Apoptosis

Frontiers in Molecular Biology

SERIES EDITORS

B. D. Hames

Department of Biochemistry and Molecular Biology University of Leeds, Leeds LS2 9JT, UK

D. M. Glover

Department of Genetics, University of Cambridge, UK

TITLES IN THE SERIES

1. **Human Retroviruses** *Bryan R. Cullen*
2. **Steroid Hormone Action** *Malcolm G. Parker*
3. **Mechanisms of Protein Folding** *Roger H. Pain*
4. **Molecular Glycobiology** *Minoru Fukuda and Ole Hindsgaul*
5. **Protein Kinases** *Jim Woodgett*
6. **RNA–Protein Interactions** *Kyoshi Nagai and Iain W. Mattaj*
7. **DNA–Protein: Structural Interactions** *David M. J. Lilley*
8. **Mobile Genetic Elements** *David J. Sherratt*
9. **Chromatin Structure and Gene Expression** *Sarah C. R. Elgin*
10. **Cell Cycle Control** *Chris Hutchinson and D. M. Glover*
11. **Molecular Immunology (Second Edition)** *B. David Hames and David M. Glover*
12. **Eukaryotic Gene Transcription** *Stephen Goodbourn*
13. **Molecular Biology of Parasitic Protozoa** *Deborah F. Smith and Marilyn Parsons*
14. **Molecular Genetics of Photosynthesis** *Bertil Andersson, A. Hugh Salter, and James Barber*
15. **Eukaryotic DNA Replication** *J. Julian Blow*
16. **Protein Targeting** *Stella M. Hurtley*
17. **Eukaryotic mRNA Processing** *Adrian Krainer*
18. **Genomic Imprinting** *Wolf Reik and Azim Surani*
19. **Oncogenes and Tumour Suppressors** *Gordon Peters and Karen Vousden*
20. **Dynamics of Cell Division** *Sharyn A. Endow and David M. Glover*
21. **Prokaryotic Gene Expression** *Simon Baumberg*
22. **DNA Recombination and Repair** *Paul J. Smith and Christopher J. Jones*
23. **Neurotransmitter Release** *Hugo J. Bellen*
24. **GTPases** *Alan Hall*
25. **The Cytokine Network** *Fran Balkwill*
26. **DNA Virus Replication** *Alan J. Cann*
27. **Biology of Phosphoinositides** *Shamshad Cockcroft*
28. **Cell Polarity** *David G. Drubin*
29. **Protein Kinase Functions** *Jim Woodgett*
30. **Molecular and Cellular Glycobiology** *Minoru Fukuda and Ole Hindsgaul*
31. **Protein–Protein Recognition** *Colin Kleanthous*
32. **Mechanisms of Protein Folding (Second Edn)** *Roger Pain*
33. **The Yeast Nucleus** *Peter Fantes and Jean Beggs*
34. **RNA Editing** *Brenda L. Bass*
35. **Chromatin Structure and Gene Expression (Second Edn)** *Sarah C. R. Elgin and Jerry L. Workman*

36. Endocytosis
Mark Marsh

37. Molecular Chaperones in the Cell
Peter Lund

38. Plant Signal Transduction
Dierk Scheel and Claus Wasternack

39. Cell Adhesion
Mary Beckerle

40. Apoptosis
Mike Jacobson and Nicola McCarthy

Apoptosis

EDITED BY

Mike Jacobson

Millennium Pharmaceuticals, Inc.
Cambridge
USA

Nicola McCarthy

Division of Cardiovascular Medicine
Addenbrookes Hospital
Cambridge
UK

This book has been printed digitally and produced in a standard specification in order to ensure its continuing availability

OXFORD
UNIVERSITY PRESS

Great Clarendon Street, Oxford OX2 6DP

Oxford University Press is a department of the University of Oxford.
It furthers the University's objective of excellence in research, scholarship,
and education by publishing worldwide in

Oxford New York

Auckland Cape Town Dar es Salaam Hong Kong Karachi
Kuala Lumpur Madrid Melbourne Mexico City Nairobi
New Delhi Shanghai Taipei Toronto

With offices in

Argentina Austria Brazil Chile Czech Republic France Greece
Guatemala Hungary Italy Japan South Korea Poland Portugal
Singapore Switzerland Thailand Turkey Ukraine Vietnam

Oxford is a registered trade mark of Oxford University Press
in the UK and in certain other countries

Published in the United States
by Oxford University Press Inc., New York

Reprinted 2011

ISBN 978-0-19-963849-9

Printed and bound in Great Britain by CPI Antony Rowe,
Chippenham and Eastbourne

Preface

Over the last ten years, the field of Programmed Cell Death (PCD/apoptosis) has changed from a relatively obscure speciality within specific areas of developmental biology, immunology and pathology, to a mainstream subject of fundamental importance to biology and medicine as a whole. Capturing on the printed page relevant and lasting ideas in any rapidly evolving field is never an easy task. Nonetheless, this is what we hope we have achieved, even if many of the details have changed as new stories have emerged. Luckily, the fundamental framework for understanding the mechanisms by which cells are programmed to die has not changed significantly over the last few years in which this book has been in production. Overall, therefore, we have included most of the important stories that have shaped our conceptual understanding of PCD through primarily focusing on the molecular mechanisms and pathways central to its regulation.

The remit of this book is to introduce the fundamentals of PCD to researchers new to this field, whether they are scientists, clinicians or undergraduates. In order to achieve this we have concentrated on the fundamental molecular processes of PCD at the expense of the more disease-associated research. The (re) discovery and background of PCD (apoptosis) is covered in chapter 1, with chapters 2-7 expanding upon individual areas of study that have aided greatly our understanding of PCD. As will be apparent throughout the book, the overall picture of PCD is gathered from many research disciplines. Arguably, much of the insight into the regulation of PCD has come from the study of less complex organisms such as the nematode worm *Caenorhabditis Elegans* (chapter 2) and the fruit fly *Drosophila* (chapter 3). The foundations laid by genetic studies in these two organisms have provided guidance for and complemented the biochemical, cellular and molecular studies conducted in mammalian cells as discussed in subsequent chapters.

The core of the death programme in animals consists of a proteolytic cascade involving a family of cysteine proteases called caspases (chapter 4). This protein family includes both the executioner caspases that target multiple cellular proteins for destruction, acting to carry out the apoptotic death of the cell, as well as death-signalling caspases that target caspases and other proteins in cell death pathways, serving to propagate or amplify the caspase death program. Two protein families play a major role in regulating this caspase cascade: the Bcl-2 family (chapter 5) and a group of proteins related to the baculovirus IAPs (Inhibitor of Apoptosis Proteins, discussed in chapters 3-5 and 8-10). The major known site of action for most of the Bcl-2 family proteins is the mitochondrion (Chapter 6), where Bcl-2 proteins regulate the release of factors from the mitochondrial inner membrane into the cytoplasm. These factors (cytochrome c, AIF, and Smac/Diablo) then activate caspases, either directly or via intermediate caspase-binding adapter/regulator proteins, and thereby trigger the death program. Crucial in both the discovery and elucidation of the biochemical function of these mitochondrial factors and their targets has been the use of cell-free systems (chapter 7).

The remainder of the book (chapters 8-11) covers the most important and best-understood pathways that regulate the intracellular death program, relaying the extracellular survival and death signals that ultimately control cell survival and death. Nearly all animal cells are programmed to die by default and are continually instructed to survive by a plethora of molecular signals (1). Recent evidence suggests that there are key proteins involved in integrating many of these signals. One such protein is the serine/threonine kinase Akt (chapter 9). Akt, a downstream target of PI-3 kinase, is implicated in the survival of many cell types and as a result any loss of function may have many repercussions in mammalian biology. In addition to these survival signals, there are also molecular signals that can specifically trigger cells to die. A particularly well-studied mechanism that has received much attention over recent years is that of the 'death receptor' pathway (chapter 8), mainly because this pathway represents a direct mechanism for activating the downstream death machinery and has been pivotal in identifying many of the key proteins that regulate PCD.

There are some disciplines in biology and medicine in which the study of PCD has been particularly important and has been so even before the decade of the 1990's that saw the emergence of apoptosis as a mainstream field of cell and molecular biology. Unfortunately, as it is not possible to cover all of these areas in just one book, we have chosen just two of the fields in which the study of PCD has made a major contribution. The first of these is virology (chapter 10). PCD is one of the cell-intrinsic mechanisms by which organisms attempt to prevent viral propagation. Essentially by committing suicide, an infected cell can prevent a virus from spreading to infect neighbouring cells and compromising the whole organism. Of course, viruses have evolved the means to counter this primitive defence, by evolving or picking up from their hosts genes that encode proteins that inhibit PCD. By understanding these viral proteins and how they work, we have been able to gain a deeper understanding of the mechanisms by which PCD operates in cells.

The second field in which PCD has had a major impact is Neurobiology. Its importance has been recognised very early on by developmental neurobiologists, and for several decades it was the main focus (although not generally recognised as such) of studies of the first described growth and survival factor, Nerve Growth Factor. Chapter 11 reviews PCD in the nervous system as an example of how PCD functions in the development of complex multicellular organisms. This chapter also reviews what we know about the role of PCD in diseases of the central nervous system, including stroke as well as more chronic degenerative diseases such as Alzheimer's disease.

There are many other areas where the discovery of PCD has revolutionised the understanding of biological function, not least in immunology, cancer and developmental biology. Regretfully, it has not been possible to cover these subjects here; however, they have been extensively reviewed in a number of books and reviews (2-7).

In closing, we would like to thank all of the contributing authors for their hard work and patience during the production of this book. This book was commissioned at a time when key researchers in PCD were being asked to contribute to books, journals and reviews on the subject every other day and many were suffering from repetitive strain injury and a lack of new plots! Our authors have managed to produce both informative and novel reviews of their specialist subjects and we are indebted to them. We would also like to thank the production team at Oxford University Press for all their help, encouragement and perseverance with this project. Finally, we hope that this book achieves its aims; to be informative, concise and clear in a subject that is forever revealing new aspects of cellular regulation that are applicable to many areas of biological research.

January 2002

Michael Jacobson, Cambridge, MA, USA
Nicola McCarthy, Cambridge, UK

References

1. Raff, M. C. (1992). Social controls on cell survival and cell death. *Nature*, **356**, 397-400.
2. Winkler, J. D. (1999). *Apoptosis in inflammation*. Progress in Inflammation Research, Birkhauser Verlag AG.
3. Sluyser, M. (1996). *Apoptosis in normal development and cancer*, Taylor & Francis, London; Bristol.
4. Martin, S. J. (ed) (1997). *Apoptosis and Cancer*, Karger.
5. Jacobson, M. D., Weil, M., and Raff, M. C. (1997). Programmed cell death in animal development. *Cell*, **88**, 347–354.
6. Vaux, D. L. and Korsmeyer, S. J. (1999). Cell death in development. *Cell*, **96**, 245–254.
7. Evan, G. I. and Vousden, K. H. (2001). Proliferation, cell cycle and apoptosis in cancer. *Nature*, **411**, 342–348.

Contents

Abbreviations

AD	Alzheimer's disease
AIF	apoptosis-inducing factor
ALS	amyotrophic lateral sclerosis
ANT	adenine nucleotide exchanger
Apaf-1	apoptotic protease activating factor 1
APP	amyloid precursor protein
BDNF	brain-derived neurotrophic factor
BH	Bcl-2 homology domain
BIR	baculoviral IAP repeat
CAD	caspase-activated deoxyribonuclease
CaM-KK	Ca^{2+}/calmodulin-dependent kinase kinase
CARD	caspase activation and recruitment domain
caspase	cysteine aspartate-specific protease
CED	cell death abnormality
CES	cell death specification
CGN	cerebellar granule neurons
CNTF	cililary neurotrophic factor
CT-1	cardiotrophin 1
CTL	cytotoxic T lymphocyte
dad-1	defender against apoptotic death
DD	death domain
DED	death effector domain
DIAP1	*Drosophila* inhibitor of cell death 1
$DiOC_{6(3)}$	3,3′dihexyloxacarbocyanine iodide
DR	death receptor
DRPLA	dentalumbralpallidoluydian atrophy
DTC	distal tip cell
EGT	early gastrulation transition
FADD	Fas-associated death domain
fem-1	feminization
FIST/HIPK3	Fas-interacting serine/threonine kinase/homeodomain-interacting protein kinase
FLICE	FADD-like ICE
FLIP	FLICE inhibitory protein
GDNF	glial-derived neurotrophic factor
gf	gain-of-function
GSK3	glycogen synthase kinase 3
HD	Huntingdons disease

her-1	hermaphrodization
HID	head involution defective
HSN	hermaphrodite-specific neurons
IAP	inhibitor of apoptosis
ICAD	inhibitor of CAD
ICE	interleukin-1β converting enzyme (caspase-1)
IGF-1	insulin-like growth factor 1
IL-3	interleukin-3
ILK	integrin-linked kinase
JIP	JNK-interacting protein
JNK	c-Jun N-terminal kinase
lf	loss-of-function
MBT	mid blastula transition
mClCCP	carbonylcyanide *m*-chlorophenylhyfrazone
MEFs	mouse embryo fibroblasts
MHC	major histocompatibility complex
NFT	neurofibrillary tangle
NGF	neuronal growth factor
NT-3	neurotrophin 3
NT-4	neurotrophin 4
NT-5	neurotrophin 5
PARP	poly (ADP-ribose) polymerase
PC12	pheochromocytoma-12 cells
PCD	programmed cell death
PDK1	phosphoinositide-dependent kinase 1
PH	pleckstrin homology domain
PI3K	phosphoinositide-3-kinase
PKA	protein kinase A
PKC	protein kinase C
POD	Pml oncogenic domain
PS-1/2	presenilin-1/2
PT	permeability transition
RANK	receptor activator of NF-κB
RNAi	RNA-mediated interference
rpr	reaper (a *Drosophila* cell death gene)
SCA	spinocerebellar ataxia
SCG	superior cervical ganglion
SMA	spinomuscular atrophy
SMBA	spinobulbarmuscular atrophy
TNF-R1	tumour necrosis factor receptor 1
TNFα	tumour necrosis factor alpha
tra-2	sexual transformer
TRADD	TNF receptor-associated death domain protein
TRAF	TNF receptor-associated factor

TRAIL	TNF-related apoptosis-inducing ligand
TRANCE	TNF-related activation-induced cytokine
TUNEL	TdT-mediated dUTP nick end-labelling
VDAC	voltage-dependent anion channel
z-VAD.fmk	*N*-benzyloxycarbonyl-Val-Ala-Asp-fluoromethyketone
$\Delta\Psi m$	mitochondrial transmembrane potential

Contributors

MARTIN R. BENNETT
Division of Cardiovascular Medicine, ACCI Level 6, PO Box 110, Addenbrookes Hospital, Cambridge CB2 2QQ, UK.

LOUISE BERGERON
Millennium Pharmaceuticals, Inc., 640 Memorial Drive Cambridge, MA 02139, USA.

ANDREAS BERGMANN
University of Texas M.D. Anderson Cancer Center, Dept. Biochemistry and Molecular Biology, 151 Holcombe Boulevard – Box 117, Houston, TX 77030, USA.

MICHAEL CARDONE
Merrimack Pharmaceuticals, 50 Church Street, Cambridge, MA 02139, USA.

PAUL R. CLARKE
Biomedical Research Centre, Level 5, Ninewells Hospital and Medical School, Dundee DD1 9SY, UK.

THOMAS F. FRANKE
Department of Pharmacology, Columbia University, 630 West 168th Street PH7-W318, New York, NY 10032, USA.

MICHAEL D. JACOBSON
Millennium Pharmaceuticals, Inc., 640 Memorial Drive, Cambridge, MA 02139, USA.

GUIDO KROEMER
CNRS UPR 420, 19 rue Guy Moquet BP 8, F-94801 Villejuif, France.

NICOLA J. MCCARTHY
Division of Cardiovascular Medicine, ACCI Level 6, PO Box 110, Addenbrookes Hospital, Cambridge CB2 2QQ, UK.

NATALIE ROY
Millennium Pharmaceuticals, Inc., 640 Memorial Drive, Cambridge, MA 02139, USA.

MANISHA S. SHAH
MCD Biology, University of Colorado, Boulder, CO 80309-0347, USA.

HERMANN STELLER
Howard Hughes Medical Institute, Rockefeller University, Weiss 1111, 1230 York Ave, New York, NY 10021, USA.

SANTOS A. SUSIN
CNRS UPR 420, 19 rue Guy Moquet BP 8, F-94801 Villejuif, France.

YOSHIHIDE TSUJIMOTO
Osaka University Medical School, Biomedical Research Center, 2-2 Yamadaoka, Suita, Osaka 565-0871, Japan.

DAVID L. VAUX
The Walter and Eliza Hall Institute, Post Office RMH, Victoria 3050, Australia.

YI-CHUN WU
Department of Zoology, National Taiwan University, Taipei 10617, Taiwan, Republic of China.

DING XUE
Campus Box 347, MCD Biology, University of Colorado, Boulder, CO 80309-0347, USA.

NAOUFAL ZAMZAMI
CNRS UPR 420, 19 rue Guy Moquet BP 8, F-94801 Villejuif, France.

1 | Why be interested in death?

NICOLA J. McCARTHY

1. Introduction

The concept that a dead cell can provide answers to several critical biological questions is a somewhat unusual one. However, this is the current opinion of many researchers working on the analogous processes of programmed cell death and apoptosis. Death of the organism is regarded as a tragic event, but at the cellular level it is, paradoxically, a prerequisite for life. Multicellular organisms have evolved specialized tissues that act in concert to enable the development and propagation of the organism. Every individual cell within each specialized tissue must function harmoniously with both its surrounding neighbours and the whole organism. It is the breakdown of these complex and integrated control systems at the cellular level, particularly loss of tissue homeostasis that is responsible for the onset of many human diseases.

Loss of tissue homeostasis can occur through over-proliferation of cells or inappropriate differentiation of cells. Significantly, the inappropriate death or survival of cells can also disrupt tissue homeostasis, a process that has hitherto been often overlooked. The study of apoptosis has revealed that the default-state of all cells in multicellular animals is death, with survival, proliferation, and differentiation all requiring positive inputs (1). Failure of this in-built suicide mechanism has drastic consequences. Aberrant surviving cells are implicated in the neoplastic process and the inappropriate death of cells can lead to, amongst others, neurodegeneration (2, 3).

Over the past twenty years research into programmed cell death has increased our knowledge of both the normal and pathological processes that can occur within tissues. This chapter serves as a broad introduction to the complex genetic and biochemical pathways of death that have been identified and which so preoccupy current apoptosis devotees.

2. Historically speaking

Wyllie, Kerr, and Currie initially used the word apoptosis in 1972 to describe a novel form of cell death (3). Cells were observed to die in a manner morphologically distinct from necrotic cell death (where the cell membrane ruptures and inflammatory cells are recruited to the scene). The 'apoptotic' deaths followed a defined morphological sequence that did not result in loss of membrane integrity, or in an inflammatory response. Moreover, these cells were disposed of rapidly by their nearest viable neighbour or passing phagocytic cell. The same morphological features had been described much earlier by developmental biologists and referred to as 'programmed cell death' (4, 5). Here, biologists had observed the neat deletion of cells that were no longer required within the developing embryo. However, the significance of this type of cell death was not realized until its description in pathological and physiological situations occurring in the adult animal. From these studies it is evident that apoptosis or programmed cell death is a regulated form of cell death that does not normally activate the immune response. Thus, throughout the life of multicellular animals, cells are regularly lost by apoptosis (6). These cells may no longer be required, or may be damaged in some way. Moreover, such cells are silently removed from the organism allowing the tissue to function as normal.

Programmed cell death or apoptosis occurs in all multicellular animals and is studied in a wide variety of animals. Indeed, many of the features of apoptosis are evolutionarily conserved in both invertebrates and vertebrates. Much of the data on the genetic regulation of apoptosis has come from the nematode worm *Caenorhabditis elegans* (7), whereas many of the biochemical pathways have been described in mammalian cell systems (8). These advances in the understanding of cell death were stimulated by the morphological description of apoptosis.

2.1 Morphological aspects

The differences between necrosis and apoptosis have been reviewed many times before hence only the salient points will be addressed here. For a more detailed description see refs 2, 9, and 10.

2.1.1 Evidence of death

Overall, necrotic and apoptotic cells are quite different. Both forms of cell death involve a characteristic number of morphological changes that can be used to classify dying cells as either apoptotic or necrotic. Necrotic cells are characterized by an overall increase in cellular size, with little change in the chromatin initially (10, 11). Organelles within the cytoplasm become disorganized and mitochondria begin to undergo distinct changes, including the accumulation of lipid-rich particles within the mitochondria and swelling, with the inner mitochondrial membrane shrinking from the outer. The cyto-architecture of the cell is lost at later stages with general release of proteases, nucleases, and lysosomal contents. The chromatin becomes flocculent and then disperses. Rupture of the cell membrane occurs leading to release of the cellular contents and pro-inflammatory cytokines leading to the recruitment of monocytes and macrophages to the site of death (see Fig. 1).

Apoptosis also follows a characteristic but distinct number of sequential morphological changes (3). Apoptotic cells shrink in size and exhibit marked alterations in their chromatin structure at an early stage. The chromatin becomes highly condensed within the nucleus and can appear concentric with the nuclear

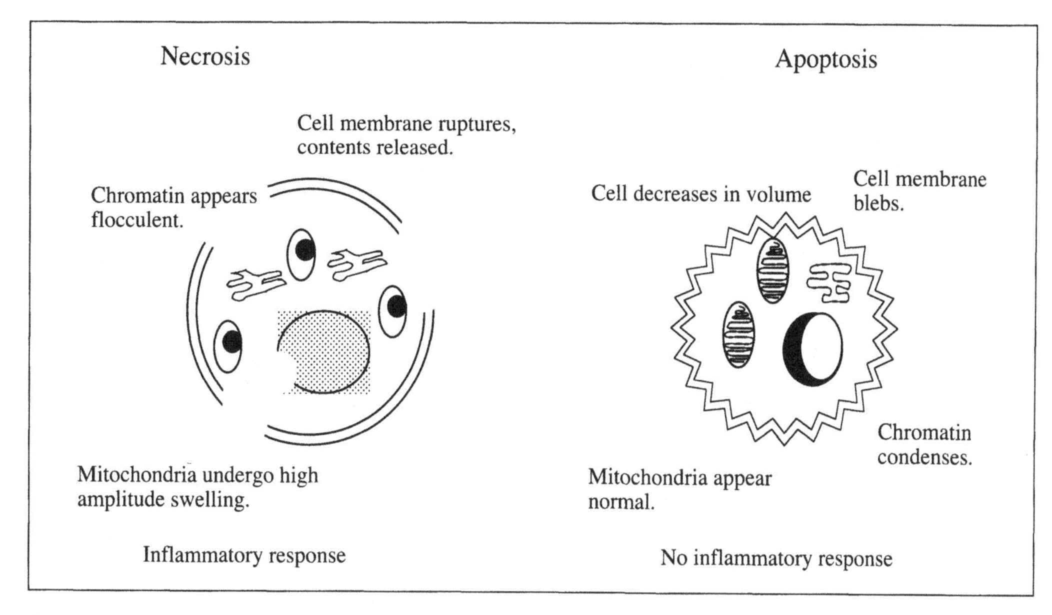

Fig. 1 Cells undergo distinct morphological changes when they progress through either necrosis or apoptosis. Necrosis is characterized by cell swelling and disruption of cellular organelles. Necrotic cells lyse releasing their contents and trigger an immune response. In contrast, apoptotic cells shrink, undergo membrane blebbing, and chromatin condensation. The cellular contents are retained and the cells are rapidly phagocytosed by viable neighbours or professional phagocytes.

membrane. The organelles within the cytoplasm remain intact and show little change apart from some swelling of the endoplasmic reticulum. Externally the cell appears to boil as the membrane becomes convoluted, this phase is often referred to as 'blebbing' and is one of the earliest morphological manifestations of imminent death (see Fig. 1). Cells *in vitro* often shed apoptotic bodies, which are membrane-bound vesicles containing intact cellular organelles and sometimes condensed chromatin. Whether or not apoptotic bodies are shed *in vivo* is still unclear. Data from time lapse video microscopy suggests that apoptotic cells can be phagocytosed prior to the shedding of apoptotic bodies. In cases where phagocytosis is delayed, breakdown of the cell into apoptotic bodies is more likely to occur (12).

Several of the morphological changes that an apoptotic cell passes through are used in the laboratory to identify apoptotic cell death. One of the best markers is still the condensation of the chromatin, observed either by light, fluorescent, or electron microscopy (13). Other biochemical methods used to identify apoptotic cells are well documented (see ref. 14) and will not be discussed here.

2.1.2 Time of death

In real time, cells can enter apoptosis and be phagocytosed within an hour, making their appearance very transient indeed. Death is also stochastic; apoptotic stimuli do not tend to induce death in all cells at precisely the same moment. Therefore, when observing a tissue section at a specific time after an apoptotic stimulus very few apoptotic cells will be observed (13, 15). Thus, evidence of apoptotic cell death is hard to determine from one time point and is one of the reasons why the importance of cell death was initially overlooked. Even a tissue that is essentially involuting may only appear to have 5% of cells in the apoptotic state at any one time, thus masking what is in fact a high level of death. One of the reasons why apoptotic cells are not often encountered in tissue sections is due to their rapid clearance by phagocytosis.

i. Disposal of the body

Despite all of the morphological changes that distinguish an apoptotic cell, perhaps one of the most important aspects of apoptosis is the rapid phagocytosis of the dying cell (9, 16). The physiological changes upon the cell surface that 'label' the cell as apoptotic and ready for phagocytosis occur very early on in the process. Current models suggest that this enables the clearance of the dying cell prior to the onset of secondary necrosis and membrane rupture, thus eliminating the involvement of the immune response (16). Invasion of a tissue by inflammatory cells can significantly disrupt the function of that tissue and hence needs to be avoided where possible (17). Importantly, the apoptosing cell does not need to attract the attentions of a passing professional phagocyte; instead it can alert its nearest viable neighbour and have them dispose of the body (18). Failure to remove an apoptotic cell will elicit an immune response since that cell will undergo secondary necrosis and lyse. In some tissues where the onset of apoptosis is acute and there are many dying cells phagocytosis is inefficient and many apoptotic cells undergo secondary necrosis. It is in these tissues that death is still classed as being necrotic, overlooking the fact that death may well have been apoptotic originally (19). This, even today, still leads to arguments about the relevance of apoptotic death. Interestingly, these uneaten secondarily necrotic cells elicit release of a specific cocktail of inflammatory mediators from the phagocytosing macrophage, the contents of which is dependent upon the time past the cells 'eat-me-by' date (20, 21). Thus, only where there is massive apoptotic cell death, inefficient phagocytosis, and secondary necrosis is the inflammatory response fully involved (16, 21).

Although treated somewhat superficially by many researchers in the field, the removal of apoptotic cells is a critical process. It is not yet clear how detrimental failure to remove apoptotic cells is to a tissue and this is one area of research that is only now coming to the fore (21). Significantly, six known genes regulate the phagocytosis of dead cells in the nematode *Caenorhabditis elegans* (see Chapter 2). Conservation of the cell death machinery is evident throughout evolution, although a greater number of genes are involved in mammalian cells (22–24). The elaborate nature of the phagocytic signals in mammalian cells is still being deciphered and is excellently reviewed in refs 25–27.

3. Where does apoptosis occur?

Apoptosis occurs in all multicellular organisms. Death of cells is seen mainly during development and is important primarily for deleting unwanted cells (6). There are many examples of where programmed cell death occurs and this has been the subject of several previous reviews (3–6). In general, cells are deleted when sculpturing tissues within the body such as formation of the digits (deletion of the inter-digital cells) (28). Cells are also deleted to remove structures that are no longer required, involution of the tadpole tail being the most cited example. Deletion of cells is also a good mechanism for precisely matching the number of cells required for a particular function. In the nervous system many more neurons and oligodendrocytes are produced than are required. Up to half are eliminated since they fail to make productive interactions with target cells (29). Finally, potentially harmful cells are deleted. A good example is the deletion of cells in the thymus during positive and negative selection (30). Thymocytes that have functionally inactive receptors, or produce potentially self-recognizing receptors undergo programmed cell death.

Programmed cell death is therefore most evident during multicellular animal development. However, apoptosis also occurs sporadically in all tissues throughout life and is a normal everyday occurrence in tissue turnover.

3.1 Primitive death

One of the most comprehensively studied examples of programmed cell death is that which occurs in the nematode *Caenorhabditis elegans* (7, 22, 31). The development of each *C. elegans* worm is invariant, 1090 cells are generated of which 131 die. The death of these 131 cells is regulated by a core of approximately 13 genes that either specify the death of the cell, carry out its suicide, or phagocytose the cell and dispose of the body (see Chapter 2). Many of these genes have been conserved throughout evolution and are functionally and structurally similar to their mammalian counterparts (Fig. 2). Hence, the function of these genes in the relatively simple nematode can be extrapolated to identify genes of similar function in higher order animals.

Evidence for programmed cell death is also seen in primitive slime moulds such as *Dictyostelium discoideum* (32). When conditions are favourable these eukaryotic cells exist as free living amoeboid cells. When food is in short supply the amoeboid cells aggregate and form a fungal-like structure with a stalk and a fruiting body. The cells within the stalk die as part of a terminal differentiation programme and these deaths exhibit morphological changes similar to those seen in eukaryotic cells.

Cell death also occurs in plants and can be seen in the development of xylem, flowers, and ovules (33). It can also occur in the hypersensitive reaction to invading pathogens. Although the dying plant cells have some similar morphological features to apoptotic animal cells, they are not phagocytosed. Presumably the presence of a cell wall prevents this. The molecular mechanisms controlling this form of cell death have yet to be fully investigated, but it would be interesting if plants also have a genetically conserved cell death programme.

The widespread occurrence of cell death and its apparent conservation throughout evolution suggests that regulation of cell death is critical in the development of multicellular organisms. Many genes have now been shown to regulate various steps within the cell death pathway and form the basis of many of the chapters in this

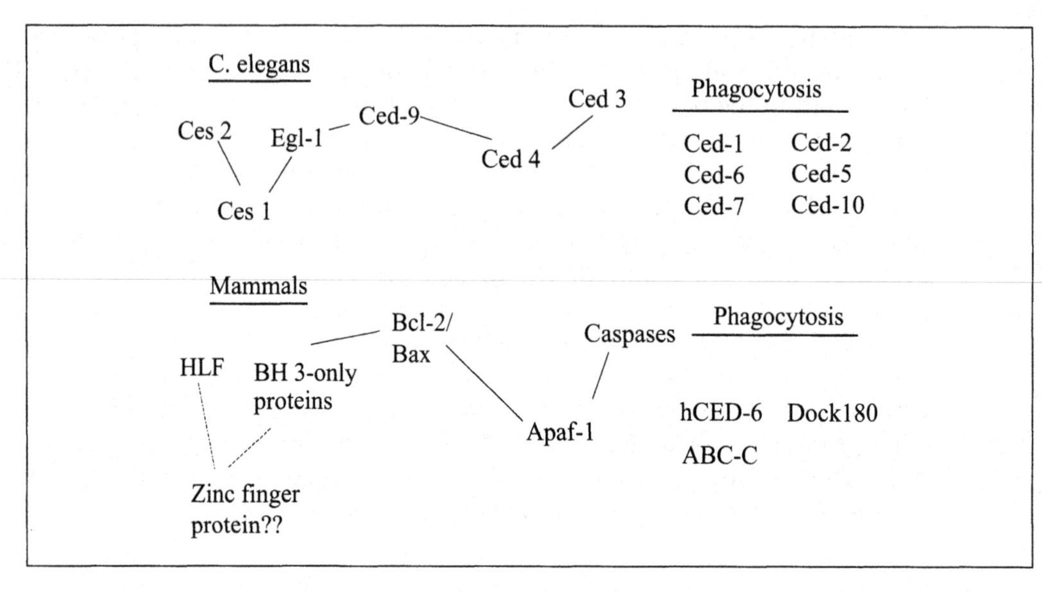

Fig. 2 Comparison of cell death pathways in nematode and man (22). Cell death is specified by Ces-1 and Ces-2 in *C. elegans* which appear to be transcription factors. Ced-9, the anti-apoptotic protein is inhibited by Egl-1, allowing activation of both Ced-4 and Ced-3. A similar pathway exists in mammalian cells with transcription factors at the start of the pathway (hepatic leukaemia factor, a basic helix-loop-helix protein), followed by the BH3-only members of the Bcl-2 family. BH3-only proteins, like Egl-1, antagonize the function of the Bcl-2-like proteins. This enables the activation of Apaf-1 (Ced-4) and caspases (Ced-3). Apoptotic cells are phagocytosed in both nematode and man and several homologous genes carry out this function in both animals.

book. However, what remains unclear is where these genes act to commit a cell to death. This commitment point is still a hotly debated topic, but for simplicity, genes can be thought of as acting in three basic phases; decision, sentence, and execution. The significance of each phase is outlined below.

4. The importance of regulated cell death

Prior to the discovery of apoptosis, cell population numbers were known to be controlled by differentiation, proliferation, and senescence. The discovery of genes that actively prevented apoptosis and the demonstration that inhibition of protein and RNA synthesis in some cell types delayed apoptosis suggested that this form of cell death was regulatable and therefore potentially important in cell population control (34–36). Indeed it is now clear that a large number of proteins act to direct whether a cell lives or dies. Moreover, this has led to the idea that all cells are programmed to die unless prevented from doing so. Thus, death is the default state of all cells.

The various fates of differentiation, proliferation, death, and senescence control cell population number (Fig. 3) (19). For cells that lose the capacity to undergo apoptosis in response to physiological stimuli the consequences can be quite dramatic. For example, one gene *bcl-2*, the first of a now large family of apoptosis-

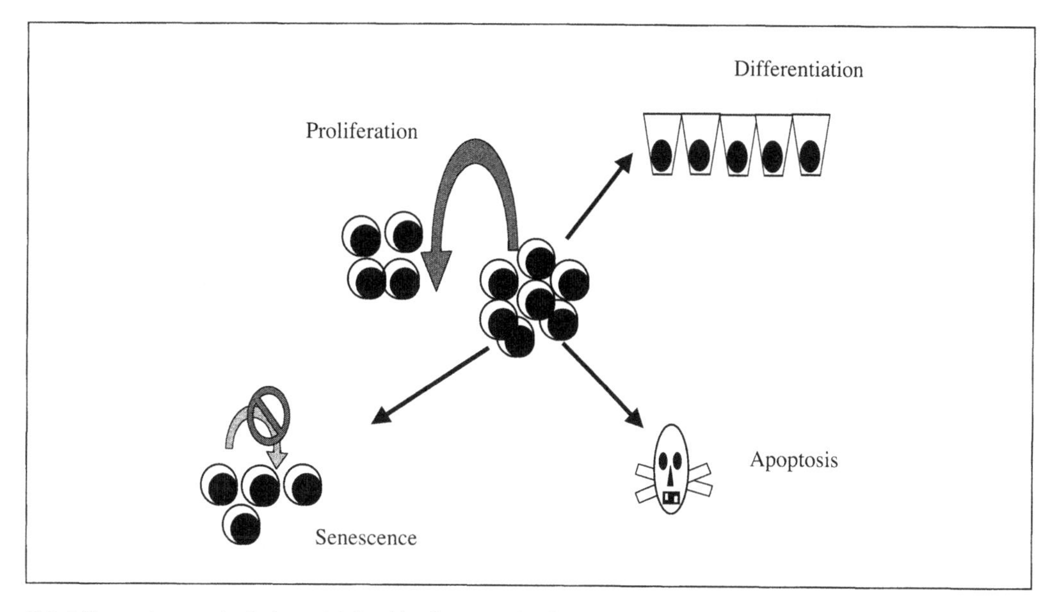

Fig. 3 Tissue homeostasis is maintained by the opposing functions of differentiation, senescence, proliferation, and death. Loss of any one of these pathways is implicated in the development of human disease.

regulating genes to be identified, causes excess accumulation of B cells when overexpressed in the B cell lineage in Eμ-*bcl-2* transgenic mice (37). These cells are not at first detrimental to the animal, however, the increased number of cells substantially increases the risk of a cell attaining a secondary and/or tertiary mutation that facilitates the clonogenic outgrowth of that cell. Thus, *bcl-2* transgenic mice develop follicular lymphoma at 12 months of age. The inappropriate survival of cells therefore, even for transient periods, is deleterious to the maintenance of tissue homeostasis.

Conversely, too much cell death can also have adverse effects. Cells can be lost from tissues due to the presence of non-physiological apoptotic stimuli or due to an over-sensitization to a particular physiological stimulus. A good example of a non-physiological stimulus of cell death would be the loss of neurons in Alzheimer's. Here cells amass aberrant forms of a specific protein that accumulates producing the characteristic disease-associated plaques that are toxic to the cell (38).

Thus, perturbations within the cell death pathway can have very profound effects on tissue physiology. However, since apoptosis is a regulated form of cell death, comprehension of the pathways that act to control whether a cell lives or dies should reveal novel targets for disease intervention. Much is now understood of the mechanisms involved in regulating cell death in several different species. Moreover, the complex system that exists in mammalian cells is paralleled by the less complex nematode and *Drosophila* cell systems. The combined research into all these animal models has identified a number of complex signalling networks involved in the apoptotic pathway.

4.1 The decision phase

One of the problems with working on a phenomenon that was initially described based upon its morphological characteristics is that many gene products do not affect these morphological changes. Indeed, the appearance of morphologically apoptotic cells characterizes the final stage of the process of regulated cellular death. The first arbitrarily assigned stage, the decision phase, is one not defined by any morphological criteria. Instead, it defines the phase that is presumed to occur when external and internal pro- and anti-apoptotic information is integrated within the cell. Acting within this phase is a whole host of known and novel gene products (see Fig. 4). Many of these gene products and their roles within the apoptotic pathway are discussed in depth throughout this book, so the following represent a brief resume only.

4.1.1 Apoptosis and genes that regulate the cell cycle

Many of the genes that are primarily associated with cell cycle regulation are also fundamentally important in the control of cell death. Products of the c-*myc*, *E2F*, c-*fos*, *ras*, and c-*abl*, genes have all been shown to exert some form of control over a cells capacity to commit suicide (39). In particular, the role of c-Myc in regulating both cell proliferation and death has underscored the intimacy of these two processes (39).

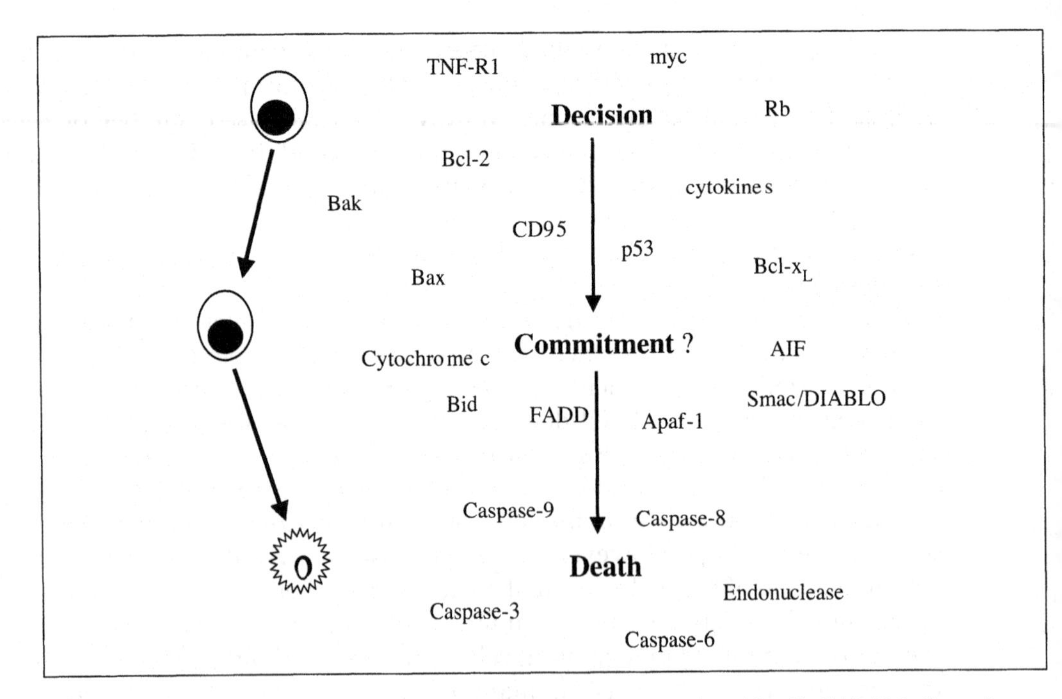

Fig. 4 Genes acting in the apoptotic pathway. The cell death pathway can be arbitrarily divided into three phases: decision, commitment, and death (execution). Where genes act within this pathway appears to be dependent upon the stimulus and the cell type being studied. Central to the understanding of apoptosis regulation at present is identifying which factors act to commit a cell to death.

The cell cycle regulatory proteins p53 and Rb can also influence a cells capacity to survive (40). Loss of Rb function produces excess apoptosis during development and slow growth and transformation in tumour cell models (41, 42). Conversely, loss of p53 enhances cellular survival in the presence of DNA damage and aides tumour progression (43, 44). Loss of both p53 and Rb leads to enhanced tumour progression, again suggesting that loss of cell cycle regulation affects apoptosis (45).

It is initially counter-intuitive that gene products which regulate proliferation can also regulate the opposing demise of the cell. However, in the context of multicellular, large, long-lived animals this is a critical 'fail-safe' mechanism. In terms of the neoplastic process, potentially the most dangerous cell within a tissue is one that replicates its DNA and then divides it between two daughter cells. Any mutation that confers a growth advantage could lead to tumour formation. However, if a cell that has the capacity to divide is also more sensitive to death, then any cell harbouring an increased capacity to replicate will also be more likely to die (46). Recent evidence widely supports this hypothesis demonstrating that cells with oncogenic lesions are more sensitive to death stimuli (reviewed in Chapter 8 and ref. 47). Moreover, genes that enhance survival, such as *bcl-2* are also cytostatic, making a long-lived cell less likely to divide (48, 49). The combined lesions of enhanced survival and proliferation are highly tumorigenic (50).

4.1.2 The Bcl-2 family

bcl-2 is often thought of as the first gene identified that primarily acted to enhance cellular survival in mammalian cells (51). Like its nematode homologue *ced-9, bcl-2* can suppress apoptosis in response to specific stimuli. Moreover, its inappropriate expression in cells is associated with enhanced survival capacity which can lead to tumour formation or autoimmune dysfunction, depending the cell lineage affected (52, 53) (see Chapter 5). Other Bcl-2-like proteins exist within mammalian cells, some of which are anti-apoptotic and others, like Bax, Bak, Bad, and Bik that are pro-apoptotic (Fig. 5). These proteins are found in many compartments within the cell, but it is their mitochondrial association that has recently been closely investigated (54). Bcl-2 family members appear to regulate the release of cytochrome *c* from mitochondria, which acts in concert with specific downstream factors (caspases) to induce apoptosis (55) (see Chapters 4, 5, and 6). In addition, members of the Bcl-2 family interact with one another and regulate each other's function (56). Binding of a pro-apoptotic member to an anti-apoptotic one can result in the loss of the anti-apoptotic function and lead to the death of the cell. Therefore, the balance of pro- and anti-apoptotic members of the Bcl-2 family regulates an important step in the decision to live or die.

4.1.3 Death receptors

The discovery of antibodies that specifically trigger the death of tumour cells lead to the identification of cell surface receptors responsible for this effect (57). These receptors are members of the tumour necrosis factor receptor (TNFR) superfamily (58). Binding of specific ligands to these receptors causes their aggregation and

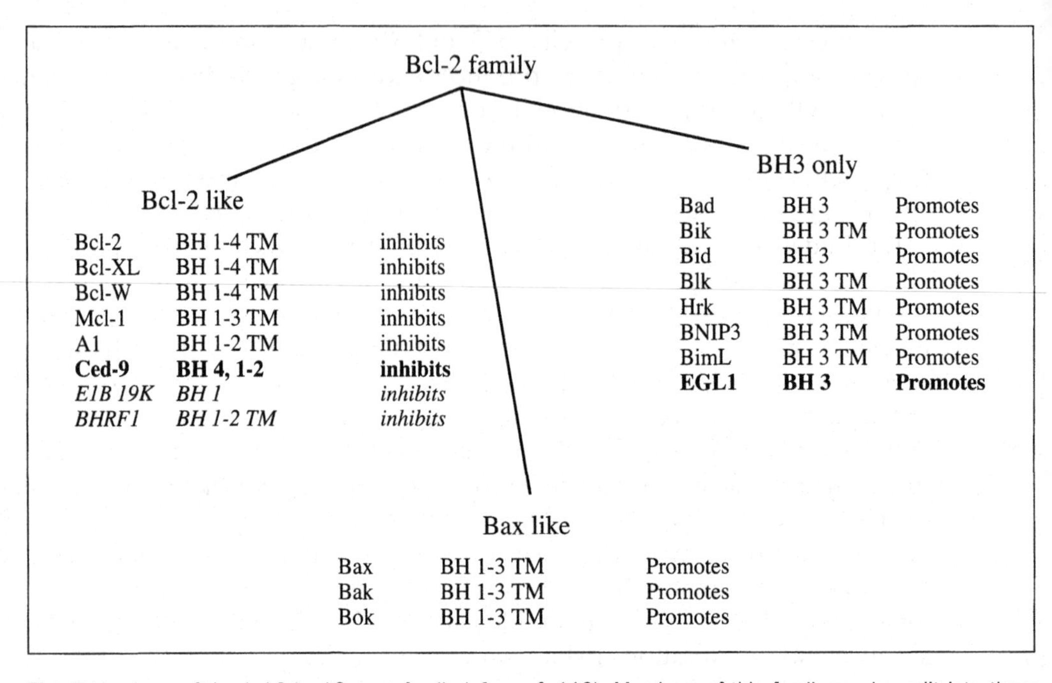

Fig. 5 Members of the *bcl-2/ced-9* gene family (after ref. 112). Members of this family can be split into three groups. The first is defined by proteins that contain four regions of homology (Bcl-2 homology-BH) and/or are anti-apoptotic in function. The second contains Bax-like proteins that do not have the 'protective' BH4 domain and activate apoptosis. The final group are BH3-only proteins that all act to activate apoptosis. Some, but not all members of the Bcl-2 family contain a transmembrane domain (TM) allowing localization to a number of membrane targets within the cell. (Bold indicates homologous genes from *C. elegans*, italics indicate viral homologues.)

activation. To the intracellular domains of these activated receptors bind several downstream adaptor proteins, some of which can activate the caspase family of proteases that carry out the systematic degradation of the cell (Fig. 6) (59–61). Other adaptor proteins initiate other signalling pathways within the cell; thus the cells will not always undergo cell death upon receptor activation (62). Death can be prevented by the presence of specific anti-apoptotic proteins that prevent the downstream caspases binding to the receptors, or in some cells the presence of survival factors such as Bcl-2 or survival cytokines will prevent cell death (63) (see Chapter 8). The importance of this family of death receptors in the immune system is particularly well understood. However, many other cell types express these receptors and moreover, they express more than one family member, giving rise to a complex network of cell death regulation. It is clear however that these receptors can either induce death or in some cells proliferation or differentiation, making them part of the decision process.

4.1.4 Survival cytokines

Cytokines have many different effects within cells that are generally cell type specific. One effect mediated by cytokines is cell survival; withdrawal of a specific survival cytokine from cells induces apoptosis (64, 65). In the presence of specific cytokines

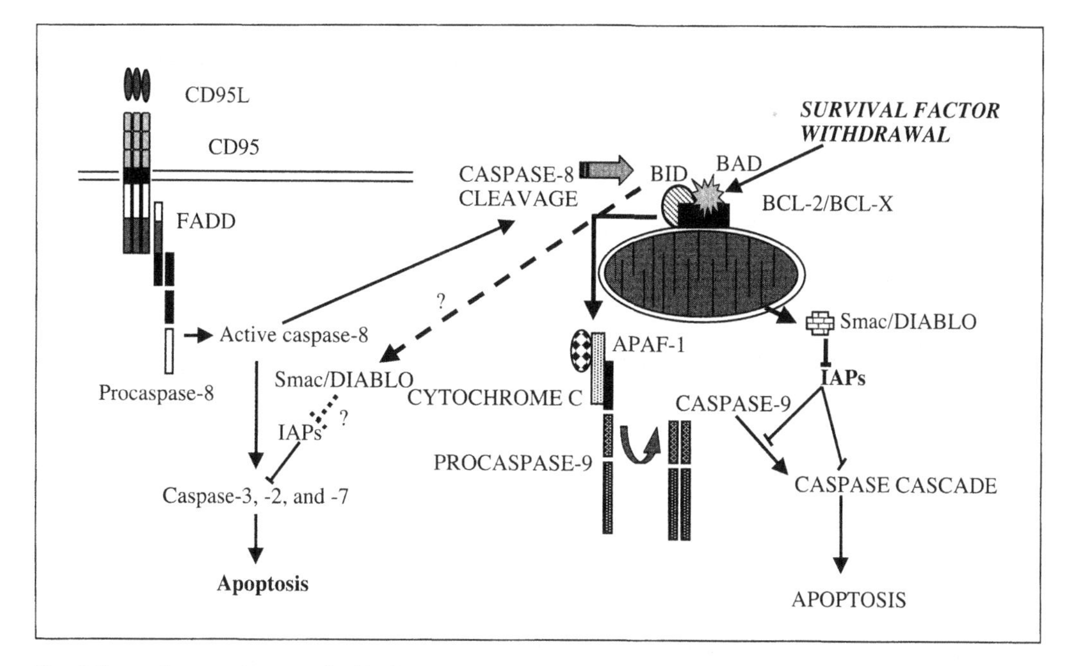

Fig. 6 Two pathways of apoptosis. Depicted are two pathways through which caspases can become activated. The first is the death receptor pathway involving activation of the CD95 receptor via its ligand CD95L. This triggers the binding of FADD to the receptor, which in turn recruits caspase-8. The presence of many caspase-8 molecules in close proximity leads to autoactivation. Once active, caspase-8 cleaves downstream caspases, such as caspases-3, -2, and -7, triggering the death of the cell. Alternatively, caspase-9 can be activated by release of cytochrome *c* from mitochondria. Cytochrome *c* along with ATP and Apaf-1 forms a complex to which caspase-9 molecules bind. Due to the close proximity of caspase-9 molecules in this complex, autoactivation of caspase-9 occurs. Active caspase-9 cleaves other downstream caspases again triggering apoptotic changes within the cell. These two pathways are linked by caspase cleavage of BH3-only proteins, such as Bid, that alter Bcl-2 family proteins at the mitochondrial membrane leading to release of cytochrome *c*. Inhibition of caspases by the IAPs is overcome by release of Smac/DIABLO from the mitochondria.

cells can survive, even if the cell happens to receive a pro-apoptotic stimulus (Fig. 6). The downstream pathways triggered by survival cytokine binding have been studied in many cell types in response to factors such as platelet-derived growth factor (PDGF), nerve growth factor (NGF), interleukin-3 (IL-3), and insulin-like growth factor 1 (IGF-1). IGF-1, for example, potentially activates both pro- and anti-apoptotic pathways, via Ras activation. The pro-apoptotic pathway is mediated by Ras-Raf signalling and the anti-apoptotic pathway by p110-mediated Akt/PKB signalling (66). Akt is a very important anti-apoptotic protein that acts to inhibit cell death at various stages within the apoptotic pathway (67–69) (see Chapter 9). Thus, the presence of survival cytokines bypasses the default-state of all cells, which is death.

4.1.5 Summary

Overall there are many genes that can influence a cells' likelihood of death. Many of these genes feed into a common pathway that involves release of cytochrome *c* from mitochondria and the activation of caspases. The release of cytochrome *c* from mitochondria is an active area of research at the current time primarily because it is one of the very early changes that occurs in almost all apoptotic cells and maybe involved in the commitment of cells to death (70).

4.2 The commitment phase

Which components of the apoptotic signalling pathway act within the commitment phase is a controversial topic. The fundamental question is precisely which molecular changes occur when a cell is condemned to death? If these particular changes can be identified then it may be possible to regulate both cell death and cell survival very effectively. Ideally, new pharmacological targets will be found that can induce death in cancer cells or prevent neuronal death in Alzheimer's.

What are the candidates for this phase of cell death? At present no known molecule that acts in the cell death pathway in mammalian cells is required absolutely for cell death to occur in all cell types. Thus, there is no one golden molecule to cure all apoptosis-associated ills. Perhaps this is not too surprising given the biological havoc that could ensue upon the mutation of this one molecule. Instead it appears that the commitment phase will vary depending upon the stimulus and the cell type concerned. A good example of this has recently been provided by the generation of Apaf-1-deficient animals (71). *apaf-1* is the mammalian homologue of *ced-4*, a gene required in the nematode for *all* somatic cell deaths to occur (72). *apaf-1*-deficient mice exhibit perinatal lethality. The defects that these mice display are mainly within the developing nervous system. Cell death is critical during neuronal development. Many more neuronal cells are produced than are needed and those that do not make connections with other neurons do not receive sufficient anti-apoptotic stimuli. Thus these cells are lost silently by apoptosis during embryogenesis. In Apaf-1-deficient animals, these cells are no longer susceptible to developmental death and remain viable. As a result, areas of the fore- and hindbrain are too big to be encapsulated by the skull bones and mice are born with neuronal protrusions. Interestingly, they also do not lose the inter-digital cells between the digits and are born with webbed paws. Apaf-1 is important in the activation of downstream caspases and the onset of the execution phase of cell death. The lack of cell death in developing neurons suggests that Apaf-1 is required for the commitment to death. Alternatively, one could argue that Apaf-1 acts in the decision phase and absence of Apaf-1 produces a potent survival signal. However, the absence of Apaf-1 in these animals does not prevent death receptor-induced cell death. Moreover, no other tissue abnormalities have been described suggesting that other death pathways are functional within these animals. Thus, Apaf-1 is not required for all apoptotic cell deaths.

Release of apoptogenic factors from mitochondria is also a prime candidate for regulation of the commitment point. Several apoptosis-inducing agents are released from mitochondria including cytochrome *c* and apoptosis-inducing factor (AIF) (73) Evidence suggests that release of cytochrome *c* however, does not necessarily induce apoptosis. Cells injected with cytochrome *c* can remain healthy in the absence of pro-apoptotic signals (74). Thus, release of factors from the mitochondria may not be the point of no return.

The recent description of a novel molecule termed either Smac or DIABLO has exemplified the role of mitochondria in the cell death process (75, 76). Smac/DIABLO also resides within the mitochondria and is released at or around the same time that cytochrome *c* release occurs, through mechanisms which are at present unclear. Smac/DIABLO is able to bind to a specific class of anti-apoptotic proteins termed IAPs (inhibitors of apoptosis). IAPs were initially described in baculovirus, but have since been characterized in nematodes, *Drosophila*, and mammals (77). IAPs have several functions, but it is their ability to inhibit active caspases that is the primary target of Smac/DIABLO (78). Smac/DIABLO binds IAPs that are bound to active caspases (-9 and -3 being prime examples) and removes them, preventing inhibition of the caspase cascade (see Fig. 6). This may explain why microinjection of cytochrome *c* does not always trigger apoptosis, since without the release of Smac/DIABLO from the mitochondria, caspases may well remain inactive due to the presence of bound IAPs (79). Thus, release of factors from the mitochondrion is certainly important for the activation of the caspase cascade. However, caspase-8 (an upstream caspase) is not targeted by IAPs (80). Hence activation of the CD95 receptor, which activates caspase-8, may not be influenced by the presence of Smac/DIABLO, again suggesting the role of this protein is context dependent (79).

Overall, which cell death molecules are included in the decision and commitment phases of cell death depends upon the model system being used and this is reflected within the chapters of this book.

4.3 The execution phase

The execution phase can be used to define the point at which all of the morphological changes that are associated with apoptosis become apparent (81, 82). The critical mediators of this phase of apoptosis are the caspases (83, 84). Caspases are proteases that cleave exclusively after aspartate residues and have a conserved active site, QACxG. There are 14 known caspases in mammalian cells with homologues in both nematode and *Drosophila* (85, 86). In mammalian cells, caspases can be broadly grouped into upstream or downstream caspases (Fig. 6). The upstream caspases are activated first, by a number of mechanisms, and once they are active they cleave and activate the downstream caspases. The caspases act in a cascade and cleave specific substrates that aid the specific breakdown and packaging of the dying cell for phagocytosis (87). Not all caspases are critical for cell death to occur (88). Indeed, in fibroblast cells treated with a broad-spectrum caspase inhibitor, the commitment to death is not suppressed, but the morphological changes of apoptosis are significantly delayed (89). However, both caspase-3 and caspase-9-deficient animals have severe abnormalities within the nervous system, that, like *apaf-1* null mice, is due to the survival of excess neuronal cells (90). Thus, both upstream and downstream caspases (-9 and -3 respectively) are critical for death to occur during development of the nervous system. In some tissues, therefore, certain caspases could be classified as acting during the commitment to cell death, whereas in others they are not required for commitment to occur. This again underscores tissue-specific differences in apoptosis regulation.

4.4 Summary

From the initial description of an intriguing morphological phenomenon has come the description of a biochemically complex, genetically regulated mechanism for the elimination of unwanted or damaged cells. Research into apoptosis is at a particularly interesting stage in that many more genes are now known to regulate death, but not all seem likely to act in the same way in every cell. Particular genes may be critical for one apoptotic signal, yet of no use for another. This maybe one reason why cells appear never to lose the capacity to die. They may exhibit resistance to one or more cell death triggers, but if pushed hard enough or with the right stimuli, cells which are apparently resistant will undergo apoptosis. So for each gene mentioned above, its role in tissue-specific death must be closely analysed before one characterizes its function within the death pathway. The arbitrary assignment of genes to the three 'phases' of apoptotic cell death is helpful when theoretically considering which gene product functions where, but the boundaries between these assignments must remain dynamic. This is a reflection of both the differences between tissues as mentioned above and the rapid progress that is made in this field. Researchers looking to find genes that act in apoptotic pathways have identified many novel genes. It is then often taken for granted that 'apoptosis regulation' is the sole function of the identified gene. However, many of the genes that function in this pathway are active in many other cellular processes and their initial role in apoptosis may turn out to be a small one indeed.

5. Models of apoptosis

Many cellular models are now used to study the process of apoptosis. This includes simple organisms such as *C. elegans* and *Drosophila* and the much more genetically complex models of mouse and man.

Animals such as *Drosophila* and *C. elegans* have been very valuable in understanding the genetic regulation of death, especially in the case of *C. elegans* whose genetic lineage is completely mapped. The function of all the cells within this organism is also known so cell type-specific functions of genes can be closely investigated. Thus, with the conservation of the death machinery throughout evolution, there is still much to be gained from genetic studies in the nematode (22). *Drosophila* is a useful system since its is amenable to the study of both genetic and biochemical consequences of gene expression (91). There is, for example, much scope within this animal for investigating the role of different caspases during cell death. However, it is interesting to note that the genes that regulate death within *Drosophila* (*reaper*, *hid*, and *grim*) have no homologues in either *C. elegans* or mammalian cells

(92). However, one function of Reaper, Hid, and Grim is to target and inhibit IAPs, suggesting some functional similarity between these *Drosophila* proteins and Smac/DIABLO (78, 79, and Chapter 3). A major limitation of both *C. elegans* and *Drosophila* is that neither suffers from the problems faced by long-lived organisms. Loss of cell death in *C. elegans* does not lead to tumour formation. Indeed, worms with excess cells appear normal (although recent data does suggest some loss of vigour). Thus, the tight regulation restricting cell growth in higher order organisms may not exist in these lower order animals. These caveats apart, the study of cell death in these animals is highly productive as demonstrated in Chapters 2 and 3.

The use of transgenic and knockout mice to study the consequences of aberrant apoptosis *in vivo* is also very valuable since such studies have elicited some unexpected findings that would not have been predicted from *in vitro* studies (93). One such example is the generation of Rb null mice (94–96). Given the important role of Rb in restricting cell cycle progression, one may have predicted that its loss would allow for unrestricted cell division resulting in an animal that was highly susceptible to cancer development. However, Rb null mice are non-viable due to massive apoptosis occurring in the central nervous system and tissues associated with red blood cell production. This is principally because these cells are unable to exit cell cycle and respond to the changing cytokine environment. Without appropriate survival factors these cells die, underlining the connection between cell proliferation and cell death.

The recent generation of mice that specifically express the oncogene c-*myc* in the skin has shown that the importance of apoptosis in the fight to suppress neoplasia depends very much on how cell numbers are controlled within a specific tissue (97). Overexpression of c-Myc in fibroblast cells *in vitro* makes them much more susceptible to cell death (98). The prediction would be that when expressed inappropriately in the skin, the cells would proliferate and also undergo apoptosis. However, although proliferation was seen in these transgenic mice, apoptosis was not. One reason for this result maybe that because skin cells are normally sloughed off once terminally differentiated, apoptosis is not required to remove the excess cells produced by Myc-induced proliferation. Instead excess cells are simply shed in greater numbers. Conversely, mice expressing a Myc transgene in β-islet cells of the pancreas develop diabetes due to Myc-induced apoptosis and loss of this population of cells (99). Thus, transgenic models are important for understanding the particular function of a pro- or anti-apoptotic gene within a particular tissue.

Another way in which pro- and anti-apoptotic proteins have been investigated is through use of cell-free systems (8, 83, 100, 101). These are generated from either mammalian cells treated with a pro-apoptotic stimulus or from using *Xenopus* egg extracts (see Chapter 7). The use of cell-free systems enables the biochemical investigation of proteins such as caspases and Bcl-2 family members. Cell-free systems have identified some of the critical pathways involved in regulating cell death, such as cytochrome *c* release and the presence of Apaf-1 and caspase-9 complexes (72). Such biochemically-based systems greatly aid the interpretation of data from both invertebrate and vertebrate models of death. However, they are somewhat limited to investigating the later stages of apoptosis such as caspase activation. Extracts derived from *Xenopus* do allow the study of several 'upstream' factors such as Bcl-2 family members (101, 102) and for the analysis of proteins from other animals involved in cell death (*reaper* from *Drosophila* being one such example).

Other models systems such as yeast, bacterial pathogens, viruses, and plants are also used to study cell death (33, 103–105). The study of viral encoded genes has identified several anti-apoptotic genes that are also conserved within mammalian cells (105–109). However, the degree to which the cell death machinery is conserved in yeast (110) and plants (111) has yet to be fully established.

Overall there are many models in which cell death is studied, all contributing their own special viewpoint on how cellular suicide is regulated.

6. Future work

Apoptosis is regulated by many genes that act in concert to ensure the demise and disposal of the condemned cell. At all of the stages of apoptosis discussed above, there are genes that act to induce and genes that act to suppress cell death. This sets up a very complex web of regulation befitting for a process that is critical in the maintenance of tissue homeostasis. Mutation of the cell death pathway is never absolute in that, although more resistant to specific triggers of apoptosis, cancer cells, for example, never completely lose their ability to undergo apoptosis. Thus, the goal of the apoptosis researcher is two-fold. First to identify a precise mechanism that regulates cell death within their chosen system, and secondly, to identify the key components within that system that will facilitate intervention in this process.

The following chapters review the critical genes involved in the regulation of apoptosis, their associated biochemical processes where known, and the diseases in which mutation of these genes is thought to be important. They provide a basic overview of research in apoptosis as it stands at the current time and review some of the controversial aspects within the field.

References

1. Raff, M. C. (1992). Social controls on cell survival and cell death. *Nature*, **356**, 397–400.
2. Wyllie, A. H. (1987). Apoptosis: cell death in tissue regulation. *J. Pathol.*, **153**, 313–16.
3. Kerr, J. F., Wyllie, A. H., and Currie, A. R. (1972). Apoptosis: a basic biological phenomenon with wide-ranging implications in tissue kinetics. *Br. J. Cancer*, **26**, 239–57.
4. Clarke, P. G. H. and Clarke, S. (1996). Nineteenth century research on naturally occurring cell death and related phenomena. *Anat. Embryol.*, **193**, 81–99.
5. Glucksmann, A. (1951). Cell deaths in normal vertebrate ontogeny. *Biol. Rev.*, **26**, 5986.
6. Jacobson, M. D., Weil, M., and Raff, M. C. (1997). Programmed cell death in animal development. *Cell*, **88**, 347–54.
7. Horvitz, H., Shaham, S., and Hengartner, M. (1994). The genetics of programmed cell-death in the nematode *caenorhabditis elegans*. *Cold Spring Harbor Symp. Quant. Biol.*, **59**, 377–85.

8. Lazebnik, Y. A., Cole, S., Cooke, C. A., Nelson, W. G., and Earnshaw, W.-C. (1993). Nuclear events of apoptosis *in vitro* in cell-free mitotic extracts a model system for analysis of the active phase of apoptosis. *J. Cell Biol.*, **123**, 7–22.
9. Morris, R. G., Hargreaves, A. D., Duvall, E., and Wyllie, A. H. (1984). Hormone-induced cell death. 2. Surface changes in thymocytes undergoing apoptosis. *Am. J. Pathol.*, **115**, 426–36.
10. Wyllie, A. H., Kerr, J. F., and Currie, A. R. (1980). Cell death: the significance of apoptosis. *Int. Rev. Cytol.*, **68**, 251–306.
11. Trump, B. F., Valigorsky, J. M., Dess, J. H., Mergner, J. W., Kim, K. M., Jones, R. T., *et al.* (1973). Cellular change in human disease: A new method of pathological analysis. *Hum. Pathol.*, **4**, 89–109.
12. Duvall, E., Wyllie, A. H., and Morris, R. G. (1985). Macrophage recognition of cells undergoing programmed cell death (apoptosis). *Immunology*, **56**, 351–8.
13. McCarthy, N. J. and Evan, G. I. (1998). Methods for detecting and quantifying apoptosis. *Curr. Top. Dev. Biol.*, **36**, 259–78.
14. Cotter, T. G. and Martin, S. J. (1996). *Techniques in apoptosis: A users guide.* Portland Press: London.
15. Arends, M. and Wyllie, A. (1991). Apoptosis: mechanisms and roles in pathology. *Int. Rev. Exp. Pathol.*, **32**, 223–54.
16. Savill, J. (1997). Recognition and phagocytosis of cells undergoing apoptosis. *Br. Med. Bull.*, **53**, 491–508.
17. Meagher, L. C., Savill, J. S., Baker, A., Fuller, R. W., and Haslett, C. (1992). Phagocytosis of apoptotic neutrophils does not induce macrophage release of thromboxane B2. *J. Leukoc. Biol.*, **52**, 269–73.
18. Hall, S. E., Savill, J. S., Henson, P. M., and Haslett, C. (1994). Apoptotic neutrophils are phagocytosed by fibroblasts with participation of the fibroblast vitronectin receptor and involvement of a mannose/fucose-specific lectin. *J. Immunol.*, **153**, 3218–27.
19. Wyllie, A. H. (1992). Apoptosis and the regulation of cell numbers in normal and neoplastic tissues: an overview. *Cancer Metas. Rev.*, **11**, 95–103.
20. Fadok, V. A., McDonald, P. P., Bratton, D. L., and Henson, P. M. (1998). Regulation of macrophage cytokine production by phagocytosis of apoptotic and post-apoptotic cells. *Biochem. Soc. Trans.*, **26**, 653–6.
21. Ren, Y. and Savill, J. (1998). Apoptosis: the importance of being eaten. *Cell Death Differ.*, **5**, 563–8.
22. Horvitz, H. R. (1999). Genetic control of programmed cell death in the nematode *Caenorhabditis elegans*. *Cancer Res.*, **59**, 1701s–6s.
23. Smits, E., Van Criekinge, W., Plaetinck, G., and Bogaert, T. (1999). The human homologue of *Caenorhabditis elegans* CED-6 specifically promotes phagocytosis of apoptotic cells. *Curr. Biol.*, **9**, 1351–4.
24. Moynault, A., Luciani, M. F., and Chimini, G. (1998). ABC1, the mammalian homologue of the engulfment gene ced-7, is required during phagocytosis of both necrotic and apoptotic cells. *Biochem. Soc. Trans.*, **26**, 629–35.
25. Gregory, C. D., Devitt, A., and Moffatt, O. (1998). Roles of ICAM-3 and CD14 in the recognition and phagocytosis of apoptotic cells by macrophages. *Biochem. Soc. Trans.*, **26**, 644–9.
26. Fadok, V. A. (1999). Clearance: the last and often forgotten stage of apoptosis. *J. Mammary Gland Biol. Neoplasia*, **4**, 203–11.
27. Aderem, A. and Underhill, D. M. (1999). Mechanisms of phagocytosis in macrophages. *Annu. Rev. Immunol.*, **17**, 593–623.

28. Zou, H. and Niswander, L. (1996). Requirement for BMP signaling in interdigital apoptosis and scale formation. *Science*, **272**, 738–41.
29. Raff, M. C., Barres, B. A., Burne, J. F., Coles, H. S., Ishizaki, Y., and Jacobson, M.-D. (1993). Programmed cell death and the control of cell survival: lessons from the nervous system. *Science*, **262**, 695–700.
30. Smith, C. A., Williams, G. T., Kingston, R., Jenkinson, E. J., and Owen, J. J. T. (1989). Antibodies to CD3/T-cell receptor complex induce death by apoptosis in immature T cells in thymic cultures. *Nature*, **337**, 181–4.
31. Ellis, H. M. and Horovitz, H. R. (1986). Genetic control of programmed cell death in the nematode *C. elegans*. *Cell*, **44**, 817–29.
32. Cornillon, S., Foa, C., Davoust, J., Buonavista, N., Gross, J. D., and Golstein, P. (1994). Programmed cell death in *Dictyostelium*. *J. Cell Sci.*, **107**, 2691–704.
33. Greenberg, J. T. (1996). Programmed cell death: A way of life for plants. *Proc. Natl. Acad. Sci. USA*, **93**, 12094–7.
34. Vaux, D. L., Aguila, H. L., and Weissman, I. L. (1992). Bcl-2 prevents death of factor-deprived cells but fails to prevent apoptosis in targets of cell mediated killing. *Int. Immunol.*, **4**, 821–4.
35. McConkey, D. J., Hartzell, P., Nicotera, P., and Orrenius, S. (1989). Calcium-activated DNA fragmentation kills immature thymocytes. *FASEB J.*, **3**, 1843–9.
36. Wyllie, A. H. (1984). Chromatin cleavage in apoptosis; association of acondensed chromatin morphology and dependence on macromolecular synthesis. *J. Pathol.*, **142**, 67–77.
37. Korsmeyer, S. J., McDonnell, T. J., Nunez, G., Hockenbery, D., and Young, R. (1990). Bcl-2: B cell life, death and neoplasia. *Curr. Top. Microbiol. Immunol.*, **166**, 203–7.
38. Uetsuki, T., Takemoto, K., Nishimura, I., Okamoto, M., Niinobe, M., Momoi, T., *et al.* (1999). Activation of neuronal caspase-3 by intracellular accumulation of wild-type Alzheimer amyloid precursor protein. *J. Neurosci.*, **19**, 6955–64.
39. Evan, G. and Littlewood, T. (1998). A matter of life and cell death. *Science*, **281**, 1317–22.
40. Lundberg, A. S. and Weinberg, R. A. (1999). Control of the cell cycle and apoptosis. *Eur. J. Cancer*, **35**, 531–9.
41. Macleod, K. F., Hu, Y. W., and Jacks, T. (1996). Loss of Rb activates both p53-dependent and independent cell-death pathways in the developing mouse nervous-system. *EMBO J.*, **15**, 6178–88.
42. Jacks, T. (1996). Tumor suppressor gene mutations in mice. *Annu. Rev. Genet.*, **30**, 603–36.
43. Attardi, L. D. and Jacks, T. (1999). The role of p53 in tumour suppression: lessons from mouse models. *Cell. Mol. Life Sci.*, **55**, 48–63.
44. Jacks, T., Remington, L., Williams, B. O., Schmitt, E. M., Halachmi, S., Bronson, R. T., *et al.* (1994). Tumor spectrum analysis in p53-mutant mice. *Curr. Biol.*, **4**, 1–7.
45. Williams, B., Morgenbesser, S., Depinho, R., and Jacks, T. (1994). Tumorigenic and developmental effects of combined germ-line mutations in RB and p53. *Cold Spring Harbor Symp. Quant. Biol.*, **59**, 449–57.
46. Hueber, A. O. and Evan, G. I. (1998). Traps to catch unwary oncogenes. *Trends Genet.*, **14**, 364–7.
47. Evan, G. I. and Vousden, K. (2001). Proliferation, cell cycle and apoptosis in cancer. *Nature*, **411**, 342–8.
48. Huang, D. C. S., Oreilly, L. A., Strasser, A., and Cory, S. (1997). The anti-apoptosis function of Bcl-2 can be genetically separated from its inhibitory effect on cell cycle entry. *EMBO J.*, **16**, 4628–38.

49. Oreilly, L. A., Huang, D. C. S., and Strasser, A. (1996). The cell death inhibitor Bcl-2 and its homologues influence control of cell cycle entry. *EMBO J.*, **15**, 6979–90.
50. Marin, M., Hsu, B., Stephens, L., Brisbay, S., and Mcdonnell, T. (1995). The functional basis of c-myc and bcl-2 complementation during multistep lymphomagenesis *in vivo*. *Exp. Cell Res.*, **217**, 240–7.
51. Reed, J. (1997). Double identity for proteins of the Bcl-2 family. *Nature*, **387**, 773–6.
52. Reed, J. C., Cuddy, M., Haldar, S., Croce, C., Nowell, P., Makover, D., *et al.* (1990). BCL2-mediated tumorigenicity of a human T-lymphoid cell line: synergy with MYC and inhibition by BCL2 antisense. *Proc. Natl. Acad. Sci. USA*, **87**, 3660–4.
53. Strasser, A., Whittingham, S., Vaux, D. L., Bath, M. L., Adams, J. M., Cory, S., *et al.* (1991). Enforced BCL2 expression in B-lymphoid cells prolongs antibody responses and elicits autoimmune disease. *Proc. Natl. Acad. Sci. USA*, **88**, 8661–5.
54. Green, D. and Reed, J. (1998). Mitochondria and apoptosis. *Science*, **281**, 1309–11.
55. Jurgensmeier, J. M., Xie, Z. H., Deveraux, Q., Ellerby, L., Bredesen, D., and Reed, J. C. (1998). Bax directly induces release of cytochrome c from isolated mitochondria. *Proc. Natl. Acad. Sci. USA*, **95**, 4997–5002.
56. Newton, K. and Strasser, A. (1998). The Bcl-2 family and cell death regulation. *Curr. Opin. Genet. Dev.*, **8**, 68–75.
57. Krammer, P. H. (1998). The CD95(APO-1/Fas)/CD95L system. *Toxicol. Lett.*, **103**, 131–7.
58. Nagata, S. (1997). Apoptosis by death factor. *Cell*, **88**, 355–65.
59. Chinnaiyan, A., Tepper, C., Seldin, M., O'Rourke, K., Kischkel, F., Hellbardt, S., *et al.* (1996). FADD/mort1 is a common mediator of CD95 (Fas/Apo-1) and tumor necrosis factor receptor induced apoptosis. *J. Biol. Chem.*, **271**, 4961–5.
60. Los, M., Van de Craen, M., Penning, L., Schenk, H., Westendorp, M., Baeuerle, P., *et al.* (1995). Requirement of an ICE/CED-3 protease for FAS/APO-1 mediated apoptosis. *Nature*, **375**, 81–3.
61. Muzio, M., Chinnaiyan, A., Kischkel, F., O'Rourke, K., Shevchenko, A., Ni, J., *et al.* (1996). FLICE, a novel FADD-homologous ice/ced-3-like protease, is recruited to the CD95 (Fas/Apo-1) death-inducing signaling complex. *Cell*, **85**, 817–27.
62. Tschopp, J., Irmler, M., and Thome, M. (1998). Inhibition of Fas death signals by FLIPs. *Curr. Opin. Immunol.*, **10**, 552–8.
63. Scaffidi, C., Fulda, S., Srinivasan, A., Friesen, C., Li, F., Tomaselli, K. J., *et al.* (1998). Two CD95 (APO-1/Fas) signaling pathways. *EMBO J.*, **17**, 1675–87.
64. Williams, G. T., Smith, C. A., Spooncer, E., Dexter, T. M., and Taylor, D. R. (1990). Haemopoietic colony stimulating factors promote cell survival by suppressing apoptosis. *Nature*, **343**, 76–9.
65. Harrington, E. A., Bennett, M. R., Fanidi, A., and Evan, G. I. (1994). c-Myc induced apoptosis in fibroblasts is inhibited by specific cytokines. *EMBO J.*, **13**, 3286–95.
66. Kauffman Zeh, A., RodriguezViciana, P., Ulrich, E., Gilbert, C., Coffer, P., Downward, J., *et al.* (1997). Suppression of c-Myc-induced apoptosis by Ras signalling through PI3K and PKB. *Nature*, **385**, 544–8.
67. Cardone, M., Roy, N., Stennicke, H., Salveson, G., Franke, T., Stanbridge, E., *et al.* (1998). Regulation of cell death protease caspase 9 by phosphorylation. *Science*, **282**, 1318–21.
68. Romashkova, J. A. and Makarov, S. S. (1999). NF-kappaB is a target of AKT in anti-apoptotic PDGF signalling. *Nature*, **401**, 86–90.
69. Kulik, G. and Weber, M. J. (1998). Akt-dependent and -independent survival signaling pathways utilized by insulin-like growth factor I. *Mol. Cell. Biol.*, **18**, 6711–18.

70. Green, D. and Kroemer, G. (1998). The central executioners of apoptosis: caspases or mitochondria? *Trends Cell Biol.*, **8**, 267–71.
71. Yoshida, H., Kong, Y. Y., Yoshida, R., Elia, A. J., Hakem, A., Hakem, R., *et al.* (1998). Apaf1 is required for mitochondrial pathways of apoptosis and brain development. *Cell*, **94**, 739–50.
72. Zou, H., Henzel, W. J., Liu, X. S., Lutschg, A., and Wang, X. D. (1997). Apaf-1, a human protein homologous to *C. elegans* CED-4, participates in cytochrome c-dependent activation of caspase-3. *Cell*, **90**, 405–13.
73. Kroemer, G., Dallaporta, B., and RescheRigon, M. (1998). The mitochondrial death/life regulator in apoptosis and necrosis. *Annu. Rev. Physiol.*, **60**, 619–42.
74. Juin, P., Hueber, A. O., Littlewood, T., and Evan, G. (1999). c-Myc-induced sensitization to apoptosis is mediated through cytochrome c release. *Genes Dev.*, **13**, 1367–81.
75. Du, C., Fang, M., Li, L., and Wang, X. (2000). Smac, a mitochondrial protein that promotes cytochrome c-dependent caspase activation by eliminating IAP inhibition. *Cell*, **102**, 33–42.
76. Verhagen, A., Ekert, P. G., Pakusch, M., Silke, J., Connolly, L. M., Reid, G. E., *et al.* (2000). Identification of DIABLO, a mammalian protein that promotes apoptosis by binding to and antagonizing IAP proteins. *Cell*, **102**, 43–53.
77. Yang, Y. L. and Li, X. M. (2000). The IAP family: endogenous caspase inhibitors with multiple biological activities. *Cell Res.*, **10**, 169–77.
78. Silke, J., Verhagen, A. M., Ekert, P. G., and Vaux, D. L. (2000). Sequence as well as functional similarity for DIABLO/Smac and Grim, Reaper and Hid? *Cell Death Differ.*, **12**, 1275.
79. Green, D. (2000). Apoptotic pathways: Paper wraps stone blunts scissors. *Cell*, **102**, 1–4.
80. Deveraux, Q. L., Roy, N., Stennicke, H. R., Van Arsdale, T., Zhou, Q., Srinivasula, S. M., *et al.* (1998). IAPs block apoptotic events induced by caspase-8 and cytochrome c by direct inhibition of distinct caspases. *EMBO J.*, **17**, 2215–23.
81. Mills, J. C., Stone, N. L., and Pittman, R. N. (1999). Extranuclear apoptosis. The role of the cytoplasm in the execution phase. *J. Cell Biol.*, **146**, 703–8.
82. Nicotera, P., Leist, M., Single, B., and Volbracht, C. (1999). Execution of apoptosis: converging or diverging pathways. *Biol. Chem.*, **380**, 1035–40.
83. Lazebnik, Y., Takahashi, A., Poirier, G., Kaufmann, S., and Earnshaw, W. (1995). Characterization of the execution phase of apoptosis *in vitro* using extracts from condemned-phase cells. *J. Cell Sci.*, **19**, 41–9.
84. Nicholson, D. and Thornberry, N. (1997). Caspases: killer proteases. *Trends Biochem. Sci.*, **8**, 299–306.
85. Yuan, J., Shaham, S., Ledoux, S., Ellis, H. M., and Horvitz, H. R. (1993). The *C. elegans* cell death gene ced-3 encodes a protein similar to mammalian interleukin-1 beta-converting enzyme. *Cell*, **75**, 641–52.
86. Fraser, A. G. and Evan, G. I. (1997). Identification of a *Drosophila melanogaster* ICE/CED3-related protease, drICE. *EMBO J.*, **16**, 2805–13.
87. Thornberry, N. and Lazebnik, Y. (1998). Caspases: Enemies within. *Science*, **281**, 1312–16.
88. Borner, C. and Monney, L. (1999). Apoptosis without caspases: an inefficient molecular guillotine? *Cell Death Differ.*, **6**, 497–507.
89. McCarthy, N. J., Whyte, M. K. B., Gilbert, C. S., and Evan, G. I. (1997). Inhibition of Ced-3/ICE-related proteases does not prevent cell death induced by oncogenes, DNA damage, or the Bcl-2 homologue Bak. *J. Cell Biol.*, **136**, 215–27.

90. Los, M., Wesselborg, S., and Schulze-Osthoff, K. (1999). The role of capsases in development, immunity and apoptotic signal transduction: Lessons from knockout mice. *Immunity*, **10**, 629–39.
91. Meier, P. and Evan, G. (1998). Dying like flies. *Cell*, **95**, 295–8.
92. Abrams, J. M. (1999). An emerging bluprint for apoptosis in *Drosophila*. *Trends Cell Biol.*, **9**, 435–40.
93. Ranger, A. M., Malynn, B. A., and Korsmeyer, S. J. (2001). Mouse models of cell death. *Nature Genet.*, **28**, 113–18.
94. Lee, E. Y., Chang, C. Y., Hu, N., Wang, Y. C., Lai, C. C., Herrup, K., *et al.* (1992). Mice deficient for Rb are non-viable and show defects in neurogenesis and haematopoiesis. *Nature*, **359**, 288–94.
95. Clarke, A. R., Maandag, E. R., van Roon, M., van de lugt, N. M., van der Valk, M., Hooper, M. L., *et al.* (1992). Requirement for a functional Rb-1 gene in murine development. *Nature*, **359**, 328–30.
96. Jacks, T., Fazeli, A., Schmitt, E. M., Bonson, R. T., Goodell, M. A., and Weinberg, R. A. (1992). Effects of an Rb mutation in the mouse. *Nature*, **359**, 295–300.
97. Pelengaris, S., Littlewood, T., Khan, M., Elia, G., and Evan, G. (1999). Reversible activation of c-Myc in skin; induction of a complex neoplastic phenotype by a single oncogenic lesion. *Mol. Cell*, **3**, 565–77.
98. Evan, G. I., Wyllie, A. H., Gilbert, C. S., Littlewood, T. D., Land, H., Brooks, M., *et al.* (1992). Induction of apoptosis in fibroblasts by c-myc protein. *Cell*, **69**, 119–28.
99. Pelengaris, S., Ruldolf, B., and Littlewood, T. D. (2000). Action of Myc *in vivo*—proliferation and apoptosis. *Curr. Opin. Genet. Dev.*, **10**, 100–5.
100. Martin, S., Newmeyer, D., Mathias, S., Farschon, D., Wang, H., Reed, J., *et al.* (1995). Cell-free reconstitution of fas-induced, uv-radiation-induced and ceramide-induced apoptosis. *EMBO J.*, **14**, 5191–200.
101. Newmeyer, D. D., Farschon, D. M., and Reed, J. C. (1994). Cell-free apoptosis in *Xenopus* egg extracts: inhibition by Bcl-2 and requirement for an organelle fraction enriched in mitochondria. *Cell*, **79**, 353–64.
102. Cosulich, S., Green, S., and Clarke, P. (1996). Bcl-2 regulates activation of apoptotic proteases in a cell-free system. *Curr. Biol.*, **6**, 997–1005.
103. Matsuyama, S., Nouraini, S., and Reed, J. C. (1999). Yeast as a tool for apoptosis research. *Curr. Opin. Microbiol.*, **2**, 18–23.
104. Weinrauch, Y. and Zychlinsky, A. (1999). The induction of apoptosis by bacterial pathogens. *Annu. Rev. Microbiol.*, **53**, 155–87.
105. Roulston, A., Marcellus, R. C., and Branton, P. E. (1999). Viruses and apoptosis. *Annu. Rev. Microbiol.*, **53**, 577–628.
106. Henderson, S., Huen, D., Rowe, M., Dawson, C., Johnson, G., and Rickinson, A. (1993). Epstein–Barr virus-coded BHRF1 protein, a viral homologue of Bcl-2, protects human B cells from programmed cell death. *Proc. Natl. Acad. Sci. USA*, **90**, 8479–83.
107. Tschopp, J., Thome, M., Hofmann, K., and Meinl, E. (1998). The fight of viruses against apoptosis. *Curr. Opin. Genet. Dev.*, **8**, 82–7.
108. Liston, P., Roy, N., Tamai, K., Lefebvre, C., Baird, S., ChertonHorvat, G., *et al.* (1996). Suppression of apoptosis in mammalian cells by NAIP and a related family of IAP genes. *Nature*, **379**, 349–53.
109. Miller, L. K. (1999). An exegesis of IAPs: salvation and surprises from BIR motifs. *Trends Cell Biol.*, **9**, 323–8.

110. Fraser, A. and James, C. (1998). Fermenting debate: do yeast undergo apoptosis? *Trends Cell Biol.*, **8**, 219–21.
111. Lam, E., Pontier, D., and del Pozo, O. (1999). Die and let live—programmed cell death in plants. *Curr. Opin. Plant Biol.*, **2**, 502–7.
112. Pellegrini, M. and Strasser, A. (1999). A portrait of the Bcl-2 protein family: life, death and the whole picture. *J. Clin. Immunol.*, **19**, 365–77.

2 | Programmed cell death in *C. elegans*: the genetic framework

DING XUE, YI-CHUN WU, and MANISHA S. SHAH

1. Introduction

Programmed cell death is an important cellular process that controls the development and homeostasis of multicellular organisms, including the nematode *Caenorhabditis elegans*. Genetic and molecular studies in *C. elegans* have played a critical part in defining a genetic pathway of programmed cell that is evolutionarily conserved between nematodes and mammals (reviewed in Chapter 1). In this chapter we will review our current understanding of programmed cell death in *C. elegans* and the insights gained from functional characterization of the *C. elegans* cell death genes.

1.1 Advantages of using *C. elegans* for the study of programmed cell death

C. elegans is a free-living worm that feeds on bacteria and can be easily maintained in the laboratory (1). Its rapid life cycle (about 2.5 days at 25#°C), self-fertilizing ability (of the hermaphrodite), and its well-defined anatomy have made it an ideal organism for sophisticated genetic manipulations. Furthermore, its transparency makes it ideal for developmental studies: cell divisions and cell deaths can be observed and followed in living animals using high magnification Nomarski DIC optics and has enabled the determination of the entire cell lineage of *C. elegans* (2–4). Of the 1090 somatic cells that are generated during the development of the adult hermaphrodite, 131 undergo programmed cell death (2–4). When observed with Nomarski microscopy, these dying cells adopt a refractile and raised button-like appearance (Fig. 1). Each of the 131 deaths occurs at a specific time and place and is essentially invariant from animal to animal. Therefore, mutants with a subtle perturbation in the cell death programme can be identified and isolated.

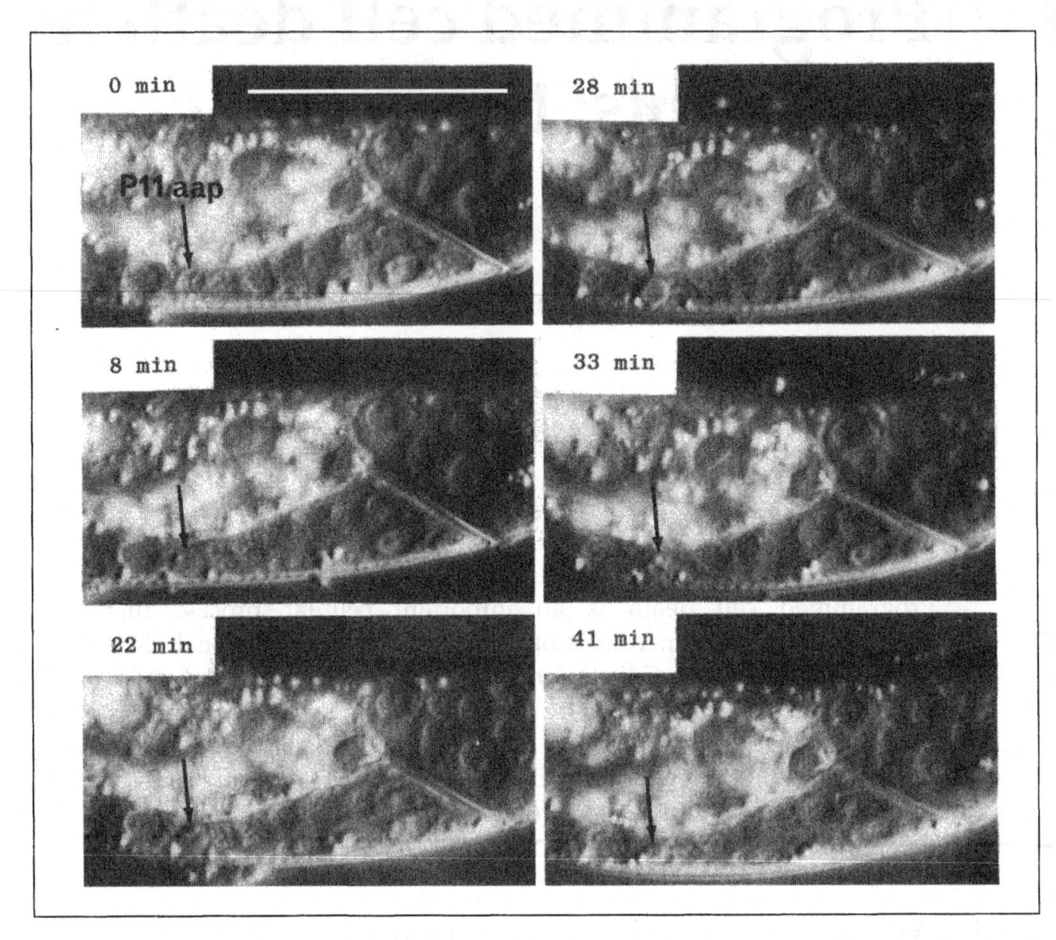

Fig. 1 Morphological changes of a dying cell during the cell death process as viewed with Nomarski optics. A dying cell, P11.aap from a L1 hermaphrodite, is indicated by an *arrow*. The cell shows maximum refractility at approximately 28 minutes after its birth (upper right panel). The scale bar represents 20 μm. (Reprinted, with permission from ref. 2.)

Genetic and phenotypic analyses of mutants that are defective in different aspects of programmed cell death have helped define a genetic pathway of programmed cell death in *C. elegans*. The combination of a detailed genetic map, a corresponding physical map (5), and the information provided by the almost complete genomic sequence (only 1% of the *C. elegans* genome left to be determined) (6) further facilitates the molecular studies of programmed cell death in *C. elegans*. Powerful molecular and genetic techniques such as germline transformation (7) and genetic mosaic analysis (8) have been developed which, in combination with detailed knowledge of the anatomy and cell lineage of *C. elegans*, have catalysed rapid progress in our understanding of the mechanisms of programmed cell death.

1.2 Origin and identity of dying cells in *C. elegans*

Most (113/131) developmental cell deaths in a hermaphrodite occur during embryogenesis, many between 250 and 450 minutes after fertilization (4). The remainder occur during early larval development (2). In males a few more cell deaths are observed in male-specific lineages and during late larval stages. In adult animals, no more somatic cell deaths are observed in either sex. However, a large number of cells in the germline of hermaphrodites, but not those of males, undergo programmed cell death (9).

Programmed cell death can be viewed as a terminal differentiation fate. The cell types programmed to die have been inferred mainly from their nuclear morphology before completion of the death process, from the fate of their lineally equivalent homologues that live and, more directly, from the cell fates they adopt when they survive in mutants that lack programmed cell deaths. It appears that most of the cell deaths in *C. elegans* are neuronal deaths: in hermaphrodites, 105/131 of the cells that die are neurons; in males it is 117/147. The other types of cells that undergo programmed cell death include neuron-associated cells, hypodermal cells, muscle cells, and pharyngeal gland cells (2, 4). No intestinal cells die in *C. elegans*, although deaths of intestinal cells have been observed in other nematode species (10).

Programmed cell deaths are not confined to a specific cell lineage during *C. elegans* development. Rather, they are asymmetrically distributed among cell lineages (Fig. 2). Deaths are found in descendants of three primary blastomeres (AB, MS, and C), and most (116/131) of these deaths occur in the AB lineage that produces much of the nervous system. Cell death appears to be a common fate in the AB lineage in which 116 of the 722 cells generated proceed to die, presumably a reflection of the many dying neurons. No cell death is found in the E lineage, which generates only intestine, or the D lineage, which generates only muscle. The germline produced from the P4 lineage gives rise to variable number of cell deaths in hermaphrodite adults (9).

1.3 Roles of programmed cell death in *C. elegans*

Programmed cell death plays several important roles in animal development (25), and the functions that have been ascribed to cell suicides in vertebrate development can also be observed clearly in *C. elegans*. The elimination of surplus cells or cells that have already fulfilled their functions and are no longer needed is one example: in *C. elegans*, the linker cell of the male gonad is generated in the second larval stage and guides the extension of the gonad during development (3). Once extension of the gonad is completed at the fourth larval stage, the linker cell is no longer needed and undergoes programmed cell death. Another function is the generation of sexual dimorphism: in *C. elegans*, cells specific to one sex die during the development of the opposite sex (4). A third example is the involvement of PCD in the sculpturing process that gives rise to species-specific organs and body structures. In *C. elegans*, two distal tip cells guide the extension of the anterior and posterior ends of the

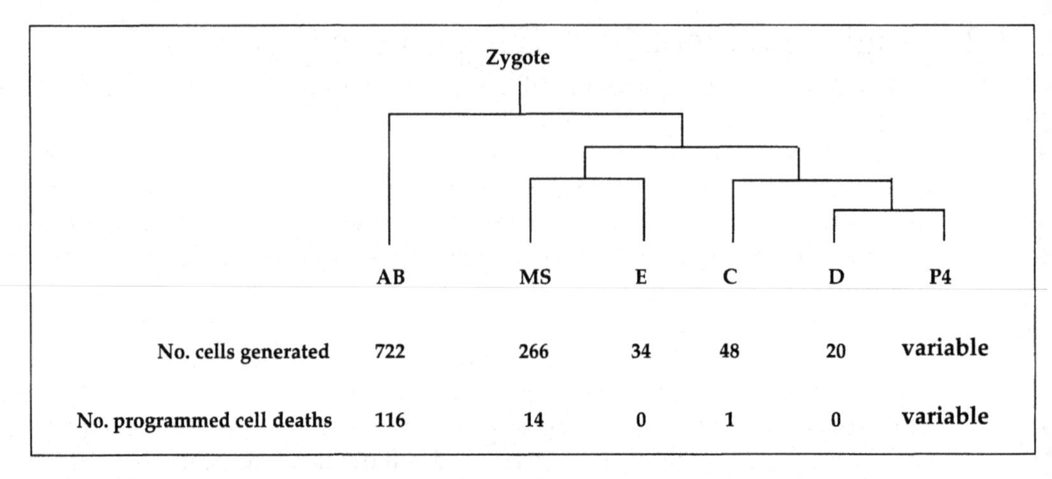

Fig. 2 Asymmetric distribution of programmed cell deaths among the lineages of the *C. elegans* hermaphrodite. Early divisions represented by horizontal lines produce six founder cells—AB, MS, E, C, D, and P4. The number of cells generated and the number of cell deaths observed from each cell lineage are indicated. The AB lineage generates hypodermis, neurons, and muscle. The MS lineage generates muscle, glands, and neurons. The C lineage generates hypodermis, neurons, and muscle. The D and E lineages produce only gut and muscle, respectively. The P4 lineage gives rise to variable number of cells and cell deaths in the germline.

developing gonads and lead to two symmetric gonadal arms, whereas in females of another nematode species, *P. redivivus*, the posterior distal tip cell undergoes programmed cell death, leading to the development of a one-armed gonad (12). Laser ablation of the posterior distal tip cell in *C. elegans* leads to a cessation of growth of the posterior gonadal arm (11), mimicking gonad development in *P. redivivus*.

1.4 Morphology and kinetics of programmed cell death in *C. elegans*

The morphological changes of cells undergoing programmed cell death have been characterized using Nomarski DIC optics and electron microscopy (2, 13). The cell programmed to die is often smaller than its sister cell after mitosis. When viewed with Nomarski DIC optics, the dying cell first appears less refractile, initially in the cytoplasm and then in the nucleus. Soon after that, both cytoplasm and nucleus become highly refractile, adopting a raised and flattened button-like appearance (Fig. 1). The mechanism underlying the refractility change during the death process is not well understood. Eventually, the refractile cell corpse disappears as it is engulfed and digested by its neighbouring cell. The entire process of cell death from division to the disappearance of the cell takes approximately one hour (Fig. 1) (2, 14).

At the ultrastructural level, as a cell undergoes programmed cell death the cytoplasm condenses and the nuclear chromatin aggregates. The engulfment process appears to occur at a very early stage of cell death: membrane processes extending from the neighbouring engulfing cell can be found even before the dying cell displays any visible morphological change (13). During the mid-stage of the cell death process, the body of the dead cell is split into several membrane-bound fragments by phagocytotic arms of the engulfing cell. Internal membranes and plasma membranes of the dead cell adopt a whorled appearance and often are enclosed within autophagic vacuoles. At the final stage, portions of the fragmented cell body fuse with vacuoles inside the engulfing cell (13).

C. elegans programmed cell death shares features with those cell deaths observed in apoptotic cell death in both invertebrates and vertebrates (15, 16). Both cell deaths show similar characteristic morphological changes at the ultrastructural level, such as cytoplasmic condensation, chromatin aggregation, and engulfment of dead cells by neighbouring cells. Recently, DNA fragmentation has also been detected during *C. elegans* programmed cell death as assayed using the TUNEL (TdT-mediated dUTP nick end-labelling) technique (Y. C. Wu, G. M. Stanfield, and H. R. Horvitz, personal communication). However, the formation of the nucleosomal DNA ladder, a hallmark of apoptosis, has yet to be demonstrated in *C. elegans*.

2. Genetic and molecular analysis of programmed cell death in *C. elegans*

2.1 The genetic pathway of programmed cell death in *C. elegans*

Over the past twenty years, genetic studies in *C. elegans* have led to the identification of more than a dozen genes that are involved in different aspects of programmed cell death. Three genes, *nuc-1* (nuclease-deficient), *ced-1*, and *ced-2* (cell death abnormality), were first identified as genes involved in the removal or degradation of cell corpses: mutations in the *nuc-1* gene block degradation of DNA from dead cells; mutations in the *ced-1* or the *ced-2* gene prevent the engulfment of many cell corpses, leading to the mutant phenotype of persistent cell corpses (Fig. 3) (14, 17). Subsequently, mutations in two additional genes, *ced-3* and *ced-4*, were isolated (18). Strikingly, mutations in both *ced-3* and *ced-4* genes prevent most, if not all, programmed cell deaths in *C. elegans*, suggesting that both genes are required for the execution of cell death. Mosaic analysis of *ced-3* and *ced-4* mutants suggests that both genes act within dying cells to cause cell death (19). This observation provided the first genetic evidence that cells die by an intrinsic suicide mechanism. Soon after that, more cell death genes were identified in a number of different genetic screens. These include four additional genes that mediate cell corpse engulfment (*ced-5, ced-6, ced-7,* and *ced-10*) (20), two genes (*ces-1* and *ces-2*; cell death specification) that specify the death of a specific set of cells (21), the *ced-9* gene that generally protects cells from programmed cell death (22), and finally, the *egl-1* gene that is also required for almost all programmed cell deaths (23).

The *ced-9* gene was initially identified by a gain-of-function (*gf*) mutation that prevents most programmed cell deaths in *C. elegans*. The normal function of *ced-9* was revealed by the phenotype of loss-of-function (*lf*) mutants in which many cells that normally live undergo programmed cell death, suggesting that *ced-9* acts to

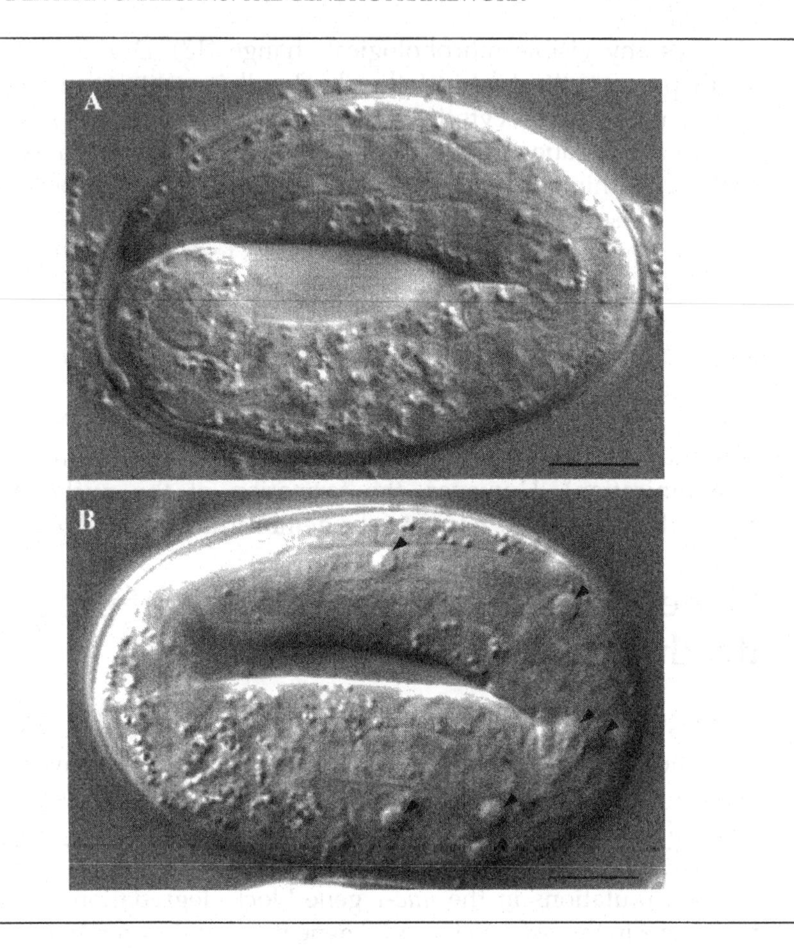

Fig. 3 An engulfment-defective mutant exhibits persistent cell corpses. (A) Nomarski micrograph of a wild-type four-fold embryo (~ 700 min) that lacks cell corpses. Only the anterior two-thirds of the embryo is in focus. (B) Nomarski micrograph of a four-fold embryo of a *ced-1* (*e1735*) mutant with many persistent cell corpses (indicated by *arrowheads*). The scale bar represents 10 μm.

protect cells from programmed cell death (22). The *egl-1* gene was originally defined by several gain-of-function mutations that cause inappropriate death of HSN neurons in hermaphrodites (24). Subsequent isolation and examination of an *egl-1* loss-of-function mutation suggest that the *egl-1* gene is required for almost all programmed cell deaths rather than just playing a role in specifying the death of HSN neurons as its *gf* mutant phenotype implicated (23).

Genetic epistasis analyses and phenotypic analyses of the above cell death mutants have placed these cell death genes into a genetic pathway that contains four sequential and genetically separable steps of cell death:

- the specification of which cell should die
- the killing process of cell death

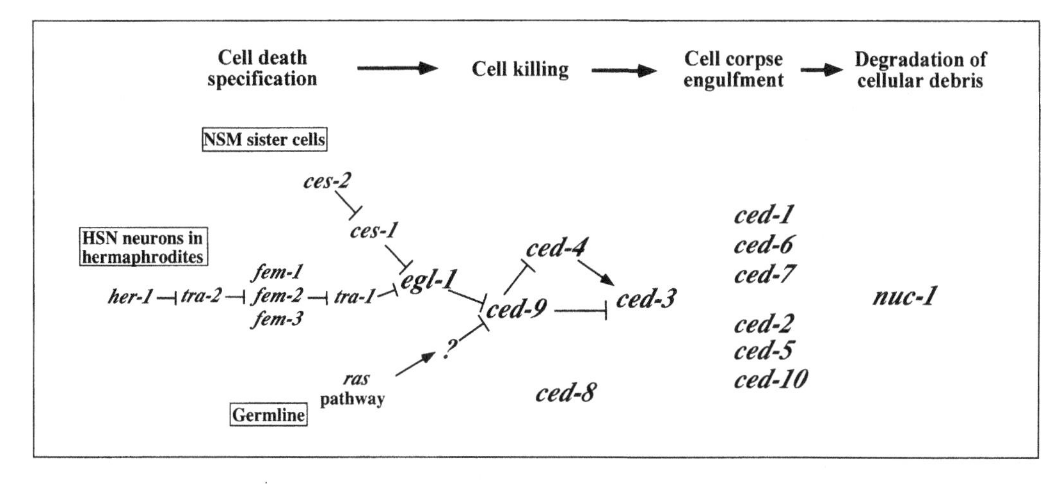

Fig. 4 The genetic pathway of programmed cell death in *C. elegans*. Four sequential steps of programmed cell death are indicated. In the cell death specification step, genes that are involved in regulating the death fates of three specific cell types (sister cells of NSM neurons, HSN neurons, and germline) are shown. In germline, activation of the *ras*/MAP kinase pathway promotes the exit of germ cells from the meiotic arrest. Some of these germ cells proceed to undergo programmed cell death. There are two partially redundant pathways (*ced-1, -6, -7* and *ced-2, -5, -10*, respectively) that mediate the engulfment of cell corpses. The position of the *ced-8* gene in the cell death pathway is not clear.

- the engulfment of cell corpses
- the degradation of cellular debris (Fig. 4) (25)

In the following section, we will review current status of the genetic and molecular characterization of genes in this pathway.

2.2 Genes involved in the killing process of programmed cell death

2.2.1 CED-3—the death protease that executes programmed cell death

The *ced-3* gene is required for most, if not all, programmed cell deaths in *C. elegans*: in strong *ced-3* loss-of-function mutants, very few, if any cell deaths occur (18). The undead cells in *ced-3* mutants never divide, but can differentiate to adopt a cell fate that is very similar to that of their sister or aunt cell or to cells at equivalent positions in a related cell lineage (18). Some of the undead cells can even function under special circumstances (26). Intriguingly, unlike higher multicellular organisms, the large number of extra undead cells in *ced-3* mutant animals, constituting 12% more somatic cells, does not seem to severely interfere with the functions of normal cells, or the animal as a whole. *ced-3* mutants are superficially indistinguishable from wild-type animals (18). However, subtle defects such as slow growth, slightly reduced brood size, and an impaired chemotaxis response have been observed with *ced-3* mutant animals (25, 27), suggesting that an increased cell number may lead to subtle differences in animal behaviour, but not the development of diseases such as cancer that are associated with increased cell number in long-lived mammals.

Genetic mosaic analysis of *ced-3* suggests that *ced-3* acts within dying cells to cause programmed cell death (19). *ced-3* encodes a protein with significant sequence similarity to a family of cysteine proteases named caspases (cysteine aspartate-specific protease), which cleave their substrates exclusively after an aspartate residue (28, 29). Several caspases have been shown to mediate apoptosis in other organisms (30–34) (see Chapter 3) and are important for the execution of programmed cell death in many species. Like many caspases, CED-3 is initially synthesized as a 56 kDa proenzyme and can be proteolytically activated to generate an active cysteine protease that is composed of p17/p15 or p17/p13 protease subunits (29, 35). The active CED-3 protease has substrate specificity similar to that of mammalian caspase-3 (35) (see Chapter 3). Several mutations that severely reduce CED-3 killing activity in nematodes affect the residues that are conserved among many caspases and which appear to be critical for the catalytic activity of the caspases (28, 36–39). *In vitro* analysis of mutant CED-3 proteases demonstrates that the reduction in protease activity correlates directly with the reduction in the capacity of CED-3 to kill *in vivo* (35). These observations indicate that the protease activity of CED-3 is essential for its role in *C. elegans* programmed cell death.

i. The regulation of ced-3

How *ced-3* killing activity is activated in the right cells and at the right time has been intensively investigated. The *ced-3* gene appears to be expressed in most, if not all, cells in *C. elegans*, since mutations in *ced-3* can block the ectopic deaths of many cells from different cell types caused by *ced-9(lf)* mutations (22). This observation argues that *ced-3* is expressed in many cells enabling it to mediate the death of normally living cells in *ced-9(lf)* mutants and in addition, that CED-3 activity is inhibited by CED-9 in living cells. Ectopic overexpression of *ced-3* in *C. elegans* can result in constitutive activation of the *ced-3* protein and the death of the cells where *ced-3* is expressed (40). This result and the observation that the CED-3 proenzyme can be proteolytically activated when overexpressed in bacteria suggest that the activation of CED-3 can be achieved by increased expression or concentration of the CED-3 proenzyme (35). The expression patterns of *ced-3* have not been determined *in C. elegans*. It is not clear whether the expression of *ced-3* is increased in cells that are fated to die.

Oligomerization of caspase zymogens that are linked to trimerized surface receptors or other oligomerized protein complexes has been shown to be important for the activation of mammalian procaspases such as procaspase-8 and procaspase-9. Presumably oligomerization brings the protease domains of these caspase zymogens into close proximity and facilitates intermolecular proteolytic cleavage (42–44) (see Chapters 4 and 8). Similarly, induced oligomerization of CED-3 protease domains can also result in the processing of CED-3 into mature protease subunits (p15/p17) *in vitro* and the activation of CED-3 killing activity in cell culture (45). Increased concentration of CED-3 in cells may enhance the possibility of CED-3 oligomerization and thus the chance of its activation. It has also been postulated that oligomerization of CED-3 can be achieved by a second mechanism: the oligomerization of another essential cell death protein CED-4 (45, 46). CED-4 has been shown to interact with CED-3 and can self-oligomerize *in vitro*. Mutant CED-4 proteins that cannot oligomerize fail to activate CED-3 killing activity in mammalian cells, suggesting that CED-4 oligomerization may be an important part of the CED-3 activation process (45).

Three additional *C. elegans* caspases have been identified, one of which can process CED-3 *in vitro* (41). However, whether these three *C. elegans* caspases have any role in programmed cell death or in CED-3 activation remains to be determined.

The activity of CED-3 can also be controlled by negative regulators. The baculovirus p35 protein, which is required to block baculovirus-infected insect cells from apoptosis, has been shown to be able to inhibit cell death in diverse organisms including *C. elegans* (47–51). Biochemical and molecular genetic studies suggest that p35 appears to do so by directly inhibiting the protease activity of CED-3 and other caspases (52, 53). So far, no homologue of p35 has been identified in *C. elegans*. However, the general cell death inhibitor CED-9 may play a similar role. CED-9 has been found to be a substrate of the CED-3 protease *in vitro* (54). Mutations that disrupt CED-3 cleavage sites in CED-9 significantly reduce the death protective activity of CED-9 in *C. elegans*, suggesting that CED-9 may function directly as a substrate inhibitor of CED-3 (54). The study of when and where *ced-3* is expressed, the level of *ced-3* expression and its subcellular localization in dying and living cells, and the expression patterns of *ced-3* in different cell death mutants will be important for understanding how CED-3 is activated.

ii. Downstream targets of ced-3

Activation of *ced-3* is critical for cell death, however, the mechanism of how activated CED-3 promotes the death of the cell is unknown. Cleavage of CED-3 targets may activate some death promoting activities and/or inactivate death inhibiting regulators and thus lead to systematic cell disassembly and the eventual recognition and engulfment of the cell corpse by its neighbouring cell. So far no genes that act downstream of *ced-3* in the genetic pathway have been found to encode a substrate of the CED-3 protease. Given that CED-3 causes cell death by cleaving multiple substrates which may lead to simultaneous activation of many facets of the cell death execution process, elimination of one of the CED-3 substrates by genetic mutations may not result in obvious cell death defects that could be detected in the genetic screens carried out to date. Instead, these mutants may partially suppress cell death or delay cell death. Using this criteria for genetic screens, several new genes have been identified that may act immediately downstream of *ced-3* to execute cell death (D. Ledwich and D. Xue, unpublished results). Genetic screens of this kind should help identify the molecular components and the genetic pathways that mediate the execution of cell death by the CED-3 protease.

2.2.2 CED-4—the regulator of CED-3 activation

Like the *ced-3* gene, the activity of *ced-4* is also required within the cell programmed to die (19). *ced-4* encodes a protein that is similar to mammalian Apaf-1 (apoptotic protease activating factor 1), a factor that is critical for the activation of mammalian caspases involved in apoptosis (46, 55–57) (see Chapters 5 and 6). By analogy, CED-4 may play a similar role in activating CED-3. Northern blot analysis demonstrates that the *ced-4* transcript is found primarily during embryo development where most programmed cell deaths occur (113 out of 131 in hermaphrodites) (46). *ced-4* expression is not altered by mutations in the *ced-3* gene, suggesting that *ced-3* does not regulate the expression of *ced-4*. Like *ced-3*, *ced-4* appears to be expressed in many cells that are fated to live, since loss-of-function mutations in *ced-4* also block the ectopic deaths of many cells in *ced-9(lf)* mutant animals (22).

Like CED-3, ectopic overexpression of CED-4 can result in the death of cells that normally survive (40). However, this cell killing caused by overexpression of CED-4 is not very efficient. The killing efficiency is greatly increased if the endogenous *ced-9* activity is eliminated by *ced-9(lf)* mutations, but markedly reduced in *ced-3(lf)* mutants (40). These observations and the finding that *ced-4(lf)* mutations completely block ectopic cell deaths in *ced-9(lf)* mutants indicate that *ced-4* acts genetically downstream of or in parallel to *ced-9*, but upstream of or in parallel to *ced-3* (Fig. 4). Consistent with this genetic ordering, cell killing mediated by overexpression of CED-3 is not obviously affected in the absence of the endogenous *ced-4* activity (40).

Biochemical studies of CED-3, CED-4, and CED-9 have provided important insights into how these three proteins may interact with one another to regulate cell death. CED-4 has been found to physically interact with both CED-3 and CED-9 *in vitro*. Moreover, CED-4 can enhance the activation of CED-3 killing activity in cultured cells when co-transfected with *ced-3*. This effect can be suppressed by co-transfection with *ced-9* (58–62). The binding of CED-9 and CED-3 to CED-4 is not mutually exclusive (45, 59). This suggests that CED-3, CED-4, and CED-9 in living cells may coexist as a ternary protein complex in which CED-3 exists as an inactive proenzyme (Fig. 5 A). In cells that are programmed to die, certain cell death factors or signals will trigger the release of the CED-4/CED-3 complex from CED-9 and the subsequent oligomerization of CED-4 will bring CED-3 proenzymes into close proximity and lead to the autoproteolytic activation of CED-3 (Fig. 5) (45, 59, 63, 64). It is also possible that CED-3 does not associate with the CED-4/CED-9 complex and exists as monomer in the cytosol. In this case, the release of CED-4 from CED-9 would allow for the formation of a CED-3/CED-4 complex and the subsequent CED-3 activation. Studies of the subcellular localization patterns of these three proteins in *C. elegans* cells should be able to address this issue.

i. CED-4 and Apaf-1

CED-4 and Apaf-1 share sequence similarity over a stretch of 320 amino acids which contain Walker's A- and B-motifs, both of which are indicative of binding and hydrolysis of a nucleotide triphosphate (46, 55). Apaf-1 *in vitro* needs the presence of both cytochrome *c* and dATP (or ATP but not any other nucleotide) to catalyse the processing of procaspase-9 (55, 65, 66). Furthermore, Apaf-1 activity is potently inhibited by ATPγS (a non-hydrolysable ATP analogue) (66), suggesting that an ATPase activity may be required for Apaf-1 to function. Indeed, purified recombin-

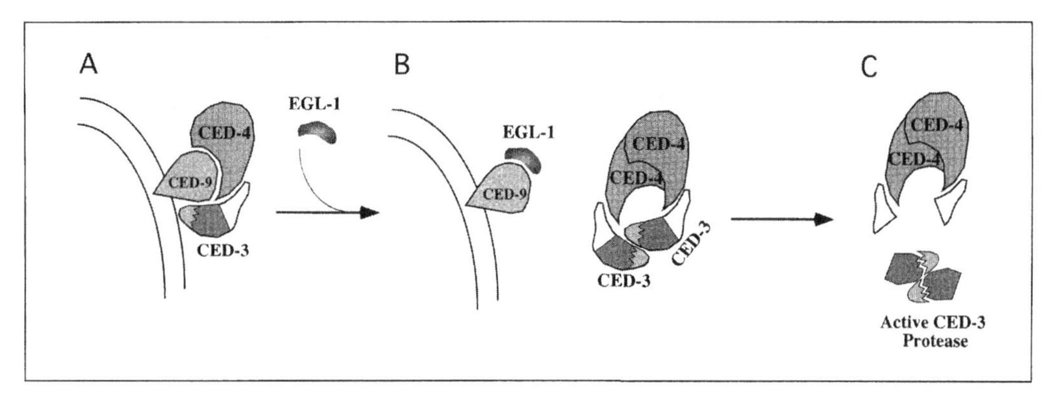

Fig. 5 Molecular model for the activation of CED-3 during programmed cell death. (A) In cells that normally live, CED-3, CED-4, and CED-9 form a ternary protein complex associated with mitochondria, in which CED-3 remains inactive. CED-3 may also directly contact CED-9 given that CED-9 has two CED-3 cleavage sites. (B) In cells that are fated to die, EGL-1 displaces the CED-4/CED-3 complex from CED-9, leading to the oligomerization of the CED-4/CED-3 complex. (C) Oligomerized CED-4/CED-3 complex facilitates the processing and activation of CED-3.

ant Apaf-1 can bind and hydrolyse ATP or dATP to ADP or dADP, respectively. The hydrolysis of ATP/dATP and the binding of cytochrome *c* promote Apaf-1 oligomerization and the subsequent recruitment and activation of procaspase-9 (67). The importance of the nucleotide binding motifs is underscored by the experiments in which mutations in the conserved residues of two Walker's motifs abolished the activity of CED-4 or Apaf-1 in activating their respective procaspases (59, 61, 67). Interestingly, one of the Walker's A-motif mutations (K165R) which inactivates CED-4 does not seem to affect the ability of CED-4 to oligomerize, indicating that oligomerization of CED-4 is not sufficient to activate CED-3 (45). So far, no ATPase activity has been demonstrated for CED-4, although CED-4 has been shown to bind ATP *in vitro* (68).

Apaf-1 has 12 or 13 WD40 repeats at its carboxyl terminus that are not found in CED-4 (67). Cytochrome *c* and dATP become dispensable if the WD40 repeat region is deleted from Apaf-1, suggesting that the WD40 repeat region acts as a negative regulatory domain for Apaf-1 (43, 69). It is not clear whether CED-4 alone is sufficient to activate CED-3 or whether it needs other cofactor(s) to do it. Since CED-4 does not contain the WD40 repeat region, cytochrome *c* may not be involved in the activation of CED-3. But this does not rule out the possibility that another factor(s) is needed.

ii. The regulation of ced-4 activity

The regulation of *ced-4* activity is also important for the appropriate control of programmed cell death in nematodes. Differential splicing of *ced-4* transcripts has been implicated in the regulation of *ced-4* activity. Specifically, *ced-4* was found to encode two transcripts. The major and shorter one, *ced-4S* transcript, causes programmed cell death, whereas the minor and longer one, *ced-4L* transcript, can act to prevent programmed cell death (70). CED-4L contains a 24 amino acid insertion right after amino acid 213 of CED-4S, between the two Walker's motifs. Nevertheless,

CED-4L and CED-4S can both self-oligomerize and can bind to CED-3 and CED-9 with similar affinities (45, 59). It is not clear whether this insertion disrupts the ATP binding ability or putative ATPase activity of CED-4S. If it does, then CED-4L may simply act in a dominant-negative manner to interfere with CED-4S. CED-4 activity may also be regulated by other factors. Recently, *C. elegans* protein MAC-1, a member of the AAA family of ATPases, has been shown to bind CED-4 (71). Overexpression of MAC-1 in *C. elegans* can prevent some physiological cell deaths in a sensitized genetic background, suggesting that MAC-1 may antagonize or negatively regulate the activity of CED-4 (71).

2.2.3 CED-9—the protector against programmed cell death

The *ced-9* gene was first defined by a gain-of-function mutation (*n1950*) that prevents almost all programmed cell deaths in *C. elegans* (22). *ced-9* loss-of-function mutations were isolated as *cis*-dominant suppressors of this *gf* allele. *ced-9(lf)* mutants have the opposite phenotype in that many cells that normally live undergo cell death (22). Two *ced-9(lf)* alleles, *n2812* and *n2077*, have nonsense mutations at codon positions 46 and 160, presumably producing truncated CED-9 proteins (72). The third *ced-9(lf)* allele, *n1653*, which causes the substitution of tyrosine 149 to glutamine, is temperature-sensitive. Intriguingly, the *ced-9(gf)* mutation, *n1950*, encodes a single amino acid change in a highly conserved residue in the Bcl-2 homology (BH) 1 domain of CED-9 (glycine 169 to glutamate) (83). It is not clear how this *gf* mutation results in increased *ced-9* death protective activity, but it does not appear to be the result of CED-9 overexpression.

Thus, *ced-9* acts as a negative regulator of programmed cell death. The extent of ectopic deaths in *ced-9(lf)* mutants is dependent on the maternal *ced-9* genotype of these animals. Homozygous *ced-9(lf)* animals derived from *ced-9(lf)*/+ mothers have much fewer cell deaths than those derived from homozygous *ced-9(lf)* mothers and usually can survive into adulthood. This is presumably because the maternal *ced-9* product is sufficient to prevent most cells from cell death. By contrast, homozygous *ced-9(lf)* animals derived from homozygous *ced-9(lf)* mothers invariably have a large number of extra cell deaths and die in early embryonic stage, leading to the maternal lethality phenotype (22). As mentioned above, the phenotypes of massive 'ectopic' cell deaths and embryonic lethality caused by *ced-9(lf)* mutations can be fully suppressed by strong *lf* mutations in either *ced-3* or *ced-4* gene, indicating that *ced-9* acts upstream of both *ced-3* and *ced-4* to negatively regulate their cell killing activities. In addition, *ced-9(lf); ced-3(lf)* animals or *ced-4(lf); ced-9(lf)* animals do not exhibit any other obvious defects beyond those observed in *ced-3(lf)* or *ced-4(lf)* single mutants, suggesting that *ced-9* may have a singular role in regulating cell death in *C. elegans* (22).

i. ced-9 and bcl-2 family members

The product of the *ced-9* gene shares significant sequence homology with that of human proto-oncogene *bcl-2* (23% identity), which plays a similar role in preventing programmed cell death in mammals (72–76) (see Chapter 5). Interestingly, over-expression of Bcl-2 can inhibit programmed cell death in nematodes and *bcl-2* can even substitute for *ced-9* to prevent ectopic cell deaths in *ced-9(lf)* animals (54, 72, 77). These results strongly suggest that *ced-9* and *bcl-2* have functional as well as sequence similarity and therefore that the cell death pathway has been evolutionarily conserved between nematodes and mammals.

ced-9 and *bcl-2* are two members of a rapidly growing gene family that play important roles in regulating programmed cell death in diverse species (for review see ref. 78 and Chapter 5). Members of this gene family can either be pro-apoptotic or anti-apoptotic (79, 80–82). Interestingly, as has been shown for Bcl-x_L and Bcl-2, *ced-9* appears to generate both death promoting and death preventing activities (83). This conclusion has come from two observations. First, the wild-type *ced-9* activity seems to be able to attenuate the death preventing activity provided by *ced-9(gf)* mutation. *ced-9(gf)*/+ animals have fewer surviving cells than *ced-9(gf)*/*Df* animals (Df: deficiency) (83). Secondly, *ced-9* appears to promote the death of those cells programmed to die. In a weak *ced-3* mutant background, which has fewer extra undead cells than strong *ced-3* mutants, *ced-9(lf)* mutations further increase the number of undead cells, an observation more consistent with the idea that *ced-9* promotes death rather than survival in dying cells (83). *ced-9(lf)* mutations cause many extra cell deaths in addition to the 131 programmed to die, suggesting that *ced-9* has to be expressed in most of the surviving cells to protect them. Moreover, overexpression of *ced-9* in nematodes can prevent most of the normally programmed cell deaths, suggesting that the level of *ced-9* expression is critical in determining whether a cell lives or dies.

Interestingly, *ced-9* is transcribed as a polycistronic message with another gene, *cyt-1*, that encodes a protein similar to cytochrome b_{560} of complex II of the mitochondrial respiratory chain (72). The ORF (open reading frame) of the *cyt-1* gene does not seem to be required for *ced-9* function, since a transgene containing the *ced-9* region but not most of the *cyt-1* region can still rescue *ced-9(lf)* mutants efficiently (72). Nevertheless, the recent findings that components of mitochondria such as cytochrome *c* and AIF (apoptosis-inducing factor) are actively involved in the activation of the death programme in mammalian cells (65, 84) (see Chapter 6), and that CED-9 associates with mitochondria when expressed in mammalian cells (60), have made this co-transcription observation rather intriguing. It is not clear, however, whether CED-9 is localized to mitochondria and other organelles in *C. elegans*, or whether CED-9 has the pore-forming ability on membranes as described for Bcl-x_L and Bcl-2 (see Chapter 5). A related and more fundamental question is whether mitochondria are involved in regulating *C. elegans* programmed cell death—and if so, whether cytochrome *c* or an equivalent molecule from mitochondria is involved. Although studies of subcellular localization of CED-9 and the nature of its interactions with other cell death proteins such as EGL-1 and CED-4 may provide some clues, so far there is no direct evidence that is suggestive of the involvement of mitochondria in *C. elegans* cell death.

ii. CED-9, EGL-1, CED-4, and CED-3

In vitro, CED-9 directly interacts with CED-4 and its upstream regulator EGL-1 (23, 58–60, 85). In mammalian cells CED-9 and CED-4 have been found to be co-localized to mitochondria (60). The interaction of CED-9 with CED-4 has been postulated as a mechanism to prevent CED-4 from activating CED-3 proenzyme (Fig. 5). *ced-9(lf)* mutation *n1653* reduces the interaction between CED-4 and CED-9, which is consistent with the hypothesis that CED-9 inhibits CED-4 activity through protein interaction (58, 61). CED-9 also interacts with EGL-1 and the binding of EGL-1 to CED-9 can effectively release CED-4 from a CED-4/CED-9 complex, thereby allowing CED-4 to activate CED-3 (23, 85) (see Section 2.2.4).

CED-9 can also directly interact with CED-3 *in vitro* and is an excellent substrate of the CED-3 protease (54). CED-3 can cleave CED-9 at two sites (aspartate 44 and aspartate 67). While mutations at one or the other of the cleavage sites do not affect CED-9 death protective activity in nematodes, mutations that disrupt both CED-3 cleavage sites markedly reduce the protective activity of CED-9, suggesting that the presence of at least one CED-3 cleavage site is important for CED-9 death protective function (54). Cleavage of CED-9 by CED-3 generates a carboxyl terminal product that resembles Bcl-2 in sequence and that has similar death protective activity to that of Bcl-2 expressed in *C. elegans*. This carboxyl terminal region of CED-9 is also sufficient to mediate interaction with CED-4 (D. Xue and H. R. Horvitz, unpublished results). Furthermore, a chimeric protein, consisting of the amino terminal region of CED-9 (amino acids 1–80) containing the two CED-3 cleavage sites fused to the entire Bcl-2 protein, protects against programmed cell death in *C. elegans* significantly better than the Bcl-2 protein alone (54). These results suggest that CED-9 may inhibit cell death in *C. elegans* through two distinct mechanisms. First, CED-9 may directly inhibit the activity of the CED-3 death protease through its two CED-3 cleavage sites, possibly by acting as a competitive substrate inhibitor. Secondly, CED-9 may indirectly inhibit the activation of CED-3 by forming a complex with CED-4 through its carboxyl terminal Bcl-2 homology regions.

2.2.4 EGL-1—the initiator of programmed cell death

The discovery that the *egl-1* gene is required for the deaths of many cells in *C. elegans* came as a pleasant surprise when a loss-of-function intragenic suppressor (*n3082*) of an *egl-1(gf)* mutation (*n1084*) was found to result in survival of almost all the cells that normally die (23). *egl-1* has long been regarded as a gene that specifies the death of HSN neurons since all *egl-1(gf)* mutations identified appear to cause no other defect except for the ectopic death of two HSN neurons in hermaphrodites (24, 86). The death of HSN neurons in *egl-1(gf)* mutants can be suppressed by *ced-9(n1950, gf)*, *ced-3(lf)*, or *ced-4(lf)* mutations, indicating that *ced-9*, *ced-3*, and *ced-4* act downstream of or in parallel to *egl-1* (18, 22, 24). Overexpression of EGL-1 can induce programmed cell death, suggesting that, like *ced-3* and *ced-4*, *egl-1* encodes a cell killing activity (23). The ectopic cell deaths caused by overexpression of *egl-1* can also be suppressed by *ced-9(gf)*, *ced-3(lf)*, or *ced-4(lf)* mutations. In contrast, the *egl-1(lf)* mutation can not suppress cell deaths caused by *ced-9(lf)* mutations (23). These results further support the conclusion that *egl-1* acts upstream of the *ced-3*, *ced-4*, and *ced-9* genes. Genetic epistasis analysis on *egl-1* and cell death specification genes *ces-1* and *ces-2* suggests that *egl-1* is likely to act downstream of these death specification genes (23). Since *egl-1* is the earliest acting gene of the general cell death pathway identified so far and acts downstream of the death specification genes, *egl-1* may play a crucial role in receiving cell death signals from the cell death specification genes and then initiating the activation of the cell death programme.

i. EGL-1 and BH3-only proteins

egl-1 encodes a relatively small protein of 91 amino acids that has no sequence similarity to any other protein except a nine amino acid stretch that resembles the BH3 motif of the Bcl-2 protein family (23). EGL-1 is thus most similar to Bcl-2 family proteins of the BH3-only subfamily that resemble each other only in the BH3 domain. Several of these proteins have been shown to be important for inducing apoptosis either by binding to pro-apoptotic proteins such as Bax and activating them, or by binding to and antagonizing anti-apoptotic proteins such as Bcl-x_L and Bcl-2 (for review see ref. 78 and Chapter 5). Similarly, EGL-1 binds CED-9 in a BH3 domain-dependent manner *in vitro* (23, 85). Furthermore, the binding of EGL-1 to CED-9 inhibits the interaction between CED-4 and CED-9 and may thus set CED-4 free to activate CED-3 (Fig. 5) (85). *n3082*, the only loss-of-function allele of *egl-1* identified so far, has a five nucleotide deletion in exon 2 that results in the formation of a truncated EGL-1 lacking the BH3 domain and the carboxyl terminus of the protein. This truncated version of EGL-1 fails to interact with CED-9 *in vitro*, providing further evidence that the interaction between EGL-1 and CED-9 is important for EGL-1 death-inducing activity *in vivo* (23).

ii. The regulation of egl-1 activity

As a gene that sits at the beginning of the central cell killing pathway, *egl-1* may play a crucial role in initiating cell death. How does *egl-1* play such a role? It is conceivable that the expression of the *egl-1* gene can be turned on or up-regulated in response to the death cues. This scenario has been supported by the recent identification of the nature of *egl-1(gf)* mutations. As mentioned above, seven *egl-1(gf)* mutations have been isolated that cause the inappropriate death of HSN neurons in hermaphrodites, a male-specific cell fate for HSN (86). These mutations have been found to alter three adjacent nucleotides in a putative binding site for the TRA-1A zinc finger protein that lies 5.6 kb downstream of the *egl-1* transcription unit (87). TRA-1A is a terminal global regulator of somatic sex determination in *C. elegans* which directs female development by repressing male-specific gene activities or by activating female-specific gene activities (Fig. 4) (88, 89). TRA-1A may thus directly repress the expression of *egl-1* in hermaphrodite HSN neurons. Indeed, *in vitro* DNA binding assays and analysis of *egl-1* expression in HSNs using an *egl-1*::GFP reporter construct carrying one of the *gf* mutations demonstrated that *egl-1(gf)* mutations disrupt the binding of TRA-1A to this site and activate the expression of the *egl-1* gene in hermaphrodite HSN neurons (87). Thus the *egl-1* gene is transcriptionally activated in male HSN neurons in response to the male sex differentiation signal. It is likely that *egl-1* is also transcriptionally regulated in other dying cells.

It is also likely that the activation of *egl-1* can be achieved at the post-translational level by modifications of EGL-1 such as phosphorylation or cleavage by a protease or by binding to other factors. Recent studies have suggested that the death-inducing activities of Bid and Bim, both of which are mammalian BH3-only proteins like EGL-1, are activated through proteolytic cleavage of Bid by caspase-8 or dissociation of Bim from the dynein motor complex (90–93). Examination of the temporal and spatial expression patterns of *egl-1* and biochemical analysis of the EGL-1 protein and the proteins it interacts with will help address these questions.

2.2.5 Others

Two genes that have not been characterized in detail but have been shown to be important for programmed cell death in *C. elegans* are *ced-8* and *dad-1*. The *ced-8* gene was first identified as a gene important for the engulfment of cell corpses. Mutations in *ced-8* result in a significant increase of the number of cell corpses in late stage embryos and young larvae, reminiscent of the engulfment-defective mutant phenotype (20). However, time course analysis of the cell corpse numbers at different embryonic stages of *ced-8* mutant animals revealed that the increased number of cell corpses in late stage embryos was due to the delay in the appearance of dying cells from early stage embryos to the late stage ones (G. M. Stanfield and H. R. Horvitz, personal communication; 27). *ced-8* may thus play a role in regulating the kinetics of cell killing in *C. elegans*.

The *dad-1* gene (defender against apopototic death) has also been implicated in regulating programmed cell death in *C. elegans* (94). The mammalian *dad-1* gene was originally discovered as a gene that could rescue the apoptotic phenotype of a temperature-sensitive mutant hamster cell line at the restrictive temperature (95). The *C .elegans dad-1* gene was then identified based on its sequence homology (> 60% identity) with mammalian *dad-1*. Overexpression of either the *C. elegans* or the human *dad-1* gene only partially inhibits programmed cell death in *C. elegans* (94), hinting that *dad-1* may function as a negative cell death regulator in *C. elegans*. *dad-1* shares significant sequence similarity (40% identity) with a yeast gene, OST2, that encodes the 16 kDa subunit of the yeast oligosaccharyltransferase, an enzyme complex that catalyses glycosylation of newly synthesized proteins in the lumen of the rough endoplasmic reticulum (96). The implication of this sequence similarity, the target(s) of *dad-1*, and the mechanism by which *C. elegans dad-1* inhibits cell death remain to be determined. So far, no genetic mutation has been identified in the *C. elegans dad-1* gene.

2.3 Genes involved in the specification of the cell death fate in specific cell types

In *C. elegans*, the deaths of 131 cells in hermaphrodite animals have been 'pre-programmed': the cells that die and the timing of their deaths are invariant from animal to animal (2–4). This pre-programming of cell deaths implies that the cell death programme must be tightly controlled such that death occurs only in the right cells and only at the right time. Genetic studies indicate that the life/death decisions of individual cells are likely to be regulated by cell type-specific regulatory genes (21). These regulatory genes may control cell deaths by modulating the activities or the expression of key components in the central cell killing pathway. In this section we review what is known about the control of cell death in specific cell types.

2.3.1 Sister cells of NSM neurons

Two genes, *ces-1* and *ces-2* (cell death specification), have been found to regulate the initiation of cell death in two cells of *C. elegans*; the sister cells of the pharyngeal serotonergic NSM neurons. In wild-type animals, the NSM sister cells undergo programmed cell death. However, a gain-of-function mutation (*n703*) in the *ces-1* gene or a reduction-of-function mutation (*n732*) in the *ces-2* gene can cause survival of these two cells (21). These two mutations were identified when a population of mutagenized animals was screened using the technique of formaldehyde-induced fluorescence for mutants that have unusual patterns of serotonin expression. In *ces-1(n703)* and *ces-2(n732)* mutants, four serotonergic cells were found in the pharynges of the mutant animals instead of the two serotonergic cells that are normally seen in wild-type animals. Analysis of the positions and the lineages of these two extra serotonergic cells indicated that they are sister cells of NSM neurons that fail to die. In addition, the *ces-1(n703gf)* mutation also prevents the death of sister cells of the pharyngeal I2 neurons (21). However, *ces-1* and *ces-2* mutations do not seem to affect the deaths of any other cells, indicating that their activities may be cell type-specific. Two additional *ces-1(gf)* alleles and two *cis*-dominant suppressors of *ces-1(gf)* mutations, which appear to be loss-of-function alleles of *ces-1*, have been isolated by using Nomarski DIC optics to screen for mutations that alter the life or death fate of NSM sister cells (21). In the two *ces-1(lf)* mutants, NSM sister cells and I2 sister cells undergo programmed cell death normally as they would in wild-type animals, suggesting that the *ces-1* activity is not required for the deaths of these two types of cells. Only one *ces-2* allele has been identified and it is hypomorphic and temperature-sensitive.

The genetic relationship between *ces-1* and *ces-2* has been carefully examined by building double mutants. The *ces-2(lf)* mutation results in the survival of NSM sister cells, suggesting that the normal function of *ces-2* is to cause the death of NSM sister cells. However, in *ces-1(lf)*; *ces-2(lf)* double mutant animals, NSM sister cells undergo programmed cell death normally as they do in wild-type or *ces-1(lf)* animals (21). This observation indicates that the activity of *ces-1* is normally inhibited to allow the death of NSM sister cells and that *ces-2* acts upstream of *ces-1* to suppress the death inhibitory activity of *ces-1* (Fig. 4).

The relationship of the *ces* genes with the genes involved in the cell killing pathway has mainly been inferred from the phenotype of a *ces-1; egl-1* double mutant. In *ces-1(lf)*; *egl-1(lf)* mutant animals, NSM sister cells survive as they do in the *egl-1(lf)* mutants, indicating that *egl-1* most likely acts downstream of *ces-1* to cause cell death and *ces-1* negatively regulates the activity of *egl-1* (23). Thus, *ces-2*, *ces-1*, and *egl-1* function in a negative regulatory chain to regulate the death fate of NSM sister cells. Since the *ces-2(lf)* mutation does not affect the death fate of I2 sister cells, there must be some other gene(s) that acts similarly to regulate the activity of *ces-1* in I2 sister cells.

The above genetic analysis has helped position *ces-1*, *ces-2*, and *egl-1* genes in the cell death pathway (see Fig. 4). However, elucidation of the mechanism by which this negative regulatory cascade works to control the death fate of NSM sister cells will require the knowledge of the molecular identities of *ces-1* and *ces-2* genes. So far, only *ces-2* has been cloned and found to encode a putative basic leucine-zipper (bZIP) transcription factor (94), suggesting that cell death specification in *C. elegans* can be regulated at the level of gene expression.

2.3.2 HSN motor neurons

The regulation of the life versus death fate of one pair of sex-specific HSN neurons (hermaphrodite-specific neurons) in *C. elegans* presents another interesting paradigm for studying cell death specification. *C. elegans* has two natural sexes: male and self-reproducing hermaphrodite (1). Two HSN motor neurons control egg laying in hermaphrodites but undergo programmed cell death in males (4). Mutations in several genes have been identified that cause the inappropriate death of HSN neurons in hermaphrodites, a male-specific fate of HSN (24, 86, 98). These include gain-of-function mutations in the *her-1* gene (hermaphroditization) and some weak loss-of-function mutations in the *tra-2* gene (sexual transformer), all of which result in partial sexual transformation of some XX hermaphrodites into male-like animals. *her-1* encodes a novel secreted molecule whose activity is high in males to promote male somatic cell fates, whereas the *tra-2* gene encodes a putative transmembrane receptor for the *her-1* protein and whose activity is inhibited by the *her-1* protein in males (98–101). Interestingly, gain-of-function mutations in the *egl-1* gene, which is required for almost all programmed cell deaths in nematodes but is not involved in sex determination, also cause ectopic HSN death in hermaphrodite animals (23, 24). Thus, *her-1* and *tra-2* may mediate a novel signal transduction pathway that integrates into the cell death pathway through the *egl-1* gene to control HSN cell death. Several other sex determining genes, *fem-1*, *fem-2*, *fem-3* (feminization), and *tra-1*, have been shown to act downstream of *tra-2* in a negative regulatory cascade to control the somatic sex determination of *C. elegans*, with *tra-1* acting at the last step of this sex determination pathway (Fig. 4) (for review see ref. 102). Since *tra-1* encodes a Zn^{2+} finger transcription factor (89), this sex determination signal transduction pathway may control HSN cell fate by turning on or off the expression of key regulatory genes such as *egl-1*. As described in Section 2.2.4.ii, the finding that all seven *egl-1(gf)* mutations alter and disrupt a putative TRA-1A binding site that lies 5.6 kb downstream of the *egl-1* transcription unit and the observation that these *gf* mutations can cause ectopic expression of *egl-1* in hermaphrodite HSN neurons, strongly suggest that this general sex determination pathway regulates the sex-specific life / death fate of HSN through transcriptional repression of *egl-1* expression by *tra-1* (87). However, this sex determination pathway can not specify the sexually dimorphic death of HSNs, since it also functions in many other somatic cells (102). The genes that truly specify the life / death fate of HSNs have yet to be identified.

2.3.3 Germline

The cells of *C. elegans* hermaphrodite germline can adopt one of four cell fates: they can undergo mitosis, enter meiosis and differentiate into sperm or oocytes, or they can undergo programmed cell death (for review see ref. 103). Germ cell deaths only occur in adult hermaphrodites and have not been observed in larval stages or in males. Early mitotic and meiotic germ cells in adult hermaphrodites are not completely enclosed by a plasma membrane and instead exist as a large syncytium of asynchronous nuclei in a common cytoplasm (103). Interestingly, direct microscopic observations suggest that nuclei of the dying germ cells rapidly cellularize and separate from the syncytium early in the cell death process, presumably to avoid the diffusion of death causing factors into the syncytium (9). Over 300 cells in the germline are estimated to die as deduced from the number of unengulfed germ cell corpses observed in engulfment-defective mutant animals (9).

Germ cell differentiation in the adult hermaphrodite corresponds to the position of the germ cell nuclei within the gonad. The gonad of the hermaphrodite animals consists of two U-shaped tubes capped by a somatic distal tip cell (DTC). The stem cell potential of germ cells in close proximity to the DTC is maintained by growth factors secreted by DTC. Germ cells farther away from the DTC enter meiosis and arrest at the pachytene stage. Near the bends of the gonad tubes where germ cell deaths are observed, germ cells receive signals from the Ras/MAPK pathway to exit meiotic arrest and progress to prophase I (103, 104). At this point germ cells that do not die increase in size to form oocytes as they migrate towards the uterus.

Phenotypic analysis of mutations that affect sexual identity or germ cell differentiation has shown that germ cell deaths are associated with oogenesis and not spermatogenesis or mitosis (9). For example, *tra-1* loss-of-function mutant XX animals, which are incapable of producing oocytes, have no observable germline programmed cell death. Furthermore, no programmed cell death is seen in *mog-1* (masculinization of germline) loss-of-function mutants, in which the germline of XX hermaphrodites is masculinized such that only sperm are produced. Interestingly, entry in meiosis and exit from meiotic arrest appear to be a requirement for germ cell death. For example, in *gld-1* (germline differentiation abnormal) mutant hermaphrodites, in which meiotic differentiation is blocked, no germ cell deaths are observed (9). As mentioned above, activation of the Ras/MAPK signalling cascade is required for exit of germ cells from meiotic arrest (103, 104). Loss-of-function mutations in several genes in the *ras* signalling pathway [*let-60* (*ras*), *lin-45* (*raf*), *mek-2* (MAPK kinase), and *mpk-1* (MAP kinase)] (105) block the ability of germ cells to complete meiosis. Additionally, the same loss-of-function mutations almost completely abolish germ cell deaths (9). Thus, exit from meiotic arrest is a prerequisite for germ cells to undergo programmed cell death.

Programmed deaths of somatic and germ cells are morphologically similar, except that germ cell corpses are usually larger in size than somatic cell corpses. Both types of cell deaths are characterized by condensation of cytoplasm and nuclei, aggregation of chromatin, and rapid recognition and phagocytosis of cell corpses by neighbouring cells (2, 9, 13). The kinetics of both cell deaths also appears to be similar. From the initiation of the germ cell death to the engulfment of the germ cell corpse by gonadal sheath cells that surround the germline, the whole process is completed in less than one hour, a time frame comparable to that of somatic cell deaths (2, 9).

Somatic and germ cell deaths share components of the cell killing machinery, as the cell killing activities of *ced-3* and *ced-4* are required for the execution of both somatic and germ cell deaths (9). Moreover, *ced-9* is required to protect germ cells from death: many more germ cells undergo programmed cell death in *ced-9(lf)* mutants. However, this general cell death programme mediated by *ced-3*, *ced-4*, and *ced-9* appears to be regulated differently in somatic and germ cells. For example, the *egl-1* gene, which is required for the activation of almost all somatic cell deaths, is not required for the activation of germ cell deaths, since in *egl-1(lf)* mutants germ cell deaths are not blocked. In addition, the *ced-9(gf)* mutation, *n1950*, that prevents almost all somatic cell deaths, has little effect on germ cell deaths (9). These observations suggest that *ced-9* may interact with a different cell death initiator to regulate the activation of cell death in germline.

One interesting question regarding germ cell death is why germ cells undergo programmed cell death at all. One possibility is that cell death is used as a mechanism to eliminate unfit germ cells. This seems unlikely since in *ced-3* or *ced-4* mutant animals there is no increase in the number of defective oocytes or embryonic lethality (9). A more feasible explanation is that many germ cells in *C. elegans* are overproduced and serve as nurse cells to synthesize cytoplasmic components for mature oocytes. Once they fulfil this function, they are eliminated by cell death. A similar process occurs in *Drosophila* and is mediated by a specific caspase DCP-1 (see Chapter 3).

2.3.4 Cell deaths in other specific cell types

Although much is known about the killing step of programmed cell death in *C. elegans*, little is known about how this step is specified and triggered in each dying cell. The cases discussed above describe only a fraction of the programmed cell deaths that occur in *C. elegans*. Elucidating the mechanisms that specify the many other cell deaths in *C. elegans* remains an important challenge.

2.4 Genes involved in the engulfment of cell corpses

Once cells undergo programmed cell death, their corpses are swiftly engulfed and degraded, usually within an hour (2, 14). Unlike *Drosophila melanogaster* and mammals, *C. elegans* does not have professional phagocytes that recognize and phagocytose dying cells. Instead, cell corpses are engulfed by their neighbouring cells. Cell corpses generated during embryogenesis are mostly engulfed by their siblings (4), whereas post-embryonic cell corpses are removed by the hypodermis, a large hypodermal syncytium that envelops most of the animal. The engulfment-inducing signal in dying cells appears to be expressed at a very early stage of the death process: pseudopodia have been observed to extend from an engulfing cell around the dying cell even before the cell division generating the dying cell has completed (13).

These phenomena beg several questions. What is the signal that marks the dying cell? How does this signal trigger phagocytosis? What is the mechanism of phagocytosis? Genetic and molecular studies of the cell corpse engulfment process in *C. elegans* have provided some valuable clues.

2.4.1 Genetic analysis of genes involved in the engulfment process

At least six genes, *ced-1, ced-2, ced-5, ced-6, ced-7,* and *ced-10* are involved in cell corpse engulfment (14, 20). Mutations in any of these genes block the engulfment of many dying cells and result in the phenotype of persistent cell corpses (Fig. 3). Since the cells destined to die still die in the engulfment-defective mutants, the engulfment process *per se* does not cause cell death. Instead, engulfment functions to remove corpses after cell death occurs. Genetic analysis suggests that these six genes fall into two groups: *ced-1, ced-6,* and *ced-7* in one group and *ced-2, ced-5,* and *ced-10* in the other (20). Single mutants or double mutants within the same group show partial engulfment defects, whereas double mutants between the two groups show more severe engulfment defects. One model consistent with these genetic data is that these two groups of genes are involved in two distinct but partially redundant pathways that lead to the recognition and subsequent phagocytosis of cell corpses by engulfing cells (20). It is possible that dying cells exhibit two different engulfment-inducing signals recognized by distinct molecules on engulfing cells and that only when both signalling systems are engaged can phagocytosis be efficient. All the engulfment genes except *ced-1* show maternal rescue for the engulfment of embryonic deaths (20), suggesting that the maternal contribution of these engulfment gene products in either embryos or germline is sufficient for engulfment to take place. However, post-embryonic somatic cell deaths do not show maternal rescue.

Distal tip cell (DTC) migration (Fig. 6A) (27, 106) is also affected by mutations in *ced-2, ced-5,* and *ced-10,* suggesting that these genes are required for cell movements underlying both migration and engulfment. Adult *ced-2, ced-5,* and *ced-10* mutant animals have abnormally shaped gonads due to DTCs making extra turns or stopping migration prematurely (Fig. 6B).

One other feature shared by *ced-2, ced-5,* and *ced-10* mutations is that they can suppress abnormal cell deaths caused by semi-dominant (*sd*) mutations in two genes *lin-24* and *lin-33* (lineage abnormal). In *lin-24*(*sd*) and *lin-33*(*sd*) mutants, P1.p-P12.p (collectively designated Pn.p) cells, which lie in the ventral midline and produce hypodermal and vulval cells, display abnormal morphologies and eventually degenerate (107–109). The death induced by *lin-24*(*sd*) and *lin-33*(*sd*) mutations does not seem to occur via a normal programmed cell death pathway since this death is morphologically different from programmed cell death, is not blocked by loss of *ced-3* or *ced-4* activity, and requires the activities of *ced-2, -5,* and *-10* (109). It is possible that *lin-24*(*sd*) and *lin-33*(*sd*) mutations cause the Pn.p cells to be recognized and engulfed by their neighbouring cells as dying cells, which requires *ced-2, ced-5,* and *ced-10* activity. If this is true, *ced-2, ced-5,* and *ced-10* genes might be involved in

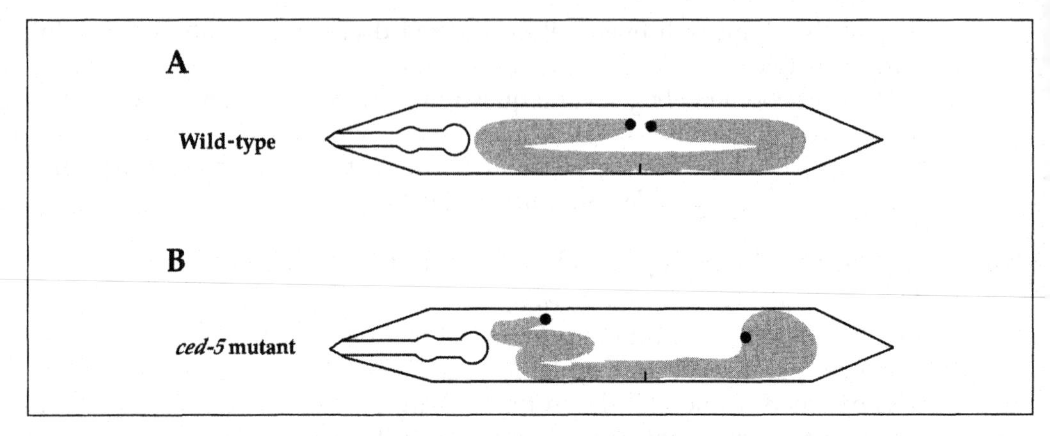

Fig. 6 Schematic drawings of the DTC locations and gonadal shapes observed in wild-type and *ced-5* adult hermaphrodites. (A) DTC locations and gonadal shape observed in the wild-type hermaphrodite. (B) An abnormally shaped gonad observed in a *ced-5* mutant, resulting from an extra turn made by the anterior DTC and the premature termination of the posterior DTC during the migration process. The gonad is shaded in grey, and the DTCs are indicated by black circles.

eliminating cells undergoing programmed cell death as well as abnormal cell deaths (27). Alternatively, the products of *ced-2*, *ced-5*, and *ced-10* may directly contribute to the abnormal death of Pn.p cells in *lin-24*(*sd*) and *lin-33*(*sd*) mutants which has nothing to do with the engulfment function of these three genes. The molecular identities of the *lin-24* and *lin-33* genes and the toxic nature of the LIN-24(*sd*) and LIN-33(*sd*) proteins have yet to be determined.

The six engulfment genes identified so far may not represent all of the genes involved in cell corpse engulfment. It has been reported that animals homozygous for some genomic deficiencies exhibit the phenotypes of increased numbers of unengulfed cell corpses and embryonic lethality (110), suggesting that other unidentified genes may also function in cell corpse engulfment.

2.4.2 Molecular analysis of genes involved in the engulfment process

Of the six engulfment genes identified so far, three (*ced-5*, *ced-6*, and *ced-7*) have been cloned. Below we will describe the molecular characterization of these genes and the potential roles they may play during cell corpse engulfment.

i. ced-5

The engulfment defect of the *ced-5* mutant can be rescued by using heat shock promoters to induce the expression of a wild-type *ced-5* transgene (106). The rescue can be achieved even when cell corpses are long dead for hours or even days in a *ced-5* mutant, suggesting that the engulfment-inducing signal(s) is stable and can stay functional for a long time with the cell corspes. The transcription/translation machinery of such late cell corpses appears to be inactive since expression of green fluorescent protein (GFP) can not be induced in these cell corpses by heat shock treatment (106). Therefore, the eventual engulfment of these late corpses by induced expression of *ced-5* is likely effected by *ced-5* expression in engulfing cells rather than in cell corpses. These observations suggest that *ced-5* functions in engulfing cells but not in dying cells. *ced-5* encodes a protein similar to human DOCK180 (111), *Drosophila* Myoblast City (MBC) (112), the predicted protein sequence from a human cDNA clone KIAA0209 (113), the yeast open reading frame L9576.7 (GeneBank accession number 664878), and a mouse expressed sequence tag (GeneBank accession number AA110899). CED-5 is most similar to DOCK180; these two proteins share 26% identity throughout their entire lengths. Both DOCK180 and MBC have been implicated in mediating the extension of cell surfaces (111, 112, 114). Interestingly, CRK, a DOCK180 interacting adaptor protein is involved in integrin-mediated signalling (115), and vitronectin, a member of the integrin superfamily, has been implicated in cell corpse engulfment in mammals (116).

Expression of human DOCK180 in *C. elegans* can rescue the cell migration defect of a *ced-5* mutant, suggesting that DOCK180 and CED-5 are, at least in this aspect, functionally interchangeable (106). However, expression of DOCK180 in *C. elegans* failed to rescue the *ced-5* engulfment defect. It is possible that DOCK180 possesses the function required for DTC migration but not that for cell corpse engulfment. It is also possible that DOCK180 does have a role in cell corpse engulfment but the sequence similarity with CED-5 is not good enough for it to interact efficiently with other components in the CED-5 pathway. Elucidation of the physiological role of DOCK180 in mammals will help distinguish these two possibilities.

The molecular characterization of the *ced-5* gene and the phenotypic analysis of *ced-5* mutants altogether suggest that *ced-5* functions in mediating the extension of cell surfaces when engulfing cells phagocytose cell corpses or when migrating DTCs move along body wall muscles. *ced-5* mutants appear to be normal in the migration of other cells (e.g. the migrations of P1-P12 precursor cells) as well as in the axonal outgrowth of neurons and in cell fusion that is involved in the development of the hypodermal syncytium (20, 106). Thus, *ced-5* is likely to function in a specific type of membrane extension that is common to both cell corpse engulfment and DTC migration.

ii. ced-7

CED-7 protein has sequence similarity to ABC (ATP binding cassette) transporters (117). ABC transporters mediate the transport of diverse substrates, including ions, sugars, vitamins, phospholipids, peptides, and proteins (118, 119). The mechanism by which each ABC transporter achieves its substrate specificity is not clear. One characteristic feature of ABC transporters is the unidirectionality of substrate transport (118, 119). The transport process appears to be export rather than import in almost all eukaryotic ABC transporters with identified substrates, except CFTR, which acts as a chloride channel (118). Whether *ced-7* can function as a transporter, and if so what its physiological substrates are, remains to be determined.

CED-7 is most similar to the ABC-C protein, a member of the ABC1 subfamily of transporter proteins (120–124). CED-7 and ABC-C exhibit 25% identity along their entire lengths. Among the ABC transporters that have been identified thus far, the

ABC1 transporter, which is 20% identical to CED-7, has been implicated in the engulfment of mammalian apoptotic cells (106, 125). The mechanism by which ABC1 mediates engulfment of apoptotic cells is still not clear.

The CED-7 protein is widely expressed during embryogenesis as determined using anti-CED-7 antibodies (106). This broad expression pattern suggests that the activity of CED-7 should be tightly regulated so that *ced-7*-mediated engulfment can specifically target dying rather than viable cells. Mosaic analysis demonstrates that CED-7 functions in both dying cells and engulfing cells during the engulfment process (106). This finding and the observation that CED-7 is localized to the plasma membrane suggest that CED-7 activity may be important for the interaction between the cell surfaces of the dying and engulfing cells. If *ced-7* functions as a transporter, identification of CED-7 substrate(s) should help understand the mechanism underlying the *ced-7*-mediated engulfment process.

iii. ced-6

ced-6 might act downstream of or in parallel to *ced-1* and *ced-7* since overexpression of *ced-6* can partially rescue the engulfment defect in *ced-1* and *ced-7* mutants but not that in *ced-2* and *ced-5* mutants (126). The CED-6 protein contains two potential functional motifs: a phosphotyrosine binding (PTB) domain and a proline/serine-rich region that may interact with the SH3 (Src homology 3) domain of signalling proteins (126–128). PTB domains can recognize a phosphorylated tyrosine residue within a NPXY consensus sequence (X: any amino acid) (126, 129). Intriguingly, the CED-7 protein contains a NPLY sequence in one of its predicted cytoplasmic domains. Whether CED-7 can be phosphorylated on this tyrosine residue and hence recruit CED-6 to the membrane remains to be determined. Genetic mosaic analysis demonstrates that *ced-6* acts within engulfing cells (126). This finding and the presence of a PTB domain and potential SH3 binding sites in CED-6 suggest that CED-6 may serve as an adaptor protein to directly or indirectly transduce a signal from a receptor to effectors or cytoskeletal proteins that initiate the rearrangement of the cytoskeleton and the phagocytic process.

Although genetic screens for engulfment mutants have not been saturated (20) and the molecular identities of three other engulfment genes (*ced-1*, *ced-2*, and *ced-10*) have yet to be determined, the molecular characterizations of *ced-5*, *ced-6*, and *ced-7* genes should open up more avenues for studying the mechanisms by which cell corpses are engulfed.

2.4.3 Unanswered questions regarding cell corpse recognition and engulfment

Although the molecular cloning of the *ced-5*, *ced-6*, and *ced-7* genes has shed some light on what molecules might be involved in the cell corpse engulfment process, several key questions remain unanswered:

(a) What are the engulfment-inducing signals and how are they generated by dying cells?

(b) What are the receptors that recognize the engulfment-inducing signals?

(c) How are the engulfment-inducing signals transduced from receptors to the phagocytosis machinery of engulfing cells to initiate the engulfment of dying cells?

Since CED-3 death protease executes cell death by cleaving critical substrates, it is possible that engulfment-inducing signals are generated by CED-3 protease cleavage. However, it is also possible that the signals could be generated by a mechanism that is not dependent on direct cleavage by CED-3, but rather, by products generated during the subsequent cell disassembly process. For example, in mammalian cells the exposure of phosphatidylserine on the outer leaflet of plasma membrane due to the loss of plasma membrane asymmetry during apoptosis has been shown to be important for recognition of apoptotic cells by macrophages *in vitro* (130, 131). Molecular characterization of the remaining engulfment genes (*ced-1*, *ced-2*, and *ced-10*) should provide additional information towards answering these questions.

2.5 Genes involved in the degradation of cell corpses

The *C. elegans* gene *nuc-1* (nuclease) gene appears to be required for DNA degradation of dead cells but not for the execution of cell death or for the engulfment of cell corpses (17). In *nuc-1* mutants, both cell death and engulfment occur, but DNA from the engulfed dead cells is not degraded and persists as a compact mass of Feulgen-reactive material (14, 17). The *nuc-1* gene is also involved in digesting the DNA of bacteria on which the animals feed, since bacterial DNA can be detected in the intestinal lumens of *nuc-1* mutants, but not in those of wild-type animals (17). The finding that an endonuclease activity present in wild-type animals is reduced to 1% in *nuc-1* mutant animals (132) suggests that *nuc-1* may encode an endonuclease or a protein that controls the activity of the endonuclease.

Although nucleosomal DNA ladders, a hallmark of apoptosis, have not been shown to occur in *C. elegans*, DNA degradation by endonuclease(s) which generate 5′-phosphate and 3′-hydroxyl ends has been detected by TUNEL staining *in situ* (Y. C. Wu, G. M. Stanfield, and H. R. Horvitz, personal communication), a technique that has been used widely to identify dying cells (133). Since the time and the position of each cell programmed to die is known in *C. elegans*, the studies of TUNEL staining patterns should help to reveal the kinetics of DNA degradation *in vivo*.

3. Conclusion and future perspectives

Programmed cell death is a widespread phenomenon in the animal kingdom and an essential cellular process in animal development and homeostasis (25, 134). Thus it has been speculated that such an important process may be evolutionarily conserved. Programmed cell deaths in *C. elegans* exhibit morphological changes that are very similar to apoptotic cell deaths in other species including humans. Moreover, many components of the *C. elegans* cell death pathway, especially those involved in the cell killing process, have homologues that play similar roles in mediating apoptosis in other species. This conclusion is underscored by the sequence and functional similarities between CED-3 and caspases, CED-4 and Apaf-1, CED-9 and Bcl-2 family proteins, and EGL-1 and BH3-only Bcl-2 family proteins. Furthermore, expression of some of these proteins in foreign species can exert similar effects on cell death. The basic mechanisms by which cell death is regulated and executed are also likely to be conserved. For example, baculovirus p35 protein can inhibit cell death in diverse species by inhibiting the activities of CED-3 and death caspases (52, 53). Also, the activation of CED-3 and caspase proenzymes is likely to be induced by conserved oligomerization mechanisms such as receptor oligomerization or adaptor protein (Apaf-1 and CED-4) oligomerization (42–45, 67). All these studies firmly establish that the cell death pathway is conserved between nematodes and mammals.

In the last few years, we have witnessed remarkable progresses towards the understanding of basic mechanisms that guide the activation and execution of programmed cell death. Genetic studies in *C. elegans* have been pivotal to the identification of a conserved pathway for programmed cell death in animals and have provided a genetic framework for studying the regulation and execution of this process. Although many important details remain to be elucidated, *C. elegans* has advantages as a model organism that enable further studies to address these unanswered questions. In addition to the relative simplicity of the cell death machinery in *C. elegans* (e.g. only four caspases identified so far in *C. elegans* compared to 14 found in mammals) (41), the completion of the genome sequence and the emergence of powerful techniques such as dsRNAi (double stranded RNA interference) (137) that allow study of the functions of virtually any *C. elegans* gene, provide powerful new tools. Genetic and molecular studies of programmed cell death in *C. elegans* will continue to contribute in a major way to deciphering the mechanisms of programmed cell death.

Acknowledgements

We thank members of D. X.'s laboratory for comments on the manuscript and P. T. Huynh for help with figures. Research in D. X.'s laboratory is supported by grants from ACS and NIH. Research in Y. C. W.'s laboratory is supported by a grant from National Science Council in Republic of China. D. X. is a recipient of Burroughs Wellcome Fund Career Award in Biomedical Sciences and the Searle Scholar Award.

References

1. Brenner, S. (1974). The genetics of *Caenorhabditis elegans*. *Genetics*, **77**, 71.
2. Sulston, J. E. and Horvitz, H. R. (1977). Post-embryonic cell lineages of the nematode, *Caenorhabditis elegans*. *Dev. Biol.*, **56**, 110.
3. Kimble, J. and Hirsh, D. (1979). The postembryonic cell lineages of the hermaphrodite and male gonads in *Caenorhabditis elegans*. *Dev. Biol.*, **70**, 396.

4. Sulston, J. E., Schierenberg, E., White, J. G., and Thomson, J. N. (1983). The embryonic cell lineage of the nematode *Caenorhabditis elegans*. *Dev. Biol.*, **100**, 64.
5. Coulson, A., Huynh, C., Kozono, Y., and Shownkeen, R. (1995). The physical map of the *Caenorhabditis elegans* genome. *Methods Cell Biol.*, **48**, 533.
6. Consortium, T. C. e. S. (1998). Genome sequence of the nematode *C. elegans*: a platform for investigating biology. *Science*, **282**, 2012.
7. Mello, C. C., Krame, J. M., Stinchcomb, D., and Ambros, V. (1992). Efficient gene transfer in *C. elegans*: Extrachromosomal maintenance and integration of transforming sequences. *EMBO J.*, **10**, 3959.
8. Hedgecock, E. M. and Herman, R. K. (1995). The *ncl-1* gene and genetic mosaics of *Caenorhabditis elegans*. *Genetics*, **141**, 989.
9. Gumienny, T. L., Lambie, E., Hartwieg, E., Horvitz, H. R., and Hengartner, M. O. (1999). Genetic control of programmed cell death in the *Caenorhabditis elegans* hermaphrodite germline. *Development*, **126**, 1011.
10. Sternberg, P. W. and Horvitz, H. R. (1982). Postembryonic nongonadal cell lineages of the nematode *Panagrellus redivivus*: description and comparison with those of *Caenorhabditis elegans*. *Dev. Biol.*, **93**, 181.
11. Kimble, J. E. and White, J. G. (1981). On the control of germ cell development in *Caenorhabditis elegans*. *Dev. Biol.*, **81**, 208.
12. Sternberg, P. W. and Horvitz, H. R. (1981). Gonadal cell lineages of the nematode *Panagrellus redivivus* and implications for evolution by the modification of cell lineage. *Dev. Biol.*, **88**, 147.
13. Robertson, A. G. and Thomson, J. N. (1982). Morphology of programmed cell death in the ventral nerve chord of *C. elegans* larvae. *J. Embryol. Exp. Morphol.*, **67**, 89.
14. Hedgecock, E. M., Sulston, J. E., and Thomson, J. N. (1983). Mutations affecting programmed cell deaths in the nematode *Caenorhabditis elegans*. *Science*, **220**, 1277.
15. Kerr, J. F., Wyllie, A. H., and Currie, A. R. (1972). Apoptosis: a basic biological phenomenon with wide-ranging implications in tissue kinetics. *Br. J. Cancer*, **26**, 239.
16. Wyllie, A. H., Kerr, J. F., and Currie, A. R. (1980). Cell death: the significance of apoptosis. *Int. Rev. Cytol.*, **68**, 251.
17. Sulston, J. E. (1976). Post-embryonic development in the ventral cord of *Caenorhabditis elegans*. *Phil. Trans. R Soc. Lond. B Biol. Sci.*, **275**, 287.
18. Ellis, H. M. and Horvitz, H. R. (1986). Genetic control of programmed cell death in the nematode *C. elegans*. *Cell*, **44**, 817.
19. Yuan, J. Y. and Horvitz, H. R. (1990). The *Caenorhabditis elegans* genes *ced-3* and *ced-4* act cell autonomously to cause programmed cell death. *Dev. Biol.*, **138**, 33.
20. Ellis, R. E., Jacobson, D. M., and Horvitz, H. R. (1991). Genes required for the engulfment of cell corpses during programmed cell death in *Caenorhabditis elegans*. *Genetics*, **129**, 79.
21. Ellis, R. E. and Horvitz, H. R. (1991). Two *C. elegans* genes control the programmed deaths of specific cells in the pharynx. *Development*, **112**, 591.
22. Hengartner, M. O., Ellis, R. E., and Horvitz, H. R. (1992). *Caenorhabditis elegans* gene *ced-9* protects cells from programmed cell death. *Nature*, **356**, 494.
23. Conradt, B. and Horvitz, H. R. (1998). The *C. elegans* protein EGL-1 is required for programmed cell death and interacts with the Bcl-2-like protein CED-9. *Cell*, **93**, 519.
24. Trent, C., Tsung, N., and Horvitz, H. R. (1983). Egg-laying defective mutants of the nematode *C. elegans*. *Genetics*, **104**, 619.
25. Ellis, R. E., Yuan, J. Y., and Horvitz, H. R. (1991). Mechanisms and functions of cell death. *Annu. Rev. Cell Biol.*, **7**, 663.

26. Avery, L. and Horvitz, H. R. (1987). A cell that dies during wild-type *C. elegans* development can function as a neuron in a *ced-3* mutant. *Cell*, **51**, 1071.
27. Hengartner, M. O. (1997). In *Cell death* (ed. D. L. Riddle, T. Blumenthal, B. J. Meyer, and J. R. Priess), p. 283. Cold Spring Harbor Laboratory Press, New York, USA.
28. Yuan, J., Shaham, S., Ledoux, S., Ellis, H. M., and Horvitz, H. R. (1993). The *C. elegans* cell death gene *ced-3* encodes a protein similar to mammalian interleukin-1 beta-converting enzyme. *Cell*, **75**, 641.
29. Alnemri, E. S., Livingston, D. J., Nicholson, D. W., Salvesen, G., Thornberry, N. A., Wong, W. W., *et al.* (1996). Human ICE/CED-3 protease nomenclature [Letter]. *Cell*, **87**, 171.
30. Kuida, K., Zheng, T. S., Na, S., Kuan, C., Yang, D., Karasuyama, H., *et al.* (1996). Decreased apoptosis in the brain and premature lethality in CPP32- deficient mice. *Nature*, **384**, 368.
31. Song, Z., McCall, K., and Steller, H. (1997). DCP-1, a *Drosophila* cell death protease essential for development. *Science*, **275**, 536.
32. Hakem, R., Hakem, A., Duncan, G. S., Henderson, J. T., Woo, M., Soengas, M. S., *et al.* (1998). Differential requirement for caspase-9 in apoptotic pathways *in vivo*. *Cell*, **94**, 339.
33. Kuida, K., Haydar, T. F., Kuan, C. Y., Gu, Y., Taya, C., Karasuyama, H., *et al.* (1998). Reduced apoptosis and cytochrome c-mediated caspase activation in mice lacking caspase-9. *Cell*, **94**, 325.
34. Varfolomeev, E. E., Schuchmann, M., Luria, V., Chiannilkulchai, N., Beckmann, J. S., Mett, I. L., *et al.* (1998). Targeted disruption of the mouse caspase-8 gene ablates cell death induction by the TNF receptors, Fas/Apo1, and DR3 and is lethal prenatally. *Immunity*, **9**, 267.
35. Xue, D., Shaham, S., and Horvitz, H. R. (1996). The *Caenorhabditis elegans* cell-death protein CED-3 is a cysteine protease with substrate specificities similar to those of the human CPP32 protease. *Genes Dev.*, **10**, 1073.
36. Walker, N. P., Talanian, R. V., Brady, K. D., Dang, L. C., Bump, N. J., Ferenz, C. R., *et al.* (1994). Crystal structure of the cysteine protease interleukin-1 beta-converting enzyme: a (p20/p10)2 homodimer. *Cell*, **78**, 343.
37. Wilson, K. P., Black, J. A., Thomson, J. A., Kim, E. E., Griffith, J. P., Navia, M. A., *et al.* (1994). Structure and mechanism of interleukin-1 beta converting enzyme [See Comments]. *Nature*, **370**, 270.
38. Rotonda, J., Nicholson, D. W., Fazil, K. M., Gallant, M., Gareau, Y., Labelle, M., *et al.* (1996). The three-dimensional structure of apopain/CPP32, a key mediator of apoptosis. *Nature Struct. Biol.*, **3**, 619.
39. Mittl, P. R., Di Marco, S., Krebs, J. F., Bai, X., Karanewsky, D. S., Priestle, J. P., *et al.* (1997). Structure of recombinant human CPP32 in complex with the tetrapeptide acetyl-Asp-Val-Ala-Asp fluoromethyl ketone. *J. Biol. Chem.*, **272**, 6539.
40. Shaham, S. and Horvitz, H. R. (1996). Developing *Caenorhabditis elegans* neurons may contain both cell-death protective and killer activities. *Genes Dev.*, **10**, 578.
41. Shaham, S. (1998). Identification of multiple *Caenorhabditis elegans* caspases and their potential roles in proteolytic cascades. *J. Biol. Chem.*, **273**, 35109.
42. Muzio, M., Stockwell, B. R., Stennicke, H. R., Salvesen, G. S., and Dixit, V. M. (1998). An induced proximity model for caspase-8 activation. *J. Biol. Chem.*, **273**, 2926.
43. Srinivasula, S. M., Ahmad, M., Fernandes-Alnemri, T., and Alnemri, E. S. (1998). Autoactivation of procaspase-9 by Apaf-1-mediated oligomerization. *Mol. Cell*, **1**, 949.
44. Steller, H. (1998). Artificial death switches: induction of apoptosis by chemically induced caspase multimerization. *Proc. Natl. Acad. Sci. USA*, **95**, 5421.

45. Yang, X., Chang, H. Y., and Baltimore, D. (1998). Essential role of CED-4 oligomerization in CED-3 activation and apoptosis. *Science*, **281**, 1355.
46. Yuan, J. and Horvitz, H. R. (1992). The *Caenorhabditis elegans* cell death gene *ced-4* encodes a novel protein and is expressed during the period of extensive programmed cell death. *Development*, **116**, 309.
47. Clem, R. J., Fechheimer, M., and Miller, L. K. (1991). Prevention of apoptosis by a baculovirus gene during infection of insect cells. *Science*, **254**, 1388.
48. Rabizadeh, S., LaCount, D. J., Friesen, P. D., and Bredesen, D. E. (1993). Expression of the baculovirus p35 gene inhibits mammalian neural cell death. *J. Neurochem.*, **61**, 2318.
49. Hay, B. A., Wolff, T., and Rubin, G. M. (1994). Expression of baculovirus P35 prevents cell death in *Drosophila*. *Development*, **120**, 2121.
50. Martinou, I., Fernandez, P. A., Missotten, M., White, E., Allet, B., Sadoul, R., *et al.* (1995). Viral proteins E1B19K and p35 protect sympathetic neurons from cell death induced by NGF deprivation. *J. Cell Biol.*, **128**, 201.
51. Sugimoto, A., Friesen, P. D., and Rothman, J. H. (1994). Baculovirus p35 prevents developmentally programmed cell death and rescues a *ced-9* mutant in the nematode *Caenorhabditis elegans*. *EMBO J.*, **13**, 2023.
52. Bump, N. J., Hackett, M., Hugunin, M., Seshagiri, S., Brady, K., Chen, P., *et al.* (1995). Inhibition of ICE family proteases by baculovirus antiapoptotic protein p35. *Science*, **269**, 1885.
53. Xue, D. and Horvitz, H. R. (1995). Inhibition of the *Caenorhabditis elegans* cell-death protease CED-3 by a CED-3 cleavage site in baculovirus p35 protein. *Nature*, **377**, 248.
54. Xue, D. and Horvitz, H. R. (1997). *Caenorhabditis elegans* CED-9 protein is a bifunctional cell-death inhibitor. *Nature*, **390**, 305.
55. Zou, H., Henzel, W. J., Liu, X., Lutschg, A., and Wang, X. (1997). Apaf-1, a human protein homologous to *C. elegans* CED-4, participates in cytochrome c-dependent activation of caspase-3 [See Comments]. *Cell*, **90**, 405.
56. Cecconi, F., Alvarez-Bolado, G., Meyer, B. I., Roth, K. A., and Gruss, P. (1998). Apaf1 (CED-4 homolog) regulates programmed cell death in mammalian development. *Cell*, **94**, 727.
57. Yoshida, H., Kong, Y. Y., Yoshida, R., Elia, A. J., Hakem, A., Hakem, R., *et al.* (1998). Apaf1 is required for mitochondrial pathways of apoptosis and brain development. *Cell*, **94**, 739.
58. Spector, M. S., Desnoyers, S., Hoeppner, D. J., and Hengartner, M. O. (1997). Interaction between the *C. elegans* cell-death regulators CED-9 and CED-4. *Nature*, **385**, 653.
59. Chinnaiyan, A. M., O'Rourke, K., Lane, B. R., and Dixit, V. M. (1997). Interaction of CED-4 with CED-3 and CED-9: a molecular framework for cell death. *Science*, **275**, 1122.
60. Wu, D., Wallen, H. D., and Nunez, G. (1997). Interaction and regulation of subcellular localization of CED-4 by CED-9. *Science*, **275**, 1126.
61. Seshagiri, S. and Miller, L. K. (1997). *Caenorhabditis elegans* CED-4 stimulates CED-3 processing and CED-3-induced apoptosis. *Curr. Biol.*, **7**, 455.
62. James, C., Gschmeissner, S., Fraser, A., and Evan, G. I. (1997). CED-4 induces chromatin condensation in *Schizosaccharomyces pombe* and is inhibited by direct physical association with CED-9. *Curr. Biol.*, **7**, 246.
63. Wu, D., Wallen, H. D., Inohara, N., and Nunez, G. (1997). Interaction and regulation of the *Caenorhabditis elegans* death protease CED-3 by CED-4 and CED-9. *J. Biol. Chem.*, **272**, 21449.
64. Hengartner, M. (1998). Apoptosis. Death by crowd control. *Science*, **281**, 1298.

65. Liu, X., Kim, C. N., Yang, J., Jemmerson, R., and Wang, X. (1996). Induction of apoptotic program in cell-free extracts: requirement for dATP and cytochrome c. *Cell*, **86**, 147.
66. Li, P., Nijhawan, D., Budihardjo, I., Srinivasula, S. M., Ahmad, M., Alnemri, E. S., *et al.* (1997). Cytochrome c and dATP-dependent formation of Apaf-1/caspase-9 complex initiates an apoptotic protease cascade. *Cell*, **91**, 479.
67. Zou, H., Li, Y., Liu, X., and Wang, X. (1999). An APAF-1.Cytochrome c multimeric complex is a functional apoptosome that activates procaspase-9. *J. Biol. Chem.*, **274**, 11549.
68. Chinnaiyan, A. M., Chaudhary, D., O'Rourke, K., Koonin, E. V., and Dixit, V. M. (1997). Role of CED-4 in the activation of CED-3. *Nature*, **388**, 728.
69. Hu, Y., Ding, L., Spencer, D. M., and Nunez, G. (1998). WD-40 repeat region regulates Apaf-1 self-association and procaspase-9 activation. *J. Biol. Chem.*, **273**, 33489.
70. Shaham, S. and Horvitz, H. R. (1996). An alternatively spliced *C. elegans ced-4* RNA encodes a novel cell death inhibitor. *Cell*, **86**, 201.
71. Wu, D., Chen, P. J., Chen, S., Hu, Y., Nunez, G., and Ellis, R. E. (1999). *C. elegans* MAC-1, an essential member of the AAA family of ATPases, can bind CED-4 and prevent cell death. *Development*, **126**, 2021.
72. Hengartner, M. O. and Horvitz, H. R. (1994). *C. elegans* cell survival gene *ced-9* encodes a functional homolog of the mammalian proto-oncogene bcl-2. *Cell*, **76**, 665.
73. Tsujimoto, Y. and Croce, C. M. (1986). Analysis of the structure, transcripts, and protein products of *bcl-2*, the gene involved in human follicular lymphoma. *Proc. Natl. Acad. Sci. USA*, **83**, 5214.
74. Seto, M., Jaeger, U., Hockett, R. D., Graninger, W., Bennett, S., Goldman, P., *et al.* (1988). Alternative promoters and exons, somatic mutation and deregulation of the Bcl-2-Ig fusion gene in lymphoma. *EMBO J.*, **7**, 123.
75. Vaux, D. L., Cory, S., and Adams, J. M. (1988). Bcl-2 gene promotes haemopoietic cell survival and cooperates with c-myc to immortalize pre-B cells. *Nature*, **335**, 440.
76. Nunez, G., London, L., Hockenbery, D., Alexander, M., McKearn, J. P., and Korsmeyer, S. J. (1990). Deregulated Bcl-2 gene expression selectively prolongs survival of growth factor-deprived hemopoietic cell lines. *J. Immunol.*, **144**, 3602.
77. Vaux, D. L., Weissman, I. L., and Kim, S. K. (1992). Prevention of programmed cell death in *Caenorhabditis elegans* by human *bcl-2*. *Science*, **258**, 1955.
78. Reed, J. C. (1998). Bcl-2 family proteins. *Oncogene*, **17**, 3225.
79. Boise, L. H., Gonzalez-Garcia, M., Postema, C. E., Ding, L., Lindsten, T., Turka, L. A., *et al.* (1993). bcl-x, a bcl-2-related gene that functions as a dominant regulator of apoptotic cell death. *Cell*, **74**, 597.
80. Cheng, E. H., Kirsch, D. G., Clem, R. J., Ravi, R., Kastan, M. B., Bedi, A., *et al.* (1997). Conversion of Bcl-2 to a Bax-like death effector by caspases. *Science*, **278**, 1966.
81. Hockenbery, D., Nunez, G., Milliman, C., Schreiber, R. D., and Korsmeyer, S. J. (1990). Bcl-2 is an inner mitochondrial membrane protein that blocks programmed cell death. *Nature*, **348**, 334.
82. Tanaka, S., Saito, K., and Reed, J. C. (1993). Structure-function analysis of the Bcl-2 oncoprotein. Addition of a heterologous transmembrane domain to portions of the Bcl-2 beta protein restores function as a regulator of cell survival. *J. Biol. Chem.*, **268**, 10920.
83. Hengartner, M. O. and Horvitz, H. R. (1994). Activation of *C. elegans* cell death protein CED-9 by an amino-acid substitution in a domain conserved in Bcl-2. *Nature*, **369**, 318.
84. Susin, S. A., Lorenzo, H. K., Zamzami, N., Marzo, I., Snow, B. E., Brothers, G. M., *et al.* (1999). Molecular characterization of mitochondrial apoptosis-inducing factor. *Nature*, **397**, 441.

85. del Peso, L., Gonzalez, V. M., and Nunez, G. (1998). *Caenorhabditis elegans* EGL-1 disrupts the interaction of CED-9 with CED- 4 and promotes CED-3 activation. *J. Biol. Chem.*, **273**, 33495.
86. Desai, C. and Horvitz, H. R. (1989). *Caenorhabditis elegans* mutants defective in the functioning of the motor neurons responsible for egg laying. *Genetics*, **121**, 703.
87. Conradt, B. and Horvitz, H. R. (1999). The TRA-1A sex determination protein of *C. elegans* regulates sexually dimorphic cell deaths by repressing the egl-1 cell death activator gene. *Cell*, **98**, 317.
88. Hodgkin, J. A. and Brenner, S. (1977). Mutations causing transformation of sexual phenotype in the nematode *Caenorhabditis elegans*. *Genetics*, **86**, 275.
89. Zarkower, D. and Hodgkin, J. (1992). Molecular analysis of the *C. elegans* sex-determining gene *tra-1*: a gene encoding two zinc finger proteins. *Cell*, **70**, 237.
90. Li, H., Zhu, H., Xu, C. J., and Yuan, J. (1998). Cleavage of BID by caspase 8 mediates the mitochondrial damage in the Fas pathway of apoptosis. *Cell*, **94**, 491.
91. Luo, X., Budihardjo, I., Zou, H., Slaughter, C., and Wang, X. (1998). Bid, a Bcl2 interacting protein, mediates cytochrome c release from mitochondria in response to activation of cell surface death receptors. *Cell*, **94**, 481.
92. Gross, A., Yin, X. M., Wang, K., Wei, M. C., Jockel, J., Milliman, C., *et al.* (1999). Caspase cleaved BID targets mitochondria and is required for cytochrome c release, while BCL-XL prevents this release but not tumor necrosis factor-R1 / Fas death. *J. Biol. Chem.*, **274**, 1156.
93. Puthalakath, H., Huang, D. C., O'Reilly, L. A., King, S. M., and Strasser, A. (1999). The proapoptotic activity of the Bcl-2 family member Bim is regulated by interaction with the dynein motor complex. *Mol. Cell*, **3**, 287.
94. Sugimoto, A., Hozak, R. R., Nakashima, T., Nishimoto, T., and Rothman, J. H. (1995). *dad-1*, an endogenous programmed cell death suppressor in *Caenorhabditis elegans* and vertebrates. *EMBO J.*, **14**, 4434.
95. Nakashima, T., Sekiguchi, T., Kuraoka, A., Fukushima, K., Shibata, Y., Komiyama, S., *et al.* (1993). Molecular cloning of a human cDNA encoding a novel protein, DAD1, whose defect causes apoptotic cell death in hamster BHK21 cells. *Mol. Cell. Biol.*, **13**, 6367.
96. Silberstein, S., Collins, P. G., Kelleher, D. J., and Gilmore, R. (1995). The essential OST2 gene encodes the 16-kD subunit of the yeast oligosaccharyltransferase, a highly conserved protein expressed in diverse eukaryotic organisms. *J. Cell Biol.*, **131**, 371.
97. Metzstein, M. M., Hengartner, M. O., Tsung, N., Ellis, R. E., and Horvitz, H. R. (1996). Transcriptional regulator of programmed cell death encoded by *Caenorhabditis elegans* gene *ces-2*. *Nature*, **382**, 545.
98. Trent, C., Wood, W. B., and Horvitz, H. R. (1988). A novel dominant transformer allele of the sex-determining gene *her-1* of *Caenorhabditis elegans*. *Genetics*, **120**, 145.
99. Kuwabara, P. E., Okkema, P. G., and Kimble, J. (1992). *tra-2* encodes a membrane protein and may mediate cell communication in the *Caenorhabditis elegans* sex determination pathway. *Mol. Biol. Cell*, **3**, 461.
100. Perry, M. D., Li, W., Trent, C., Robertson, B., Fire, A., Hageman, J. M., *et al.* (1993). Molecular characterization of the *her-1* gene suggests a direct role in cell signaling during *Caenorhabditis elegans* sex determination. *Genes Dev.*, **7**, 216.
101. Kuwabara, P. E. (1996). A novel regulatory mutation in the *C. elegans* sex determination gene *tra-2* defines a candidate ligand / receptor interaction site. *Development*, **122**, 2089.
102. Meyer, B. J. (1997). In *Sex determination and X chromosome dosage compensation* (ed. D. L. Riddle, T. Blumenthal, B. J. Meyer, and J. R. Priess), p. 209. Cold Spring Harbor, New York, USA.

103. Schedl, T. (1997). In *Developmental genetics of the germline* (ed. D. L. Riddle, T. Blumenthal, B. J. Meyer, and J. R. Priess), p. 241. Cold Spring Harbor Laboratory Press, New York, USA.
104. Church, D. L., Guan, K. L., and Lambie, E. J. (1995). Three genes of the MAP kinase cascade, *mek-2, mpk-1/sur-1* and *let-60 ras*, are required for meiotic cell cycle progression in *Caenorhabditis elegans*. *Development*, **121**, 2525.
105. Sternberg, P. W. and Han, M. (1998). Genetics of RAS signaling in *C. elegans*. *Trends Genet.*, **14**, 466.
106. Wu, Y. C. and Horvitz, H. R. (1998). *C. elegans* phagocytosis and cell-migration protein CED-5 is similar to human DOCK180 [See Comments]. *Nature*, **392**, 501.
107. Ferguson, E. L. and Horvitz, H. R. (1985). Identification and characterization of 22 genes that affect the vulval cell lineages of the nematode *Caenorhabditis elegans*. *Genetics*, **110**, 17.
108. Ferguson, E. L., Sternberg, P. W., and Horvitz, H. R. (1987). A genetic pathway for the specification of the vulval cell lineages of *Caenorhabditis elegans*. *Nature*, **326**, 259.
109. Kim, S. (1994). Two *C. elegans* genes that can mutate to cause degenerative cell death. Ph.D. Thesis. Massachusetts Institute of Technology. Cambridge.
110. Ahnn, J. and Fire, A. (1994). A screen for genetic loci required for body-wall muscle development during embryogenesis in *Caenorhabditis elegans*. *Genetics*, **137**, 483.
111. Hasegawa, H., Kiyokawa, E., Tanaka, S., Nagashima, K., Gotoh, N., Shibuya, M., *et al.* (1996). DOCK180, a major CRK-binding protein, alters cell morphology upon translocation to the cell membrane. *Mol. Cell. Biol.*, **16**, 1770.
112. Rushton, E., Drysdale, R., Abmayr, S. M., Michelson, A. M., and Bate, M. (1995). Mutations in a novel gene, myoblast city, provide evidence in support of the founder cell hypothesis for *Drosophila* muscle development. *Development*, **121**, 1979.
113. Nagase, T., Seki, N., Ishikawa, K., Ohira, M., Kawarabayasi, Y., Ohara, O., *et al.* (1996). Prediction of the coding sequences of unidentified human genes. VI. The coding sequences of 80 new genes (KIAA0201-KIAA0280) deduced by analysis of cDNA clones from cell line KG-1 and brain. *DNA Res.*, **3**, 321.
114. Erickson, M. R., Galletta, B. J., and Abmayr, S. M. (1997). *Drosophila* myoblast city encodes a conserved protein that is essential for myoblast fusion, dorsal closure, and cytoskeletal organization. *J. Cell Biol.*, **138**, 589.
115. Clark, E. A. and Brugge, J. S. (1995). Integrins and signal transduction pathways: the road taken. *Science*, **268**, 233.
116. Savill, J., Dransfield, I., Hogg, N., and Haslett, C. (1990). Vitronectin receptor-mediated phagocytosis of cells undergoing apoptosis. *Nature*, **343**, 170.
117. Wu, Y. C. and Horvitz, H. R. (1998). The *C. elegans* cell corpse engulfment gene *ced-7* encodes a protein similar to ABC transporters. *Cell*, **93**, 951.
118. Higgins, C. F. and Gottesman, M. M. (1992). Is the multidrug transporter a flippase? *Trends Biochem. Sci.*, **17**, 18.
119. Ruetz, S. and Gros, P. (1994). Phosphatidylcholine translocase: a physiological role for the mdr2 gene. *Cell*, **77**, 1071.
120. Luciani, M. F., Denizot, F., Savary, S., Mattei, M. G., and Chimini, G. (1994). Cloning of two novel ABC transporters mapping on human chromosome 9. *Genomics*, **21**, 150.
121. Klugbauer, N. and Hofmann, F. (1996). Primary structure of a novel ABC transporter with a chromosomal localization on the band encoding the multidrug resistance-associated protein. *FEBS Lett.*, **391**, 61.
122. Connors, T. D., Van Raay, T. J., Petry, L. R., Klinger, K. W., Landes, G. M., and Burn, T. C. (1997). The cloning of a human ABC gene (ABC3) mapping to chromosome 16p13.3. *Genomics*, **39**, 231.

123. Allikmets, R. (1997). A photoreceptor cell-specific ATP-binding transporter gene (ABCR) is mutated in recessive Stargardt macular dystrophy. *Nature Genet.*, **17**, 122.
124. Illing, M., Molday, L. L., and Molday, R. S. (1997). The 220-kDa rim protein of retinal rod outer segments is a member of the ABC transporter superfamily. *J. Biol. Chem.*, **272**, 10303.
125. Luciani, M. F. and Chimini, G. (1996). The ATP binding cassette transporter ABC1, is required for the engulfment of corpses generated by apoptotic cell death. *EMBO J.*, **15**, 226.
126. Liu, Q. A. and Hengartner, M. O. (1998). Candidate adaptor protein CED-6 promotes the engulfment of apoptotic cells in *C. elegans*. *Cell*, **93**, 961.
127. Ren, R., Mayer, B. J., Cicchetti, P., and Baltimore, D. (1993). Identification of a ten-amino acid proline-rich SH3 binding site. *Science*, **259**, 1157.
128. Kavanaugh, W. M., Turck, C. W., and Williams, L. T. (1995). PTB domain binding to signaling proteins through a sequence motif containing phosphotyrosine. *Science*, **268**, 1177.
129. Zhou, S., Margolis, B., Chaudhuri, M., Shoelson, S. E., and Cantley, L. C. (1995). The phosphotyrosine interaction domain of SHC recognizes tyrosine-phosphorylated NPXY motif. *J. Biol. Chem.*, **270**, 14863.
130. Fadok, V. A., Savill, J. S., Haslett, C., Bratton, D. L., Doherty, D. E., Campbell, P. A., *et al.* (1992). Different populations of macrophages use either the vitronectin receptor or the phosphatidylserine receptor to recognize and remove apoptotic cells. *J. Immunol.*, **149**, 4029.
131. Fadok, V. A., Voelker, D. R., Campbell, P. A., Cohen, J. J., Bratton, D. L., and Henson, P. M. (1992). Exposure of phosphatidylserine on the surface of apoptotic lymphocytes triggers specific recognition and removal by macrophages. *J. Immunol.*, **148**, 2207.
132. Hevelone, J. and Hartman, P. S. (1988). An endonuclease from *Caenorhabditis elegans*: partial purification and characterization. *Biochem. Genet.*, **26**, 447.
133. Gavrieli, Y., Sherman, Y., and Ben-Sasson, S. A. (1992). Identification of programmed cell death *in situ* via specific labeling of nuclear DNA fragmentation. *J. Cell Biol.*, **119**, 493.
134. Steller, H. (1995). Mechanisms and genes of cellular suicide. *Science*, **267**, 1445.
135. Miura, M., Zhu, H., Rotello, R., Hartwieg, E. A., and Yuan, J. (1993). Induction of apoptosis in fibroblasts by IL-1 beta-converting enzyme, a mammalian homolog of the *C. elegans* cell death gene *ced-3*. *Cell*, **75**, 653.
136. Fire, A., Xu, S., Montgomery, M. K., Kostas, S. A., Driver, S. E., and Mello, C. C. (1998). Potent and specific genetic interference by double-stranded RNA in *Caenorhabditis elegans*. *Nature*, **391**, 806.

3 | Genetic and molecular analysis of programmed cell death in *Drosophila*

ANDREAS BERGMANN and HERMANN STELLER

1. Introduction

During development of multicellular organisms large numbers of cells die by a specialized form of programmed cell death (PCD), termed apoptosis. Apoptosis is characterized by a number of characteristic morphological and ultrastructural events including membrane blebbing, cytoplasmic condensation, nuclear fragmentation, and engulfment of dying cells by phagocytes (1). Cell loss as a normal feature of animal development has long been recognized in the vertebrate nervous system where about 50% of all neurons and other cell types die by apoptosis (2, 3). It is now clear that programmed cell death is an important process for proper development and tissue homeostasis.

Many different signals originating from within or outside a cell influence the decision of a cell to live or die. These include cell–cell contact, steroid hormones, lack of peptide survival factors, activation of cell surface death receptors, oxidative stress, excitotoxicity, ischaemia, unfolded proteins, and unrepaired DNA strand breaks. Misregulation of cell death is implicated in many diseases that are associated with inappropriate activation or inactivation of cell death including neurodegenerative disorders, ischaemic injury, cancer, autoimmune disorders, and viral infections (reviewed in ref. 4).

Genetically amenable invertebrate model systems have been extremely useful for our understanding of many biological processes. These include signal transduction, basic transcription, behaviour, development of the nervous system, and PCD. Traditionally, the nematode *Caenorhabditis elegans* has been the genetic model organism of choice to study programmed cell death (see Chapter 2). However, although the lineage-restricted development of *C. elegans* was indispensable for the identification of the core components of the cell death programme, it is somewhat limited for the identification of the non-autonomous regulators of cell death because PCD is genetically predetermined in this organism.

In *Drosophila*, a large number of cells die during embryonic and imaginal development, as well as during metamorphosis, with all the characteristics of apoptotic cell death (5–7). However, unlike *C. elegans*, this cell death is not genetically predetermined in a lineage-restricted manner, but is dependent on environmental circumstances. Initially, a large number of cells are formed prior to cell differentiation. At this point, a given cell does not 'know' its final cell fate. During and after cell fate determination, extra, misplaced, and unwanted cells are eliminated by PCD. This appears to be a common strategy for the generation of complex patterns such as neuronal connectivity that must be imposed on populations of cells whose numbers cannot be precisely specified by the genome. Thus, *Drosophila* shares the genetic accessibility with *C. elegans* and the developmental plasticity with vertebrates. Therefore, molecular genetic studies in *Drosophila* promise considerable hope for advancing our understanding of the basic control mechanisms involved in the regulation of PCD and will provide a conceptual framework for elucidating the mechanisms and control of apoptosis in humans.

This chapter summarizes the key findings and concepts which have emerged from studies of apoptosis in *Drosophila*. After a description of PCD during *Drosophila* development (Section 2), we present the identification, characterization, and regulation of the cell death inducers *reaper*, *hid*, and *grim* (Sections 3–5). Another class of important cell death regulators are the IAP molecules (Section 6). The basic executioners of apoptosis, the caspases, are covered in Section 7. In Section 8 we present a potential model how *reaper*, *hid*, and *grim* induce apoptosis. Given the importance of *reaper*, *hid*, and *grim* for the regulation of cell death in *Drosophila* we discuss the existence of mammalian homologues in Section 9. However, *reaper*, *hid*, and *grim* do not account for all cell death in *Drosophila*. Thus, in Section 10 we consider alternative cell death paradigms in *Drosophila*. The clearance of apoptotic cells by phagocytosis is presented in Section 11. Before closing with our final considerations (Section 13), we present several human disorders that can be studied readily in *Drosophila* (Section 12).

2. Description of programmed cell death during *Drosophila* development

The life cycle of holometabolous insects such as *Drosophila* consists of distinct phases of development during which the animal has a body pattern specialized for its behaviour at that stage. The *Drosophila* embryo which develops in 17 distinct developmental stages hatches into a larva that has little physical resemblance with the adult fly it will eventually become. During embryonic development, the primordia of the adult structures are segregated from the larval tissues, and form the imaginal discs during larval stages. (Note: The adult insect is also known as the imago. Although this name is not in current use anymore, the precursors of the adult structures continue to be referred to as 'imaginal' discs). There are three larval stages followed by pupariation. In a process called metamorphosis, the imaginal discs undergo dramatic morphogenetic changes to take on the shape of the adult fly. During each of these developmental phases a large number of cells die by PCD (5–7). In the following, we describe the pattern of PCD during these developmental phases. Cell death regulation by trophic interactions as well as by intracellular factors will also be presented. During the description, different modes of PCD regulation by a variety of stimuli will become apparent.

A large number of cells undergo PCD during *Drosophila* embryogenesis, and these cells display the characteristic ultrastructural features described for apoptosis (5). To visualize and to follow the cell death pattern during embryogenesis, live embryos were stained with the vital dyes Acridine orange (AO) or Nile blue. AO positive cell death starts at about stage 11, and thereafter becomes widespread, affecting many different tissues and regions of the embryo (Fig. 1; see Plate 1) (5). Although the cell death pattern is highly dynamic, it is fairly reproducible for any given stage during development. Moreover, detailed analysis of the pattern in the CNS showed that the exact number and position of dying cells on either side of the midline varies (5). This is surprising for a bilateral organism such as *Drosophila* but indicates that the onset of PCD in the embryonic CNS is somewhat variable and suggests that the decision

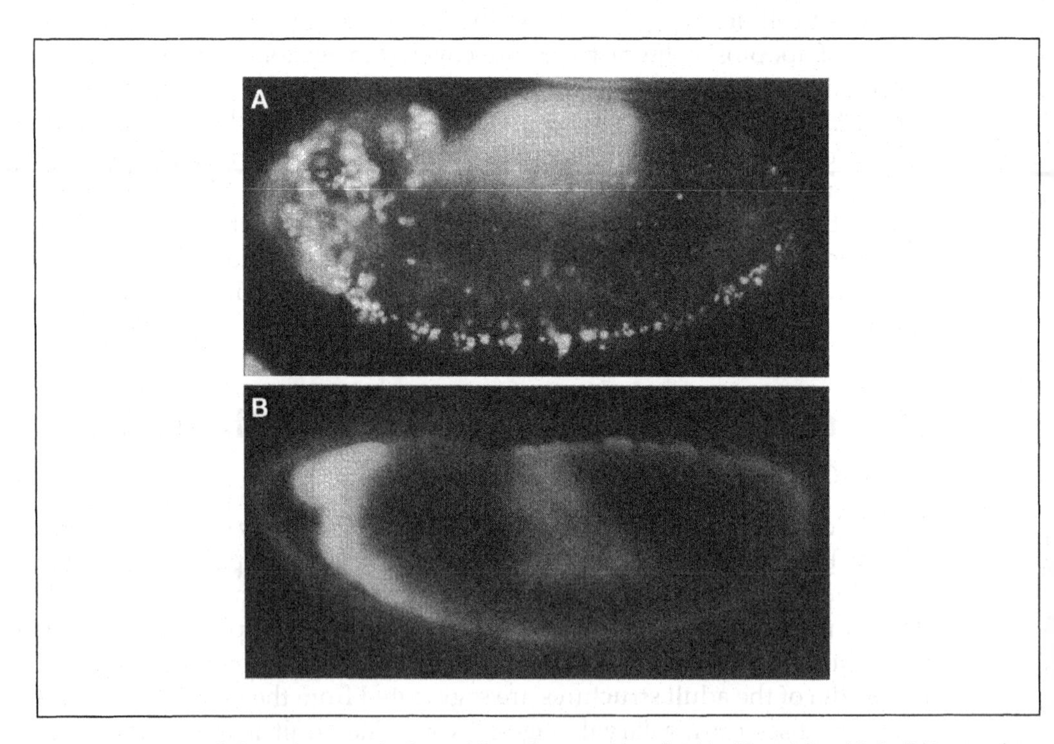

Fig. 1 (See Plate 1) Cell death pattern in *Drosophila* embryos as visualized by AO staining. (A) A wild-type embryo at stage 12 shows widespread cell death that is particularly pronounced in the head region and along the ventral nerve cord where the CNS develops. (B) A *H99* homozygous mutant embryo, completely lacking AO positive cell death.

to die is not genetically predetermined but rather influenced by local cell–cell interactions.

Cell death occurring during imaginal development has been best described for the eye imaginal disc (but see ref. 8 for cell death in the wing imaginal disc). In the early larval stages there is virtually no detectable PCD in the eye imaginal disc (6). Late in larval development (third instar larval stage), a low amount of PCD is present in two bands of dead cells, one ahead of the morphogenetic furrow and one near the posterior margin of the eye disc (6). The significance of this cell death, which affects only about 30–40 cells, is unknown. However, much later in pupal development, between 35–50 hours after pupation, about 2000 cells are eliminated (6). This amount of cell death corresponds to about two to three cells per ommatidium that ultimately consists of 19 cells. The dying cells are usually undifferentiated, interommatidial cells or surplus secondary and tertiary pigment cells which have to be removed to achieve the highly regular, reiterative lattice of the compound eye. The pupal cell death appears to be distributed uniformly over the entire eye disc. Interestingly, most of the dying cells have bristle cells as their neighbours suggesting a form of positional bias, perhaps contact-mediated, by the preferential elimination of cells adjacent to bristle cells (6). Finally, a unique population of ommatidia at the periphery of the retina is removed in a further round of cell death between 60–70 hours. The death of these ommatidia occurs simultaneously, rather than being scattered over a period of 20 hours (6).

Metamorphosis, the transformation of the worm-like larva into the mature adult fly, involves large scale cell death. Metamorphosis occurs in two major phases. During the initial phase, the transition from larva to pupa, in which a crude adult form is generated, larval structures that are no longer needed are removed by PCD. For example, muscles and neurons that are needed for larval locomotion are eliminated by PCD (7, 9). The second round of PCD occurs much later, once the adult has emerged. Here, PCD is used to remove any larval cells that were maintained during metamorphosis, as well as those structures required only for adult eclosion (10–12). Both early and late metamorphic cell death is controlled by changes in the titre of the steroid hormone ecdysone, though in opposite ways. The initial wave of PCD is triggered by the surge of ecdysone at pupation (7, 13). In this case, ecdysone acts as a death-inducing signal. Interestingly, changes in ecdysone levels have the opposite effect on late metamorphic PCD. After eclosion of the adult, it is a decline of the ecdysone levels that induces the late wave of cell death. Injection of 20-hydroxyecdysone can block the post-metamorphic cell death (12, 13).

The effects of ecdysteroids are mediated through heterodimeric nuclear hormone receptors encoded by two genes that are members of the steroid receptor superfamily, the *Drosophila Ecdysone receptor* (*EcR*, NR1H1) gene and the RXR homologue *ultraspiracle* (*usp*, NR2B4) gene (14–16). Thus, it is attractive to speculate that the binding of ecdysone to the EcR/USP complex induces the transcription of one or several genes required to initiate the cell death programme. The observation that treatment with the protein synthesis inhibitors actinomycin D and cycloheximide can reduce the level of PCD in the abdomen (17) is consistent with the postulated requirement of *de novo* RNA and protein synthesis to execute PCD. Excellent candidates for downstream targets of the ecdysone receptor are the cell death inducers *reaper* and *hid* (see below). The *EcR* gene encodes three different isoforms, *EcR-A*, *EcR-B1*, and *EcR-B2* which have distinct N-termini (18). *The EcR-A* and *EcR-B1* isoforms are expressed with distinct spatial and temporal profiles during metamorphosis (18). Interestingly, almost all cells in the CNS that will undergo post-metamorphic cell death express high levels of the *EcR-A* isoform (12). This strong expression of the EcR-A protein in doomed neurons helps explain why only some neurons die in response to a decline in ecdysone. However, two neurons, n6 and n7, survive despite high *EcR-A* expression levels suggesting that this explanation is not sufficient for determining which neurons die. Finally, decapitation of adults soon after eclosion will prevent post-metamorphic death in the CNS, even though ecdysone titres will decline under these conditions (11). This has led to the postulation of a 'head factor' that is required for death. The molecular nature of the head factor remains to be determined but it is possible that neuronal activity is required for the death-inducing activity.

Regulation of PCD in *Drosophila* also appears to involve trophic-like interactions. The concept of trophic survival mechanisms was originally developed to explain the massive neuronal cell loss in the developing vertebrate CNS and was later extended to include most, if not all, animal cells (reviewed in refs 19–21). This concept is based on the assumption that the cell intrinsic suicide programme operates by default unless it is suppressed by trophic factors such as the neurotrophins secreted by neighbouring cells (20, 21). This social control of cell survival ensures the functional integrity of a given tissue or organ by matching the number of different cell types to each other. For example, in *Drosophila* it is striking that the size of the optic ganglia, the CNS portion of the insect visual system, is always matched to the variable size of the retina (22). This results from the adjustment of both cell proliferation and cell death through competitive interactions between cells in the retina and optic ganglia (23–25). Apparently, the continued survival of neurons in the optic ganglia depends on retinal input. Only those neurons that are innervated by retinal photoreceptors survive in the optic ganglia. Neurons that do not participate in the neuronal circuit are eliminated by PCD.

PCD in *Drosophila* can also be induced by a number of intracellular signals that act autonomously within the dying cell. For example, DNA damage induced by ionizing radiation efficiently triggers apoptosis in affected cells (5, 26). Also, numerous mutations that block cellular differentiation cause the death of cells that fail to complete their developmental programme (5, 27–30). Apparently, an unknown intrinsic mechanism monitors a cells' ability to terminally differentiate and activates the death programme if it fails to do so. Another interesting class of mutants cause light-dependent retinal degeneration, presumably as a result of 'stress' that results from executing a defective phototransduction cascade (31, 32). At this point, little is known regarding the molecular nature of any of the underlying cell autonomous signalling pathways in *Drosophila*.

3. The *H99* cell death genes: *reaper, hid,* and *grim*

A genetic approach was used to screen a large fraction of the *Drosophila* genome for genes that are required for embryonic cell death. Embryos homozygous mutant for previously identified chromosomal deletions were stained with AO and examined for their cell death pattern (26). Several deletions which are deficient for the chromosomal interval 75C1,2 were found to be essential for all embryonic cell death. Embryos mutant for the smallest of these deficiencies, *Df(3R)H99* (from hereon referred to as *H99*) completely lack cell death (Fig. 1; see Plate 1) and have a large excess of cells resulting for instance in an enlarged CNS (26). These embryos die at the end of embryogenesis with numerous morphological defects due to the presence of extra cells demonstrating the importance of PCD for proper development. In addition to developmental PCD, cell death induced in response to X-ray irradiation or caused by developmental defects is also inhibited in *H99* mutant embryos (26). However, extremely high doses of X-ray are able to induce a small amount of cell death in *H99* mutant embryos, indicating that the basic cell death machinery is functionally intact in these embryos (26). Thus, the *H99* gene(s) encode important inducers, but not effectors of cell death.

Molecular genetic analysis of the *H99* region has led to the identification of three cell death genes: *reaper* (*rpr*), *head involution defective* (*hid*), and *grim* (Fig. 2) (26, 33, 34). The three genes encode novel proteins without any significant homology to other proteins in the database that would indicate their biochemical function. However, the first 14 N-terminal amino acid residues of *rpr, hid,* and *grim* share some degree of homology (Fig. 3) (33, 34). As discussed in more detail below, the N-terminus of RPR, HID, and GRIM is required for protein–protein interaction with another class of cell death regulators, the IAP molecules (see Section 6). RPR also has some weak homology to the death domain of Fas and TNF-R1 (35). The significance of this homology is controversial since mutations in conserved residues between RPR and Fas do not affect the killing activity of RPR (36, 37). *rpr* and *grim* are specifically expressed in cells which are going to die (26, 34, 38). Their expression pattern is remarkably similar to the cell death pattern known from AO staining. This observation supports the notion that *rpr* and *grim* are important regulators of PCD, in contrast to the core components of the cell death machinery which are expected to be present ubiquitously (20, 39). Induced ectopic cell death (in response to X-ray irradiation or developmental defects) is accompanied by ectopic *rpr* expression (40–42). This provides evidence that the integration of incoming death-inducing stimuli occurs—at least in part—by a transcriptional mechanism (see next section). As outlined in Section 5 in more detail, the activity of the third gene, *hid,* appears to be regulated differently and includes both transcriptional and post-translational mechanisms.

Ectopic expression of any one of the cell death regulators, *rpr, hid,* or *grim,* in various *Drosophila* tissues and cell lines is sufficient for the induction of PCD, even in the absence of the other two genes (i.e. in an *H99* mutant background), indicating that they are able to induce apoptosis independently of each other (26, 33, 34, 43).

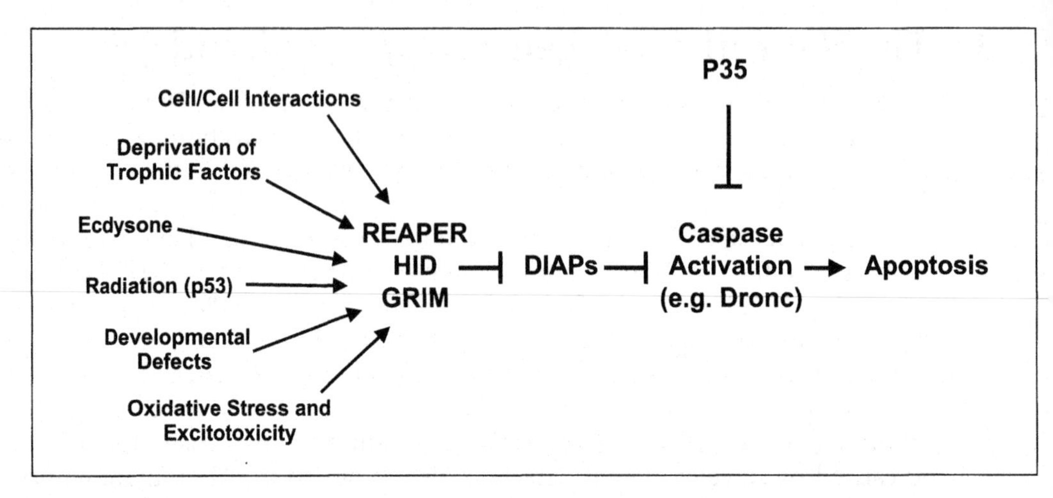

Fig. 2 The cell death pathway in *Drosophila*. During *Drosophila* embryogenesis, the cell death inducers REAPER, HID, and GRIM are central for the regulation of apoptosis. They integrate a large number of different incoming signals to activate a common cell death pathway directly or indirectly by activating caspases. *Drosophila* IAPs (DIAPs) counteract REAPER-, HID-, and GRIM-induced cell death. In this chapter, we propose a mechanistic model for how REAPER, HID, and GRIM induce apoptosis. The baculoviral cell death inhibitor p35 is known to inhibit apoptosis by directly binding to caspases (45, 46).

Expression of either gene in the fly compound eye using the eye-specific promoter GMR results in eye ablation (for instance *GMR-hid* in Fig. 5A; see Plate 2A). The eye ablation phenotype is restored to wild-type if the baculoviral cell death inhibitor *p35* is co-expressed in the compound eye (33, 34, 43, 44). Since p35 exerts its anti-apoptotic activity by inhibiting caspase activation (45, 46), this finding suggests that RPR, HID, and GRIM kill by activating a caspase pathway (Fig. 2).

Although RPR, HID, and GRIM have been shown to induce apoptosis independently of each other there is increasing evidence that they may act in a combinatorial fashion to regulate cell death. For instance, during metamorphosis in response to the steroid hormone ecdysone, *rpr* and *hid* transcripts accumulate in salivary glands and the larval midgut, whereas *grim* transcripts do not (47). Prior to the onset of apoptosis of a certain subset of neurons, the n4 cluster, expression of *rpr* and *grim* is up-regulated whereas *hid* expression is not (41, 48). More strikingly, Zhou *et al.* (49)

HID	2 **AVPFYLPE**GGADD14
REAPER	2 **AVAFYIPDQATLL** 14
GRIM	2 **AIAYFIPDQA**QLL 14

Fig. 3 The N-terminus of REAPER, HID, and GRIM contains a conserved motif. Alignment of the first 14 amino acid residues reveals identical and conserved residues, printed in bold. The analysis reveals that this stretch of residues is more closely related between REAPER and GRIM, whereas HID is a little more distant. Outside this region, the three proteins do not show any obvious similarity to each other nor to any other known proteins in the database.

demonstrated that apoptosis of the midline glia in the embryonic CNS occurs only if *rpr* and *hid* are expressed together. In contrast, *grim* appears to act in a distinct fashion from *rpr* and *hid*, as targeted expression of *grim* alone was sufficient to induce midline glia cell death (50). These distinct killing properties might be—at least in part—due to the failure of certain anti-apoptotic proteins such as DIAP2 (see below) to block GRIM-induced apoptosis whereas cells are well protected from RPR- and HID-induced cell death by DIAP2 (50). This work is very informative as it shows that cells might respond differently to several apoptotic stimuli, probably by expression of distinct inherent anti-apoptotic functions.

4. Transcriptional regulation of the *reaper* gene

Work in mammalian tissue culture cells and in *Drosophila* has demonstrated that many different signals can influence the decision of a cell to die or to live. These signals originate either internally or externally and include inhibition of cell differentiation, viral infection, unfolded proteins, DNA breaks, lack of peptide survival factors, steroid hormones (ecdysone), activation of death receptors, oxidative stress, excitotoxicity, and ischaemia (Fig. 2). Depending on the cell type studied and the signal applied, this cell death either requires or does not require *de novo* RNA and protein synthesis. Studies in *Drosophila* have been very informative to delineate these two modes of cell death regulation. As mentioned before, expression of the *rpr* and *grim* genes precedes the onset of apoptosis by several hours and is restricted to cells that are doomed to die (26, 34). Their expression is also induced in cells when they are exposed to DNA damage inducing stimuli like X-ray irradiation or when they have a developmental defect (26, 40, 42). Thus, PCD by the *rpr* and *grim* genes requires their *de novo* gene expression.

The regulation of *rpr* gene expression was investigated by analysis of the *rpr* promoter. Early work had indicated that the *rpr* promoter is very large and comprises at least 11 kb genomic DNA (40). A *lacZ* reporter gene fused to 11 kb of genomic DNA upstream of the *rpr* translation start site mimicked most but not all of the expression pattern of the endogenous gene (40). However, the 11 kb genomic fragment contained all elements which are necessary to induce *rpr* expression in response to X-ray irradiation, ecdysone stimulation, and developmental defects (40, 51, A. F. Lamblin and H. S., unpublished). To identify the DNA sequences in the *rpr* promoter which mediate *rpr* expression in response to X-ray, reporter transgenes with smaller 5′*rpr* genomic sequences were generated and analysed. This analysis resulted in the identification of a discrete 150 bp genomic fragment approximately 5 kb upstream of the *rpr* transcriptional start site which is required for X-ray-induced *rpr* expression (Fig. 4) (52). Within the 150 bp enhancer a putative 20 bp sequence was identified that strongly resembles the consensus for the human p53 DNA binding site. In a yeast one-hybrid assay, it was shown that the *Drosophila* homologue of p53, Dmp53, is able to bind to this motif and induces gene expression (52). In transgenic flies, *rpr* induction in response to X-ray was largely mediated through binding of Dmp53 to this *cis*-regulatory element (Fig. 4). Consistently, transgenic expression of a

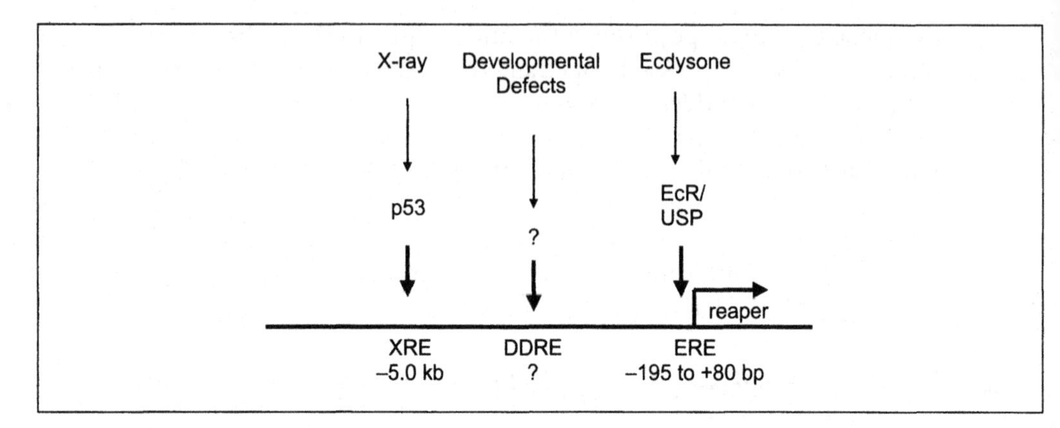

Fig. 4 Integration of death-inducing signals by the *reaper* promoter. The *reaper* promoter is complex and contains more than 11 kb of genomic DNA upstream of the *reaper* transcription unit. Dissection of the promoter into smaller genomic fragments revealed that distinct *cis*-regulatory elements respond to different cell death-inducing stimuli. X-ray-induced DNA damage results in the activation of Dmp53 which binds the promoter at a X-ray response element (XRE) located at around –5 kb of *reaper* upstream genomic DNA (52). The ecdysone response element (ERE) is located immediately upstream of the transcriptional start site of the *reaper* promoter (51). The response element for developmental defects (DDRE) is distinct from the XRE and ERE (52; A. F. Lamblin and H. Steller, unpublished); its exact position is not known.

dominant-negative Dmp53 blocked irradiation-induced apoptosis (52, 53) although it has not been demonstrated that this is due to a reduction in *rpr* expression. In addition, overexpression of Dmp53 in the fly eye (*GMR-Dmp53*) causes massive apoptosis and a reduction in the size of the adult eye (53). Although it has not been directly shown, it is possible that *GMR-Dmp53*-mediated eye ablation is caused by transcriptional activation of *rpr* expression.

An analogous study was performed to identify sequence elements in the *rpr* promoter that are required for the ecdysone response (51). This study showed that *rpr* expression is induced directly by the ecdysone/EcR/USP complex through an essential response element in the *rpr* promoter. The ecdysone response element is distinct from the X-ray response element (the Dmp53 binding site) and is located between −195 bp to +80 bp relative to the *rpr* transcription start site (Fig. 4). In addition, many developmental mutants in which cells have failed to appropriately develop or differentiate display an increase in apoptosis (5, 27–30), which often is accompanied by an up-regulation of *rpr* gene expression (26, 40, 42). The analysis of the *rpr* promoter for the response element which is required to trigger cell death in developmental mutants revealed that it is different from both the X-ray and the ecdysone response element (52, A. F. Lamblin and H. S., unpublished).

Thus, in response to different death-inducing stimuli, distinct elements in the *rpr* promoter are required. This work confirmed the *rpr* gene as a convergence point of a variety of death-inducing signals that activate a common death programme. The regulation of *rpr* expression is very complex and requires at least 11 kb of genomic DNA upstream of the translation start site. Different *cis*-regulatory elements in the *rpr* promoter are recognized by different transcription factors which are activated in response to various stimuli (Fig. 4).

However, not all death-inducing stimuli work through up-regulation of *rpr* transcription. For instance, mutations in the eye gene *sine oculis* and the wing gene *vestigial* did not induce *rpr* expression (A. F. Lamblin and H. S., unpublished), although these mutations cause massive ectopic cell death (27). Apparently, not all cells undergoing PCD require *rpr* expression. The expression of other genes such as *grim* might be up-regulated instead. Alternatively, other mechanisms that do not require gene expression and protein synthesis might be involved. The following section offers such an alternative mechanism.

5. The *Drosophila* gene *hid* links RAS-dependent cell survival and apoptosis

In contrast to *rpr* and *grim*, the regulation of the third cell death inducer, *hid*, appears to be more complex. It is not only expressed in dying cells, but also in cells which live (33; unpublished observation). Thus, it seems likely that HID activity is regulated at a post-translational level which would not necessarily require new protein synthesis. In addition, ectopic expression of *hid* can very efficiently induce apoptosis (33, 54). Thus, in *hid*-expressing cells which survive, very potent mechanisms must operate which prevent these cells from undergoing HID-induced apoptosis.

Recent experiments revealed a potential mechanism through which extracellular survival signalling regulates HID activity. This work took advantage of the strong eye ablation phenotype caused by expression of *hid* under control of the eye-specific GMR promoter (*GMR-hid*) (Fig. 5; see Plate 2). In a genetic screen for modifiers of the *GMR-hid*-induced eye phenotype, mutations in genes encoding negative regulators of the EGFR/RAS/MAPK pathway were recovered as strong suppressors (54). Subsequent analysis showed that activation of EGFR/RAS/MAPK results in strong inhibition of HID-induced apoptosis (Fig. 5; see Plate 2) (54–56). An activated allele of *Ras*, Ras^{V12}, which is associated with many human tumours, is a very potent inhibitor of HID-induced apoptosis (Fig. 5; see Plate 2). This suppression appears to be specific for HID since Ras^{V12} showed little or no inhibition of *CMR rpr* or *GMR-grim*-induced eye phenotypes, respectively (54, 55).

Evidence for an anti-apoptotic function of the RAS pathway is apparent from mammalian cell culture (reviewed in ref. 57) and the role of RAS signalling in regulating cell proliferation and cell differentiation is well established genetically (reviewed in ref. 58). However, it was difficult to study genetically apoptotic phenotypes caused by *EGFR/RAS/MAPK* mutants because of a simultaneous lack of cell proliferation in these mutants. Animals homozygous for *Ras*, *Raf*, and *MAPK* die as third instar larvae completely lacking imaginal discs, which constitute the bulk of proliferating tissue during larval development. Strong loss-of-function alleles of *Ras*, *Raf*, and *MAPK* are cell lethal in homozygous clones in imaginal discs (59). Thus, to study a potential anti-apoptotic function of EGFR/RAS/MAPK signalling it is

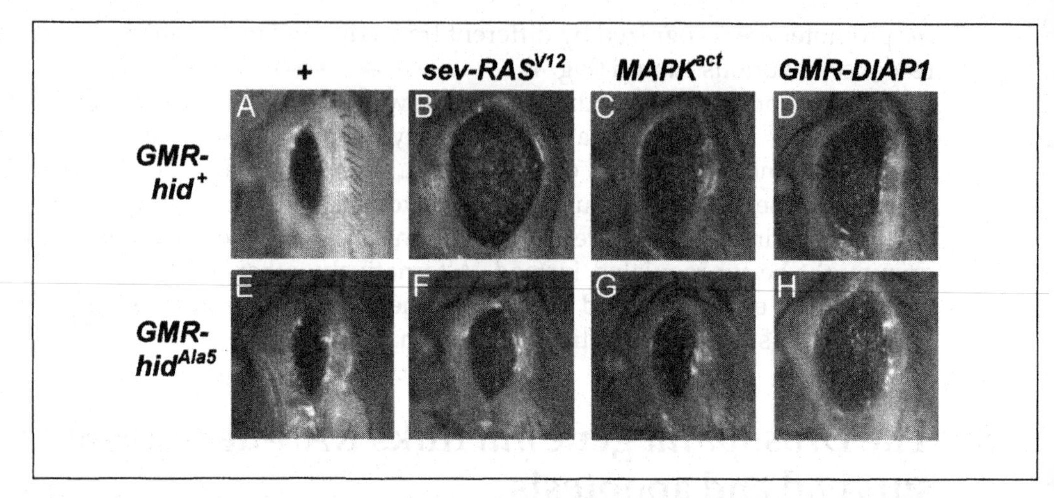

Fig. 5 (See Plate 2) Activation of the RAS/MAPK signalling pathway inhibits HID-induced apoptosis. (A) Wild-type *hid* expressed under control of the eye-specific GMR promoter (*GMR-hid*$^+$) causes a strong eye ablation phenotype. (B) This phenotype is efficiently suppressed by the oncogenic *RAS*^{V12} allele which is under control of the eye-specific *sevenless* (*sev*) promoter. (C) An activated MAPK allele (*rl*^{Sem}) leads to a weaker but clearly detectable suppression of *GMR-hid*$^+$. (E) The MAPK site deficient mutant of HID (*GMR-hid*^{Ala5}) causes an eye ablation phenotype similar to *GMR-hid*$^+$ (compare to Fig. 5A; Plate 2A). (F, G) However, activation of the RAS/MAPK pathway largely fails to suppress the *GMR-hid*^{Ala5} phenotype. This failure suggests that phosphorylation of HID by activated MAPK is critical for the execution of the anti-apoptotic effects of the RAS/MAPK pathway. (D, H) The inhibitory protein DIAP1 suppresses *GMR-hid*$^+$ and *GMR-hid*^{Ala5} to a similar extent indicating that the failure of RAS/MAPK to suppress *GMR-hid*^{Ala5} is due to the change of the MAPK phosphorylation sites and not to any unrelated effect. Please note that *RAS*^{V12} suppresses *GMR-hid*$^+$ much more potently than does *GMR-DIAP1*, a specialized cell death inhibitor (compare Figs 5B and 5D; Plates 2B and 2D).

necessary to uncouple the cell proliferation activity from the ability to inhibit PCD. This problem has been overcome by expressing a dominant-negative allele of *EGFR* (*DN-EGFR*) in the developing eye after cell proliferation is complete (in post-mitotic cells), so that any role of EGFR/RAS/MAPK signalling in cell division is irrelevant (60). Expression of the *DN-EGFR* mutant in the developing eye induces ectopic cell death, and the resulting eye phenotype is remarkably similar to the eye ablation phenotype caused by *GMR-hid* (Fig. 5; see Plate 2) (60). Moreover, expression of *argos*, an extracellular inhibitor of EGFR (61) under GMR promoter control (*GMR-argos*) induces ectopic PCD during eye development and results in a smaller eye (56). Strikingly, the *GMR-argos* eye phenotype can be dominantly suppressed by the deficiency *H99* (56), which deletes the gene *hid* as well as *reaper* and *grim*. This experiment is complementary to the finding that mutations in *argos* dominantly suppress the *GMR-hid*-induced eye phenotype (54). In this context, it is important to note that *hid* is expressed during eye development (33). In addition, in an elegant set of cell ablation studies, Miller and Cagan (62) demonstrated the important anti-apoptotic function of RAS during eye development. For instance, laser ablation of the primary pigment cells of the compound eye results in apoptosis of secondary and tertiary pigment cells which is effectively blocked by expression of the activated *Ras* allele, *Ras*^{V12} (62). Taken together, these results strongly support the notion that the EGFR/RAS/MAPK signalling pathway delivers its anti-apoptotic signal via inhibition of the cell death regulator *hid*.

By which mechanism does the RAS pathway influence the intrinsic cell death machinery? A recent report (54) showed that the MAPK phosphorylation sites in HID are critical for the anti-apoptotic response of RAS/MAPK signalling. Mutation of the phospho-acceptor sites of HID to Ala abolished the inhibitory effect of RAS/MAPK signalling suggesting that phosphorylation of HID by activated MAPK leads to its inactivation (54). Furthermore, Kurada and White (55) demonstrated that activation of RAS also down-regulates *hid* transcription. Thus, HID activity appears to be regulated at least at two levels: after the decision of a *hid*-expressing cell to live has been made, existing HID protein pools are immediately inactivated by MAPK phosphorylation. In a second response, the cell ensures that HID is no longer produced by down-regulation of its transcription.

This mechanism provides a conceptual framework for how extracellular survival signalling interferes with the intrinsic cell death programme by activating the RAS/MAPK pathway. Significantly, cell–cell communication, as outlined for the trophic theory for cell survival in vertebrates in Section 2 (19–21), may make use of such a mechanism to regulate cell death in the nervous system and other tissues. Cells which require trophic signalling to survive may contain an analogous function to the *Drosophila* HID protein which is active by default unless it is suppressed by survival factors. Mammalian trophic factors such as the neurotrophins NGF, BDNF, NT3, and NT4/6 are known to activate the RAS pathway by binding to the *trk* family of receptor tyrosine kinases (63). Two *Drosophila* genes have been identified that are related to the *trk* family of neurotrophin receptors, *Dtrk* and *Dsor* (64, 65). Both genes are expressed in the CNS, but their precise function is not known. *Drosophila* homologues of mammalian neurotrophins have not been reported. However, the EGF ligand encoded by the *spitz* locus, might be used under certain circumstances as a trophic factor (62, 66, 67). This is consistent with the aforementioned finding that overexpression of *argos*, an inhibitor of *spitz/EGFR* activity results in induction of apoptosis in post-mitotic cells (56).

This work revealed a striking difference between regulation of RPR- and HID-induced apoptosis. Induction of RPR-mediated apoptosis is an active gene-directed process that requires *rpr* gene transcription and protein synthesis. In contrast, HID activity has to be suppressed for cell survival since it is present in cells that die as well as live. Trophic survival factors may provide a signal for cells to suppress HID-induced apoptosis. A cell that does not receive this signal is committed to die.

In summary, inhibition of HID by activation of the RAS/MAPK pathway links extracellular survival signalling and the intrinsic cell death machinery. Significantly, about 30% of human tumours carry activated forms of the *Ras* oncogene, such as *Ras*^{V12} (68). These oncogenes have a very strong anti-apoptotic activity. In fact, in *Drosophila*, inhibition of HID by *Ras*^{V12} is much more potent than inhibition by DIAP1, a direct inhibitor of the cell death programme (see Fig. 5; Plate 2; and next section). Tumour cells are less sensitive to apoptotic stimuli (69). Inactivation of an intrinsic component of the cell death machinery, such as HID, is a logical target of RAS-dependent survival signalling during oncogenesis. Thus, a mechanism that is used during normal development to adjust cell number may be deregulated by oncogenic mutations favouring tumour formation.

6. *Drosophila* inhibitors of apoptosis (DIAPs)

Another class of proteins which appears to regulate RPR, HID, and GRIM activity are the IAP molecules (inhibitor of apoptosis protein) (Fig. 2). IAPs were originally discovered due to their ability to functionally substitute for the baculoviral p35 protein to inhibit apoptosis (70–72; see recent review ref. 73) (see Chapter 10). It is believed that baculoviruses express IAPs to escape the apoptotic host response, which would otherwise limit viral replication. Subsequently, IAP molecules from a wide variety of organisms including humans, flies, worms, and yeast were identified. There are at least five human IAP genes. Mutations in the NAIP (neuronal apoptosis inhibitor protein) gene are thought to contribute to spinal muscular atrophy (SMA) which involves inappropriate apoptosis of motor neurons (74). Cellular IAPs (cIAP1 and cIAP2) were identified based on their ability to bind to TRAF1 and TRAF2 proteins which are associated with the TNF type 2 receptor (75) (see Chapter 8). XIAP (X-linked IAP) was identified in a protein database search using baculoviral IAP sequences (76). Expression of Survivin, another human IAP gene, correlates with oncogenic transformation (77; reviewed in ref. 78). In addition, several mammalian viruses contain IAP genes (79).

IAPs share several structural motifs: they contain at least one but usually two or three tandem baculoviral IAP repeat (BIR) motifs, and some have a carboxy terminal RING finger domain (Fig. 6). Ectopic expression of human IAPs can suppress

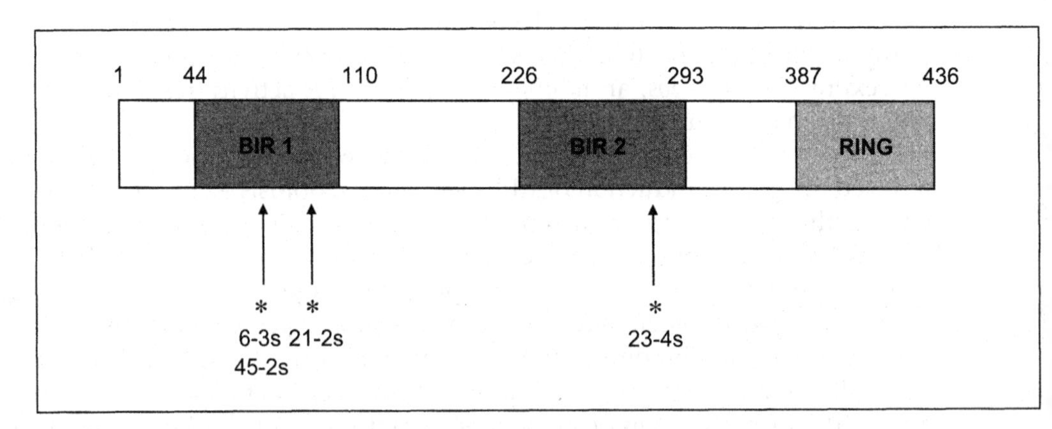

Fig. 6 Schematic outline of the structure of DIAP1 and relative location of the gain-of-function mutations. DIAP1 contains two BIR domains, termed BIR1 and BIR2, and a carboxy terminal RING finger domain. The mutations in DIAP1 which convert the protein into a much stronger suppressor of REAPER- and HID-induced cell death are located in the BIR domains. The two suppressors lines 6-3s and 45-2s affect the same residue in DIAP1. These mutations reduce the binding properties of DIAP1 towards REAPER and HID (87).

apoptosis in several paradigms (76, 77, 80–83). However, not all IAPs inhibit apoptosis. For instance, it is unlikely that the two yeast IAP genes encode anti-apoptotic proteins since no apoptotic programme has been identified in this organism (84). For that reason, we follow a suggestion by Uren *et al.* (79) to rename this protein family BIRP (for BIR-containing proteins), with the anti-apoptotic IAPs as a subfamily.

Four BIRP genes have been identified in *Drosophila* two of which are IAP-like: *diap1* and *diap2* (85). Heterozygous mutants for *diap1* are strong enhancers of RPR- and HID-induced cell death (85, 86). Homozygous *diap1* embryos exhibit developmental arrest and die by massive ectopic cell death starting at stage 5 (86–88) which is much earlier than in wild-type embryos (stage 11). Overexpression of DIAP1 suppresses RPR-, HID-, and GRIM-induced cell death (85, 86, 89, 90). DIAP2 was isolated based on its sequence homology to DIAP1 (85). Since no *diap2* mutants have been identified, it is not known whether *diap2* encodes an anti-apoptotic protein. Overexpression of DIAP2 inhibits RPR- and HID-induced apoptosis but is ineffective with GRIM-induced cell death in the *Drosophila* CNS and compound eye (50). In contrast, Vucic *et al.* (90) demonstrated that DIAP2 inhibits GRIM-induced apoptosis and that the two proteins can directly associate with each other in the heterologous lepidopteran cell line SF-21.

The inhibition of RPR, HID, and GRIM activity by baculoviral and *Drosophila* IAPs appears to be mediated by direct interaction altering the subcellular localization of RPR, HID, and GRIM to punctuate perinuclear staining which coincides with IAP localization in SF-21 cells (89–91). This interaction requires the N-terminal 15 amino acids of RPR, HID, and GRIM, the region of similarity of the proteins (Fig. 3) (89–91). The N-terminal 37 residues of HID were found to be necessary and sufficient for its binding to IAPs and for induction of apoptosis (88, 90). As far as IAPs are concerned, the BIR domains are required for the interaction with and inhibition of RPR, HID, and GRIM; the removal of the RING domain of IAPs did not show any effects (85, 89–92). Vucic *et al.* (91) determined that the second of the two BIR motifs of baculoviral opIAP and *Drosophila* DIAP1 can bind HID and inhibit HID-induced apoptosis. However, most of the conserved residues of the second BIR of opIAP were found not to be required for interaction with HID. Instead, the least conserved part of the BIR, a carboxy terminal stretch of 17 residues, is essential for binding and HID inhibition (91).

As mentioned above, loss-of-function alleles of *diap1* strongly enhance RPR-, HID-, and GRIM-induced cell death (85, 86). However, in the *GMR-rpr* and *GMR-hid* cell death screens a large number of suppressors have been isolated which appear to be gain-of-function alleles of *diap1* (86, 87). These alleles were very informative for the characterization of the interaction of RPR/HID/GRIM with DIAP1 and allowed the ordering of these genes in a genetic hierarchy. Sequencing of the gain-of-function alleles revealed amino acid substitutions in the BIR domains of DIAP1 (Fig. 5; see Plate 2). Since the BIR domains are necessary for binding to RPR/HID/GRIM (89–91), binding studies with the gain-of-function mutants and the *H99* gene products were performed. These studies show a diminished binding affinity of RPR and HID with the mutant DIAP1 proteins (87). How then does diminished binding between DIAP1 and RPR/HID account for the strong suppression of RPR- and HID-induced cell death? The answer to this question allowed the determination of the genetic order of the *H99* genes and *diap1*. If we assume that IAPs act upstream of *rpr/hid/grim* and exert their anti-apoptotic function simply by blocking RPR/HID/GRIM activation, then a diminished interaction would be expected to result in an enhancement of RPR/HID/GRIM-induced apoptosis. However, the strong suppression observed for the gain-of-function alleles is not consistent with this assumption.

The alternative model—IAPs acting downstream of *rpr/hid/grim*—helps explain the suppression of RPR/HID/GRIM-induced cell death by the gain-of-function alleles of *diap1*. In this model, we assume that RPR/HID/GRIM exert their pro-apoptotic activity by binding to and inactivating IAPs, thus blocking their anti-apoptotic function. Impaired binding of RPR/HID/GRIM to IAPs results in less efficient transduction of the cell death signal by RPR/HID/GRIM. Consequently, their cell killing activities are suppressed by the gain-of-function alleles of *diap1*.

Another very informative way to determine the hierarchical relationship between two genes is the analysis of the double mutant phenotype if the phenotypes of the single mutants are different from each other. If the double mutant phenotype resembles or is identical to the phenotype caused by mutations in one gene, then this indicates that the function of that gene is required downstream or in parallel to the other one. The *H99* mutant phenotype is characterized by the lack of PCD (26), whereas loss-of-function mutations in *diap1* cause massive cell death (86–88). The loss-of-function *diap1-H99* double mutant has the same phenotype as the loss-of-function *diap1* mutant alone placing *diap1* genetically downstream of or in parallel to the *H99* genes *rpr*, *hid*, and *grim* (87, 88). The biochemical observation that DIAP1 interacts with caspases (93, 94), makes it even more likely that it acts downstream of the *H99* genes. Furthermore, the strong apoptosis phenotype caused by the *diap1-H99* double mutant indicates that the *H99* genes are not required for the induction of apoptosis if the inhibitory *diap1* function is missing.

In summary, the available data strongly support the notion that *diap1* acts genetically downstream of *rpr/hid/grim*. In Section 8, we will propose a model to explain how IAPs exert an anti-apoptotic function and how RPR/HID/GRIM overcome it.

Another protein which has been reported to interact with IAP molecules is DOOM, an alternative splice product of the *Drosophila mod*(*mdg4*) gene (95). DOOM was isolated in a yeast two-hybrid screen using the baculoviral opIAP as bait. Overexpression of DOOM in SF-21 cells induces apoptosis and changes the subcellular localization of opIAP from cytoplasmic to nuclear (95). The apoptotic activity of DOOM and the interaction to opIAP is mediated by the C-terminus of DOOM which is unique among the splice variants of the *mod*(*mdg4*) gene. In fact, the C-terminus on its own is able to induce apoptosis (95). Simultaneous overexpression of baculoviral IAPs or p35 blocks DOOM-induced apoptosis (95). The products of the *mod*(*mdg4*) gene are involved in the regulation of chromatin structure. A role in apoptosis has not been anticipated and further genetic studies will be needed to assess the extent of involvement of DOOM in the apoptotic pathways in *Drosophila*.

Not all BIRPs suppress apoptosis, i.e. are IAPs (79). What is/are the function(s) of non-apoptotic BIRPs. A first clue comes from studies in *C. elegans*. The *C. elegans* genome encodes two BIRP genes, termed *bir1* and *bir2* (79). These genes contain one and two BIR motifs, respectively, and lack the RING finger domain. Overexpression of BIR1 and BIR2 failed to block programmed cell death in worms (96). Elimination of *bir1* and *bir2* activity by RNA-mediated interference (RNAi) did not generate a cell death phenotype either (96). Hence, the *C. elegans* BIRPs appear not to encode anti-apoptotic proteins. Further analysis revealed that elimination of *bir1* activity by RNAi causes cytokinesis defects which can partially be rescued by the human BIRP Survivin (96). Bir1 and Survivin are most similar to each other. The partial rescue of *bir1* (RNAi) embryos by Survivin indicates a conserved role of BIRPs in the regulation of cytokinesis.

7. *Drosophila* caspases

The importance of caspases as essential components of the cell death programme became clear with the cloning of *ced-3*, a *C. elegans* caspase gene (97) (see Chapters 2 and 4). During the apoptotic process, it is thought that caspases are activated in an amplifying proteolytic cascade, cleaving one another in sequence. Thus, a caspase cascade in which class I ('initiator') cleave class II ('executioner') caspases has been proposed (98–100). Class I caspases are characterized by a long prodomain, whereas class II caspases have a short or no prodomain (reviewed in ref. 101). The long prodomain of class I caspases usually contains protein–protein interaction motifs (CARD, DED, see below) that helps recruit class I caspases to specific death complexes (101). Class II caspases cleave a growing number of cellular proteins such as nuclear lamins, and they cause most of the changes that are characteristic of apoptotic cell death.

The *Drosophila* genome contains eight caspase encoding genes (102) five of which have been further characterized: *dcp-1, drICE, dcp-2/dredd, dronc,* and *decay* (103–108). DCP-1, DRICE, and DECAY are members of class II caspases most similar to mammalian caspase-3 and caspase-7, the class I caspase DCP-2/DREDD is most similar to caspase-8 and caspase-10 but has some unusual properties, and the class I caspase DRONC is most similar to caspase-2. DCP-1, DRICE, and DECAY induce apoptosis in both insect and mammalian cell lines, DRONC has been tested for the ability to induce apoptosis in yeast and mammalian cells. DCP-1 and DRONC also induce apoptosis when overexpressed in the fly eye (109, 110). With the exception of DRONC the cell death induced by these caspases is inhibited by known caspase inhibitors such as p35, IAPs, and various synthetic peptide inhibitors. The promiscuous caspase inhibitor p35 fails to rescue DRONC-induced cell death *in vivo* and is not cleaved by DRONC *in vitro*, making DRONC the first identified p35-resistent caspase (110). In contrast to p35, expression of the poxvirus caspase inhibitor CrmA in a yeast assay and expression of DIAP1 *in vivo* (see below) does inhibit DRONC-induced cell death (110).

At this point, loss-of-function mutants have only been described for *dcp-1*. These mutants allowed the investigation of *dcp-1* function *in vivo*. Animals homozygous mutant for *dcp-1* arrest during larval development and contain melanotic tumours indicating an essential developmental function for *dcp-1* (103). However, the cell death pattern in these mutants appears to be normal overall. This might be due to redundant caspase function (*drICE* is very similar to *dcp-1*), or to the strong maternal contribution which might be sufficient to compensate for the zygotic loss of *dcp-1* during embryogenesis. To eliminate the maternal *dcp-1* contribution, McCall and Steller (111) generated females with homozygous *dcp-1* germlines in otherwise heterozygous animals. These females are sterile, and only a few embryos are laid, which arrest early in development with severe mitotic defects. However, the main defect occurs during oogenesis. The *Drosophila* egg chamber consists of a cluster of 16 germline cells. One cell in this cluster becomes the oocyte while the other 15 develop into nurse cells that transcriptionally and translationally support the oocyte in its development. At stage 12 of oogenesis, the nurse cells start degenerating with apoptotic characteristics (112, 113). During degeneration, the nurse cells transport their contents into the oocyte, a process called 'dumping'. Finally, they are cleared from the egg chamber. This process is a clear example of how a single cell, the oocyte, uses the death of its sister cells, the nurse cells, to develop properly. In *dcp-1* mutants, the onset of nurse cell death is delayed to about stage 14 (111). As a consequence of this delay, the dumping process does not occur or only to a limited extent. Further analysis showed that the nurse cells of *dcp-1* mutants are defective in nuclear breakdown, cytoskeletal reorganization, and membrane contraction, all hallmarks of apoptotic cell death (111). This study indicated that *dcp-1* has an essential function in nurse cell death during oogenesis in *Drosophila*.

Based on the prodomain structure, DCP-1, DRICE, and DECAY belong to class II executioner caspases, most similar to caspase-3 and -7. These caspases are thought to require cleavage by class I caspases for activation. DCP-2/DREDD and DRONC are members of the class I caspase subfamily. DCP-2/DREDD contains two putative DED (death effector domain) motifs (106). DED-containing caspases such as caspase-8 and -10 are activated on FADD adaptor-mediated recruitment to death receptors (see Chapter 8) (reviewed in ref. 114). The presence of two putative DED domains in DCP-2/DREDD implies that death receptor pathways may exist in *Drosophila*. However, DCP-2/ DREDD has some unusual properties which makes it distinct from all known caspases (106). Its catalytic site is unique among caspases since it contains a glutamic acid residue at a position which is normally occupied by a glycine residue (106). This change might impose a novel cleavage specificity for DCP-2/DREDD and might reflect a preference for novel substrates that have yet to be identified. Furthermore, DCP-2 is not ubiquitously expressed like most other caspases. Instead it is specifically expressed in cells that are going to die and requires the *H99* cell death genes for this expression (106). In addition, unlike most other caspases, ectopic expression of DCP-2/DREDD is not sufficient to induce apoptosis on its own, and processing of the protein was not observed (106). Co-expression with RPR or GRIM resulted in the processing of DCP-2/DREDD. However, this processing is insensitive to p35 inhibition and other anti-caspase peptides. Therefore, *dcp-2/dredd* may encode a novel caspase-like protein with interesting properties. Unfortunately, a *dcp-2/dredd* mutant has not been reported which would allow the examination of its precise role *in vivo*.

DRONC is most similar to caspase-2 and contains a CARD (caspase recruitment domain) motif (107, 110). CARD motifs mediate homotypic protein–protein interactions between caspases (such as CED-3, caspase-1, -2, -9) and adaptor molecules (such as CED-4, CARDIAC, RAIDD, and APAF-1, respectively) (see Chapters 4 & 8). The presence of a CARD motif in DRONC implies that similar DRONC-interacting proteins exist in flies (see Section 10). DRONC was identified in a database search for CARD-containing molecules (107) as well in a yeast two-hybrid screen using DRICE as bait (110). Similar to DCP-2/DREDD, the catalytic Cys in the active site of DRONC is encompassed by a sequence distinct from all other known caspases. However, its cleavage specificity is most similar to caspase-2. During development, *dronc* is highly expressed in early embryonic and early pupal stages (107). Significantly, *dronc* expression is particularly high in midgut and salivary glands, tissues which are known to die during metamorphosis in response to release of the steroid hormone ecdysone (47). In fact, Dorstyn *et al.* (107) showed that *dronc* expression is up-regulated in response to ecdysone.

Activation of caspases is a critical event in the apoptotic programme. As such, it is a tightly regulated process. Caspases are the target of several viral and cellular protein inhibitors of apoptosis including p35, crmA, FLIPs, ARC, and IAPs (reviewed in ref. 115). IAPs function by directly binding to and inhibiting caspases from activation (83, 88, 93, 94). Interestingly, IAPs appear to interact with only a certain subset of caspases. For instance, human XIAP as well as cIAP1 and cIAP2 can bind to and inhibit caspase-3, -7, and -9 but fail to interact with caspase-8, -1, and -6 (83, 116). Survivin binds specifically to caspase-3 and -7, but not to caspase-8 (117). Conversely, NAIP fails to bind to any of these caspases. The interaction between IAPs and caspases requires the BIR domains of IAPs. A detailed analysis showed that the second domain of the three BIR domains of XIAP is sufficient for inhibiting caspases (118). The RING domain might modify this interaction.

In *Drosophila*, the eye ablation phenotype caused by expression of *dronc* under eye-specific GMR promoter control (*GMR-dronc*) is effectively blocked by *GMR-diap1* (110). This inhibition requires the CARD prodomain of DRONC since *GMR-diap1* fails to block the cell death caused by a mutant form of DRONC which lacks the CARD domain (110). In fact, in yeast two-hybrid assays the physical interaction between the CARD motif of DRONC and the BIR2 domain of DIAP1 was demonstrated (110). This is particularly intriguing since the BIR2 domain of DIAP1 is also known to physically interact with, and block the pro-apoptotic activity of, RPR, HID, and GRIM (89–91; see also next section). Similarly, the expression of *dcp-1* both in yeast and in the fly eye is efficiently inhibited by co-expression of DIAP1 (94, 109). This effect is presumably due to direct binding of DIAP1 to DCP-1 since GST-DIAP1 inhibits DCP-1 caspase activity *in vitro* (94). DIAP1 also suppresses apoptosis induced by DRICE and physically interacts with the active form, but not the procaspase of DRICE in the lepidopteran cell line SF-21 (93). Thus, the available evidence strongly suggests that IAPs bind to caspases and inhibit their biochemical activity.

8. How do the cell death regulators *rpr, hid,* and *grim* induce cell death in *Drosophila*?

Genetic evidence suggests that *diap1* acts genetically downstream of *rpr/hid/grim* (see Section 6) (87, 88) and that it inhibits RPR/HID/GRIM-induced apoptosis (85). In addition, RPR/HID/GRIM are known to induce apoptosis by activation of caspases (33, 34, 43, 109). Biochemical data has shown that IAPs physically interact with both RPR/HID/GRIM and caspases (83, 89–91, 93, 110, 116, 117). Furthermore, in a yeast system RPR/HID/GRIM block DIAP1s ability to suppress caspase-dependent cell death (88). All the available genetic and biochemical data are formally consistent with the model shown in Fig. 7 (87, 88, 119).

In surviving healthy cells the *Drosophila* IAP molecules, especially DIAP1, bind to caspases such as DRONC, DCP-1, and DRICE and prevent them from being activated. In response to a death-inducing signal, cells express *rpr, grim,* and/or *hid,* or activate HID. RPR/HID/GRIM bind directly to IAP molecules and release the block of caspase activation. The mechanistic details of this process are not entirely clear but since RPR/HID/GRIM as well as caspases bind directly to the BIR domains of IAPs, it appears attractive to speculate that RPR/HID/GRIM compete with caspases for binding to the BIR domains of IAPs. Thus, RPR/HID/GRIM may dissociate the IAP/caspase protein complex simply by competition for binding to the BIR domains of IAPs. How caspases are activated after IAP release is not known. This might involve association with activation factors such as APAF-1-like proteins (Fig. 7) (see Section 10).

Song *et al.* (109) provided evidence that DCP-1 is activated by RPR and GRIM, but not by HID. A similar observation was made for activation of DRICE. Thus, RPR- and GRIM-induced cell death is mechanistically different from HID-induced cell death (Fig. 7). Since HID-induced apoptosis is blocked by p35 (33), a broad range caspase inhibitor (45, 46), we postulate the existence of one or more HID-specific caspases, termed caspase-X in Fig. 7. This difference between RPR and HID is surprising since loss-of-function alleles of *diap1* enhance RPR-, HID-, and GRIM-induced apoptosis equally well, whereas certain gain-of-function alleles of *diap1* and GMR-DIAP1 suppress them (85–87). Furthermore, genetic interaction studies indicate that *dronc* is required for both RPR- and HID- (and presumably GRIM-) induced cell death (110). Somehow in an unknown manner, RPR and GRIM release the DIAP1-mediated block of caspase activation by a mechanism distinct from HID (Fig. 7). In fact, there is genetic support for this assumption. In the cell death screens, a number of *diap1* alleles were isolated which enhance the *GMR-rpr-*, but suppress

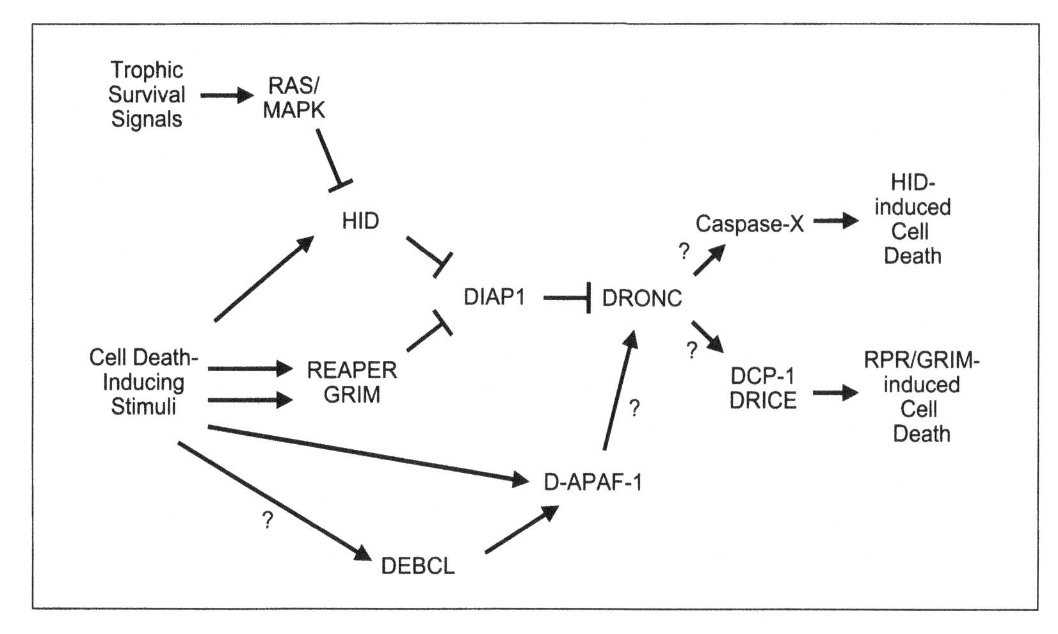

Fig. 7 Regulation of apoptosis via dual control of caspase activation in *Drosophila*. This model accommodates the available genetic and biochemical data. IAP molecules, such as DIAP1, inhibit caspases (most notably DRONC, but also DCP-1, DRICE, and an unknown HID-specific caspase termed caspase-X) from being activated. The expression of *reaper*, *hid*, and *grim* by cell death-inducing stimuli (see Fig. 2), or the derepression of HID by trophic survival factor withdrawal, releases the inhibitory block of caspase activation by DIAP1. REAPER, HID, and GRIM may dissociate the DIAP1/caspase protein complex simply by competing for binding to the BIR domains. However, in an unknown manner, REAPER- and GRIM-induced apoptosis results specifically in the activation of the caspases DCP-1 and DRICE (109). HID appears to induce apoptosis by releasing a different caspase of unknown identity (caspase-X) (109). The caspase release by REAPER, HID, or GRIM might allow access of Apaf-1-like molecules (see Section 10) to the CARD domain of the class I caspase DRONC (131) resulting in full activation of DRONC and subsequently downstream class II caspases DCP-1, DRICE, and caspase-X. D-APAF-1 activity may be promoted by pro-apoptotic members of the Bcl-2 family like DEBCL (see Section 10). The mechanistic details of this model might be more complex than shown. See Section 10 for a more detailed discussion.

the *GMR-hid*-induced eye phenotype (86; J. Agapite, K. McCall, and H. S., unpublished observation). How specificity is conferred in these pathways is currently unknown. However, the molecular analysis of these very interesting *diap1* alleles will provide important information about the regulation of these parallel pathways.

Other groups have proposed that the cell death inducers RPR, HID, and GRIM bind directly to and activate caspases (93). However, there is no biochemical evidence that these proteins bind to and activate caspases (88). Rather, the strong apoptosis phenotype caused by the *diap1-H99* double mutant (see above) (87, 88) indicates that the *H99* genes are not required for the induction of apoptosis if the inhibitory *diap1* function is missing. Thus, binding of RPR, HID, and GRIM to caspases is genetically not required for caspase activation. Kaiser *et al.* (93) also proposed in the 'apopstat model' that changes in the levels of IAP molecules are responsible for the regulation of caspase activation. This model is based on overexpression studies in the heterologous lepidopteran cell line SF-21 in which expression levels can easily be manipulated. Although this is an attractive idea, there is no genetic evidence for this mechanism. Rather, all the genetic evidence points towards the level of the cell death inducers RPR, HID, and GRIM as being critical for the control of caspase activation (26, 33, 34, 43). Clearly, more extensive genetic and biochemical investigations are needed to resolve these controversies and to provide a comprehensive detailed understanding of the major cell death pathway in *Drosophila*.

9. Where are the mammalian *rpr/hid/grim* homologues?

The *Drosophila* genes *rpr / hid / grim* are the key regulatory components of programmed cell death during embryogenesis. Mutants which lack these genes are completely devoid of PCD. Despite their central importance for *Drosophila* PCD, mammalian homologues of *rpr / hid / grim* have not been reported to date.

However, there are a number of reasons to believe that genes similar to *Drosophila rpr/hid/grim* exist in vertebrates. RPR / HID / GRIM have been shown to interact with IAPs, an evolutionarily conserved class of genes. Furthermore, HID is the target of RAS / MAPK anti-apoptotic signalling, and an anti-apoptotic function of the RAS / MAPK pathway has been described in vertebrate / mammalian model systems (reviewed in ref. 57).

Consistent with a conserved pro-apoptotic role of *rpr / hid / grim* are a number of reports stating that their expression in vertebrate / mammalian systems induces cell death with apoptotic morphology. Evans *et al.* (120) used addition of recombinant RPR protein to *Xenopus* oocyte extracts to induce many hallmarks of apoptosis including mitochondrial cytochrome *c* release, caspase activation, nuclear fragmentation, and the characteristic 'DNA laddering' seen in apoptotic cells of diverse origin. Subsequently, a 150 kDa RPR interacting protein named SCYTHE was purified and molecularly cloned. The characterization of SCYTHE showed that it is an essential component of RPR-induced caspase activation in the *Xenopus* system (121). The *Drosophila* genome contains one predicted *scythe* homologue indicating that the mechanisms of RPR-induced cell death may be conserved between flies and vertebrates.

Several groups have transfected *rpr*, *hid*, and *grim* cDNAs into mammalian / human cell lines. RPR and GRIM induce apoptosis in human MCF7 breast carcinoma cells (92), GRIM can also kill mouse fibroblasts (122), and HID can cause apoptosis in human HeLa cells (123). Known caspase inhibitors such as p35, IAPs, Bcl-2 family members, and small peptide inhibitors block this cell death, indicating that RPR / HID / GRIM induce apoptosis by activation of caspases. Interestingly, not all human cell lines are susceptible to induction of cell death by the *Drosophila* genes. For instance, human 293 cells are resistant to HID-induced killing (123). This finding argues strongly against a general toxic effect of RPR / HID / GRIM expression in mammalian cells.

These studies indicate that the pro-apoptotic function of *Drosophila* RPR / HID / GRIM is conserved in mammalian cells. However, efforts by different laboratories have failed to identify homologues using standard molecular cloning techniques. Maybe the primary sequence of these genes is not conserved enough for these approaches. It might still be possible to obtain homologues using interaction techniques such as the yeast two-hybrid system taking advantage of the well defined interaction between RPR / HID / GRIM and the conserved IAP molecules. In addition, completion of the genome sequencing projects should reveal any mammalian homologues. (see Chapter 10 for further information).

10. Alternative cell death paradigms in *Drosophila*

Although the *H99* cell death genes are absolutely required for PCD during embryogenesis, there is indication that they do not account for all PCD occurring in *Drosophila*. Foley and Cooley (124) analysed the requirement of these genes for nurse cell death during oogenesis (see Section 7). The *H99* genes *rpr* and *hid* start being expressed during oogenesis at stage 9 and continue throughout the rest of oogenesis. *grim* transcripts are also present at lower, but clearly detectable levels (124). However, egg chambers deficient for the *H99* genes apparently develop normally without the lack of apoptotic events (124). This finding indicates that the *H99* genes are not required for apoptosis in oogenesis. Therefore, other as yet unidentified genes appear to be responsible for the nurse cell death during oogenesis.

Potential candidates for these genes are *Drosophila* homologues of the *C. elegans ced-4* and mammalian Apaf-1 genes. CED-4 and APAF-1 directly bind procaspases and activate their self-processing. Evidence for the existence of these genes was provided as it was shown that expression of CED-4 in the *Drosophila* compound eye induces massive ectopic cell death suggesting a conserved function of this class of genes (125). Recently, several groups reported the cloning of a *Drosophila ced-4* / Apaf-1 homologue, termed *hac-1* (126), *dark* (127), or *D-apaf-1* (128). (Until a common consensus is reached about the name of this gene we refer to it as *D-apaf-1*.) This gene was identified in a database search and encodes two distinct isoforms, *D-apaf-1L* and *D-apaf-1S* (128). The N-terminal 400 amino acids of D-APAF-1L share homology with both CED-4 and mammalian APAF-1, including a putative CARD motif (see Section 7) and two ATP binding domains (126–128). D-APAF-1L also contains at least eight WD repeats which are found in mammalian APAF-1 but not in CED-4. The other isoform, D-APAF-1S lacks the WD repeats and is therefore similar in structure to CED-4 (128). WD repeats bind to cytochrome *c*, an important cofactor for mammalian APAF-1 activation (129). Consistently, D-APAF-1L has been shown to bind cytochrome *c* (127, 128), and addition of cytochrome *c* to wild-type embryonic lysates resulted in activation of caspases, whereas lysates from *D-apaf-1* mutants did not respond in this assay (128). In support of an important function of cytochrome *c* for triggering apoptosis in *Drosophila* is the recent finding that altered cytochrome *c* display precedes apoptotic cell death in flies (130).

Homozygous *D-apaf-1* mutants show a decreased amount of apoptosis in embryogenesis and an enlarged larval nervous system (126–128). Also, these mutants partially suppress *GMR-rpr-*, *GMR-hid-*, and *GMR-grim*-induced eye phenotypes

(127, 128) indicating a function of *D-apaf-1* in RPR/HID/GRIM-mediated apoptosis. Strikingly, similar to *rpr* expression (see Section 4) *D-apaf-1* is expressed in response to X-ray and UV irradiation (126). Also, heterozygous *D-apaf-1* flies suppress the *GMR-dcp-1*-induced eye phenotype (126) suggesting an important function of *D-apaf-1* in caspase activation. Consistently, D-APAF-1 binds to the class I (initiator) caspase DCP-2/DREDD (127) in a process similar to activation of caspase-9 by Apaf-1 (131). However, it remains to be seen whether D-APAF-1 activates DCP-2/DREDD *in vivo* since this caspase does not contain a CARD motif (105, 106) which is required for binding of APAF-1 and caspase-9 in the mammalian system (131). Another *Drosophila* caspase, DRONC, however, possesses a CARD motif (107, 110) and thus might represent an additional binding partner for D-APAF-1 (Fig. 7). Taking together, the available genetic and biochemical data strongly support an important pro-apoptotic function of *D-apaf-1*in *Drosophila* similar to its homologues *ced-4* and Apaf-1.

The activity of CED-4/APAF-1 proteins is tightly regulated by CED-9/Bcl-2 family members (reviewed in ref. 132). There are both pro- and anti-apoptotic members of the Bcl-2 family. The *Drosophila* genome contains at least two Bcl-2 genes, though more divergent family members may exist (102, 133–136). In overexpression studies, both in insect cell lines and in the fly, these genes were found to be pro-apoptotic members of the Bcl-2 gene family. Interestingly, the cell death induced by one of the Bcl-2 homologues, termed *Drob1/dBorg1/Dbok* is insensitive to p35 inhibition indicating that it is inducing PCD independent of caspases (133). The second homologue, termed *Debcl* induces p35-inhibitable apoptosis (134). This gene also shows a genetic interaction with *D-apaf-1* suggesting that *Debcl* and *D-apaf-1* are involved in the same apoptotic pathway (Fig. 7). However, without loss-of-function mutants of the Bcl-2 homologues it is difficult to assess the relative importance of these genes for the apoptotic programme. RNAi-mediated loss-of-function analysis, however, resulted in reduction of apoptosis and extra glia cells indicating a requirement of both genes for embryonic apoptosis (134, 135).

It appears that *Drosophila* uses two distinct mechanisms simultaneously to control the activation of a caspase-based death programme (Fig. 7). Factors such as *ced-4*/Apaf-1 promote caspase activation, and this step is simultaneously inhibited by IAPs. Both regulatory inputs are controlled by upstream factors (Bcl-2 family versus *reaper, hid,* and *grim*), which in turn are the targets for complex control by signalling pathways (Fig. 7). It is very likely that a similar dual control of caspase activation operates during mammalian apoptosis. The proposed dual control of caspase activation would offer increased security and flexibility for selecting cells that are fated to die. For example, conditions where the *ced-4*/Apaf-1 pathway is only weakly active may be insufficient to overcome protection by IAPs. On the other hand, modest levels of RPR/HID/GRIM may be inadequate to kill in the absence of active CED-4/APAF-1-like proteins. Therefore, under physiological conditions, the successful induction of cell death might either require that both pathways are co-ordinately regulated, or that one of them is induced very strongly. Significantly, at least some death-inducing signals, such as ionizing radiation, can indeed activate both pathways simultaneously (126). Working out how these pathways are interconnected and how they are utilized *in situ* will almost certainly remain an interesting challenge during the coming years.

Another alternative cell death paradigm might involve the *Drosophila* homologue of the mammalian anti-apoptotic protein kinase Akt (137). In mammalian cell culture, Akt phosphorylates BAD, a pro-apoptotic member of the Bcl-2 family, resulting in its binding to 14-3-3 proteins as an inactive protein complex (138, 139). There is a *Drosophila* Akt homologue, called DAKT, which is ubiquitously expressed (140, 141). A mutant analysis of *dakt* revealed increased levels of apoptosis during embryogenesis (142). The *H99: dakt* double mutant has the same phenotype as the *dakt* mutant alone indicating that DAKT exerts its anti-apoptotic function either downstream of the *H99* genes or independent of them (142). However, *dakt* is not required for regulation of the nurse cell death during oogenesis, since *dakt* mutant egg chambers apparently develop normally (142). In addition, cells in homozygous *dakt* mutant clones in imaginal discs do not undergo apoptosis, but are smaller in size (143). Consistently, overexpression of wild-type or activated forms of DAKT in imaginal discs fails to block programmed cell death but results in cells which are larger in size (143). Thus, these findings indicate a requirement of *dakt* for cell size control in imaginal discs.

Finally, the existence of death domain-containing receptors such as Fas and TNF-R1 in *Drosophila* cannot be ruled out. There is one report stating that expression of Fas in *Drosophila* cell lines causes apoptosis with a caspase activation profile that is distinct from the one obtained for RPR-induced apoptosis (144). The presence of two putative DED domains in DCP-2/DREDD implies that death receptor pathways may exist in *Drosophila* (105, 106). However, no death domain-containing receptors are known in flies. The completion of the *Drosophila* sequencing project will be informative in this regard.

11. Phagocytosis—the final fate

After initiation of the apoptotic programme, dying cells are removed from the organism by scavenger cells. These can be neighbouring cells acting as semi-professional phagocytes (as in *C. elegans*) or specialized cells such as macrophages (as in *Drosophila* and vertebrates). Phagocytic removal of apoptotic cells is essential in protecting tissue from inflammatory injury caused by leakage of harmful contents from dying cells. Phagocytosis is a complex cellular event in which apoptotic bodies are recognized, engulfed, and eliminated. The engagement of apoptotic cells and phagocytes requires cell movement or anatomical proximity of target and effector cells. It appears that phagocytes taking up apoptotic cells may employ one or more of a range of possible recognition mechanisms, while in turn apoptotic cells may display their 'eat me' status in a number of different ways (for review see ref. 145). For instance, apoptotic cells display at their cell surface a number of different molecules which are in healthy cells components of the inner leaflet of the lipid bilayer such as phosphatidylserine, the carbohydrate *N*-acetylglucosamine, and its dimer *N,N'*-diacetylchitolase, vitronectin, and fibronectin. These exposed molecules are recognized by cognate receptors at the surface of macrophages. The recognition event results in movement of the phagocyte membrane (guided by changes in the cytoskeleton) engulfing the target cell and the formation of the phagosome. After fusion of the phagosome with a lysosome, the contents of the resulting phagolysosome are digested.

In *C. elegans*, six genes have been identified which are required for phagocytosis of apoptotic cells. Three of them have been molecularly characterized. The protein encoded by *ced-7* has sequence similarities to ABC (ATP binding cassette) transporters (146) which are known to transport a variety of substrates across membranes. *ced-7* is broadly expressed during embryogenesis and localized to plasma membranes. Mosaic analysis indicates that its function is required both in dying and in engulfing cells (146). Thus, based on its biochemical function, it appears that CED-7 may be involved in the recognition event of dying and engulfing cells by translocating 'eat me' signals and their cognate receptors to create a link between the cell surfaces.

CED-5 is most similar to human DOCK180 and to the *Drosophila* gene *myoblast city* (147), both of which appear to function by extension of cell surfaces. DOCK180 was isolated based on its interaction with the cytoskeleton-associated protein CRK (148) which has been implicated in integrin-mediated signalling and cell movement. *Myoblast city* is necessary for myoblast fusion and for migration of some epithelial cells, both of which require the extension of cell surfaces, presumably through reorganization of the cytoskeleton (149, 150). Thus, Wu and Horvitz (147) proposed that *ced-5* is required to envelop apoptotic cells by extension of the cell surface of engulfing cells presumably through effecting a reorganization of the cytoskeleton.

Finally, Liu and Hengartner (151) report the cloning of *ced-6*. It encodes a protein with a phosphotyrosine binding domain at its N-terminus and a Pro/Ser-rich domain at its C-terminus. Mosaic analysis indicates that *ced-6* acts within engulfing cells. Based on these findings, it was proposed that CED-6 is an adaptor molecule acting in a signal transduction pathway that mediates the recognition and engulfment of apoptotic cells (151).

In *Drosophila*, macrophages represent a subpopulation of blood cells, or haemocytes, which are contained within the haemolymph space. Haemocytes can be identified at late stage 10 (shortly before the onset of PCD which starts at stage 11) as a discrete subpopulation of mesodermal cells located in the head of the embryo (152). At early stage 12, the haemocytes start migrating throughout the embryo. From stage 12 onwards, many haemocytes show phagocytic activity. They contain one or more vacuoles filled with dark inclusions which represent shrunken cytoplasm and pycnotic nuclei of ingested cells that have undergone apoptotic cell death (152). Cell death occurs independently of macrophages. Mutant embryos that do not form haemocytes show abundant cell death indicating that macrophages do not play an active role in the initiation of PCD. As a consequence of the absence of macrophages, cellular debris is located extracellularly in the haemolymph space (152).

All embryonic macrophages express a receptor for apoptotic cells, termed *croquemort* (*crq*) (153). CRQ is a member of the CD36 superfamily. Human CD36 acts as a scavenger receptor (154) and also binds apoptotic cells in combination with the macrophage vitronectin receptor and thrombospondin (155). Expression of *crq* is detected immediately after the onset of PCD (stage 11), and is restricted to migrating phagocytic haemocytes, i.e. macrophages (153). Cos7 cells transfected with *crq* bind to and ingest apoptotic thymocytes within four hours suggesting that CRQ may participate in the removal of apoptotic cells during embryogenesis (153). In embryos homozygous deficient for *crq*, haemocytes largely fail to engulf apoptotic cells. Macrophages in wild-type embryos contain on average about four apoptotic bodies. In contrast, in *crq*-deficient macrophages this number is reduced to about 0.25 apoptotic bodies (156). This result demonstrates that *crq* is required for efficient phagocytosis of apoptotic targets. However, phagocytosis of apoptotic cells is not completely abolished in *crq*-deficient embryos. The recent completion of the *Drosophila* genome sequencing project revealed the presence of nine members of the CD36 family as well as three macrophage receptors of the scavenger receptor dSR-C1 class (102, 157). These genes might share overlapping functions with *crq* in the engulfment process.

Franc *et al.* (156) also noted a correlation in the amount of PCD and the relative expression of the *crq* gene. In *H99* mutant embryos, in which no PCD occurs (see Section 3) (26), *crq* expression is decreased by 75%. Consistently, under conditions with increased cell death such as X-ray irradiated embryos, macrophages show several-fold increase in *crq* expression (156). Thus, signals generated by dying cells cause increased expression of *crq*, which would facilitate the clearance of the cell corpses.

These studies demonstrate that *Drosophila* provides a powerful tool to study phagocytosis genetically. Further studies will reveal the molecular nature of the ligand for CRQ, and will elucidate the phagocytic pathways for apoptotic cells during development.

12. Inhibition of pathological cell death

A variety of diseases, including cancer, AIDS, stroke, myopathies, and various neurodegenerative disorders, are associated with the abnormal regulation of apoptosis (4). Consequently, there has been considerable interest in exploring the possibility of using apoptosis-based technologies and drugs for therapeutic purposes. However, it is uncertain whether the prevention of apoptosis in degenerative disorders provides a benefit for the organism, since rescued cells may no longer be functional. *Drosophila* offers unique opportunities to study the specific contribution of apoptosis in disease and provides interesting models for human diseases that are associated with abnormal apoptosis (see ref. 158 for review). For example, mutations in rhodopsin or associated proteins can cause retinitis pigmentosa in humans and mice, and a very similar condition in flies (31, 159, 160). In the *Drosophila* model, it has been shown that the retinal degeneration, due to mutations in either rhodopsin or a rhodopsin phosphatase gene, occurred by apoptosis and could be blocked by caspase inhibitors (32). Significantly, this suppression of apoptosis restored visual function to otherwise blind flies. These findings demonstrate that 'undead cells' can serve a useful purpose for the organism, and these results provide a strong rationale to further explore anti-apoptotic therapies in mammalian models for retinitis pigmentosa, as well as other neurodegenerative disorders.

In a different model, it has been shown that inactivation of the *Drosophila* homologue of the tumour suppressor APC (adenomatous polyposis coli) causes retinal neuronal degeneration which results from apoptotic cell death, a condition remarkably similar to that found in humans with germline APC mutations (161). The degenerating neurons in APC mutants can be rescued by expression of the caspase inhibitor p35 without any obvious change in cell fate (161). However, it was not demonstrated that the rescued neurons are functional.

Mutations in the α-synuclein gene are linked to familial Parkinson's disease. Transgenic expression of wild-type and mutant forms of α-synuclein in *Drosophila* recapitulates the essential features of the human disorder: adult-onset loss of dopaminergic neurons, intraneuronal inclusions, and locomotor dysfunction (162). These parallels indicate that the three abnormalities may be causally related.

Finally, Jackson *et al.* (163) and Warrick *et al.* (164) described *Drosophila* models for the glutamine repeat neurological disorders Huntington's disease and Machado–Joseph disease, respectively. In both studies, polyglutamine-expanded alleles of the human genes expressed in the *Drosophila* compound eye cause massive degeneration of photoreceptor neurons by cell death. The induced neuronal degeneration appears to be independent of both the *H99* cell death genes and of caspases, as co-expression of the p35 caspase inhibitor fails to inhibit the neuronal degeneration caused by the human disease genes. However, co-expression of the chaperone HSP70 suppresses polyglutamine-mediated neurodegeneration *in vivo* (165). This study indicates that HSP70 or related molecular chaperones may provide a means of treating these and other neurodegenerative diseases associated with abnormal protein conformation and toxicity. These examples illustrate that the powerful genetic system of *Drosophila* combined with many molecular tools including the complete genomic sequence will help improve our conceptual understanding of human disease mechanisms.

13. Final considerations

The regulation and execution of programmed cell death is a very complex biological phenomenon. During the last few years, there has been considerable progress in understanding the basic cellular mechanisms of apoptosis. However, we still know very little about how specific cells are selected for death and how cells that should live are protected from inappropriate activation of the death programme. Cell death is regulated by a large number of different intracellular and extracellular signals. Studies in vertebrate animal models are time-consuming and hampered by the fact that many genes are present as redundant functions. Thus, gene targeting very often does not show any phenotype or an unexpected one. Fortunately, the basic mechanism of apoptosis has been conserved throughout animal evolution from *C. elegans* to *Drosophila* to humans. *Drosophila* is open to a wide variety of technical approaches and provides an extremely powerful combination of techniques to study apoptosis in an organism that is intermediate between that of nematodes and vertebrates. Therefore, studies on the mechanism and regulation of apoptosis in *Drosophila* in the context of a developing organism will have a substantial impact upon the understanding of mammalian apoptosis. The completion of the *Drosophila* genome sequencing project reveals many more conserved factors and will allow now their genetic analysis (102). In addition, genetic screens for mutants affecting cell death and the molecular analysis of the corresponding genes will help closing the gaps of the apoptotic pathways. Thus, future genetic studies in *Drosophila* combined with molecular and biochemical techniques will greatly advance our knowledge in these areas and provide a conceptual framework for elucidating the mechanisms and control of apoptosis in humans. That is, after all, what model organisms are good for.

Acknowledgements

We are grateful to Dr Anne-Françoise Lamblin and Dr Masayuki Miura for providing information prior to publication. We would like to thank our colleagues in the Steller laboratory for advice and comments on the manuscript, especially Scott Waddel, Julie Agapite, Zhiwei Song, Lei Zhou, Lakshmi Goyal, and Lutz Kockel. A. B. is supported by post-doctoral fellowships from the Human Frontier Science Program Organization (HFSPO), the Anna Fuller Fund in Molecular Oncology, and the MIT/Merck Collaboration program. H. S. is an investigator of the Howard Hughes Medical Institute (HHMI).

References

1. Kerr, J. F. R., Wyllie, A. H., and Currie, A. R. (1972). Apoptosis: a basic biological phenomenon with wide ranging implications in tissue kinetics. *Br. J. Cancer*, **26**, 239–57.
2. Ernst, M. (1926). Über Untergang von Zellen während der normalen Entwicklung bei Wirbeltieren. *Z. Anat. Entwicklungsgesch.*, **79**, 228–62.
3. Glücksmann, A. (1951). Cell deaths in normal vertebrate ontogeny *Biol. Rev.*, **26**, 59–86.
4. Thompson, C. B. (1995). Apoptosis in the pathogenesis and treatment of disease. *Science*, **267**, 1456–62.
5. Abrams, J. M., White, K., Fessler, L. I., and Steller, H. (1993). Programmed cell death during *Drosophila* embryogenesis. *Development*, **117**, 29–43.
6. Wolff, T. and Ready, D. F. (1991). Cell death in normal and rough eye mutants of *Drosophila*. *Development*, **113**, 825–39.
7. Truman, J. W., Thorn, R. S., and Robinow, S. (1992). Programmed neuronal death in insect development. *J. Neurobiol.*, **23**, 1295–311.
8. Milan, M., Campuzano, S., and Garcia-Bellido, A. (1997). Developmental parameters of cell death in the wing disc of *Drosophila*. *Proc. Natl. Acad. Sci. USA*, **94**, 5691–6.
9. Truman, J. W. (1992). Insect systems for the study of programmed neuronal death. *Exp. Gerontol.*, **27**, 17–28.
10. Truman, J. W. (1983). Programmed cell death in the nervous system of an adult insect. *J. Comp. Neurol.*, **216**, 445–52.

11. Kimura, K. I. and Truman, J. W. (1990). Postmetamorphic cell death in the nervous and muscular systems of *Drosophila melanogaster*. *J. Neurosci.*, **10**, 403–10.
12. Robinow, S., Talbot, W. S., Hogness, D. S., and Truman, J. W. (1993). Programmed cell death in the *Drosophila* CNS is ecdysone-regulated and coupled with a specific ecdysone receptor isoform. *Development*, **119**, 1251–9.
13. Truman, J. W. and Schwartz, L. M. (1984). Steroid regulation of neuronal death in the moth nervous system. *J. Neurosci.*, **4**, 274–80.
14. Koelle, M. R., Talbot, W. S., Segraves, W. A., Bender, M. T., Cherbas, P., and Hogness, D. S. (1991). The *Drosophila EcR* gene encodes an ecdysone receptor, a new member of the steroid receptor superfamily. *Cell*, **67**, 59–77.
15. Yao, T. P., Segraves, W. A., Oro, A. E., McKeown, M., and Evans, R. M. (1992). *Drosophila ultraspiracle* modulates ecdysone receptor function via heterodimer formation. *Cell*, **71**, 63–72.
16. Thomas, H. E., Stunnenberg, H. G., and Stewart, A. F. (1993). Heterodimerization of the *Drosophila* ecdysone receptor with retinoid X receptor and *ultraspiracle*. *Nature*, **362**, 471–5.
17. Fahrbach, S. E., Choi, M. K., and Truman, J. W. (1994). Inhibitory effects of actinomycin D and cycloheximide on neuronal death in adult Manducasexta. *J. Neurobiol.*, **25**, 59–69.
18. Talbot, W. S., Swyryd, E. A., and Hogness, D. S. (1993). *Drosophila* tissues with different metamorphic responses to ecdysone express different ecdysone receptor isoforms. *Cell*, **73**, 1323–37.
19. Oppenheim, R. W. (1991). Cell death during development of the nervous system. *Annu. Rev. Neurosci.*, **14**, 453–501.
20. Raff, M. C., Barres, B. A., Burne, J. F., Coles, H. S., Ishizaki, Y., and Jacobson, M. D. (1993). Programmed cell death and the control of cell survival: lessons from the nervous system. *Science*, **262**, 695–700.
21. Raff, M. C. (1992). Social controls on cell survival and cell death. *Nature*, **356**, 397–400.
22. Power, M. E. (1943). The effect of reduction in numbers of ommatidia upon the brain of *Drosophila melanogaster*. *J. Exp. Zool.*, **94**, 33–72.
23. Steller, H., Fischbach, K.-F., and Rubin, G. M. (1987). *Disconnected*: A locus required for neuronal pathway formation in the visual system of *Drosophila*. *Cell*, **50**, 1139–53.
24. Selleck, S. and Steller, H. (1991). The influence of retinal innervation on neurogenesis in the first optic ganglion of *Drosophila*. *Neuron*, **6**, 83–99.
25. Campos, A. R., Fischbach, K.-F., and Steller, H. (1992). Survival of photoreceptor neurons in the compound eye of *Drosophila* depends on connections with the optic ganglia. *Development*, **114**, 355–66.
26. White, K., Grether, M., Abrams, J., Young, L., Farrell, K., and Steller, H. (1994). Genetic control of cell death in *Drosophila*. *Science*, **264**, 677–83.
27. Fristrom, D. (1969). Cellular degeneration in the production of some mutant phenotypes in *Drosophila melanogaster*. *Mol. Gen. Genet.*, **103**, 363–79.
28. Dura, J. M., Randsholt, N. B., Deatrick, J., Erk, I., Santamaria, P., Freeman, J. D., *et al.* (1987). A complex genetic locus, *polyhomeotic*, is required for segmental specification and epidermal development in *D. melanogaster*. *Cell*, **51**, 829–39.
29. Magrassi, L. and Lawrence, P. A. (1998). The pattern of cell death in *fushi tarazu*, a segmentation gene of *Drosophila*. *Development*, **104**, 447–51.
30. Bonini, N. M., Leiserson, W. M., and Benzer, S. (1993). The *eyes absent* gene: genetic control of cell survival and differentiation in the developing *Drosophila* eye. *Cell*, **72**, 379–95.

31. Kurada, P. and O'Tousa, J. E. (1995). Retinal degeneration caused by dominant mutations in *Drosophila. Neuron*, **14**, 571–9.
32. Davidson, F. F. and Steller, H. (1998). Blocking apoptosis prevents blindness in *Drosophila* retinal degeneration mutants. *Nature*, **391**, 587–91.
33. Grether, M. E., Abrams, J. M., Agapite, J., White, K., and Steller, H. (1995). The *head involution defective* gene of *Drosophila melanogaster* functions in programmed cell death. *Genes Dev.*, **9**, 1694–708.
34. Chen, P., Nordstrom, W., Gish, B., and Abrams, J. M. (1996). *grim*, a novel cell death gene in *Drosophila. Genes Dev.*, **10**, 1773–82.
35. Golstein, P., Marguet, D., and Depraetere, V. (1995). Homology between Reaper and the cell death domains of Fas and TNFR1. *Cell*, **81**, 185–6.
36. Chen, P., Lee, P., Otto, L., and Abrams, J. M. (1996). Apoptotic activity of REAPER is distinct from signaling by the tumor necrosis factor receptor 1 death domain. *J. Biol. Chem.*, **247**, 25735–7.
37. Vucic, D., Seshagiri, S., and Miller, L. K. (1997). Characterization of reaper- and FADD-induced apoptosis in a lepidopteran cell line. *Mol. Cell. Biol.*, **17**, 667–76.
38. Nassif, C., Daniel, A., Lengyel, J. A., and Hartenstein, V. (1998). The role of morphogenetic cell death during *Drosophila* embryonic head development. *Dev. Biol.*, **197**, 170–86.
39. Weil, M., Jacobson, M. D., Coles, H. S., Davies, T. J., Gardner, R. L., Raff, K. D., *et al.* (1996). Constitutive expression of the machinery for programmed cell death. *J. Cell Biol.*, **133**, 1053–9.
40. Nordstrom, W., Chen, P., Steller, H., and Abrams, J. M. (1996). Activation of the *reaper* gene defines an essential function required for both naturally-occuring apoptosis and induced cell killing in *Drosophila. Dev. Biol.*, **180**, 227–41.
41. Robinow, S., Draizen, T. A., and Truman, J. W. (1997). Genes that induce apoptosis: transcriptional regulation in identified, doomed neurons of the *Drosophila* CNS. *Dev. Biol.*, **190**, 206–13.
42. Frank, L. H. and Rushlow, C. (1996). A group of genes required for maintenance of the amnioserosa tissue in *Drosophila. Development*, **122**, 1343–52.
43. White, K., Tahaoglu, E., and Steller, H. (1996). Cell killing by *Drosophila reaper. Science*, **271**, 805–7.
44. Hay, B. A., Wolff, T., and Rubin, G. M. (1994). Expression of baculovirus P35 prevents cell death in *Drosophila. Development*, **120**, 2121–9.
45. Xue, D. and Horvitz, H. R. (1995). Inhibition of the *Caenorhabditis elegans* cell-death protease CED-3 by a CED-3 cleavage site in baculovirus p35 protein. *Nature*, **377**, 248–51.
46. Bump, N. J., Hackett, M., Hugunin, M., Seshagiri, S., Brady, K., Chen, P., *et al.* (1995). Inhibition of ICE family proteases by baculovirus antiapototic protein p35. *Science*, **269**, 1885–8.
47. Jiang, C., Baehrecke, E. H., and Thummel, C. S. (1997). Steroid regulated programmed cell death during *Drosophila* metamorphosis. *Development*, **124**, 4673–83.
48. Draizen, T. A., Ewer, J., and Robinow, S. (1999). Genetic and hormonal regulation of the death of peptidergic neurons in the *Drosophila* central nervous system. *J. Neurobiol.*, **38**, 455–65.
49. Zhou, L., Schnitzler, A., Agapite, J., Schwartz, L. M., Steller, H., and Nambu, J. R. (1997). Cooperative functions of the *reaper* and *head involution defective* genes in programmed cell death of *Drosophila* CNS midline cells. *Proc. Natl. Acad. Sci. USA*, **94**, 5131–6.

50. Wing, J. P., Zhou, L., Schwartz, L. M., and Nambu, J. R. (1998). Distinct cell killing properties of the *Drosophila reaper, head involution defective,* and *grim* genes. *Cell Death Differ.*, **5**, 930–9.
51. Jiang, C., Lamblin, A.-F. J., Steller, H., and Thummel, C. S. (2000). A steroid-triggered transcriptional hierarchy controls salivary gland cell death during *Drosophila* metamorphosis. *Mol. Cell*, **5**, 445–55.
52. Brodsky, M. H., Nordstrom, W., Tsang, G., Kwan, E., Rubin, G. M., and Abrams, J. M. (2000). *Drosophila* p53 binds a damage response element at the *reaper* locus. *Cell*, **101**, 103–13.
53. Ollmann, M., Young, L. M., Di Como, C. J., Karim, F., Belvin, M., Robertson, S., *et al.* (2000). *Drosophila* p53 is a structural and functional homolog of the tumor suppressor p53. *Cell*, **101**, 91–101.
54. Bergmann, A., Agapite, J., McCall, K. A., and Steller, H. (1998). The *Drosophila* gene *hid* is a direct molecular target of Ras-dependent survival signaling. *Cell*, **95**, 331–41.
55. Kurada, P. and White, K. (1998). Ras promotes cell survival in *Drosophila* by downregulating *hid* expression. *Cell*, **95**, 319–29.
56. Sawamoto, K., Taguchi, A., Yamada, C., Jin, M., and Okano, H. (1998). Argos induces cell death in the developing *Drosophila* eye by inhibition of the Ras pathway. *Cell Death Differ.*, **5**, 262–70.
57. Downward, J. (1998). Ras signaling and apoptosis. *Curr. Opin. Genet. Dev.*, **8**, 49–54.
58. Wassarman, D. A. and Therrien, M. (1997). RAS1-mediated photoreceptor development in *Drosophila*. *Adv. Dev. Biol.*, **5**, 1–41.
59. Diaz-Benjumea, F. J. and Hafen, E. (1994). The *sevenless* signalling cassette mediates *Drosophila* EGF receptor function during epidermal development. *Development*, **120**, 569–78.
60. Freeman, M. (1996). Reiterative use of the EGF receptor triggers differentiation of all cell types in the *Drosophila* eye. *Cell*, **87**, 651–60.
61. Schweitzer, R., Shaharabany, M., Seger, R., and Shilo, B. Z. (1995). Secreted Spitz triggers the DER signaling pathway and is a limiting component in embryonic ventral ectoderm determination. *Genes Dev.*, **9**, 1518–29.
62. Miller, D. T. and Cagan, R. L. (1998). Local induction of patterning and programmed cell death in the developing *Drosophila* retina. *Development*, **125**, 2327–35.
63. Kaplan, D. R. and Miller, F. D. (1997). Signal transduction by the neurotrophin receptors. *Curr. Opin. Cell Biol.*, **9**, 213–21.
64. Pulido, D., Campuzano, S., Koda, T., Modolell, J., and Barbacid, M. (1992). *Dtrk*, a *Drosophila* gene related to the trk family of neurotrophin receptors, encodes a novel class of neural cell adhesion molecule. *EMBO J.*, **11**, 391–404.
65. Wilson, C., Goberdhan, D. C., and Steller, H. (1993). *Dror*, a potential neurotrophic receptor gene, encodes a *Drosophila* homolog of the vertebrate Ror family of Trk-related receptor tyrosine kinases. *Proc. Natl. Acad. Sci. USA*, **90**, 7109–13.
66. Freeman, M. (1997). Cell determination strategies in the *Drosophila* eye. *Development*, **124**, 261–70.
67. Huang, Z., Shilo, B. Z., and Kunes, S. (1998). A retinal axon fascicle uses *spitz*, an EGF receptor ligand, to construct a synaptic cartridge in the brain of *Drosophila*. *Cell*, **95**, 693–703.
68. Bos. J. L. (1989). *ras* oncogenes in human cancer: a review. *Cancer Res.*, **49**, 4682–9.
69. Hoffman, B. and Liebermann, D. A. (1994). Molecular controls of apoptosis: differentiation/growth arrest primary response genes, proto-oncogenes, and tumor suppressor genes as positive and negative modulators. *Oncogene*, **9**, 1807–12.

70. Crook, N. E., Clem, R. J., and Miller, L. K. (1993). An apoptosis-inhibiting baculovirus gene with a zinc finger-like motif. *J. Virol.*, **67**, 2168–74.
71. Birnbaum, M. J., Clem, R. J., and Miller, L. K. (1994). An apoptosis-inhibiting gene from nuclear polyhedrosis virus encoding a polypeptide with Cys/His sequence motifs. *J. Virol.*, **68**, 2521–8.
72. Clem, R. J. and Miller, L. K. (1994). Control of programmed cell death by the baculovirus genes p35 and iap. *Mol. Cell. Biol.*, **14**, 5212–22.
73. Deveraux, Q. L. and Reed, J. C. (1999). IAP family proteins–suppressors of apoptosis. *Genes Dev.*, **13**, 239–52.
74. Roy, N., Mahadevan, M. S., McLean, M., Shutler, G., Yaraghi, Z., Farahani, R., *et al.* (1995). The gene for neuronal apoptosis inhibitory protein is partially deleted in individuals with spinal muscular atrophy. *Cell*, **80**, 167–78.
75. Rothe, M., Pan, M. G., Henzel, W. J., Ayres, T. M., and Goeddel, D. V. (1995). The TNFR2-TRAF signaling complex contains two novel proteins related to baculoviral inhibitor of apoptosis proteins. *Cell*, **83**, 1243–52.
76. Liston, P., Roy, N., Tamai, K., Lefebvre, C., Baird, S., Cherton-Horvant, G., *et al.* (1996). Suppression of apoptosis in mammalian cells by NAIP and a related family of IAP genes. *Nature*, **379**, 349–53.
77. Ambrosini, G., Adida, C., and Altieri, D. C. (1997). A novel anti-apoptosis gene, survivin, expressed in cancer and lymphoma. *Nature Med.*, **3**, 917–21.
78. LaCasse, E. C., Baird, S., Korneluk, R. G., and MacKenzie, A. E. (1998). The inhibitors of apoptosis (IAPs) and their emerging role in cancer. *Oncogene*, **17**, 3247–59.
79. Uren, A. G., Coulson, E. J., and Vaux, D. L. (1998). Conservation of baculovirus inhibitor of apoptosis repeat proteins (BIRPs) in viruses, nematodes, vertebrates and yeasts. *Trends Biochem. Sci.*, **23**, 159–62.
80. Duckett, C. S., Nava, V. E., Gedrich, R. W., Clem, R. J., Van Dongen, J. L., Gilfillan, C., *et al.* (1996). A conserved family of cellular genes related to the baculovirus iap gene and encoding apoptosis inhibitors. *EMBO J.*, **15**, 2685–94.
81. Uren, A. G., Pakusch, M., Hawkins, C. J., Puls, K. L., and Vaux, D. L. (1996). Cloning and expression of apoptosis inhibitory protein homologs that function to inhibit apoptosis and/or bind tumor necrosis factor receptor-associated factors. *Proc. Natl. Acad. Sci. USA*, **93**, 4974–8.
82. Chu, Z. L., McKinsey, T. A., Liu, L., Gentry, J. J., Malim, M. H., and Ballard, D. W. (1997). Suppression of tumor necrosis factor-induced cell death by inhibitor of apoptosis c-IAP2 is under NF-kB control. *Proc. Natl. Acad. Sci. USA*, **94**, 10057–62.
83. Deveraux, Q. L., Takahashi, R., Salvesen, G. S., and Reed, J. C. (1997). X-linked IAP is a direct inhibitor of cell-death proteases. *Nature*, **388**, 300–4.
84. Fraser, A. and James, C. (1998). Fermenting debate: do yeast undergo apoptosis? *Trends Cell Biol.*, **8**, 219–21.
85. Hay, B. A., Wassarman, D. A., and Rubin, G. M. (1995). *Drosophila* homologs of baculovirus inhibitor of apoptosis proteins function to block cell death. *Cell*, **83**, 1253–62.
86. Lisi, S., Mazzon, I., and White, K. (2000). Diverse domains of THREAD/DIAP1 are required to inhibit apoptosis induced by REAPER and HID in *Drosophila*. *Genetics*, **154**, 669–78.
87. Goyal, L., McCall, K., Agapite, J., Hartwieg, E., and Steller, H. (2000). Induction of apoptosis by *Drosophila reaper, hid* and *grim* through inhibition of IAP function. *EMBO J.*, **19**, 589–97.

88. Wang, S. L., Hawkins, C. J., Yoo, S. J., Muller, H. A., and Hay, B. A. (1999). The *Drosophila* caspase inhibitor DIAP1 is essential for cell survival and is negatively regulated by HID. *Cell*, **98**, 453–63.
89. Vucic, D., Kaiser, W. J., Harvey, A. J., and Miller, L. K. (1997). Inhibition of Reaper-induced apoptosis by interaction with inhibitor of apoptosis proteins (IAPs). *Proc. Natl. Acad. Sci. USA*, **94**, 10183–8.
90. Vucic, D., Kaiser, W. J., Harvey, A. J., and Miller, L. K. (1998). IAPs physically interact with and block apoptosis induced by *Drosophila* proteins HID and GRIM. *Mol. Cell. Biol.*, **18**, 3300–9.
91. Vucic, D., Kaiser, W. J., and Miller, L. K. (1998). A mutational analysis of the baculovirus inhibitor of apoptosis Op-IAP. *J. Biol. Chem.*, **273**, 33915–21.
92. McCarthy, J. V. and Dixit, V. M. (1998). Apoptosis induced by *Drosophila reaper* and *grim* in a human system. Attenuation by inhibitor of apoptosis proteins (cIAP). *J. Biol. Chem.*, **273**, 37–41.
93. Kaiser, W. J., Vucic, D., and Miller, L. K. (1998). The *Drosophila* inhibitor of apoptosis D-IAP1 suppresses cell death induced by the caspase drICE. *FEBS Lett.*, **440**, 243–8.
94. Hawkins, C. J., Wang, S. L., and Hay, B. A. (1999). A cloning method to identify caspases and their regulators in yeast: identification of *Drosophila* IAP1 as an inhibitor of the *Drosophila* caspase DCP-1. *Proc. Natl. Acad. Sci. USA*, **96**, 2885–90.
95. Harvey, A. J., Bidwai, A. P., and Miller, L. K. (1997). Doom, a product of the *Drosophila mod(mdg4)* gene, induces apoptosis and binds to baculovirus inhibitor-of-apoptosis proteins. *Mol. Cell. Biol.*, **17**, 2835–43.
96. Fraser, A. G., James, C., Evan, G. I., and Hengartner, M. O. (1999). *Caenorhabditis elegans* inhibitor of apoptosis protein (IAP) homologue BIR-1 plays a conserved role in cytokinesis. *Curr. Biol.*, **9**, 292–301.
97. Yuan, J., Shaham, S., Ellis, H. M., and Horvitz, H. R. (1993). The *C. elegans* cell death gene *ced-3* encodes a protein similar to mammalian interleukin-1beta-converting enzyme. *Cell*, **75**, 641–52.
98. Takahashi, A. and Earnshaw, W. C. (1996). ICE-related proteases in apoptosis. *Curr. Opin. Genet. Dev.*, **6**, 50–5.
99. Salvesen, G. S. and Dixit, V. M. (1997). Caspases: intracellular signaling by proteolysis. *Cell*, **91**, 443–6.
100. Thornberry, N. A., Rano, T. A., Peterson, E. P., Rasper, D. M., Timkey, T., Garcia-Calvo, M., *et al.* (1997). A combinatorial approach defines specificities of members of the caspase family and granzyme B. Functional relationships established for key mediators of apoptosis. *J. Biol. Chem.*, **272**, 17907–11.
101. Kumar, S. and Colussi, P. A. (1999). Prodomains–adaptors–oligomerization: the pursuit of caspase activation in apoptosis. *Trends Biochem. Sci.*, **24**, 1–4.
102. Rubin, G. M., Yandell, M. D., Wortman, J. R., Gabor, G. L., Miklos, G. L., Nelson, C. R., *et al.* (2000). Comparative genomics of the eukaryotes. *Science*, **287**, 2204–15.
103. Song, Z., McCall, K. A., and Steller, H. (1997). DCP-1, a *Drosophila* cell death protease essential for development. *Science*, **275**, 536–40.
104. Fraser, A. G. and Evan, G. I. (1997). Identification of a *Drosophila melanogaster* ICE/CED-3-related protease, drICE. *EMBO J.*, **16**, 2805–13.
105. Inohara, N., Koseki, T., Hu, Y., Chen, S., and Nunez, G. (1997). CLARP, a death effector domain-containing protein interacts with caspase-8 and regulates apoptosis. *Proc. Natl. Acad. Sci. USA*, **94**, 10717–22.

106. Chen, P., Rodriguez, A., Erskine, R., Thach, T., and Abrams, J. M. (1998). *Dredd*, a novel effector of the apoptosis activators *Reaper, Grim*, and *Hid* in *Drososphila. Dev. Biol.*, **201**, 202–16.
107. Dorstyn, L., Colussi, P. A., Quinn, L. M., Richardson, H., and Kumar, S. (1999). DRONC, an ecdysone-inducible *Drosophila* caspase. *Proc. Natl. Acad. Sci. USA*, **96**, 4307–12.
108. Dorstyn, L., Read, S. H., Quinn, L. M., Richardson, H., and Kumar, S. (2000). DECAY, a novel *Drosophila* caspase related to mammalian caspase-3 and caspase-7. *J. Biol. Chem.*, **274**, 30778–83.
109. Song, Z., Guan, B., Bergman, A., Nicholson, D. W., Thornberry, N. A., Peterson, E. P., *et al.* (2000). Biochemical and genetic interactions between *Drosophila* caspases and the proapoptotic genes *rpr, hid*, and *grim*. *Mol. Cell. Biol.*, **20**, 2907–14.
110. Meier, P., Silke, J., Leevers, S. J., and Evan, G. I. (2000). The *Drosophila* caspase DRONC is regulated by DIAP1. *EMBO J.*, **19**, 598–611.
111. McCall, K. A. and Steller, H. (1998). Requirement of DCP-1 caspase during *Drosophila* oogenesis. *Science*, **279**, 230–4.
112. Waddington, C. H. and Okada, E. (1960). Some degenerative phenomena in *Drosophila* ovaries. *J. Embryol. Exp. Morphol.*, **8**, 341–8.
113. Giorgi, F. and Deri, P. (1976). Cell death in ovarian chambers of *Drosophila melanogaster*. *J. Embryol. Exp. Morphol.*, **35**, 521–33.
114. Ashkenazi, A. and Dixit, V. M. (1998). Death receptors: signaling and modulation. *Science*, **281**, 1305–8.
115. Thornberry, N. A. and Lazebnik, Y. (1998). Caspases: enemies within. *Science*, **281**, 1312–16.
116. Roy, N., Deveraux, Q. L., Takahashi, R., Salvesen, G. S., and Reed, J. C. (1997). The cIAP-1 and cIAP-2 proteins are direct inhibitors of specific caspases. *EMBO J.*, **16**, 6914–25.
117. Tamm, I., Wang, Y., Sausville, E., Scudiero, D. A., Vigna, N., Oltersdorf, T., *et al.* (1998). IAP-family protein survivin inhibits caspase activity and apoptosis induced by Fas (CD95), Bax, caspases, and anticancer drugs. *Cancer Res.*, **58**, 5315–20.
118. Takahashi, R., Deveraux, Q., Tamm, I., Welsh, K., Assa-Munt, N., Salvesen, G. S., *et al.* (1998). A single BIR domain of XIAP sufficient for inhibiting caspases. *J. Biol. Chem.*, **273**, 7787–90.
119. Bergmann, A., Agapite, J., and Steller, H. (1998). Mechanisms and control of programmed cell death in invertebrates. *Oncogene*, **17**, 3215–23.
120. Evans, E. K., Kuwana, T., Strum, S. L., Smith, J. J., Newmeyer, D. D., and Kornbluth, S. (1997). Reaper-induced apoptosis in a vertebrate system. *EMBO J.*, **16**, 7372–81.
121. Thress, K., Henzel, W., Shillinglaw, W., and Kornbluth, S. (1998). Scythe: a novel reaper-binding apoptotic regulator. *EMBO J.*, **17**, 6135–43.
122. Claveria, C., Albar, J. P., Serrano, A., Buesa, J. M., Barbero, J. L., Martinez, A. C., *et al.* (1998). *Drosophila grim* induces apoptosis in mammalian cells. *EMBO J.*, **17**, 7199–208.
123. Haining, W. N., Carboy-Newcomb, C., Wei, C. L., and Steller, H. (1999). The pro-apoptotic function of *Drosophila* Hid is conserved in mammalian cells. *Proc. Natl. Acad. Sci. USA*, **96**, 4936–41.
124. Foley, K. and Cooley, L. (1998). Apoptosis in late stage *Drosophila* nurse cells does not require genes within the *H99* deficiency. *Development*, **125**, 1075–82.
125. Kanuka, H., Hisahara, S., Sawamoto, K., Shoji, S., Okano, H., and Miura, M. (1999). Proapoptotic activity of *Caenorhabditis elegans* CED-4 protein in *Drosophila*: implicated mechanisms for caspase activation. *Proc. Natl. Acad. Sci. USA*, **96**, 145–50.

126. Zhou, L., Song, Z., Tittel, J., and Steller, H. (1999). HAC-1, a *Drosophila* homolog of APAF-1 and CED-4, functions in developmental and radiation-induced apoptosis. *Mol. Cell*, **4**, 745–55.
127. Rodriguez, A., Oliver, H., Zou, H., Chen, P., Wang, X., and Abrams, J. M. (1999). Dark is a *Drosophila* homologue of Apaf-1/CED-4 and functions in an evolutionarily conserved death pathway. *Nature Cell Biol.*, **1**, 272–9.
128. Kanuka, H., Sawamoto, K., Inohara, N., Matsuno, K., Okano, H., and Miura, M. (1999). Control of cell death pathway by *Dapaf-1*, a *Drosophila* Apaf-1/CED-4 related caspase activator. *Mol. Cell*, **4**, 757–69.
129. Zou, H., Henzel, W. J., Liu, X., Lutschg, A., and Wang, X. (1997). Apaf-1, a human protein homologous to *C. elegans* CED-4, participates in cytochrome c-dependent activation of caspase-3. *Cell*, **90**, 405–13.
130. Varkey, J., Chen, P., Jemmerson, R., and Abrams, J. M. (1999). Altered cytochrome c display precedes apoptotic cell death in *Drosophila*. *J. Cell Biol.*, **144**, 701–10.
131. Li, P., Nijhawan, D., Budihardjo, I., Srinivasula, S. M., Ahmad, M., Alnemri, E. S., *et al.* (1997). Cytochrome c and dATP-dependent formation of Apaf-1/Caspase-9 complex initiates an apoptotic protease cascade. *Cell*, **91**, 479–89.
132. Gross, A., McDonnell, J. M., and Korsmeyer, S. J. (2000). BCL-2 family members and the mitochondria in apoptosis. *Genes Dev.*, **13**, 1899–911.
133. Igaki, T., Kanuka, H., Inohara, N., Sawamoto, K., Nunez, G., Okano, H., *et al.* (2000). *Drob-1*, a *Drosophila* member of the Bcl-2/CED-9 family that promotes cell death. *Proc. Natl. Acad. Sci. USA*, **97**, 662–7.
134. Colussi, P. A., Quinn, L. M., Huang, D. C., Coombe, M., Read, S. H., Richardson, H., *et al.* (2000). *Debcl*, a proapoptotic Bcl-2 homologue, is a component of the *Drosophila melanogaster* cell death machinery. *J. Cell Biol.*, **148**, 703–14.
135. Brachmann, C. B., Jassim, O. W., Wachsmuth, B. D., and Cagan, R. L. (2000). The *Drosophila* bcl-2 family member *dBorg-1* functions in the apoptotic response to UV-irradiation. *Curr. Biol.*, **10**, 547–50.
136. Zhang, H., Huang, Q., Ke, N., Matsuyama, S., Hammock, B., Godzik, A., *et al.* (2000). *Drosophila* pro-apoptotic Bcl-2/Bax homologue reveals evolutionary conservation of cell death mechanisms. *J. Biol. Chem.*, **275**, 27303–6.
137. Kennedy, S. G., Wagner, A. J., Conzen, S. D., Jordan, J., Bellacosa, A., Tsichlis, P. N., *et al.* (1997). The PI 3-kinase/Akt signaling pathway delivers an anti-apoptotic signal. *Genes Dev.*, **11**, 701–13.
138. Datta, S. R., Dudek, H., Tao, X., Masters, S., Fu, H., Gotoh, Y., *et al.* (1997). Akt phosphorylation of BAD couples survival signals to the cell-intrinsic death machinery. *Cell*, **91**, 231–41.
139. del Peso, L., Gonzalez-Garcia, M., Page, C., Herrera, R., and Nunez, G. (1997). Interleukin-3-induced phosphorylation of BAD through the protein kinase Akt. *Science*, **278**, 687–9.
140. Andjelkovic, M., Jones, P. F., Grossniklaus, U., Cron, P., Schier, A. F., Dick, M., *et al.* (1995). Developmental regulation of expression and activity of multiple forms of the *Drosophila* RAC protein kinase. *J. Biol. Chem.*, **270**, 4066–75.
141. Franke, T. F., Tartof, K. D., and Tsichlis, P. N. (1994). The SH2-like Akt homology (AH) domain of c-akt is present in multiple copies in the genome of vertebrate and invertebrate eucaryotes. Cloning and characterization of the *Drosophila melanogaster c-akt* homolog *Dakt1*. *Oncogene*, **9**, 141–8.
142. Staveley, B. E., Ruel, L., Jin, J., Stambolic, V., Mastronardi, F. G., Heitzler, P., *et al.* (1998). Genetic analysis of protein kinase B (AKT) in *Drosophila*. *Curr. Biol.*, **8**, 599–602.

143. Verdu, J., Buratovich, M. A., Wilder, E. L., and Birnbaum, M. J. (2000). Cell-autonomous regulation of cell and organ growth in *Drosophila* by Akt / PKB. *Nature Cell Biol.*, **1**, 500–6.
144. Kondo, T., Yokokura, T., and Nagata, S. (1997). Activation of distinct caspase-like proteases by Fas and *reaper* in *Drosophila* cells. *Proc. Natl. Acad. Sci. USA*, **94**, 11951–6.
145. Ren, Y. and Savill, J. (1998). Apoptosis: the importance of being eaten. *Cell Death Differ.*, **5**, 563–8.
146. Wu, Y. C. and Horvitz, H. R. (1998). The *C. elegans* cell corpse engulfment gene *ced-7* encodes a protein similar to ABC transporters. *Cell*, **93**, 951–60.
147. Wu, Y. C. and Horvitz, H. R. (1998). *C. elegans* phagocytosis and cell-migration protein CED-5 is similar to human DOCK180. *Nature*, **392**, 501–4.
148. Hasegawa, H., Kiyokawa, E., Tanaka, S., Nagashima, K., Gotoh, N., Shibuya, M., *et al.* (1996). DOCK180, a major CRK-binding protein, alters cell morphology upon translocation to the cell membrane. *Mol. Cell. Biol.*, **16**, 1770–6.
149. Rushton, E., Drysdale, R., Abmayr, S. M., Michelson, A. M., and Bate, M. (1995). Mutations in a novel gene, *myoblast city*, provide evidence in support of the founder cell hypothesis for *Drosophila* muscle development. *Development*, **121**, 1979–88.
150. Erickson, M. R., Galletta, B. J., and Abmayr, S. M. (1997). *Drosophila myoblast city* encodes a conserved protein that is essential for myoblast fusion, dorsal closure, and cytoskeletal organization. *J. Cell Biol.*, **138**, 589–603.
151. Liu, Q. A. and Hengartner, M. O. (1998). Candidate adaptor protein CED-6 promotes the engulfment of apoptotic cells in *C. elegans*. *Cell*, **93**, 961–72.
152. Tepass, U., Fessler, L. I., Aziz, A., and Hartenstein, V. (1994). Embryonic origin of hemocytes and their relationship to cell death in *Drosophila*. *Development*, **120**, 1829–37.
153. Franc, N. C., Dimarcq, J. L., Lagueux, M., Hoffmann, J., and Ezekowitz, R. A. (1996). *Croquemort*, a novel *Drosophila* hemocyte / macrophage receptor that recognizes apoptotic cells. *Immunity*, **4**, 431–43.
154. Endemann, G., Stanton, L. W., Madden, K. S., Bryant, C. M., White, R. T., and Protter, A. A. (1993). CD36 is a receptor for oxidized low density lipoprotein. *J. Biol. Chem.*, **268**, 11811–16.
155. Savill, J., Hogg, N., Ren, Y., and Haslett, C. (1992). Thrombospondin cooperates with CD36 and the vitronectin receptor in macrophage recognition of neutrophils undergoing apoptosis. *J. Clin. Invest.*, **90**, 1513–22.
156. Franc, N. C., Heitzler, P., Ezekowitz, R. A., and White, K. (1999). Requirement for *croquemort* in phagocytosis of apoptotic cells in *Drosophila*. *Science*, **284**, 1991–4.
157. Pearson, A., Lux, A., and Krieger, M. (1995). Expression cloning of dSR-CI, a class C macrophage-specific scavenger receptor from *Drosophila melanogaster*. *Proc. Natl. Acad. Sci. USA*, **92**, 4056–60.
158. Fortini, M. E. and Bonini, N. M. (2000). Modeling human neurodegenerative diseases in *Drosophila*: on a wing and a prayer. *Trends Genet.*, **16**, 161–7.
159. Chang, G. Q., Hao, Y., and Wong, F. (1993). Apoptosis: final common pathway of photoreceptor death in rd, rds and rhodopsin mutant mice. *Neuron*, **11**, 595–605.
160. Gregory-Evans, K. and Bhattacharya, S. (1998). Genetic blindness: current concepts in the pathogenesis of human outer retinal dystrophies. *Trends Genet.*, **14**, 103–8.
161. Ahmed, Y., Hayashi, S., Levine, A., and Wieschaus, E. (1998). Regulation of Armadillo by a *Drosophila* APC inhibits neuronal apoptosis during retinal development. *Cell*, **93**, 1171–82.
162. Feany, M. B. and Bender, W. W. (2000). A *Drosophila* model of Parkinson's disease. *Nature*, **404**, 394–8.

163. Jackson, G. R., Salecker, I., Dong, X., Yao, X., Arnheim, N., Faber, P. W., *et al.* (1998). Polyglutamine-expanded human huntingtin transgenes induce degeneration of *Drosophila* photoreceptor neurons. *Neuron*, **21**, 633–42.
164. Warrick, J. M., Paulson, H. L., Gray-Board, G. L., Bui, Q. T., Fischbeck, K. H., Pittman, R. N., *et al.* (1998). Expanded polyglutamine protein forms nuclear inclusions and causes neural degeneration in *Drosophila*. *Cell*, **93**, 939–49.
165. Warrick, J. M., Chan, H. Y., Gray-Board, G. L., Chai, Y., Paulson, H. L., and Bonini, N. M. (2000). Suppression of polyglutamine-mediated neurodegeneration in *Drosophila* by the molecular chaperone HSP70. *Nature Genet.*, **23**, 425–8.

4 The caspases: consequential cleavage

NATALIE ROY and MICHAEL H. CARDONE

1. Introduction

The regulation of cellular pathways by specific proteolytic systems is not unusual. The most noted example would perhaps be the cell cycle which is regulated by the co-ordinated degradation of cyclins and of cyclin inhibitors. Over recent years it has become apparent that a key feature of apoptosis is the proteolysis of specific substrates which enable several morphological and biochemical characteristics of apoptosis to be manifest. The proteases primarily responsible for the cleavage of specific substrates during apoptosis are known as caspases. Caspases are cysteine proteases that cleave their substrates after aspartic acid residues (cysteine *aspartase*) and are highly selective proteases giving them distinct biological roles in the cell as mediators of apoptosis and inflammation (1). Evidence suggests that caspases do not simply lead to the degradation of their substrates but instead serve to modulate the function of their substrates and have emerged as critical signalling molecules in the apoptotic pathway.

2. Caspase involvement in apoptosis

Genetic studies in the nematode *C. elegans* were critical in the initial characterization of caspases as key effectors of apoptosis (2–5) (see Chapter 2). Loss-of-function mutations in, or deletion of, a single gene *ced-3* (cell death defective) prevents the death of the 131 cells programmed to die during the development of this organism. Conversely, overexpression of CED-3 induces these cells to die (6). Sequence analysis unveiled the homology of Ced-3 to the mammalian protease, interleukin-1β converting enzyme (ICE). ICE (or caspase-1 as it was later renamed) was initially identified as the enzyme responsible for cleaving the pro-inflammatory cytokine, interleukin-1 beta, to its mature active form (4). This protease did not share any homology with any known protease family and hence was the founder member of a unique class of cysteine proteases (7, 8). The homology of ICE with CED-3 provided the first evidence that this class of protease may be critical in executing mammalian apoptosis and that the mammalian cell death machinery is conserved across species (4, 9). Overexpression of either CED-3 or caspase-1 induces apoptosis in mammalian cells substantiating this hypothesis (10). After the identification of caspase-1 numerous caspases were discovered in mammalian cells defining a family of homologous cysteine proteases. The rapid discovery of these proteases by several laboratories led to inconsistent nomenclature which has now been replaced by the uniform terminology, caspase and a designated number (1). Currently 14 members have been identified and can be classified based upon their structural and functional homology as depicted in Fig. 1.

Various studies have established the role of caspases as integral mediators of apoptosis. Initially several caspases were observed to be activated early in the apoptotic programme and their overexpression in cells induced death (10–15). In addition, key regulatory and structural proteins are cleaved during apoptosis and the proteases responsible for the hydrolysis of these proteins were identified as the caspases (16–18). Peptide inhibitors of the caspases were shown to inhibit apoptosis *in vitro*, confirming their role as executioners of this programme (9, 19, 20). Complementing these studies, ectopic expression of mutant caspases in which key residues of their catalytically active sites were mutated were ineffective at inducing cell death and furthermore could inhibit apoptosis by acting in a dominant-negative fashion (21). These studies established that active caspase molecules are required for classical apoptotic death with distinct morphological characteristics. However, the most striking evidence substantiating the role that caspases play in programmed cell death comes from the generation of mice or nematodes deficient in particular

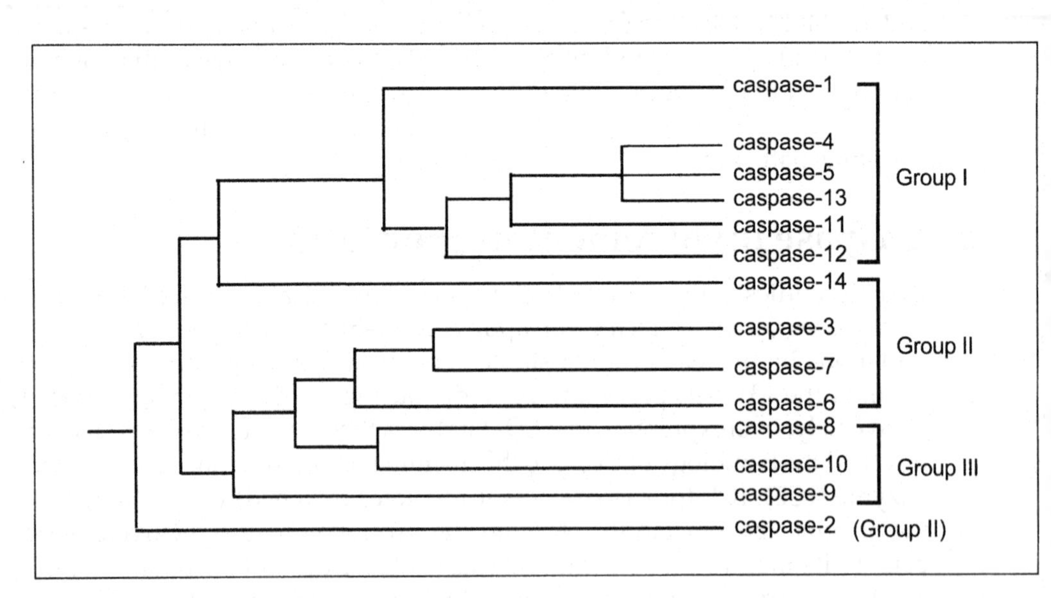

Fig. 1 Caspase family evolutionary relationship. The structural evolutionary relationship of the caspases is shown. The functional groups that the caspases have been classified into is indicated at the right. With the exception of caspase-2, the caspases are most closely related to those within their functional class.

caspases. Mutant worms that lack CED-3 have a complete absence of developmental cell death while mice deficient in caspase-3 or caspase-9 have extreme defects in neuronal programmed cell death (22–24). Both these mice have increased brain cell number that is evident as ectopic cell masses or protrusions of cranial brain tissue. Apoptosis occurs normally in other tissues from these mice suggesting a complex system of caspase regulation that may be cell type and tissue specific.

3. Class and structure

Peptidases are classified into five families; aspartic, cysteine, metallo, serine, and threonine based upon the chemical group responsible for catalysis (Merops classification). These families are further classified based upon the order of catalytic site residues. The caspases are thus distinguished from the papain-like, the viral 'papain-like', the viral 'chymotrypsin-like', and the adenovirus cysteine proteases within their family in that they contain catalytic residues in the order His, Cys. Plant, bacterial, and fungi peptidases exist with the same catalytic motif however, they share no sequence or functional homology with the caspases. Distant homologues, identified by sequence similarity, in *Dictyostelium discoideuor*, *Sreptomyces colicolor*, and *Rhizobium* have been reported although their involvement in apoptosis remains to be demonstrated (25). The caspases are conserved across animal species with functional homologues (i.e. orthologues) in *Drosophila*, *C. elegans*, and *Xenopus*, suggesting evolution from a conserved protease superfamily (see Chapters 2 & 3). Granzyme B and calpain are the closest to functional analogues that share no sequence homology but may have some similar functional roles as both induce cell death via activation of the caspases (26, 27). Granzyme B, a serine protease, is most similar in that it shares a preference for aspartic acid residues and mediates cell death in the immune response (28, 29).

Within their family caspases can be grouped in various ways. Most commonly they are grouped according to their substrate specificity or their functional roles. However, they can also be grouped according to their prodomain structure. Grouping the caspases in these various different ways allows us to make predictions about their importance in the regulation of the apoptotic pathway.

3.1 Three-dimensional structure

Caspases are expressed in the cell as inactive enzymes or zymogens and are proteolytically activated to form the mature protein, similar to other protease zymogens. The structure of the caspase is depicted in Fig. 2. Processing of the proenzyme at specific aspartate residues involves removal of the variable length N-terminal peptide or prodomain (2–25 kDa) and cleavage between the large and small subunits. In some cases a short linker exists between the large and small subunit that is also removed. These cleavage events result in a heterodimeric active enzyme consisting of a large subunit of approximately 17–21 kDa and a smaller subunit of approximately 10–13 kDa. The three-dimensional structure of caspase-1 and caspase-

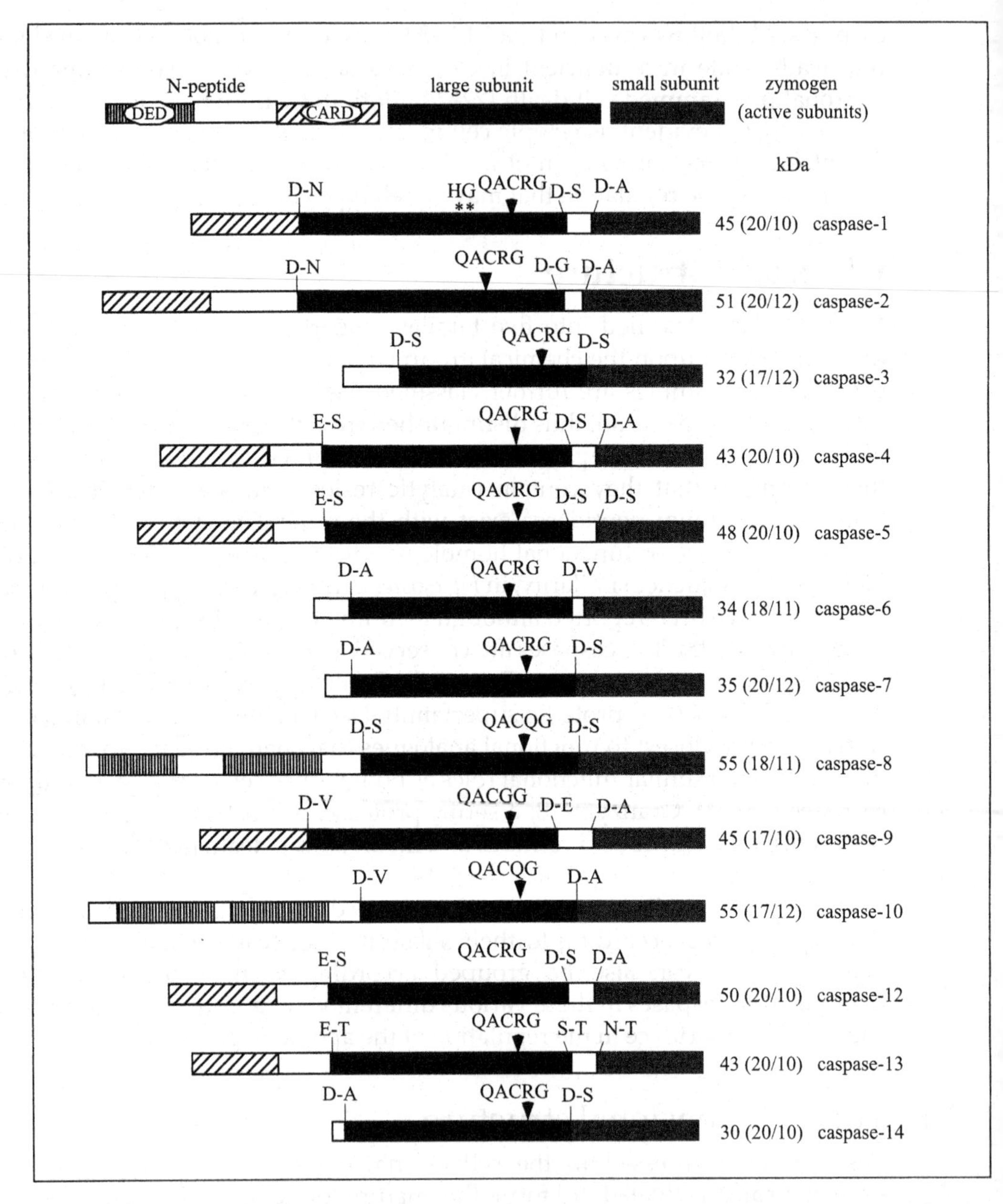

Fig. 2 Caspase zymogen structure. Caspases are synthesized in their zymogen form containing prodomains encoding CARD (angled lines) or DED (vertical lines) motifs. The mature active heterodimeric enzyme results from cleavage between the large (black) and small (grey) subunits. The sites of cleavage are indicated by lines. The active site (*arrow*) and the histidine and glycine (*asterisks*) residues that make up the catalytic diad are indicated (caspase-1 numbering). The molecular weight of the zymogen and active subunits (in brackets) is given.

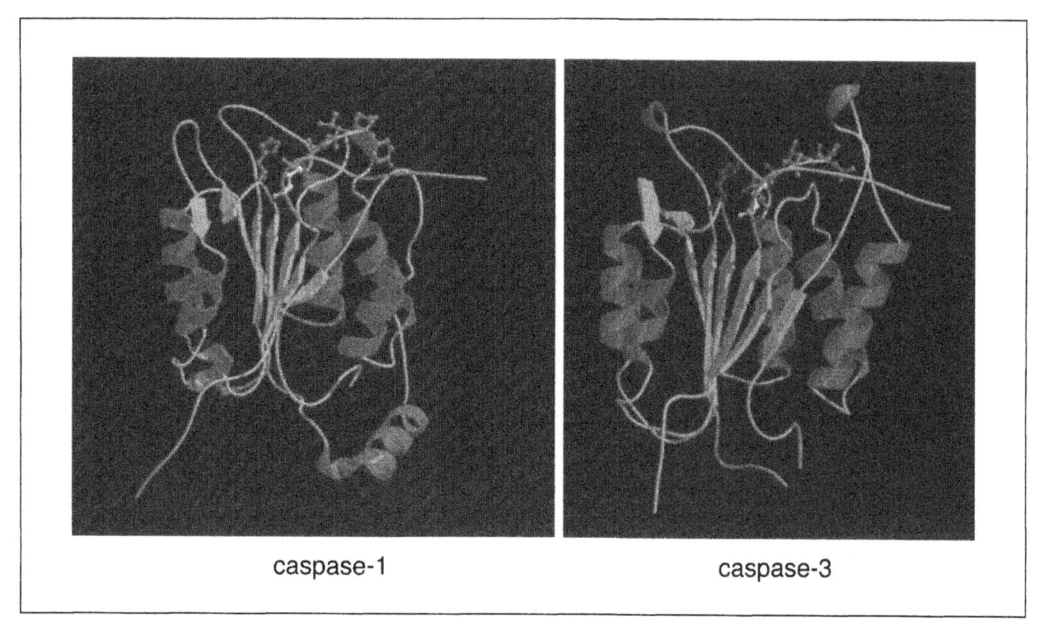

Fig. 3 (See Plate 3) Three-dimensional structure of caspase-1 (*left*) and caspase-3 (*right*). One molecule of the dimer is shown. Catalytic residues are shown in ball-and-stick representation. For caspase-1 residue His237 is in purple and Cys285 is in yellow. For caspase-3 His121 and Cys163 are marked. The bound inhibitor is shown in grey in ball-and-stick representation. Caspase-1 is in complex with acetyl-Tyr-Val-Asp acetyl-Asp-Val-Ala-Asp.

3 has revealed an active enzyme in which two heterodimers associate to form a tetramer (Fig. 3; see Plate 3) (30–32). These structures were determined in complex with tetrapeptide inhibitors, thus enabling the identification of essential residues for substrate binding and catalysis within both subunits. The two subunits bind to form a central six-strand β-sheet core, flanked on either side by α-helices. Characteristic of all caspases is a conserved QACXG pentapeptide encompassing the active site cysteine residue. A single catalytic domain or diad is formed by a cysteine nucleophile (Cys285, caspase-1 numbering) in close proximity to a histidine imidazole group (His237). Hydrogen bond interactions by the backbone amide protons of the active site cysteine and Gly238 lead to stabilization of the oxyanion of the tetrahedral intermediates.

The primary substrate recognition pocket (S1) interacts with the aspartic acid of the substrate and is stabilized by two residues from each subunit (Arg179, Gln283 and Arg341, Ser347). Additional pockets (S2–S4) distinguish the caspases from one another with the S4 subsite largely determining the substrate specificity. This can be observed when comparing caspase-1 and caspase-3 structures as an additional surface loop is present in caspase-3 that forms one side of the S4 subsite 30. The loop folds onto the S4 binding site and appears to confer steric hindrance on large apolar residues conferring a rigid preference for aspartic acid residues. In contrast, caspase-1 is able to recognize various aromatic P4 residues. The loop is formed by an additional 10 amino acids that are absent in caspase-1.

3.2 Substrate specificity

The caspase requirement for an aspartate residue at the P1 site is highly specific so much so that more than a 100-fold decrease in the second order rate constant (k_{cat}/K_m) results if this residue is changed (7, 33, 34). Efficient catalysis however requires a tetrapeptide sequence where the P4 subsite requirement differs considerably between caspase members and thus largely determines substrate specificity (7). One notable study performed an exhaustive examination of caspase substrate preferences using a combinatorial peptide library which shuffled amino acids in the P4, P3, and P2 positions (35). These tetrapeptides linked to amino-methyl-coumarin (AMC) are processed efficiently by the caspases and used as fluorogenic substrates to identify optimal substrate cleavage sights and catalytic activity (7, 36). Three classes of caspases can thus be distinguished based upon their optimal substrate cleavage site as shown in Table 1 (35, 37). Group I, consisting of caspase-1, -4, and -5 favours hydrophobic residues in the P4 site with the optimal sequence WEHD. The effector or downstream caspases (Group II) have a firm requirement for an aspartic acid with a preference for DEXD sequences in their substrates. Processing of the procaspases occurs at similar sites suggesting that autocatalysis occurs and/or implies a caspase cascade. Amino acids with branched aliphatic side chains, optimally V/LEXD, are favoured by the initiator or upstream caspases (Group III). Peptide substrates can thus be used to determine the class of caspase which is activated in a given cell extract by careful measurement of the k_{cat}/K_m. It has been suggested that caspase-2 may have a slightly extended binding pocket as pentapeptides are cleaved 10-fold more efficiently by this caspase (38).

Several modifications have been made to these ideal tetrapeptide substrates resulting in inhibitors that either bind reversibly, in the case of aldehydes, nitriles, or ketones or bind irreversibly in the case of the flouro-diazomethyl ketones (FMK) derived peptides (39). The FMK coupled and benzyloxycarbonyl (Z) modified peptide Z-VAD-FMK acts as a general inhibitor of all of the classes of caspases and is widely used as a membrane permeable inhibitor of apoptosis in cell culture. The reversible aldehyde peptides have variable potencies for inhibition, which are predicted from their substrate specificity. For instance, as Group II caspases have a highly stringent requirement for the Asp in position P4 with a k_{cat}/K_m of $1.6\text{-}10^6$, they are potently inhibited by DEVD-CHO with an inhibition constant (K_i) of 230 pM in contrast to a significantly higher K_i of 1 μM for YVAD (35, 39).

3.3 Prodomain structure

The N-terminal peptide or prodomain of the zymogen is divergent structurally which serves to classify the procaspases into two broad groups. Long prodomains are found in caspase-1, -2, -4, -5, -8, -9, -10, -12, and -13 while short or no prodomains are present in caspase-3, -6, -7, -11, and -14 (Fig. 2). The prodomain is removed during processing to the active form however, it plays a crucial role in activation. These domains can serve either to target procaspases to death receptors at the cell

Table 1 Three classes of caspases have been distinguished based upon their optimal substate cleavage sites (35, 37)

		Prodomain motif[a]	Adaptor		$k_{cat}M^{-1s}/K_m^{-1\times10^6}$ [b]	Optimal tetrapeptide cleavage site (P_4-P_1)[c]	
			Activator	Inhibitor			
Group I[d]	**Mediators of inflammation**						
	Caspase-1	CARD	CARDIAK	Unknown	3.3	W/Y/ FEHD	W/LEHD consensus Favour hydrophobic residues in S_4
	Caspase-4(11)	CARD	Unknown	Unknown	1.2	W/L/ FEHD	
	Caspase-5	CARD	Unknown	Unknown	0.3	W/L/ FEHD	
	Caspase-12	CARD	Unknown	Unknown	nd	Unknown	
	Caspase-13	CARD	Unknown	Unknown	nd[e]	Unknown	
Group II[f]	**Effectors**						
	Caspase-2	CARD	RAIDD	ARC	nd	DEHD	DEXD[h] consesus Aspartic acid requirements in S_4
	Caspase-3	–[g]	–	–	14.0	DEVD	
	Caspase-7	–	–	–	6.3	DEVD	
	Caspase-14	–	–		nd	DEVD	
Group III[i]	**Upstream activators**						Cleave non DXXD substrates with preference for amino acids containing branched aliphatic side chains Cleave Group II zymogens contain long prodomains with adaptor domains
	Caspase-6	–	–	–	nd	V/T/ IEHD	
	Caspase-8	DED/DRD	FADD	Unknown	0.37	LETD	
	Caspase-9	CARD	APAF-1	Unknown	0.10	LEHD	
	Caspase-10	DED	FADD	Unknown	nd	Unknown	
Invertebrate caspases							
	CED-3	CARD	CED-4	Unknown	nd	DEXD	
	DREDD	DED	DARK	Inknown	nd	nd	

[a] CARD = xaspases activation and recruitment domain, DED = death effector domain.
[b] Adapted from ref. 38.
[c] Standard single amino acid code is used to represent the optimum coding sequence for sites P4–P1.
[d] Group I, consisting of caspase-1, -4, and -5 favour hydrophobic residues in the P4 site with the optimal sequence WEHD.
[e] nd, not determined.
f Group II caspases (the effector or downstream caspases) have a firm requirement for an aspartic acid at P4, with a preference for DEXD sequences in their substrates.
[g] –, not applicable, short prodomain with no motif present.
[h] X, broad amino acid specificity.
i Group III caspases (initiator or upstream caspases) favour amino acids with branched aliphatic side chains, optimally V/LEXD, at the P4 site.

membrane or to mediate assembly of a caspase-activating complex within the cell. By targeting the caspases to different cellular compartments these domains serve to regulate their activity. The effector caspases lie within the class that either lack or only contain a short prodomain and appear to require processing by those with long prodomains, the initiator or apical caspases.

Two unique protein–protein interaction domains are found within the long prodomains; death effector domains (DEDs) and caspase recruitment domains (CARDs) (40, 41). Both are approximately 90 amino acids in length and although not significantly similar at the sequence level, they adopt a similar fold consisting of approximately six α-helices (42, 43). This observation suggests that these domains could interact with one another and implies cross-regulation of pathways. The CARDs mediate electrostatic interactions while DEDs mediate hydrophobic associations (43). Both domains nonetheless serve to recruit caspases to specific complexes though adaptor molecules via their CARD and DED domains.

Caspase-8 and -10 each have two DED domains and through their association with the DED domain of their adaptor molecule FADD, are targeted to the ligand-activated receptors Fas, TNF-R1, DR3, and DR5 of the TNFR superfamily (44–48) (see Chapter 8). It is postulated that one of the DED domains may be required for homodimerization while the other recruits the caspase to the adaptor protein (49). This class of adaptor molecules, in addition to their DED domain, contain the third characteristic protein–protein interaction domain present in several molecules active in the apoptotic pathway, the death domain (DD) that serves to tether these complexes to the receptor (50) (see Chapter 8). The *Drosophila* caspase DREDD resembles caspase-8 and -10 in that it contains two DED domains. This implies that death receptor pathways maybe present in *Drosophila* although the identity of such receptors has not been ascertained (51).

CARD domains are found in caspase-1, -2, -9, and CED-3 mediating their association with CARD containing adaptor molecules CARDIAK, RAIDD, Apaf-1, and CED-4 respectively (52–54). Analogous to FADD, CARDIAK and RAIDD target their bound procaspase to activated receptors. A DD in RAIDD mediates the assembly of a receptor complex with caspase-2 via a DD in RIP, which directly binds the Fas or TNF-R1 receptors. Tethering of caspase-1 to receptor complexes via CARDIAK is mediated by its interaction with the TNF-associated factors TRAF1 and 2. However, the CARD containing adaptors, Apaf-1 and CED-4, mediate pathways downstream of the cell membrane by assembly of a cytoplasmic caspase complex. A CARD domain is also present in the *Drosophila* caspase DRONC, implying the presence of a partner CARD containing, CED-4-like molecule (55). Ironically the CED-4/Apaf homologue DARK has been reported to bind the DED containing caspase DREDD as opposed to DRONC (56).

4. Caspase activation

Activation of the caspases represents a pivotal step in the cell death signalling cascade. Upon initiation of an apoptotic signal multiple caspases are activated within the cell and this has been shown to occur via proteolytic caspase cascades that ensue in the cell (14, 16–18). Activation of multiple caspases serves to amplify the apoptotic signal and their sequential activation results in a pathway that may be regulated at many points. To date the mechanisms observed for caspase activation are: transactivation, proteolysis by other classes of proteases, or autoactivation. Procaspase zymogens can be cleaved by the catalytically active mature forms of the caspases themselves resulting in transactivation. Alternatively caspases are substrates of other active proteases in the dying cell. The last mechanism involves aggregation or oligimerization of the procaspases that results in autoprocessing. The route to caspase activation is dependent upon the apoptotic stimulus and the subcellular localization of the caspases. The proteolytic caspase cascades can be initiated either at the cell membrane by receptor complexes or in the cytosol by other death stimuli. In general, once initiated, two signalling pathways may be utilized, one that is mediated through the mitochondria and one that bypasses mitochondria. Mitochondria have emerged as central organelles mediating many of the signalling events in apoptosis (57, 58) (see Chapters 5 and 6). These signalling events are discussed below in the context of caspase activation.

4.1 Transactivation

Most latent proteases are directly cleaved and activated or serve as substrates for another protease and indeed this was the first mechanism described for caspase activation. The observation that the proteolytic specificity of caspases corresponds to the target site of cleavage between the large and small subunits within procaspases suggested a proteolytic signalling cascade (35). This type of regulation is reminiscent of the complement proteins within the blood coagulation cascade that cleave one another with strict specificity for Arg residues. One can predict a sequential caspase cascade based upon individual substrate specificity. *In vitro* experiments first demonstrated that certain caspases such as caspase-8 and -10 could activate procaspase-3 and -7 (40, 59–61). These studies on the sequential activation of caspases are now being confirmed *in vivo* (62). It is well established that upstream or apyical caspases with long prodomains cleave the downstream or effector caspases, which either have or lack short prodomains. The downstream or executioner caspases serve to amplify the death signal and cause the demise of the cell. For example active caspase-9 activates caspases-7 and -3 which for the latter caspase results in the further activation of caspases-2 and -6 (21, 63–65). Moreover caspase-3 can cleave and activate procaspase-9 providing a feed-forward amplification loop (62). Caspase-6 acts in a similar fashion by cleaving caspase-3. The job of the downstream caspases such as caspse-3, -6, and -7 is to cleave key regulatory and structural proteins enabling the collapse of the cell (see section below) (60, 66). Activation of the pro-inflammatory caspases, caspase-1, -4, and -5 is less well understood. Caspase-11-deficient mice however have shown that activation of caspase-1 is dependent on active caspase-11 (67).

In addition to the caspases themselves other classes of proteases may cleave and activate the procaspases. Granzyme B, a serine protease with specificity for aspartic residues, directly cleaves and activates caspase-3 leading to the destruction of the cell (26, 68). This is observed in cytotoxic T lymphocyte (CTL)-mediated cytotoxicity, the bodies main defence against tumorigenic and virus-infected cells. The elimination of cells is mediated in part by the perforin–granzyme B pathway whereby perforin facilitates the entry of granzymes into the target cell, that serve to subsequently induce apoptosis (29). In addition to CTLs, lysosomes were identified by fractionation studies as a source of a protease that could activate procaspase-11 (69). This protease, cathepsin B, can also activate procaspase-1. Possibly other, as yet unidentified, proteases can activate the zymogen forms of the caspases.

4.2 Aggregation leading to autoprocessing

An alternative mechanism employed to induce caspase activation is aggregation or oligomerization facilitating intermolecular proteolysis. Aggregation occurs by a number of distinct mechanisms at particular sites in the cell. Dependent upon the death stimulus aggregation may ensue at the cell membrane, via receptor-mediated interactions, or within the cytosol. Regardless, the control of caspase complex formation is largely mediated by adaptor molecules. The first indication of autocatalytic processing via aggregation emerged from recombinant expression systems where procaspases were observed to autocatalytically process to their mature active forms. In their zymogen forms the caspases have low intrinsic enzymatic activity, however upon cleavage and assembly into tetramers, an increase in catalytic activity occurs. The high concentration of the caspases produced in recombinant systems allows the low catalytic activity of one molecule to cleave a second forming an active tetramer.

Oligomerization as a mode of activation was first proposed based upon *in vivo* observations that activation of the receptors Fas and TNF-R1 led to formation of signalling complexes that induced activation of specific procaspases (40, 70, 71) (see Chapter 8). Several groups recreated the *in vivo* model of Fas oligomerization by inducing dimerization by artificial means (72, 73). Forced oligomerization of membrane targeted procaspase-8 resulted in its proteolytic processing and cell death. Removal of the prodomain however abolished the apoptotic activity of procaspase-8 indicating that prodomain dimerization is likely to be necessary for processing and autoactivation to occur. The initiator caspases have all been shown to be activated in a similar fashion supporting the idea that the principle role of adaptor molecules is to bring the procaspases into close proximity to allow their activation. Forced dimerization/oligomerization of downstream procaspases that lack long CARD or DED containing prodomains does not lead to their processing supporting the idea that long prodomains are essential for oligomerized-induced activation (73). This result is also consistent with the observation that overexpression of the downstream caspases, such as caspase-3, does not lead to high levels of cell death, presumably due to their inability to oligomerize. Fusion of the prodomain of caspase-2, an initiator caspase, to caspase-3 leads to its autoactivation providing direct evidence that the prodomains are required for dimerization and autoactivation (74).

In addition to adaptor molecules, polyglutamine inclusions have also been found to function in the same manner to activate procaspases (see Chapter 11). Trinucleotide repeat expansions that encode polyglutamine tracts occur in proteins involved in several neurodegenerative disorders. The genes involved in X-linked spinal and bulbar atrophy (Kennedy's disease), Huntingtons, Dentatorubral-palidoluysian atrophy (DRPLA), and spinocerebellar ataxia (SCA) all harbour this type of mutation (see last section) (75–80). Caspase cleavage of these expanded polyglutamine tracts modulates the formation of protein aggregates or inclusions within neuronal cells that are toxic to the cell (81, 82). The presence of plaques themselves was not thought to be entirely responsible for the death of the affected neuronal cells. Consistent with this hypothesis, work by Yuan and colleagues demonstrated that procaspase-8 and adaptor molecules such as FADD are recruited to these aggregates (83). Recruitment of procaspase-8 induces its oligomerization and activation, which is deleterious to the cell. Aggregate formation thus further enhances toxicity by the activation of caspases creating a positive feedback loop. Evidence for the recruitment of procaspase-3 to these aggregates could not be found suggesting that procaspases with long prodomains are preferentially recruited or may be dependent upon recruitment of adaptor molecules. However more data is still needed to support this hypothesis.

4.3 Caspase activation at the mitochondria

Activation of procaspases by oligomerization within the cytosol occurs by various mechanisms. At present the most studied is caspase activation events occurring at the mitochondria (21, 53, 84). Many death stimuli induce mitochondrial swelling, membrane depolarization, and cytochrome *c* release from the mitochondria into the cytosol (58) (see Chapters 5 and 6). This event promotes the formation of a caspase-activating complex of cytochrome *c*, Apaf-1, and procaspase-9 that triggers activation of caspase-9 and leads to a proteolytic cascade. The release of cytochrome *c* from the mitochondria, in the presence of ATP, enables cytochrome *c* to bind APAF-1 in the cytosol. This interaction is mediated via protein–protein interaction domains within Apaf-1, the WD repeats. These repeats are essential for an oligomeric complex of at least eight Apaf-1 subunits to form (85). Conformational changes occur upon cytochrome *c* binding allowing Apaf-1 to bind procaspase-9 via CARD domains present in both molecules. Aggregation of procaspase-9 leads to autoproteolysis and activation of downstream caspases (85). Alternatively, signalling downstream from the membrane receptor Fas can bypass the mitochondria resulting in a modified signalling cascade whereby caspase-8 can directly cleave caspase-3 (86) (see Chapter 8). Recent evidence suggests that caspases are also located within the intermitochondrial space. Release of these caspases, along with factors such as cytochrome *c* and apoptosis-inducing factor (AIF), is subject to a number of regulatory events including binding of pro- or anti-apoptotic members of the Bcl-2 family (96–98) (see Chapters 5 and 6). Bax, a pro-apoptotic member of the Bcl-2 family inhibits the action of Bcl-2, enabling release of proteins from the mitochondria. AIF release from mitochondria may amplify the apoptotic signal by in turn releasing both cytochrome *c* and caspase-9 from the mitochondria. Thus, caspase-9 activation is subject, indirectly, to regulation by proteins such as Bax, BID, and AIF (99, 100). Most recently the mitochondrial protein SMAC/DIABLO has been shown to be released during apoptosis promoting caspase activation (87, 88).

The mitochondrial pathway in part, is evolutionarily conserved with similar events evident in *C. elegans* (see Chapter 2). CED-3 and CED-4, the functional equivalents of caspase-9 and Apaf-1 respectively, form an activating complex in an analogous, albeit less complex, manner. Cytochrome *c* release is not observed and is hence not required for assembly of the complex. Oligomerization of CED-4 is induced by a region located in the putative ATPase domain, clustering CED-3 molecules and leading to CED-3 cleavage and activation (73, 89). The interaction of procaspases to molecules containing ATP binding sites first suggested that ATP hydrolysis induces conformational changes in the caspases resulting in activation. This has been confirmed by studies demonstrating that dATP is required for oligomerization of procaspase-9 to occur.

Procaspase-9 may also be activated directly by other CARD containing molecules that reside in the cytoplasm. The death-inducing signal of caspase-9 may be enhanced by way of its interaction with CARD4/Nod (90, 91). CARD-4, similar to CED-4, does not induce death when expressed ectopically in cells. However, although the data is conflicting with regards to the association of these two molecules, it has been suggested that CARD-4 induces the oligomerization and thus activation of caspase-9. A second CARD containing molecule, mE10, functions in a comparable, yet slightly different, manner. mE10 contains an N-terminal CARD domain that shares homology with a CARD sequence from the E10 gene of equine herpesvirus type 2. When overexpressed in cells mE10 induces death (92). Surprisingly, mE10 binds to caspase-9 via its unique C-terminal domain and not its CARD domain. The CARD domain induces mE10 self-oligomerization thus enabling mE10 to activate capsase-9 by an induced proximity mechanism.

4.4 Conformational changes leading to activation

The speculation that conformational changes may induce activation of the caspases was first introduced when ATP was observed to be required for activation of caspase-9 in complex with Apaf-1. This mechanism is supported by other studies. In an effort to discern the components leading to caspase activation, Xanthoudakis *et al.* (93) purified a multiprotein activation complex from cells treated with the topoisomerase inhibitor, campothecin. Caspase-3 and -8 were shown to be components of this complex by Western blotting while mass spectroscopy revealed one of the unknown components to be Hsp60. *In vitro* reconstitution experiments demonstrated that Hsp60 stimulated the proteolytic processing of procaspase-3, but only in the presence of the upstream caspase-6, -8, and -9. Hsp60 enhances the activation by these upstream caspases and this can be further stimulated by ATP. Concurrently, a second study demonstrated that cytosolic fractions from Fas-activated cells induce the release of cytochrome *c*, Hsp60, and Hsp10 from mitochondria (94). Hsp60 and

Hsp10 could be co-immunoprecipitated with procaspase-3 from mitochondrial fractions in untreated cells and in agreement with Xanthoudakis *et al.*, both proteins had a stimulatory effect on the rate of caspase activation that occurred in an ATP-dependent manner. Hsp60 is a chaperonin that induces the folding of proteins in an ATP-dependent manner while Hsp10 binds to Hsp60 and regulates its substrate binding in addition to its ATPase activity. These studies suggest that the Hsps induce conformational changes in procaspase-3 that promote its activation. Given the requirement for an upstream caspase suggests that the caspase-3 prodomain may need to be cleaved in order for these conformational changes to occur.

The direct activation of caspases by specific peptides has also recently been described. Arginine–glycine–aspartate (RGD) peptides have been found to induce apoptosis in cells by directly inducing autoprocessing and activity of caspase-3 (95). The RGD motif is a well-characterized integrin-recognition motif located in many ligands hence RGD peptides have been used as inhibitors of integrin–ligand interactions. Integrin activation is normally triggered by RGD-DDX interactions. Caspase-3 contains both these motifs suggesting that RGD peptides induce conformational changes that trigger processing and activation. Presumably oligomerization is increased resulting in autoproteolytic processing.

5. Caspase inhibition

Studies on the molecular mechanisms of viral pathogenesis have broadened our understanding of the components involved in the regulation of cell death. Apoptosis is a critical defence mechanism for host cells once infected and viruses have evolved scrupulous strategies to avoid such a fate. Viruses encode anti-apoptotic genes that prolong the life of the infected cell and in some instances are required for establishing their persistence and allowing them to evade immune responses.

Inflammatory responses against viral infection can be avoided by viral-mediated inhibition of apoptosis. This phenomenon was first seen in cowpox virus, mediated by the gene CrmA, a cytokine response modifier. CrmA prevents the release and processing of the pro-inflammatory cytokine interleukin-1 beta of the host cell by inhibiting the activation of caspase-1 (ICE) (96). CrmA can also inhibit caspase-8 potentially protecting against Fas-, TNF-, and TRAIL-mediated apoptosis (61) and is a weak inhibitor of caspase-3, -7, and granzyme B (97, 98). CrmA is thus a selective caspase inhibitor and can be distinguished from the baculovirus inhibitor p35. p35 is a broad-spectrum caspase inhibitor potently inhibiting Group I, II, and III caspases (99–101). Spontaneous mutations within p35 were found in a mutant baculovirus that resulted in plasma membrane blebbing and a rapid cytolytic phenotype upon infection in insect cells (102). Thus, p35 is required during infection to inhibit host cell apoptosis. p35 itself is cleaved at the sequence DQMD, an event that correlates with caspase inhibition (99). Thus p35 is a suicide inhibitor. The cleavage products disable the active forms of the caspases by remaining tightly and irreversibly bound, blocking the active site within the caspase (100). The association and inhibition of the caspase requires a reactive site loop within p35 that mediates association and confers stability to the complex (103). The molecular and/or structural features of the caspases that enable p35 to be a pan inhibitor of caspases remains to be clarified.

Evidence that caspase activation could be inhibited or negatively regulated by endogenous mammalian proteins, first came from viruses. In addition to p35, baculovirus also has a second anti-apoptotic gene; the inhibitor of apoptosis or IAP (104, 105). The IAPs are a family of anti-apoptotic proteins that are found in mammalian cells and are conserved across several species including the nematode, *Drosophilia*, and yeast (106–111). Several mammalian IAPs have been shown to be caspase inhibitors when overexpressed and were the first mammalian caspase inhibitors to be identified (112–114). The IAPs, unlike the broad-spectrum inhibitor p35, inhibit select caspases and as such potentially serve as potent regulators that can repress different apoptotic pathways. The IAPs target the active mature forms of the distal caspase-3 and -7 enabling them to block many stimuli. In addition, these inhibitors can target caspase-9, modulating its activity in two ways. Similar to their regulation of caspase-3 and -7, the IAPs can directly inhibit the active form of caspase-9. Moreover, they can bind the zymogen form of caspase-9, preventing processing of the enzyme to its active form. Association of the IAPs to procaspase-9 serves as a two-fold inhibitory mechanism. The association blocks the interaction of caspase-9 with Apaf-1 in complex with cytochrome *c* and dATP, preventing its auto-activation and secondly it prevents the cleavage of procaspase-9 by caspase-3. By targeting procaspase-9, the IAPs block mitochondrial apoptotic pathways and by targeting caspase-3 and -7 block those stimuli that can bypass the mitochondria and directly target the distal caspases. The caspase activator Smac DIABLO however binds the IAPs and blocks their inhibitory function presumably by disabling their interaction with caspase-9 (88). Recent evidence from both nematode and man suggests that not all IAPs function solely within the apoptotic pathway. BIR-2 is required for cytokinesis during meiosis in the nematode. Moreover, the human IAP survivin is required for mitotic spindle checkpoint control, preventing the onset of apoptosis during G2/M phase. These recent results suggest that IAPs have a much more complex role in cellular survival than simply serving to inhibit caspases.

DED domains previously described in molecules that positively regulate caspase activation are additionally found in molecules that negatively regulate the caspases. The first examples were identified in herpesvirus and poxvirus and contained two DEDs which could bind the DED of FADD thereby disabling the recruitment of procaspase-8 and -10 to FADD binding death receptors (115–117). A homologous mammalian protein was identified with a similar function termed FLIP/Casper/I-FLICE/FLAME-1/CASH/CLARP/MRIT (117–122). This protein resembles capase-8 and -10 in its structure however lacks enzymatic activity as the active site cysteine residue is replaced by a tyrosine in the caspase-like domain. FLIP competes with the caspases for binding to FADD and displacement of the caspase by FLIP prevents the onset of the proteolytic caspase cascade. Direct inhibition of caspase-8 however may also occur by the formation of catalytically inactive heterodimers (123). FLIP antagonizes primarily receptor-mediated apoptosis. Some cells are resistant to Fas-mediated apoptosis and this may be largely due to high levels of FLIP expression.

Potentially, modulating the levels of FLIP in the cell may sensitize cells to apoptosis. For example, vascular endothelial cells are resistant to Fas-mediated apoptosis, however, upon treatment with oxidized low density lipoprotein they are sensitized to death due to a down-regulation of FLIP expression (124).

Like proteins containing DED domains, CARD domain-containing proteins also act as dominant-negative regulators of caspase activation. The inhibitor ARC contains an N-terminal CARD domain allowing it to bind procaspase-2 and surprisingly procaspase-8 (125). The interaction with procaspase-8 occurs through its DED domain indicating that the CARD and DED domains, although not similar at the amino acid level, are similar at the structural level and can form complexes. Resembling FLIP, ARC represses receptor-mediated death signals acting as a competitive inhibitor.

Caspase activity may also be regulated by protein modifications. This type of regulation is reversible and can occur by two mechanisms; phosphorylation or nitrosylation. Akt is a pro-survival kinase and is activated in response to growth factors such as insulin-like growth factor 1 (IGF-1) and platelet-derived growth factor (PDGF). Growth factors promote cell survival by suppressing the apoptotic machinery via pathways that activate PI3K and its downstream target Akt (126). Regulated by the oncoproteins Ras, ILK, and PI3 kinase, Akt plays a key role in the onset of tumorigenesis and inappropriate survival. Akt phosphorylates the pro-apoptotic molecule BAD, altering its binding affinity and activity (127, 128). Subsequently direct phosphorylation of the active, mature human caspase-9 enzyme by Akt was shown, resulting in inactivation of caspase-9 *in vitro* and *in vivo* (129, 130). This mechanism of regulation is not likely to be conserved among mammalian species as Akt does not phosphorylate mouse, dog, or hamster caspase-9. These species do not share Akt phosphorylation sites on caspase-9. One speculation is that longer-lived mammalian species have evolved this additional protection against caspase-9 activation. How phosphorylation of caspase-9 inhibits its function is unknown. The phosphorylated serine is distal to the substrate binding pocket, therefore inhibition may involve a mechanism that affects subunit dimerization or alters the catalytic machinery of the substrate cleft through conformational changes. In addition, the pro-form of caspase-9 is phosphorylated *in vivo* by Akt, but as yet it is unclear if this alters the association of procaspase-9 to Apaf-1 or other binding partners such as e10 or CARD4.

The second type of caspase modification that may occur is the formation of thiol adducts or nitrosylation. Regulation of a minority of proteins in the cell by nitrosylation has been demonstrated by nitric oxide. Nitric oxide inhibits apoptosis and NO synthase activity leads to caspase inhibition (131). Caspases can be *S*-nitrosylated by NO donors specifically at the active site, thus rendering the caspase inactive (132, 133). This occurs in a reversible manner and upon Fas stimulation an increase in caspase denitrosylation can be observed.

Finally, caspases as with their activation may be inhibited by upstream events. As described above for the nematode, CED-4 binds CED-3 promoting its aggregation and activation (134, 135). CED-9 regulates cell death by forming a multimeric complex with CED-4 and CED-3 and repressing activation of CED-3 (136). An analogous situation occurs in mammalian cells with homologous counterparts. It was originally shown that overexpression of Bcl-2 or Bcl-x_L blocks the release of cytochrome *c* stopping the apoptotic cascade (137, 138). The mammalian counterpart of *ced-9*, *bcl-x_l*, can bind Apaf-1 via its CED-4 domain inhibiting the association between Apaf-1 and caspase-9 (139). Bcl-x_L thus regulates caspase-9 activation indirectly by preventing its oligomerization. A second member of the Bcl-2 family, Boo, that is specifically expressed in the ovary and epididymis, regulates apoptosis in a similar manner (140). Boo forms a multimeric complex with Apaf-1 and caspase-9 through its interaction with Apaf-1. Given that caspase-9 can be detected in this ternary complex may indicate that Boo is preventing conformational changes and/or oligomerization of caspase-9. Regulation of caspase-9 activity may also occur in the mitochondria directly, concordant with the mechanism observed in *C. elegans* that is independent of cytochrome *c* release. Support for this hypothesis has only recently emerged with caspase-3 activation in the mitochondrial membrane being independent of cytochrome *c* and Bcl-2 inhibitable (141).

6. Caspase substrates

The execution of apoptosis entails the disassembly of the cell and the cessation of cellular function in a programmed manner. Characteristic apoptotic morphology such as condensed nuclei with nucleosome cleaved DNA fragments, loss of plasma leaflet asymmetry, and disassembly of organelles, facilitate the demise of the cell and the removal of the cell carcass by phagocytosis. It was originally proposed that caspase proteolysis regulates these events based upon the findings that expression of exogenous active caspases induced apoptosis. Conversely, the active site caspase mutants failed to induce death and in some cases acted in a dominant-negative fashion to block the apoptotic stimulus. Furthermore, peptides that correspond to the substrate cleavage sites disrupt the morphological changes associated with cell death. These observations suggest that caspase substrates are either required for cell survival and must be inhibited by cleavage or cause cell death and must be activated by cleavage. Consistent with this hypothesis some proteins in the cell are cleaved as apoptosis occurs. Thus many groups set about identifying which substrates were cleaved by caspases and whether or not such cleavage had to occur for apoptosis to be manifest. The substrates identified to date are listed in Table 2.

6.1 Substrate discovery

The development of an *in vitro* system was crucial for the identification of the caspases responsible for the cleavage of proteins during apoptosis and for determining the mechanism and purpose of this proteolysis. Cell extracts taken from cells undergoing apoptosis were found to promote condensation of nuclei isolated from healthy cells indicating that late stage apoptotic events were occurring. These extracts also supported the cleavage of purified poly (ADP-ribose) polymerase (PARP), a DNA binding protein involved in base excision repair. PARP is cleaved from a 125 kDa to an 85 kDa protein during apoptosis (142, 143). *In vitro* PARP cleavage was inhibited by the cysteine alkylating reagents, iodoacetamide and *N*-ethylmaleimide suggesting that a cysteine protease was mediating this cleavage event.

A second protease activity was later identified in apoptotic cell lysates, which cleaves the nuclear lamins A and B1 (144). The kinetics of cleavage were compared to PARP cleavage. The lamins took 10 times longer to be cleaved compared to PARP and more importantly, their sensitivity to the peptide inhibitor of ICE/caspase-1 (Z-YVAD-CHO) was dissimilar (145). The differing sensitivity to this inhibitor suggested that distinct caspases were required for the apoptotic events in the cell-free lysates. Interestingly, one consequence of inhibiting the lamin protease was the inhibition of packaging condensed chromatin into apoptotic bodies but not inter-nucleosomal DNA cleavage, nor apoptosis. It thus become clear that apparently different proteases had unique roles to play in the apoptotic signalling cascade, cleaving different substrates and affecting different aspects of apoptosis. The identification and purification of the PARP protease, caspase-3, at that time called CPP-32/YAMA, and the development of its specific inhibitor, Z-DEVD-CHO allowed this hypothesis to evolve. Z-DEVD-CHO was found to be 10^3 times more potent than Z-YVAD-CHO at inhibiting PARP cleavage (36). Purification of the lamin protease, caspase-6, allowed a direct demonstration of the differing substrate specificity (13, 145).

6.2 Proteolysis regulating function

The cellular functions affected by the caspases include the apoptotic pathway, cell cycle and growth regulating pathways, and the maintenance of cell structure. Proteolysis of caspase substrates within these pathways modifies protein function in distinct ways. Substrates can be directly activated, directly inactivated, or can modulate the function of other proteins as a result of cleavage. Activation subsequent to proteolysis occurs in the case of caspases themselves, the actin polymer regulating protein gelsolin (146), and a growing list of kinases involved in diverse cellular growth and stress response signalling cascades. Examples include the p21-activated kinase PAK-2, protein kinase Cδ, the stress pathway kinase MEKK-1, and the p34cdc2 related kinase PITSLRE1 (147–152). In contrast, caspase cleavage can directly inactivate protein function. Examples are the kinases, Akt, Raf-1, DNA protein kinase (DNA-PK) (17, 153, 154), and the sterol regulatory element binding protein (SREBP) (155). Cleavage of structural proteins affects their ability to assemble into structures, and thereby also inactivates them. Caspase proteolysis of proteins that function to regulate other molecules in the cell also occurs. For example the NF-κB regulating protein IκB is cleaved by caspase-3 allowing IκB to remain bound to NF-κB. The caspases have cleverly targeted the proteins within signalling pathways that will assist in their purpose to destroy the cell.

The primary effect of caspase proteolysis is eventual apoptosis. It is not surprising that many of the proteins that function to activate or regulate apoptosis have thus been identified as caspase substrates. As described in the earlier section the activa-

Table 2 Substrates cleaved by the caspases which have been identified in recent years

Caspase substrate	Substrate function	Caspase Substrate Transcription/translation	Substrate function
Apoptosis regulation			
Procaspases (activators)	Activate activator and effector caspases		
Procaspases (effectors)	Cleave procaspases and the proteins listed below	NF-κB	Transcription regulator of inflammation and apoptosis proteins
Bcl-2	Anti-apoptotic regulator		
Bcl-Xl	Anti-apoptotic regulator	IκB	Inhibits nuclear localization/function of NF κB
Bld	Pro-apoptotic interacts with BAX		
Bap31	Pro-apoptotic Bcl-2 interator	Sterol regulatory element binding protein (SREBP)	Cholsterol metabolism
AKT	Signalling receptor regulated surivial kinase		
PKCδ/PKCθ	Pro-apoptotic membrane associated	STAT1	Cytokine signalling
Dap-kinase	Death inducing calcium/calmodulin regulated serine/threonine kinase localized to the cytoskeleton	Sp-1	Trascription factor
		U1-70 kD sRNP	Pre-mRNA splicing
		Heteronuclear ribonuclear proteins(hnRNPs)	Pre-mRNA splicing
Cytoskeletal proteins		**Survival and Growth**	
Lamins	Nuclear matrix structural protein		
β-catenin	Cadherin complex and LEF binding	Raf kinase	Regulates MAPK
	Transcription factor	MEKK-1	Stress kinase
Ademous poloposis cancer (APC)	Regulation of B catenin cytosolic levels	PAK-2	p21 (cell cycle) Activated kinase
Actin	Actin filament subunits	Ras GAP	Ras GTPase activating protein
Gelsolin	Actin filament cleavage	Protein phosphokinase-2A (PP2A)	Signalling kinase
Plakaglobin	Mediates cell adhesion via α-catenin		
Fodrin	Stabilize actin filament ends	Mst-1	SAPK and p38 pathway activator
Keratin-18 and 19	Intermediate filament subunits	PLA2 (cytosolic)	Phospholipid metabolism

Gas2	Microfilament assembly
Focal adhesion kinase (FAK)	Regulates actin cytoskeleton
Vimentin	Intermediate filament protein
Rabaptin-5	endosome fusion
D4-GDI	Rho GDP dissociation inhibitor
Cell cycle regulation	
p21Cip1/Waf1	Cdk2 inhibitor
p27(KIP)	Cdk inhibitor
Wee 1	Cdc2 inhibitor
Cdc 27	Anaphase promoting complex
Cyclin A	Mitosis
Retinoblastoma protein (Rb)	Negative regulator of S phase
PITSLRE kinase	p34cdc2 related kinase
MDM2	Mediates degradation of p53
NuMa	Nucleur matrix protein/associated with spindle poles during mitosis
Topoisomerase I	DNA replication
MCM3	DNA replication
DNA associated proteins	
Poly(ADP-ribose) polymerase (PARP)	DNA repair
Inhibitor of caspase activated DNase(ICAD,DFF)	DNA cleavage
DNA-dependent protein kinase (DNA-PK)	DNA repair
Ataxia telangiectascia (ATM)	DNA damage
PKC-related kinase 2 (PRK2)	Signalling kinase
Calmodulin-dependent kinase II	Calcium channel ion regulator
Calpastatin	Calpain inhibitor
Cytokine processing	
Pro-interleukin-1β	Immune regulation
Pro-interleukin-16	Immune regulation
Pro-interleukin-18 (IFN-γ-inducing factor)	Immune regulation
Group 1 caspases	Inflammation regulators
Disease associated proteins	
Androgen receptor	Involved in Kennedys disease (X-linked spinal and bulbar atrophy)
Huntingtin	Involved in Huntingdon's disease
Atrophin-1	Involved in Dentatorubral-palidoluysian atrophy (DRPLA)
Ataxin-3	Involved in spinocerebellar ataxia (SCA)
Presenilins	Involved in Alzheimer's disease
β-amyloid precusor protein (βAPP)	Involved in Alzheimer's disease

Adapted from Stroh and Osthoff (1999) (224).

tion of caspases in response to pro-apoptotic signals involves initial cleavage and activation of the apical caspases. The localization of the procaspases to the regions of the cell where they assemble into complexes allowing processing relies on the prodomains that interact with adaptor molecules targeting the caspase to the activating complex. The ensuing activation develops into an amplified pro-apoptotic signal, which can travel in several directions; downstream to activate the effector caspases that directly target a range of substrates, or to cleave other molecules involved in parallel pathways of apoptosis regulation/activation. In the case of the death agonist BID, caspase-mediated proteolysis results in its activation. Fas activated caspase-8 cleavage of BID causes BID to translocate from the cytosol, where it resides as a full-length protein, to the mitochondria where it interacts with BAX and causes the loss of membrane potential and release of cytochrome *c* (156–158). Thus death receptor-mediated caspase cleavage of substrates can engage the mitochondria to amplify the pro-apoptotic signal (159).

In addition to the activation of apoptotic molecules required to sustain the death signal, the caspases activate molecules engaged in other signalling pathways to trigger death. An interesting example is the mitogen-activated kinase/ERK kinase, kinase-1 (MEKK-1), which functions in the cell primarily in the stress and growth pathways, and has also been implicated as a survival kinase in embryonic stem cells (160) and in cells that have disrupted microtubules. Cleavage of MEKK-1, by caspase-3, -7, or -8, removes its N-terminal regulatory and membrane-anchoring region (150, 152, 154). Expression of the cleavage product causes apoptosis in a number of cell lines and furthermore the caspase-activated MEKK-1 can induce the activation of the caspases creating a positive feedback amplification loop (150). This finding suggests that the role of MEKK-1 changes to a cell death promoting kinase after cleavage by the caspases. One speculative proposal on this shift in function comes from the observation that the subcellular localization of MEKK-1 changes from membrane to cytosolic following cleavage that is the likely result of removing the membrane anchoring N-terminal fragment (161, 162). It seems that membrane association of the active MEKK-1 in the full-length form prevents the pro-apoptotic function. In keeping, anchoring the active caspase cleavage product to the membrane by adding a CAAX membrane targeting sequence also serves to block its apoptotic function. Further, MEKK-1 seems to be required for the onset of apoptosis as expression of a dominant-negative, kinase dead, version of MEKK-1 blocks apoptosis caused by removal of MDCK cells from their matrix or apoptosis caused by DNA damage following UV irradiation, cisplatin, etoposide, or mitomycin C treatment (150, 163).

The protein kinase C family members PKCδ and PKCτ are also cleaved by the caspases as observed in keratinocytes exposed to ionizing radiation or after Fas, Ara-c etoposide, or cisplatin treatment (164, 165). Normally these kinases become activated after translocating to the cell membrane where they interact with activating polar lipids and associate with specific receptors for the activated kinases (RACKS) (166). Activation of these kinases in this manner causes them to function as signalling molecules for a wide range of cellular functions. Proteolytic cleavage of PKCδ or

PKCτ by caspase-3 yields a 40 kDa catalytic fragment that is present in the soluble fraction of apoptotic cells which, similar to MEKK-1, induces apoptosis (149, 167). Here again it seems transferring the activity of this kinase from the membrane where it normally resides, to the cytosol, causes a shift to an apoptotic function.

To further aid the execution of the apoptotic programme, proteins that function primarily to protect against the onset of apoptosis are negated by cleavage. The Bcl-2 family members block apoptosis primarily by regulating membrane events at the mitochondria such as the release of cytochrome *c*, AIF, and reactive oxygen species into the cytosol. The interaction between the pro-apoptotic and anti-apoptotic members of this protein family determines the likelihood of the above events occurring. Interactions between members of the Bcl-2 family and other proteins are mediated by highly conserved regions. These regions are termed Bcl-2 homology (BH) 1–4, with the BH3 region being central in promoting apoptosis (168). Caspase-3 cleavage of Bcl-2 removes the N-terminal membrane anchor region and leaves the BH3 region intact, allowing the protein to interact with the pro-apoptotic molecule BAX (169). Thus Bcl-2 cleavage by caspase-3 converts it from an anti-apoptotic protein to a pro-apoptotic protein via its modified ability to associate with death agonists.

In most cell types some form of survival signalling, mediated from growth factors, cell–matrix interactions, or cell–cell contact is required to maintain the non-apoptotic state. The kinase Raf-1 mediates growth and survival signalling by activating the ERK/MAP kinase pathway and by interacting with members of the Bcl-2 family . A mechanism for Raf-1-mediated survival signalling was uncovered by observations that Bcl-2 targets Raf-1 to mitochondria, where it then phosphorylates and inactivates BAD (170). This survival function is abrogated when Raf is cleaved and then degraded by proteolysis following apoptotic stimuli. This cleavage is Ac-YVAD-CMK sensitive indicating that a Group I caspase maybe responsible. The significance of the serine threonine kinase Akt as a key survival signalling molecule has come to light recently. Akt blocks apoptosis by phosphorylating and inactivating the pro-apoptotic molecules BAD and caspase-9 (127–129). Akt, similar to Raf is cleaved in an Ac-YVAD-CMK sensitive manor although the caspase responsible for proteolysis and the sites of cleavage have yet to be identified (154). The inactivation of these kinases occurs late on in the programme and allows the apoptotic pathway to progress smoothly.

Caspase activity enhances apoptosis by targeting transcription factors, or their regulating proteins. At present the best characterized examples are the cell cycle and apoptosis regulating protein p53 and the inflammation and apoptosis regulating protein NF-κB (171, 172). Caspase cleavage mediates the inactivation of NF-κB, and thereby blocks transcription of anti-apoptotic genes, by two distinct mechanisms. The first mechanism involves the cleavage of the inhibitor of NF-κB, IκB (173, 174). IκB forms a complex in the cytosol with NF-κB, masking its NLS and thus preventing its translocation to the nucleus (175). Subsequent to stimulation, phosphorylation of IκB at specific serine residues results in its ubiquitination and degradation by the proteosome enabling NF-κB to translocate to the nucleus and promote transcription. Cleavage of IκB at its N-terminus removes these serine residues preventing its degradation and IκB remains bound to NF-κB blocking its translocation to the nucleus (173, 174). The second manner in which NF-κB is inactivated by cleavage involves a direct mechanism. The p65 subunit of NF-κB forms a complex with the cofactor CBP/300 and upon phosphorylation of p65, a conformational change occurs in this complex, stimulating transcription. p65 may be cleaved by caspase-3, -6, or -7 and inactivated as the N-terminal DNA binding domain is separated from the C-terminal transactivation domains. The cleaved form of p65 may act in a dominant-negative fashion to inhibit transcription as it lacks transcriptional activity, yet may compete with full-length active p65 for DNA binding.

The balance between cell growth, progression through the cell cycle, and apoptosis maintains cellular homeostasis. Cell cycle checkpoints at G1/S and G2/M protect against genetic instability by affecting cell cycle arrest or apoptosis in response to DNA damage. The transcription activator p53 and the E2F transcription regulating protein pRB are important primarily in the G1/S checkpoint and in apoptosis (176). Absence of p53 results in cells being predisposed to cancer, while exogenous over-expression of p53 can induce apoptosis. In contrast, pRB null mice undergo massive apoptosis and die *in utero* (177–179) while cells expressing ectopic pRB are resistant to apoptosis induced by ionizing radiation (180). Taken together these findings indicate that pRB and p53 play opposing roles in controlling the onset of apoptosis. It follows that the caspases affect these two proteins differently.

During genotoxic stress (DNA damage) p53 is activated and stabilized. Normally the oncogene MDM2 destabilizes p53 by targeting it to be degraded by the ubiquitin–proteosome pathway. MDM2 is cleaved to a 60 kDa form by the caspases, inactivating it (181, 182). p53 thus persists in the cell and induces apoptosis or growth arrest. The caspase responsible for cleavage of MDM2 has a unique substrate recognition sequence, DVPD, but its identity is not known. Activity of this DVPD cleaving caspase is up-regulated by p53 creating a positive feedback mechanism that leads to p53 stabilization and p53-mediated cell death (183). In contrast to p53, pRB is cleaved by caspases directly after being dephosphorylated during DNA damage induced apoptosis (176). The role of pRB as an essential anti-apoptotic protein is evident given the increased cell death in developing Rb null mice. Proteolysis of Rb by caspases yields different cleavage fragments in response to different apoptotic stimuli. Fas-induced apoptosis results in the cleavage of pRB to a 68 and 48 kDa fragment (184). The 68 kDa fragment maintains the ability to bind to E2F making the reason for cleavage unclear. Tumour necrosis factor (TNFα)- and staurosporine-induced apoptosis however, causes the caspase-mediated removal of the C-terminal 42 amino acids. This cleavage also leaves pRB function intact except for its ability to bind MDM2, possibly affecting its regulation of apoptosis (185). Why RB is differentially cleaved depending on stimulus is unclear and awaits confirmation *in vivo*.

Active p53 promotes the transcription of the WAF1/CIP1 gene product, p21, and the apoptosis-inducing protein BAX. p21 is an inhibitor of cyclin-dependent kinases (CDKs) that are required for the G1/S transition. The ectopic expression of full-length p21 increases G1 arrest and desensitizes cells to apoptotic stimuli. During

DNA damage induced apoptosis or growth factor removal in endothelial cells p21 and a second CDK inhibitor p27 are cleaved by caspase-3 (186–188). The caspase-3 cleaved p21 fails to cause G1 arrest or block apoptosis (189). Caspase cleavage affects the subcellular localization of p21 by removing the N-terminal nuclear localization sequence and preventing its translocation into the nucleus where it normally functions to inhibit CDKs. Consistent with this finding, expression of dominant-negative CDK2 in endothelial cells suppresses apoptosis. These observations suggest that caspase cleavage of the CDK inhibitors plays a role in the sensitization of the cell cycle transitional cell to apoptosis. Catastrophic CDK activation is predicted to contribute to the physiological cell death as is the case for the mitosis-promoting kinase Cdc-2, which causes mitotic catastrophe when its expression is ill-timed (190, 191).

Cell proliferation is also linked to the apoptotic programme through the cytoskeleton. The anchorage requirement for survival and growth is well documented. Attachment of cells to extracellular matrix components such as collagen, fibronectin, and vitronectin, modulate growth receptor signalling and protects the cell against apoptosis (192). Detachment of cells from their extracellular matrix causes apoptosis in many cell types and this phenomenon has been termed 'anoikis', the Greek word for homelessness (193, 194). Matrix-mediated survival signalling pathways rely on the integrity of the cell cytoskeleton, and the function of the kinases that transduce matrix-mediated survival signals; PI3 kinases, the integrin-linked kinase (ILK), Akt, and the focal adhesion kinase (FAK). FAK mediates survival signals from the ECM, apparently downstream of integrin $\alpha V\beta 5$ binding to fibronectin and upstream of PI3 kinase and Akt. The intact 125 kDa FAK protein is cleaved during Fas-induced apoptosis in Jurkat T cells (195, 196) first to a 85 kDa form, by caspase-3 or -7, and then to a 77 kDa fragment by caspase-6. The resulting 77 kDa fragment contains the kinase domain but has decreased kinase activity (195). This loss of activity is presumed to contribute to the disassembly of the focal adhesions by promoting the detachment of the dying cell from the extracellular matrix. Loss of FAK activity also translates into a loss of survival signalling which may be p53 dependent (197).

The caspases orchestrate disassembly of the actin cytoskeleton during apoptosis. In addition to cleaving and inactivating FAK, the actin monomers, the actin-linking proteins fodrin, β-catenin, plakaglobin, and VE cadherin, the microfilament stabilizing protein Gas-2, and the actin severing protein gelsolin are all cleaved by caspases during apoptosis (146, 198–202). Caspase-mediated morphological changes result from the disassembly of the actin microfilaments, the loss of intact adherent junctions and apparently, focal adhesions. As would be expected, the cells round up, detach from their ECM, and eventually begin to bleb as a consequence. Less apparent, but clearly significant consequences of the changes in cell shape, are the loss of cell survival and growth signalling pathways due to desegregation of surface signalling complexes. It is unclear at present whether a detaching cell is already committed to die, if so such survival signalling pathways become surplus. Microtubules are not cleaved by caspases during apoptosis suggesting that they may be required for the dismantling of the cell. Intact microtubules may facilitate the trafficking of apoptosis signalling molecules to the appropriate intracellular domains where they can execute their apoptotic function.

7. Functional diversity

At present 14 distinct mammalian caspases have been identified and at least 70 substrates described. Several caspases cleave the same proteins (as illustrated in Table 2), indicating overlapping specificity. So why are there so many caspases? Substrate overlap suggests functional redundancy that underscores the need to ensure a thorough execution of the cell death programme. However, this suggestion is based largely on analysis of caspase function *in vitro*. *In vivo*, diverse caspases are likely to serve a number of unique functions. This hypothesis is supported by studies with caspase-deficient mice. Studies over the past year implicate both redundant and non-redundant roles for the caspases. Foremost the caspases differentially respond to diverse apoptotic stimuli. One example is the different caspase-mediated cleavage of pRB during different apoptotic stimuli in different cell types. Multiple signalling pathways are activated by distinct stimuli that in turn activate specific caspases. Further, the caspases are activated in a tissue-specific manner as their levels and the presence or absence of their inhibitors and/or activators may differ. Finally, caspases are localized in different subcellular compartments, which determine their target substrates and function.

Combinatorial peptide-based studies and *in vitro* studies using a number of full-length substrates have shown that caspase-3 and -7 have virtually identical substrate specificity. However, the catalytic efficiencies for the hydrolysis of their optimal tetrapeptide substrate differs 10-fold, suggesting subtle regulation. The generation of caspase-3-deficient mice suggests that this enzyme plays a major role in neuronal apoptosis, yet functionally redundant caspases compensate for its absence in other tissues (22, 203). Caspase-3-deficient mice are smaller than littermate controls and only survive until three weeks of age. The pronounced phenotype is in the nervous system and is characterized by massive hyperplasia and ectopic cell masses of the brain in addition to protrusions of neuroepithelium in the retina. Caspase-3 had been perceived as the principle enzyme responsible for cleaving PARP, however proteolysis of PARP is maintained in these mice. Independently, the breast carcinoma cell line MCF-7 has been utilized to study the function of caspase-3 as it lacks a functional copy of this enzyme (204). Although many apoptotic stimuli cause death of these cells, the characteristic nuclear changes associated with apoptosis, such as DNA strand breakage, condensation, and membrane blebbing, are not seen. This discovery suggests that caspase-3 is the primary member of this family required for nuclear morphological events in apoptosis. Indeed, restoration of caspase-3 activity in these cells by ectopic expression of caspase-3 re-establishes the classic nuclear changes one observes during apoptosis. In support of this concept, cleavage of gelsolin, fodrin, and DFF/ICAD is delayed in caspase-3 null cells. Caspase-7 presumably acts to compensate in the absence of caspase-3 on account of their similarity and may account for the lack of perceivable abnormalities in tissues such as the heart, liver, kidney, or lung. A salient observation however is that caspase-7 expression is lacking in brain thus, in its absence a compensatory caspase may not exist for caspase-3 and one might anticipate the phenotype observed in the caspase-3-deficient mice.

Several caspase-deficient mice have been generated that similarly exhibit tissue and cell type-specific or stimulus-dependent defects in apoptosis. Caspase-9-deficient mice resemble caspase-3-deficient mice in that they have defective brain development but also show embryonic lethality (23, 24). These mice were compared in an attempt to identify caspase-3 and -9 dependent pathways. Caspase-9 and -3 dependent pathways were observed in brain and in ES cells that were resistant to UV irradiation. Caspase-9-independent, but caspase-3-dependent apoptosis was observed in activated T cells that were sensitive to CD95 and CD3 stimulation in caspase-9 null cells. This is consistent with previous data that suggested the presence of two pathways downstream of CD95, one utilizing the mitochondria and one bypassing it. Surprisingly caspase-9-dependent but caspase-3-independent apoptosis was observed in thymocytes and splenocytes treated with γ-irradiation. Apoptotic pathways independent of both these caspases were also observed in thymocytes and splenocytes in response to UV irradiation. These studies clearly illustrate the differential pathways elicited by different apoptotic stimuli that occur in a tissue and cell type-specific manner.

Knockout mice of the adaptor molecules FADD and APAF-1 have been generated confirming the pathways regulated by caspase-8 and -9 respectively. Similar to caspase-3 and -9 knockout mice, APAF-1-deficient mice are characterized by profound defects of the central nervous system (205). Caspase-8 and FADD-deficient mice however are critical for embryonic development since both have profound cardiac defects and die *in utero* (48, 206). Although these molecules are recruited to the Fas receptor they exhibit a unique phenotype compared with mice that have inactive Fas or Fas ligand which are characterized by severe immune defects. However caspase-10 is also recruited to this receptor and mutations within caspase-10 that decrease the activity of the enzyme result in immune dysfunction suggesting that this caspase is predominantly involved in negative selection within the immune system downstream of Fas. Taken together, these mice clearly demonstrate the distinct tissue and cellular specific roles that the caspases have evolved to play.

The caspase substrates are located in different intracellular compartments, thus functional diversity determined by subcellular localization seems evident (Fig. 4; see Plate 4). During Fas-induced apoptosis active caspase-3 is found exclusively in the cytosol of mouse liver cells, while the active caspase-7 is located exclusively in the mitochondrial and microsomal fractions (207). Concurrently, the endoplasmic reticulum resident protein SREP is cleaved in a ZVAD inhibitable manner, most likely by caspase-7. The implication was that caspase-7 had a unique function based on its localization to the ER and first suggested distinct functions for caspases with overlapping substrate specificity based on cellular compartmentalization. These observations furthermore suggested that the ER may confer a significant role in apoptosis regulation. The localization of the anti-apoptotic proteins Bcl-2 and Bcl-x_L

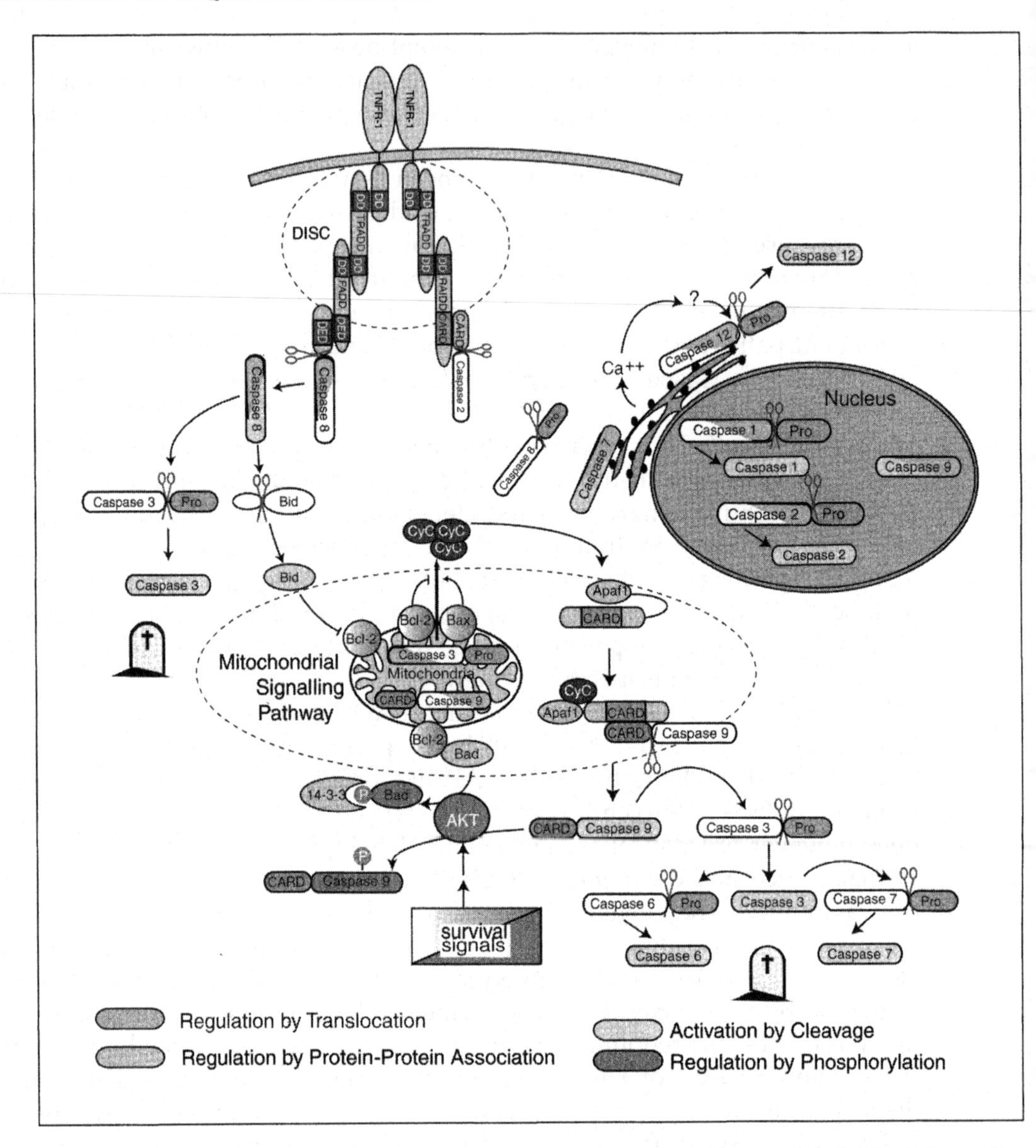

Fig. 4 (See Plate 4) A simplified schematic of pro- and anti-apoptotic signals generated by TNF-R1. Colour coding indicates different types of signalling events: pink denotes the assembly of macromolecular protein complexes; blue denotes translocated proteins; yellow indicates proteins activated by cleavage; red indicates phosphorylation.

to the outer membrane of both the ER and mitochondria suggests cross-talk between these two organelles possibly by calcium signals transmitted from the ER to the mitochondria (208). Supporting this notion, inhibition of the type I inositol 1,4,5-triphosphate receptor calcium release channel in the ER provides resistance to apoptosis (209). It is notable that activity of caspase-7 specifically is dramatically increased by calcium possibly implying its regulation and amplification at the ER (38). In keeping with this hypothesis, recently it was demonstrated that caspase-12 activity is regulated by the Ca^{2+} concentration near the endoplasmic reticulum microenvironment in response to ER stress (210). Procaspase-8 has also been shown to be targeted to the ER via the integral membrane protein p28Bap31 that forms a complex with Bcl-2 or Bcl-x_L (211). Substrates for caspase-8 that reside in the ER have not been identified to date.

Although nuclear substrates were the first to be identified for this family of proteases, the localization of caspases to this compartment has only recently been described. Both the pro-forms of caspases-1 and -2 have been observed to translocate to the nucleus where processing and activation occurs. The prodomain mediates this event by the presence of a nuclear localization signal (NLS) that is released upon processing. The mode of activation within the nucleus is unknown as are the substrates for these particular caspases. Despite this, the localization of caspase-1 and -2 to this subcellular compartment implies a functional role.

Immunohistochemistry and immunogold electron microscopy have localized procaspase-9 to the inner membrane space (IMS) of the mitochondria in several cell types including neuronal cells. However, when neuronal cells are exposed to different apoptosis-inducing drugs or to ischaemic insult the active form of caspase-9 is found largely in the nucleus. This finding suggests that caspase-9 functions within the nucleus, possibly as an effector caspase. How cleavage of procaspase-9 enables its translocation to the nucleus is not known at present. It is possible that targeting signals are present in caspase-9 that are exposed after the pro-form is cleaved. It is also not clear whether procaspase-9 is processed in the mitochondria or whether it moves out of the mitochondria before it is cleaved. The precise role of caspase-9 in the mitochondria has not been elucidated at this time. It may simply be that caspase-9 is sequestered in this compartment such that it is utilized primarily for mitochondrial-mediated death pathways.

8. Caspases in disease

Apoptosis plays a pivotal role in normal tissue turnover and maintenance of homeostasis in the cell, therefore it is now widely believed that deregulation of the apoptotic pathway results in several pathologies. Insufficient apoptosis may contribute to autoimmune disorders and tumour formation while excessive or inappropriate activation of cell death underlies ischaemia, neurodegeneration, and AIDS. Caspases have been implicated in the aetiology of these disorders.

In theory, caspases could be involved in tumour development due to an up-regulation of caspase inhibitors or mutations within the caspases themselves. In addition to contributing to tumour development however alterations of the caspases and their regulators may render tumours resistant to chemotherapeutic drugs that normally mediate tumour regression by the induction of apoptosis. The caspases regulate tumour suppressor function by several mechanisms. In some cases the caspases act downstream of tumour suppressors as mediators in these signal transduction pathways or they may bind tumour suppressors, directly regulating their activity.

p53 inhibits oncogenic transformation via its ability to induce apoptosis or growth arrest. Accordingly, defects in the proteins mediating p53-dependent apoptosis, such as Apaf-1 or caspase-9, have been shown to enable cellular transformation *in vivo* (212). This implies that loss of these effectors can aid tumour formation, however, at present it is unclear whether human tumours exhibit defects in these effector proteins. In addition to acting downstream, the caspases can directly bind certain tumour suppressors. The most common subset of non-Hodkins lymphomas arising in extranodal sites are the B cell lymphomas of mucosa-associated lymphoid tissue (MALT lymphomas) which are most frequently found in the gastrointestinal tract. Many of these lymphomas contain (1;14) (p22;q32) breakpoints that lie upstream of the promoter of the CARD containing gene mE10 also named Bcl10 (213, 214). A frameshift mutation in Bcl10 results in a truncated protein that terminates distal to the CARD domain resulting in gain-of-function transforming activity. The C-terminal end that is absent in these lymphomas normally binds caspase-9 resulting in oligomerization, activation, and subsequent death. Bcl10 may normally suppress tumour formation by activation of caspase-9 while mutations prevent this allowing aberrant cell survival.

Inappropriate activation or excessive apoptosis mediated by the caspases plays a considerable role in the regulation of various neuronal disorders. Alzheimer's disease (AD) is distinguished by the invariant accumulation of senile plaques and the prominent loss of particular neurons in the brain. Senile plaques are comprised principally of amyloid β-peptide (Aβ), the proteolytic cleavage product of β amyloid precursor protein (βAPP) (215). Increased production of Aβ thus leads to an augmentation of plaques. βAPP can be directly cleaved by caspase-3 with a preference for the $VEVD^{720}$ cleavage site (216, 217). Caspase-3 is up-regulated in dying pyramidal neurons of the hippocampus in Alzheimer's patients and the caspase-cleaved form of APP co-localizes in these neurons. In cultured neuronal cells the rate of Aβ formation is increased due to the caspase-mediated proteolysis of APP. An inherited form of AD with an earlier onset is associated with the 'Swedish' mutation of APP that changes the β-secretase site, $VKMD^{653}$ to $VNLD^{653}$ (218). This results in a preferred caspase recognition site. Indeed the mutation increases the rate of cleavage specifically by caspase-6 and results in enhanced Aβ peptide formation (217). The caspases thus play a dual role whereby caspase-3 cleaves APP resulting in enhanced Aβ peptide formation that is cytotoxic and caspase-6 cleavage of APP is increased specifically in patients with the Swedish mutation further exacerbating the progression of the disorder, possibly contributing to early onset of the disease. Caspase-3 cleavage resulting in an augmentation of senile plaques may amplify the condition by aggregating the caspases within plaques resulting in their activation however this remains to be determined. The caspases furthermore contribute to the pathogenesis of AD by cleaving the presenilins (PS), PS-1 and PS-2 (219, 220). Mutations within PS-1 and PS-2 are a frequent cause of familial AD and similar to mutations in βAPP result in increased Aβ production. PS-2 is cleaved by the caspases and resulting in an augmentation of Aβ peptide by activating the cleavage of APP. Particular mutations within PS-2 enhance its pro-apoptotic activity and patients which carry these mutations correspondingly have an increase in the PS-2 caspase fragment in their affected neurons.

The inherited autoimmune lymphoproliferative syndromes (ALPS) are a genetic disorder in which patient's manifest autoimmune disorders and lymphocytosis. Mutations in the Fas receptor and Fas ligand have been identified in patients with type 1a and 1b of the disorder respectively. Mutations in type 2 have now been detected in caspase-10 (221). These caspase-10 mutant enzymes have defective autocatalytic processing and reduced catalytic activity and act as dominant-negatives to inhibit Fas, TNF-RI, and TRAIL apoptosis in cells. Dendritic cell apoptosis mediated via TRAIL was reduced in patients with ALPS and a concordant accumulation of dendritic cells was observed. These data directly implicate the caspases in this disease, although their role in the development of other diseases has yet to be fully understood.

9. Conclusion

It has become clear that a metazoan cell comes equipped with a highly evolved mechanism to dismantle itself given the proper cues. The caspases provide a signalling pathway and effector machinery to eliminate unwanted cells. Once started the signalling mediated by the caspases cascades into a wide range of activities, which affects literally all of the cell functions. There is a very tight logic in the selection of caspase substrates, which in some cases are converted from their primary function to pro-apoptotic function, or which are simply disabled, in their non-apoptotic function. The thoroughness of the effects of caspase activity underscores the importance of completing the apoptotic task. The evolution of the cell has seen to it that this process is completely effective and the caspases are the highly evolved implementers.

It follows that having such a highly potent mechanism for cell removal there should be a highly developed mechanism for ensuring appropriate exercise and control of the system. We have described the events, which lead to irreversible activation of the caspases and the cells commitment to carrying out apoptosis. The sequence of these events have been described in terms of time, but it is becoming apparent that the subcellular location of the caspases can determine their ability to function. This along with the finding that, at least in the case of the caspase-9 being inactivated by Akt phosphorylation suggests that there are layers of regulation which can occur after caspase processing occurs.

One question which is still to be addressed is whether caspase have other cellular functions that do not induce apoptosis. It seems that the overwhelming purpose is to signal and carry out apoptosis and yet the first caspase identified is involved in cytokine signalling, as are all of the Group I caspases. Observations have been made describing the roles of activated caspases in development of the eye and in maintaining cytoskeletal structure in non-apoptosing cells (222, 223). These observations suggest that further examination of the non-apoptotic role of the caspases may be warranted. If, as has been proposed in many forms in the literature, death is not a consequence of life, but is a part of life, then examining the interplay between the different roles of the caspases is likely to underline further the complex interaction between proteins that regulate both cell life and cell death.

References

1. Alnemri, E. S., Livingston, D. J., Nicholson, D. W., Salvesen, G., Thornberry, N. A., Wong, W. W. *et al.* (1996). Human ICE/CED-3 protease nomenclature [Letter]. *Cell*, **87**, 171.
2. Hengartner, M. O., Ellis, R. E., and Horvitz, H. R. (1992). *Caenorhabditis elegans* gene ced-9 protects cells from programmed cell death. *Nature*, **356**, 494–9.
3. Yuan, J. and Horvitz, H. R. (1992). The *Caenorhabditis elegans* cell death gene ced-4 encodes a novel protein and is expressed during the period of extensive programmed cell death. *Development*, **116**, 309–20.
4. Yuan, J., Shaham, S., Ledoux, S., Ellis, H. M., and Horvitz, H. R. (1993). The *C. elegans* cell death gene ced-3 encodes a protein similar to mammalian interleukin-1 beta-converting enzyme. *Cell*, **75**, 641–52.
5. Shaham, S. and Horvitz, H. R. (1996). Developing *Caenorhabditis elegans* neurons may contain both cell-death protective and killer activities. *Genes Dev.*, **10**, 578–91.
6. Yuan, J. Y. and Horvitz, H. R. (1990). The *Caenorhabditis elegans* genes ced-3 and ced-4 act cell autonomously to cause programmed cell death. *Dev. Biol.*, **138**, 33–41.
7. Thornberry, N. A., Bull, H. G., Calaycay, J. R., Chapman, K. T., Howard, A. D., Kostura, M. J., and Miller, D. K. *et al.* (1992). A novel heterodimeric cysteine protease is required for interleukin-1 beta processing in monocytes. *Nature*, **356**, 768–74.
8. Cerretti, D. P., Kozlosky, C. J., Molsey, B., Nelson, N., Van Ness, K., Greenstreet, T. A., *et al.* (1992). Molecular cloning of the interleukin-1 beta converting enzyme. *Science*, **256**, 97–100.
9. Thornberry, N. A. and Molineaux, S. M. (1995). Interleukin-1 beta converting enzyme: a novel cysteine protease required for IL-1 beta production and implicated in programmed cell death. *Protein Sci.*, **4**, 3–12.
10. Miura, M., Zhu, H., Rotello, R., Hartwieg, E. A., and Yuan, J. (1993). Induction of apoptosis in fibroblasts by IL-1 beta-converting enzyme, a mammalian homolog of the *C. elegans* cell death gene ced-3. *Cell*, **75**, 653–60.
11. Fernandes-Alnemri, T., Litwack, G., and Alnemri, E. S. (1994). CPP32, a novel human apoptotic protein with homology to *Caenorhabditis elegans* cell death protein Ced-3 and mammalian interleukin-1 beta-converting enzyme. *J. Biol. Chem.*, **269**, 30761–4.
12. Wang, L., Miura, M., Bergeron, L., Zhu, H., and Yuan, J. (1994). Ich-1, an Ice/ced-3-related gene, encodes both positive and negative regulators of programmed cell death. *Cell*, **78**, 739–50.
13. Orth, K., Chinnaiyan, A. M., Garg, M., Froelich, C. J., and Dixit, V. M. (1996). The CED-3/ICE-like protease Mch2 is activated during apoptosis and cleaves the death substrate lamin A. *J. Biol. Chem.*, **271**, 16443–6.
14. MacFarlane, M., Cain, K., Sun, X. M., Alnemri, E. S., and Cohen, G. M. (1997). Processing/activation of at least four interleukin-1beta converting enzyme-like proteases occurs during the execution phase of apoptosis in human monocytic tumor cells. *J. Cell Biol.*, **137**, 469–79.

15. Duan, H., Chinnaiyan, A. M., Hudson, P. L., Wing, J. P., He, W. W., and Dixit, V. M. (1996). ICE-LAP3, a novel mammalian homologue of the *Caenorhabditis elegans* cell death protein Ced-3 is activated during Fas- and tumor necrosis factor-induced apoptosis. *J. Biol. Chem.*, **271**, 1621–5.
16. Casciola-Rosen, L. A., Miller, D. K., Anhalt, G. J., and Rosen, A. (1994). Specific cleavage of the 70-kDa protein component of the U1 small nuclear ribonucleoprotein is a characteristic biochemical feature of apoptotic cell death. *J. Biol. Chem.*, **269**, 30757–60.
17. Casciola-Rosen, L. A., Anhalt, G. J., and Rosen, A. (1995). DNA-dependent protein kinase is one of a subset of autoantigens specifically cleaved early during apoptosis. *J. Exp. Med.*, **182**, 1625–34.
18. Casciola-Rosen, L., Nicholson, D. W., Chong, T., Rowan, K. R., Thornberry, N. A., Miller, D. K., *et al.* (1996). Apopain/CPP32 cleaves proteins that are essential for cellular repair: a fundamental principle of apoptotic death [See Comments]. *J. Exp. Med.*, **183**, 1957–64.
19. Boudreau, N., Sympson, C. J., Werb, Z., and Bissell, M. J. (1995). Suppression of ICE and apoptosis in mammary epithelial cells by extracellular matrix. *Science*, **267**, 891–3.
20. Armstrong, R. C., Aja, T., Xiang, J., Gaur, S., Krebs, J. F., Hoang, K., Bai, X., *et al.* (1996). Fas-induced activation of the cell death-related protease CPP32 Is inhibited by Bcl-2 and by ICE family protease inhibitors. *J. Biol. Chem.*, **271**, 16850–5.
21. Li, P., Nijhawan, D., Budihardjo, I., Srinivasula, S. M., Ahmad, M., Alnemri, E. S., *et al.* (1997). Cytochrome c and dATP-dependent formation of Apaf-1/caspase-9 complex initiates an apoptotic protease cascade. *Cell*, **91**, 479–89.
22. Kuida, K., Zheng, T. S., Na, S., Kuan, C., Yang, D., Karasuyama, H., *et al.* (1996). Decreased apoptosis in the brain and premature lethality in CPP32- deficient mice. *Nature*, **384**, 368–72.
23. Kuida, K., Haydar, T. F., Kuan, C. Y., Gu, Y., Taya, C., Karasuyama, H., *et al.* (1998). Reduced apoptosis and cytochrome c-mediated caspase activation in mice lacking caspase 9. *Cell*, **94**, 325–37.
24. Hakem, R., Hakem, A., Duncan, G. S., Henderson, J. T., Woo, M., Soengas, M. S., *et al.* (1998). Differential requirement for caspase 9 in apoptotic pathways *in vivo*. *Cell*, **94**, 339–52.
25. Aravind, L., Dixit, V. M., and Koonin, E. V. (1999). The domains of death: evolution of the apoptosis machinery. *Trends Biochem. Sci.*, **24**, 47–53.
26. Darmon, A. J., Nicholson, D. W., and Bleackley, R. C. (1995). Activation of the apoptotic protease CPP32 by cytotoxic T-cell-derived granzyme B. *Nature*, **377**, 446–8.
27. Quan, L. T., Tewari, M., O'Rourke, K., Dixit, V., Snipas, S. J., Poirier, G. G., *et al.* (1996). Proteolytic activation of the cell death protease Yama/CPP32 by granzyme B. *Proc. Natl. Acad. Sci. USA*, **93**, 1972–6.
28. Harris, J. L., Peterson, E. P., Hudig, D., Thornberry, N. A., and Craik, C. S. (1998). Definition and redesign of the extended substrate specificity of granzyme B. *J. Biol. Chem.*, **273**, 27364–73.
29. Krahenbuhl, O. and Tschopp, J. (1990). Involvement of granule proteins in T-cell-mediated cytolysis. *Nature Immun. Cell Growth Regul.*, **9**, 274–82.
30. Rotonda, J., Nicholson, D. W., Fazil, K. M., Gallant, M., Gareau, Y., Labelle, M., *et al.* (1996). The three-dimensional structure of apopain/CPP32, a key mediator of apoptosis. *Nature Struct. Biol.*, **3**, 619–25.
31. Wilson, K. P., Black, J. A., Thomson, J. A., Kim, E. E., Griffith, J. P., Navia, M. A., *et al.* (1994). Structure and mechanism of interleukin-1 beta converting enzyme [See Comments]. *Nature*, **370**, 270–5.

32. Walker, N. P., Talanian, R. V., Brady, K. D., Dang, L. C., Bump, N. J., Ferenz, C. R., *et al.* (1994). Crystal structure of the cysteine protease interleukin-1 beta-converting enzyme: a (p20/p10)2 homodimer. *Cell*, **78**, 343–52.
33. Howard, A. D., Kostura, M. J., Thornberry, N., Ding, G. J., Limjuco, G., Weidner, J., *et al.* (1991). IL-1-converting enzyme requires aspartic acid residues for processing of the IL-1 beta precursor at two distinct sites and does not cleave 31- kDa IL-1 alpha. *J. Immunol.*, **147**, 2964–9.
34. Sleath, P. R., Hendrickson, R. C., Kronheim, S. R., March, C. J., and Black, R. A. (1990). Substrate specificity of the protease that processes human interleukin- 1 beta. *J. Biol. Chem.*, **265**, 14526–8.
35. Thornberry, N. A., Rano, T. A., Peterson, E. P., Rasper, D. M., Timkey, T., Garcia-Calvo, M., *et al.* (1997). A combinatorial approach defines specificities of members of the caspase family and granzyme B. Functional relationships established for key mediators of apoptosis. *J. Biol. Chem.*, **272**, 17907–11.
36. Nicholson, D. W., Ali, A., Thornberry, N. A., Vaillancourt, J. P., Ding, C. K., Gallant, M., *et al.* (1995). Identification and inhibition of the ICE/CED-3 protease necessary for mammalian apoptosis [See Comments]. *Nature*, **376**, 37–43.
37. Talanian, R. V., Quinlan, C., Trautz, S., Hackett, M. C., Mankovich, J. A., Banach, D., *et al.* (1997). Substrate specificities of caspase family proteases. *J. Biol. Chem.*, **272**, 9677–82.
38. Garcia-Calvo, M., Peterson, E. P., Rasper, D. M., Vaillancourt, J. P., Zamboni, R., Nicholson, D. W., *et al.* (1999). Purification and catalytic properties of human caspase family members. *Cell Death Differ.*, **6**, 362–9.
39. Garcia-Calvo, M., Peterson, E. P., Leiting, B., Ruel, R., Nicholson, D. W., and Thornberry, N. A. (1998). Inhibition of human caspases by peptide-based and macromolecular inhibitors. *J. Biol. Chem.*, **273**, 32608–13.
40. Muzio, M., Chinnaiyan, A. M., Kischkel, F. C., O'Rourke, K., Shevchenko, A., Ni, J., *et al.* (1996). FLICE, a nóvel FADD-homologous ICE/CED-3-like protease, is recruited to the CD95 (Fas/APO-1) death-inducing signaling complex. *Cell*, **85**, 817–27.
41. Hofmann, K., Bucher, P., and Tschopp, J. (1997). The CARD domain: a new apoptotic signalling motif. *Trends Biochem. Sci.*, **22**, 155–6.
42. Eberstadt, M., Huang, B., Chen, Z., Meadows, R. P., Ng, S. C., Zheng, L., *et al.* (1998). NMR structure and mutagenesis of the FADD (Mort1) death-effector domain. *Nature*, **392**, 941–5.
43. Chou, J. J., Matsuo, H., Duan, H., and Wagner, G. (1998). Solution structure of the RAIDD CARD and model for CARD/CARD interaction in caspase-2 and caspase-9 recruitment. *Cell*, **94**, 171–80.
44. Boldin, M. P., Varfolomeev, E. E., Pancer, Z., Mett, I. L., Camonis, J. H., and Wallach, D. (1995). A novel protein that interacts with the death domain of Fas/APO1 contains a sequence motif related to the death domain. *J. Biol. Chem.*, **270**, 7795–8.
45. Chinnaiyan, A. M., O'Rourke, K., Tewari, M., and Dixit, V. M. (1995). FADD, a novel death domain-containing protein, interacts with the death domain of Fas and initiates apoptosis. *Cell*, **81**, 505–12.
46. Chaudhary, P. M., Eby, M., Jasmin, A., Bookwalter, A., Murray, J., and Hood, L. (1997). Death receptor 5, a new member of the TNFR family, and DR4 induce FADD- dependent apoptosis and activate the NF-kappaB pathway. *Immunity*, **7**, 821–30.
47. Chinnaiyan, A. M., Tepper, C. G., Seldin, M. F., O'Rourke, K., Kischkel, F. C., Hellbardt, S., *et al.* (1996). FADD/MORT1 is a common mediator of CD95 (Fas/ APO-1) and tumor necrosis factor receptor-induced apoptosis. *J. Biol. Chem.*, **271**, 4961–5.

48. Yeh, W. C., Pompa, J. L., McCurrach, M. E., Shu, H. B., Elia, A. J., Shahinian, A., *et al.* (1998). FADD: essential for embryo development and signaling from some, but not all, inducers of apoptosis. *Science*, **279**, 1954–8.
49. Ashkenazi, A. and Dixit, V. M. (1998). Death receptors: signaling and modulation. *Science*, **281**, 1305–8.
50. Tartaglia, L. A., Ayres, T. M., Wong, G. H., and Goeddel, D. V. (1993). A novel domain within the 55 kd TNF receptor signals cell death. *Cell*, **74**, 845–53.
51. Chen, P., Rodriguez, A., Erskine, R., Thach, T., and Abrams, J. M. (1998). Dredd, a novel effector of the apoptosis activators reaper, grim, and hid in *Drosophila. Dev. Biol.*, **201**, 202–16.
52. Duan, H. and Dixit, V. M. (1997). RAIDD is a new 'death' adaptor molecule. *Nature*, **385**, 86–9.
53. Zou, H., Henzel, W. J., Liu, X., Lutschg, A., and Wang, X. (1997). Apaf-1, a human protein homologous to *C. elegans* CED-4, participates in cytochrome c-dependent activation of caspase-3 [See Comments]. *Cell*, **90**, 405–13.
54. Seshagiri, S., Chang, W. T., and Miller, L. K. (1998). Mutational analysis of *Caenorhabditis elegans* CED-4. *FEBS Lett.*, **428**, 71–4.
55. Dorstyn, L., Colussi, P. A., Quinn, L. M., Richardson, H., and Kumar, S. (1999). DRONC, an ecdysone-inducible *Drosophila* caspase. *Proc. Natl. Acad. Sci. USA*, **96**, 4307–12.
56. Rodriguez, A., Oliver, H., Zai, H., Chen, P., Wang, X., and Abrams, J. M. (1999). Dark is a *Drosophila* homologue of Apaf-1/CED-4 and functions in an evolutionarily conserved death pathway. *Nature Cell Biol.*, **1**, 30798–83.
57. Kroemer, G. (1998). The mitochondrion as an integrator/coordinator of cell death pathways. *Cell Death Differ.*, **5**, 547.
58. Kroemer, G., Zamzami, N., and Susin, S. A. (1997). Mitochondrial control of apoptosis. *Immunol. Today*, **18**, 44–51.
59. Fernandes-Alnemri, T., *et al.* (1996). *In vitro* activation of CPP32 and Mch3 by Mch4, a novel human apoptotic cysteine protease containing two FADD-like domains. *Proc. Natl. Acad. Sci. USA*, **93**, 7464–9.
60. Srinivasula, S. M., Fernandes-Alnemri, T., Zangrilli, J., Robertson, N., Armstrong, R. C., Wang, L., *et al.* (1996). The Ced-3/interleukin 1beta converting enzyme-like homolog Mch6 and the lamin-cleaving enzyme Mch2alpha are substrates for the apoptotic mediator CPP32. *J. Biol. Chem.*, **271**, 27099–106.
61. Srinivasula, S. M., Ahmad, M., Fernandes-Alnemri, T., Litwack, G., and Alnemri, E. S. (1996). Molecular ordering of the Fas-apoptotic pathway: the Fas/APO-1 protease Mch5 is a CrmA-inhibitable protease that activates multiple Ced-3/ICE- like cysteine proteases. *Proc. Natl. Acad. Sci. USA*, **93**, 14486–91.
62. Slee, E. A., Harte, M. T., Kluck, R. M., Wolf, B. B., Casiano, C. A., Newmeyer, D. D., *et al.* (1999). Ordering the cytochrome c-initiated caspase cascade: hierarchical activation of caspases-2, -3, -6, -7, -8, and -10 in a caspase-9-dependent manner. *J. Cell Biol.*, **144**, 281–92.
63. Srinivasula, S. M., Ahmad, M., Fernandes-Alnemri, T., and Alnemri, E. S. (1998). Autoactivation of procaspase-9 by Apaf-1-mediated oligomerization. *Mol. Cell*, **1**, 949–57.
64. Pan, G., Humke, E. W., and Dixit, V. M. (1998). Activation of caspases triggered by cytochrome c *in vitro* [Published erratum appears in *FEBS Lett.* (1998) May 29, 428(3), 309]. *FEBS Lett.*, **426**, 151–4.
65. Hirata, H., Takahashi, A., Kobayashi, S., Yonehara, S., Sawai, H., Okazaki, T., *et al.* (1998).

Caspases are activated in a branched protease cascade and control distinct downstream processes in Fas-induced apoptosis. *J. Exp. Med.*, **187**, 587–600.

66. Takahashi, A., Alnemri, E. S., Lazebnik, Y. A., Fernandes-Alnemri, T., Litwack, G., Moir, R. D., *et al.* (1996). Cleavage of lamin A by Mch2 alpha but not CPP32: multiple interleukin 1 beta-converting enzyme-related proteases with distinct substrate recognition properties are active in apoptosis. *Proc. Natl. Acad. Sci. USA*, **93**, 8395–400.
67. Wang, S., Miura, M., Jung, Y. K., Zhu, H., Li, E., and Yuan, J. (1998). Murine caspase-11, an ICE-interacting protease, is essential for the activation of ICE. *Cell*, **92**, 501–9.
68. Andrade, F., Roy, S., Nicholson, D., Thornberry, N., Rosen, A., and Casciola-Rosen, L. (1998). Granzyme B directly and efficiently cleaves several downstream caspase substrates: implications for CTL-induced apoptosis. *Immunity*, **8**, 451–60.
69. Schotte, P., Van Criekinge, W., Van de Craen, M., Van Loo, G., Desmedt, M., Grooten, J., *et al.* (1998). Cathepsin B-mediated activation of the proinflammatory caspase-11. *Biochem. Biophys. Res. Commun.*, **251**, 379–87.
70. Boldin, M. P., Goncharov, T. M., Goltsev, Y. V., and Wallach, D. (1996). Involvement of MACH, a novel MORT1/FADD-interacting protease, in Fas/APO-1- and TNF receptor-induced cell death. *Cell*, **85**, 803–15.
71. Medema, J. P., Scaffidi, C., Kischkel, F. C., Shevchenko, A., Mann, M., Krammer, P. H., *et al.* (1997). FLICE is activated by association with the CD95 death-inducing signaling complex (DISC). *EMBO J.*, **16**, 2794–804.
72. Muzio, M., Stockwell, B. R., Stennicke, H. R., Salvesen, G. S., and Dixit, V. M. (1998). An induced proximity model for caspase-8 activation. *J. Biol. Chem.*, **273**, 2926–30.
73. Yang, X., Chang, H. Y., and Baltimore, D. (1998). Autoproteolytic activation of procaspases by oligomerization. *Mol. Cell*, **1**, 319–25.
74. Colussi, P. A., Harvey, N. L., Shearwin-Whyatt, L. M., and Kumar, S. (1998). Conversion of procaspase-3 to an autoactivating caspase by fusion to the caspase-2 prodomain. *J. Biol. Chem.*, **273**, 26566–70.
75. The Huntington's Disease Collaborative Research Group. (1993). A novel gene containing a trinucleotide repeat that is expanded and unstable on Huntington's disease chromosomes [See Comments]. *Cell*, **72**, 971–83.
76. La Spada, A. R., Wilson, E. M., Lubahn, D. B., Harding, A. E., and Fischbeck, K. H. (1991). Androgen receptor gene mutations in X-linked spinal and bulbar muscular atrophy. *Nature*, **352**, 77–9.
77. Koide, R., Ikeuchi, T., Onodera, O., Tanaka, H., Igarashi, S., Endo, K., Takahashi, H., *et al.* (1994). Unstable expansion of CAG repeat in hereditary dentatorubral- pallidoluysian atrophy (DRPLA). *Nature Genet.*, **6**, 9–13.
78. Imbert, G., Saudou, F., Yvert, G., Devys, D., Trottier, Y., Garnier, J. M., *et al.* (1996). Cloning of the gene for spinocerebellar ataxia 2 reveals a locus with high sensitivity to expanded CAG/glutamine repeats [See Comments]. *Nature Genet.*, **14**, 285–91.
79. Banfi, S., Servadio, A., Chung, M. Y., Kwiatkowski, T. J., Jr., McCall, A. E., Duvick, L. A., *et al.* (1994). Identification and characterization of the gene causing type 1 spinocerebellar ataxia. *Nature Genet.*, **7**, 513–20.
80. Zhuchenko, O., Bailey, J., Bonnen, P., Ashizawa, T., Stockton, D. W., Amos, C., *et al.* (1997). Autosomal dominant cerebellar ataxia (SCA6) associated with small polyglutamine expansions in the alpha 1A-voltage-dependent calcium channel. *Nature Genet.*, **15**, 62–9.
81. Goldberg, Y. P., Nicholson, D. W., Rasper, D. M., Kalchman, M. A., Koide, H. B., Graham,

R. K., *et al.* (1996). Cleavage of huntingtin by apopain, a proapoptotic cysteine protease, is modulated by the polyglutamine tract [See Comments]. *Nature Genet.*, **13**, 442–9.

82. Martindale, D., Hackam, A., Wieczorek, A., Ellerby, L., Wellington, C., McCutcheon, K., *et al.* (1998). Length of huntingtin and its polyglutamine tract influences localization and frequency of intracellular aggregates. *Nature Genet.*, **18**, 150–4.
83. Sanchez, I., Xu, C. J., Juo, P., Kakizaka, A., Blenis, J., and Yuan, J. (1999). Caspase-8 is required for cell death induced by expanded polyglutamine repeats [See Comments]. *Neuron*, **22**, 623–33.
84. Zou, H., Li, Y., Liu, X., and Wang, X. (1999). An APAF-1.cytochrome c multimeric complex is a functional apoptosome that activates procaspase-9. *J. Biol. Chem.*, **274**, 11549–56.
85. Saleh, A., Srinivasula, S. M., Acharya, S., Fishel, R., and Alnemri, E. S. (1999). Cytochrome c and dATP-mediated oligomerization of apaf-1 is a prerequisite for procaspase-9 activation [In Process Citation]. *J. Biol. Chem.*, **274**, 17941–5.
86. Scaffidi, C., Fulda, S., Srinivasan, A., Friesen, C., Li, F., Tomaselli, K. J., *et al.* (1998). Two CD95 (APO-1/Fas) signaling pathways. *EMBO J.*, **17**, 1675–87.
87. Verhagen, A., Ekert, P. G., Pakush, M., Silke, J., Connolly, L. M., Reid, G. E., *et al.* (2000). Identification of DIABLO, a mammalian protein that promotes apoptosis by binding to and antagonizing IAP proteins. *Cell*, **102**, 43–53.
88. Du, C., Fang, M., Li, L., and Wang, W. (2000). Smac, a mitochondrial protein that promotes cytochrome c-dependent caspase activation by eliminating IAP inhibition. *Cell*, **102**, 33–42.
89. Yang, X., Chang, H. Y., and Baltimore, D. (1998). Essential role of CED-4 oligomerization in CED-3 activation and apoptosis [See Comments]. *Science*, **281**, 1355–7.
90. Inohara, N., Koseki, T., del Peso, L., Hu, Y., Yee, C., Chen, S., *et al.* (1999). Nod1, an Apaf-1-like activator of caspase-9 and nuclear factor-kappaB. *J. Biol. Chem.*, **274**, 14560–7.
91. Bertin, J., Nir, W. J., Fischer, C. M., Tayber, O. V., Errada, P. R., Grant, J. R., *et al.* (1999). Human CARD4 protein is a novel CED-4/Apaf-1 cell death family member that activates NF-kappaB. *J. Biol. Chem.*, **274**, 12955–8.
92. Yan, M., Lee, J., Schilbach, S., Goddard, A., and Dixit, V. (1999). mE10, a novel caspase recruitment domain-containing proapoptotic molecule. *J. Biol. Chem.*, **274**, 10287–92.
93. Xanthoudakis, S., Roy, S., Rasper, D., Hennessey, T., Aubin, Y., Cassady, R., *et al.* (1999). Hsp60 accelerates the maturation of pro-caspase-3 by upstream activator proteases during apoptosis. *EMBO J.*, **18**, 2049–56.
94. Samali, A., Cai, J., Zhivotovsky, B., Jones, D. P., and Orrenius, S. (1999). Presence of a pre-apoptotic complex of pro-caspase-3, Hsp60 and Hsp10 in the mitochondrial fraction of jurkat cells. *EMBO J.*, **18**, 2040–8.
95. Buckley, C. D., Pilling, D., Henriquez, N. V., Parsonage, G., Threlfall, K., Scheel-Toellner, D., *et al.* (1999). RGD peptides induce apoptosis by direct caspase-3 activation [See Comments]. *Nature*, **397**, 534–9.
96. Ray, C. A., Black, R. A., Kronheim, S. R., Greenstreet, T. A., Sleath, P. R., Salvesen, G. S., *et al.* (1992). Viral inhibition of inflammation: cowpox virus encodes an inhibitor of the interleukin-1 beta converting enzyme. *Cell*, **69**, 597–604.
97. Zhou, Q., Snipas, S., Orth, K., Muzio, M., Dixit, V. M., and Salvesen, G. S. (1997). Target protease specificity of the viral serpin CrmA. Analysis of five caspases. *J. Biol. Chem.*, **272**, 7797–800.
98. Quan, L. T., Caputo, A., Bleackley, R. C., Pickup, D. J., and Salvesen, G. S. (1995). Granzyme

B is inhibited by the cowpox virus serpin cytokine response modifier A. *J. Biol. Chem.*, **270**, 10377–9.

99. Bertin, J., Mendrysa, S. M., LaCount, D. J., Gaur, S., Krebs, J. F., Armstrong, R. C., *et al.* (1996). Apoptotic suppression by baculovirus P35 involves cleavage by and inhibition of a virus-induced CED-3/ICE-like protease. *J. Virol.*, **70**, 6251–9.
100. Bump, N. J., Hackett, M., Hugunin, M., Seshagiri, S., Brady, K., Chen, P., *et al.* (1995). Inhibition of ICE family proteases by baculovirus antiapoptotic protein p35. *Science*, **269**, 1885–8.
101. Xue, D. and Horvitz, H. R. (1995). Inhibition of the *Caenorhabditis elegans* cell-death protease CED-3 by a CED-3 cleavage site in baculovirus p35 protein. *Nature*, **377**, 248–51.
102. Clem, R. J., Fechheimer, M., and Miller, L. K. (1991). Prevention of apoptosis by a baculovirus gene during infection of insect cells. *Science*, **254**, 1388–90.
103. Fisher, A. J., Cruz, W., Zoog, S. J., Schneider, C. L., and Friesen, P. D. (1999). Crystal structure of baculovirus P35: role of a novel reactive site loop in apoptotic caspase inhibition. *EMBO J.*, **18**, 2031–9.
104. Birnbaum, M. J., Clem, R. J., and Miller, L. K. (1994). An apoptosis-inhibiting gene from a nuclear polyhedrosis virus encoding a polypeptide with Cys/His sequence motifs. *J. Virol.*, **68**, 2521–8.
105. Crook, N. E., Clem, R. J., and Miller, L. K. (1993). An apoptosis-inhibiting baculovirus gene with a zinc finger-like motif. *J. Virol.*, **67**, 2168–74.
106. Roy, N., Mahadevan, M. S., McLean, M., Shutler, G., Yaraghi, Z., Farahani, R., *et al.* (1995). The gene for neuronal apoptosis inhibitory protein is partially deleted in individuals with spinal muscular atrophy [See Comments]. *Cell*, **80**, 167–78.
107. Liston, P., Roy, N., Tamai, K., Lefebvre, C., Baird, S., Cherton-Horvat, G., *et al.* (1996). Suppression of apoptosis in mammalian cells by NAIP and a related family of IAP genes. *Nature*, **379**, 349–53.
108. Duckett, C. S., Nava, V. E., Gedrich, R. W., Clem, R. J., Van Dongen, J. L., Gilfillan, M. C., *et al.* (1996). A conserved family of cellular genes related to the baculovirus iap gene and encoding apoptosis inhibitors. *EMBO J.*, **15**, 2685–94.
109. Rothe, M., Pan, M. G., Henzel, W. J., Ayres, T. M., and Goeddel, D. V. (1995). The TNFR2-TRAF signaling complex contains two novel proteins related to baculoviral inhibitor of apoptosis proteins. *Cell*, **83**, 1243–52.
110. Uren, A. G., Pakusch, M., Hawkins, C. J., Puls, K. L., and Vaux, D. L. (1996). Cloning and expression of apoptosis inhibitory protein homologs that function to inhibit apoptosis and/or bind tumor necrosis factor receptor-associated factors. *Proc. Natl. Acad. Sci. USA*, **93**, 4974–8.
111. Lu, C. D., Altieri, D. C., and Tanigawa, N. (1998). Expression of a novel antiapoptosis gene, survivin, correlated with tumor cell apoptosis and p53 accumulation in gastric carcinomas. *Cancer Res.*, **58**, 1808–12.
112. Roy, N., Deveraux, Q. L., Takahashi, R., Salvesen, G. S., and Reed, J. C. (1997). The c-IAP-1 and c-IAP-2 proteins are direct inhibitors of specific caspases. *EMBO J.*, **16**, 6914–25.
113. Deveraux, Q. L., Takahashi, R., Salvesen, G. S., and Reed, J. C. (1997). X-linked IAP is a direct inhibitor of cell-death proteases. *Nature*, **388**, 300–4.
114. Deveraux, Q. L., Roy, N., Stennicke, H. R., Van Arsdale, T., Zhou, Q., Srinivasula, S. M., *et al.* (1998). IAPs block apoptotic events induced by caspase-8 and cytochrome c by direct inhibition of distinct caspases. *EMBO J.*, **17**, 2215–23.

115. Bertin, J., Armstrong, R. C., Ottilie, S., Martin, D. A., Wang, Y., Banks, S., *et al.* (1997). Death effector domain-containing herpesvirus and poxvirus proteins inhibit both Fas- and TNFR1-induced apoptosis. *Proc. Natl. Acad. Sci. USA*, **94**, 1172–6.
116. Hu, S., Vincenz, C., Buller, M., and Dixit, V. M. (1997). A novel family of viral death effector domain-containing molecules that inhibit both CD-95- and tumor necrosis factor receptor-1-induced apoptosis. *J. Biol. Chem.*, **272**, 9621–4.
117. Thome, M., Schneider, P., Hofmann, K., Fickenscher, H., Meinl, E., Neipel, F., *et al.* (1997). Viral FLICE-inhibitory proteins (FLIPs) prevent apoptosis induced by death receptors. *Nature*, **386**, 517–21.
118. Srinivasula, S. M., Ahmad, M., Ottilie, S., Bullrich, F., Banks, S., Wang, Y., *et al.* (1997). FLAME-1, a novel FADD-like anti-apoptotic molecule that regulates Fas/TNFR1-induced apoptosis. *J. Biol. Chem.*, **272**, 18542–5.
119. Rasper, D. M., Vaillancourt, J. P., Hadano, S., Houtzager, V. M., Seiden, I., Keen, S. L., *et al.* (1998). Cell death attenuation by 'Usurpin', a mammalian DED-caspase homologue that precludes caspase-8 recruitment and activation by the CD-95 (Fas, APO-1) receptor complex. *Cell Death Differ.*, **5**, 271–88.
120. Inohara, N., Koseki, T., Hu, Y., Chen, S., and Nunez, G. (1997). CLARP, a death effector domain-containing protein interacts with caspase-8 and regulates apoptosis. *Proc. Natl. Acad. Sci. USA*, **94**, 10717–22.
121. Shu, H. B., Halpin, D. R., and Goeddel, D. V. (1997). Casper is a FADD- and caspase-related inducer of apoptosis. *Immunity*, **6**, 751–63.
122. Han, D. K., Chaudhary, P. M., Wright, M. E., Friedman, C., Trask, B. J., Riedel, R. T., *et al.* (1997). MRIT, a novel death-effector domain-containing protein, interacts with caspases and BclXL and initiates cell death. *Proc. Natl. Acad. Sci. USA*, **94**, 11333–8.
123. Irmler, M., Thome, M., Hahne, M., Schneider, P., Hofmann, K., Steiner, V., *et al.* (1997). Inhibition of death receptor signals by cellular FLIP. *Nature*, **388**, 190–5.
124. Sata, M. and Walsh, K. (1998). Endothelial cell apoptosis induced by oxidized LDL is associated with the down-regulation of the cellular caspase inhibitor FLIP. *J. Biol. Chem.*, **273**, 33103–6.
125. Koseki, T., Inohara, N., Chen, S., and Nunez, G. (1998). ARC, an inhibitor of apoptosis expressed in skeletal muscle and heart that interacts selectively with caspases. *Proc. Natl. Acad. Sci. USA*, **95**, 5156–60.
126. Alessi, D. R. and Cohen, P. (1998). Mechanism of activation and function of protein kinase B. *Curr. Opin. Genet. Dev.*, **8**, 55–62.
127. del Peso, L., Gonzalez-Garcia, M., Page, C., Herrera, R., and Nunez, G. (1997). Interleukin-3-induced phosphorylation of BAD through the protein kinase Akt. *Science*, **278**, 687–9.
128. Datta, S. R., Dudek, H., Tao, X., Masters, S., Fu, H., Gotoh, Y., *et al.* (1997). Akt phosphorylation of BAD couples survival signals to the cell-intrinsic death machinery. *Cell*, **91**, 231–41.
129. Cardone, M. H., Roy, N., Stennicke, H. R., Salvesen, G. S., Franke, T. F., Stanbridge, E., *et al.* (1998). Regulation of cell death protease caspase-9 by phosphorylation. *Science*, **282**, 1318–21.
130. Hermann, C., Assmus, B., Urbich, C., Zeiher, A. M., and Dimmeler, S. (2000). Insulin-mediated stimulation of protein kinase Akt: A potent survival signaling cascade for endothelial cells. *Arterioscler. Thromb. Vasc. Biol.*, **20**, 402–9.
131. Mannick, J. B., Hausladen, A., Liu, L., Hess, D. T., Zeng, M., Miao, Q. X., *et al.* (1999). Fas-induced caspase denitrosylation. *Science*, **284**, 651–4.

132. Dimmeler, S., Haendeler, J., Nehls, M., and Zeiher, A. M. (1997). Suppression of apoptosis by nitric oxide via inhibition of interleukin-1beta-converting enzyme (ICE)-like and cysteine protease protein (CPP)- 32-like proteases. *J. Exp. Med.*, **185**, 601–7.
133. Rossig, L., Fichtlscherer, B., Breitschopf, K., Haendeler, J., Zeiher, A. M., Mulsch, A., *et al.* (1999). Nitric oxide inhibits caspase-3 by S-nitrosylation *in vivo*. *J. Biol. Chem.*, **274**, 6823–6.
134. Chinnaiyan, A. M., O'Rourke, K., Lane, B. R., and Dixit, V. M. (1997). Interaction of CED-4 with CED-3 and CED-9: a molecular framework for cell death [See Comments]. *Science*, **275**, 1122–6.
135. Wu, D., Wallen, H. D., Inohara, N., and Nunez, G. (1997). Interaction and regulation of the *Caenorhabditis elegans* death protease CED-3 by CED-4 and CED-9. *J. Biol. Chem.*, **272**, 21449–54.
136. Spector, M. S., Desnoyers, S., Hoeppner, D. J., and Hengartner, M. O. (1997). Interaction between the *C. elegans* cell-death regulators CED-9 and CED- 4. *Nature*, **385**, 653–6.
137. Kluck, R. M., Bossy-Wetzel, E., Green, D. R., and Newmeyer, D. D. (1997). The release of cytochrome c from mitochondria: a primary site for Bcl-2 regulation of apoptosis [See Comments]. *Science*, **275**, 1132–6.
138. Yang, J., Liu, X., Bhalla, K., Kim, C. N., Ibrado, A. M., Cai, J., *et al.* (1997). Prevention of apoptosis by Bcl-2: release of cytochrome c from mitochondria blocked [See Comments]. *Science*, **275**, 1129–32.
139. Hu, Y., Benedict, M. A., Wu, D., Inohara, N., and Nunez, G. (1998). Bcl-XL interacts with Apaf-1 and inhibits Apaf-1-dependent caspase-9 activation. *Proc. Natl. Acad. Sci. USA*, **95**, 4386–91.
140. Song, Q., Kuang, Y., Dixit, V. M., and Vincenz, C. (1999). Boo, a novel negative regulator of cell death, interacts with Apaf-1. *EMBO J.*, **18**, 167–78.
141. Krebs, J. F., Armstrong, R. C., Srinivasan, A., Aja, T., Wong, A. M., Aboy, A., *et al.* (1999). Activation of membrane-associated procaspase-3 is regulated by Bcl-2. *J. Cell Biol.*, **144**, 915–26.
142. Lazebnik, Y. A., Kaufmann, S. H., Desnoyers, S., Poirier, G. G., and Earnshaw, W. C. (1994). Cleavage of poly(ADP-ribose) polymerase by a proteinase with properties like ICE. *Nature*, **371**, 346–7.
143. Tewari, M., Quan, L. T., O'Rourke, K., Desnoyers, S., Zeng, Z., Beidler, D. R., *et al.* (1995). Yama / CPP32 beta, a mammalian homolog of CED-3, is a CrmA-inhibitable protease that cleaves the death substrate poly(ADP-ribose) polymerase. *Cell*, **81**, 801–9.
144. Lazebnik, Y. A., Takahashi, A., Poirier, G. G., Kaufmann, S. H., and Earnshaw, W. C. (1995). Characterization of the execution phase of apoptosis *in vitro* using extracts from condemned-phase cells. *J. Cell Sci. Suppl.*, **19**, 41–9.
145. Lazebnik, Y. A., Takahashi, A., Moir, R. D., Goldman, R. D., Poirier, G. G., Kaufmann, S. H., *et al.* (1995). Studies of the lamin proteinase reveal multiple parallel biochemical pathways during apoptotic execution. *Proc. Natl. Acad. Sci. USA*, **92**, 9042–6.
146. Kothakota, S., Azuma, T., Reinhard, C., Klippel, A., Tang, J., Chu, K., *et al.* (1997). Caspase-3-generated fragment of gelsolin: effector of morphological change in apoptosis. *Science*, **278**, 294–8.
147. Bokoch, G. M. (1998). Caspase-mediated activation of PAK2 during apoptosis: proteolytic kinase activation as a general mechanism of apoptotic signal transduction? *Cell Death Differ.*, **5**, 637–45.
148. Rudel, T. and Bokoch, G. M. (1997). Membrane and morphological changes in apoptotic cells regulated by caspase-mediated activation of PAK2. *Science*, **276**, 1571–4.

149. Denning, M. F., Wang, Y., Nickoloff, B. J., and Wrone-Smith, T. (1998). Protein kinase Cdelta is activated by caspase-dependent proteolysis during ultraviolet radiation-induced apoptosis of human keratinocytes. *J. Biol. Chem.*, **273**, 29995–30002.
150. Cardone, M. H., Salvesen, G. S., Widmann, C., Johnson, G., and Frisch, S. M. (1997). The regulation of anoikis: MEKK-1 activation requires cleavage by caspases. *Cell*, **90**, 315–23.
151. Beyaert, R., Kidd, V. J., Cornelis, S., Van de Craen, M., Denecker, G., Lahti, J. M., *et al.* (1997). Cleavage of PITSLRE kinases by ICE/CASP-1 and CPP32/ CASP-3 during apoptosis induced by tumor necrosis factor. *J. Biol. Chem.*, **272**, 11694–7.
152. Widmann, C., Gerwins, P., Johnson, N. L., Jarpe, M. B., and Johnson, G. L. (1998). MEK kinase 1, a substrate for DEVD-directed caspases, is involved in genotoxin-induced apoptosis. *Mol. Cell. Biol.*, **18**, 2416–29.
153. Wang, X., Pai, J. T., Wiedenfeld, E. A., Medina, J. C., Slaughter, C. A., Goldstein, J. L., *et al.* (1995). Purification of an interleukin-1 beta converting enzyme-related cysteine protease that cleaves sterol regulatory element-binding proteins between the leucine zipper and transmembrane domains. *J. Biol. Chem.*, **270**, 18044–50.
154. Widmann, C., Gibson, S., and Johnson, G. L. (1998). Caspase-dependent cleavage of signaling proteins during apoptosis. A turn-off mechanism for anti-apoptotic signals. *J. Biol. Chem.*, **273**, 7141–7.
155. Wang, X., Zelenski, N. G., Yang, J., Sakai, J., Brown, M. S., and Goldstein, J. L. (1996). Cleavage of sterol regulatory element binding proteins (SREBPs) by CPP32 during apoptosis. *EMBO J.*, **15**, 1012–20.
156. Desagher, S., Osen-sand, A., Nichols, A., Eskes, R., Montessuit, S., Lauper, S., *et al.* (1999). Bid-induced conformational change of Bax is responsible for mitochondrial cytochrome c release during apoptosis. *J. Cell Biol.*, **144**, 891–901.
157. Li, H., Zhu, H., Xu, C. J., and Yuan, J. (1998). Cleavage of BID by caspase 8 mediates the mitochondrial damage in the Fas pathway of apoptosis. *Cell*, **94**, 491–501.
158. Gross, A., Yin, X. M., Wang, K., Wei, M. C., Jockel, J., Milliman, C., *et al.* (1999). Caspase cleaved BID targets mitochondria and is required for cytochrome c release, while BCL-XL prevents this release but not tumor necrosis factor-R1/Fas death. *J. Biol. Chem.*, **274**, 1156–63.
159. Bossy-Wetzel, E. and Green, D. R. (1999). Caspases induce cytochrome c release from mitochondria by activating cytosolic factors. *J. Biol. Chem.*, **274**, 17484–90.
160. Yujiri, T., Sather, S., Fanger, G. R., and Johnson, G. L. (1998). Role of MEKK1 in cell survival and activation of JNK and ERK pathways defined by targeted gene disruption. *Science*, **282**, 1911–14.
161. Schlesinger, T. K., Fanger, G. R., Yujiri, T., and Johnson, G. L. (1998). The TAO of MEKK. *Front. Biosci.*, **3**, D1181–6.
162. Deak, J. C., Cross, J. V., Lewis, M., Qian, Y., Parrott, L. A., Distelhorst, C. W., *et al.* (1998). Fas-induced proteolytic activation and intracellular redistribution of the stress-signaling kinase MEKK1. *Proc. Natl. Acad. Sci. USA*, **95**, 5595–600.
163. Widmann, C., Johnson, N. L., Gardner, A. M., Smith, R. J., and Johnson, G. L. (1997). Potentiation of apoptosis by low dose stress stimuli in cells expressing activated MEK kinase 1. *Oncogene*, **15**, 2439–47.
164. Reyland, M. E., Anderson, S. M., Matassa, A. A., Barzen, K. A., and Quissell, D. O. (1999). Protein kinase C delta is essential for etoposide-induced apoptosis in salivary gland acinar cells. *J. Biol. Chem.*, **274**, 19115–23.
165. Frutos, S., Moscat, J., and Diaz-Meco, M. T. (1999). Cleavage of zetaPKC but not lambda/iotaPKC by caspase-3 during UV-induced apoptosis. *J. Biol. Chem.*, **274**, 10765–70.

166. Mochly-Rosen, D., Khaner, H., Lopez, J., and Smith, B. L. (1991). Intracellular receptors for activated protein kinase C. Identification of a binding site for the enzyme. *J. Biol. Chem.*, **266**, 14866–8.
167. Datta, R., Kojima, H., Yoshida, K., and Kufe, D. (1997). Caspase-3-mediated cleavage of protein kinase C theta in induction of apoptosis. *J. Biol. Chem.*, **272**, 20317–20.
168. Kelekar, A. and Thompson, C. B. (1998). Bcl-2-family proteins: the role of the BH3 domain in apoptosis. *Trends Cell Biol.*, **8**, 324–30.
169. Cheng, E. H., Kirsch, D. G., Clem, R. J., Ravi, R., Kastan, M. B., Bedi, A., *et al.* (1997). Conversion of Bcl-2 to a Bax-like death effector by caspases. *Science*, **278**, 1966–8.
170. Wang, H. G., Rapp, U. R., and Reed, J. C. (1996). Bcl-2 targets the protein kinase Raf-1 to mitochondria. *Cell*, **87**, 629–38.
171. Miyashita, T. and Reed, J. C. (1995). Tumor suppressor p53 is a direct transcriptional activator of the human bax gene. *Cell*, **80**, 293–9.
172. Baichwal, V. R. and Baeuerle, P. A. (1997). Activate NF-kappa B or die? *Curr. Biol.*, **7**, R94–6.
173. Reuther, J. Y. and Baldwin, A. S., Jr. (1999). Apoptosis promotes a caspase-induced amino-terminal truncation of IkappaBalpha that functions as a stable inhibitor of NF-kappaB. *J. Biol. Chem.*, **274**, 20664–70.
174. Barkett, M., Xue, D., Horvitz, H. R., and Gilmore, T. D. (1997). Phosphorylation of IkappaB-alpha inhibits its cleavage by caspase CPP32 *in vitro*. *J. Biol. Chem.*, **272**, 29419–22.
175. Siebenlist, U., Franzoso, G., and Brown, K. (1994). Structure, regulation and function of NF-kappa B. *Annu. Rev. Cell Biol.*, **10**, 405–55.
176. Kasten, M. M. and Giordano, A. (1998). pRb and the cdks in apoptosis and the cell cycle. *Cell Death Differ.*, **5**, 132–40.
177. Clarke, A. R., Maandag, E. R., van Roon, M., van der Lugt, N. M., van der Valk, M., Hooper, M. L., *et al.* (1992). Requirement for a functional Rb-1 gene in murine development. *Nature*, **359**, 328–30.
178. Jacks, T., Fazeli, A., Schmitt, E. M., Bronson, R. T., Goodell, M. A., and Weinberg, R. A. (1992). Effects of an Rb mutation in the mouse. *Nature*, **359**, 295–300.
179. Lee, E. Y., Chang, C. Y., Hu, N., Wang, Y. C., Lai, C. C., Herrup, K., *et al.* (1992). Mice deficient for Rb are nonviable and show defects in neurogenesis and haematopoiesis [See Comments]. *Nature*, **359**, 288–94.
180. Haas-Kogan, D. A., Kogan, S. C., Levi, D., Dazin, P., T'Ang, A., Fung, Y. K., *et al.* (1995). Inhibition of apoptosis by the retinoblastoma gene product. *EMBO J.*, **14**, 461–72.
181. Chen, L., Marechal, V., Moreau, J., Levine, A. J., and Chen, J. (1997). Proteolytic cleavage of the mdm2 oncoprotein during apoptosis. *J. Biol. Chem.*, **272**, 22966–73.
182. Erhardt, P., Tomaselli, K. J., and Cooper, G. M. (1997). Identification of the MDM2 oncoprotein as a substrate for CPP32-like apoptotic proteases. *J. Biol. Chem.*, **272**, 15049–52.
183. Pochampally, R., Fodera, B., Chen, L., Lu, W., and Chen, J. (1999). Activation of an MDM2-specific caspase by p53 in the absence of apoptosis. *J. Biol. Chem.*, **274**, 15271–7.
184. Dou, Q. P., An, B., Antoku, K., and Johnson, D. E. (1997). Fas stimulation induces RB dephosphorylation and proteolysis that is blocked by inhibitors of the ICE protease family. *J. Cell. Biochem.*, **64**, 586–94.
185. Hsieh, J. K., Chan, F. S., O'Connor, D. J., Mittnacht, S., Zhong, S., and Lu, X. (1999). RB regulates the stability and the apoptotic function of p53 via MDM2. *Mol. Cell*, **3**, 181–93.

186. Zhang, Y., Fujita, N., and Tsuruo, T. (1999). Caspase-mediated cleavage of p21Waf1/Cip1 converts cancer cells from growth arrest to undergoing apoptosis. *Oncogene*, **18**, 1131–8.
187. Levkau, B., Koyama, H., Raines, E. W., Clurman, B. E., Herren, B., Orth, K., *et al.* (1998). Cleavage of p21Cip1/Waf1 and p27Kip1 mediates apoptosis in endothelial cells through activation of Cdk2: role of a caspase cascade. *Mol. Cell*, **1**, 553–63.
188. Loubat, A., Rochet, N., Turchi, L., Rezzonico, R., Far, D. F., Auberger, P., *et al.* (1999). Evidence for a p23 caspase-cleaved form of p27[KIP1] involved in G1 growth arrest. *Oncogene*, **18**, 3324–33.
189. Suzuki, A., Tsutomi, Y., Miura, M., and Akahane, K. (1999). Caspase 3 inactivation to suppress Fas-mediated apoptosis: identification of binding domain with p21 and ILP and inactivation machinery by p21. *Oncogene*, **18**, 1239–44.
190. Harvey, K. J., Blomquist, J. F., and Ucker, D. S. (1998). Commitment and effector phases of the physiological cell death pathway elucidated with respect to Bcl-2 caspase, and cyclin-dependent kinase activities. *Mol. Cell. Biol.*, **18**, 2912–22.
191. King, K. L. and Cidlowski, J. A. (1995). Cell cycle and apoptosis: common pathways to life and death. *J. Cell. Biochem.*, **58**, 175–80.
192. Ruoslahti, E. and Reed, J. C. (1994). Anchorage dependence, integrins, and apoptosis. *Cell*, **77**, 477–8.
193. Meredith, J. E., Jr., Fazeli, B., and Schwartz, M. A. (1993). The extracellular matrix as a cell survival factor. *Mol. Biol. Cell*, **4**, 953–61.
194. Frisch, S. M. and Francis, H. (1994). Disruption of epithelial cell-matrix interactions induces apoptosis. *J. Cell Biol.*, **124**, 619–26.
195. Wen, L. P., Fahrni, J. A., Troie, S., Guan, J. L., Orth, K., and Rosen, G. D. (1997). Cleavage of focal adhesion kinase by caspases during apoptosis. *J. Biol. Chem.*, **272**, 26056–61.
196. Crouch, D. H., Fincham, V. J., and Frame, M. C. (1996). Targeted proteolysis of the focal adhesion kinase pp125 FAK during c-MYC-induced apoptosis is suppressed by integrin signalling. *Oncogene*, **12**, 2689–96.
197. Ilic, D., Almeida, E. A., Schlaepfer, D. D., Dazin, P., Aizawa, S., and Damsky, C. H. (1998). Extracellular matrix survival signals transduced by focal adhesion kinase suppress p53-mediated apoptosis. *J. Cell Biol.*, **143**, 547–60.
198. Martin, S. J., O'Brien, G. A., Nishioka, W. K., McGahon, A. J., Mahboubi, A., Saido, T. C., *et al.* (1995). Proteolysis of fodrin (non-erythroid spectrin) during apoptosis. *J. Biol. Chem.*, **270**, 6425–8.
199. Cryns, V. L., Bergeron, L., Zhu, H., Li, H., and Yuan, J. (1996). Specific cleavage of alpha-fodrin during Fas- and tumor necrosis factor-induced apoptosis is mediated by an interleukin-1beta converting enzyme/Ced-3 protease distinct from the poly(ADP-ribose) polymerase protease. *J. Biol. Chem.*, **271**, 31277–82.
200. Herren, B., Levkau, B., Raines, E. W., and Ross, R. (1998). Cleavage of beta-catenin and plakoglobin and shedding of VE-cadherin during endothelial apoptosis: evidence for a role for caspases and metalloproteinases. *Mol. Biol. Cell*, **9**, 1589–601.
201. Schmeiser, K. and Grand, R. J. (1999). The fate of E- and P-cadherin during the early stages of apoptosis. *Cell Death Differ.*, **6**, 377–86.
202. Brancolini, C., Benedetti, M., and Schneider, C. (1995). Microfilament reorganization during apoptosis: the role of Gas2, a possible substrate for ICE-like proteases. *EMBO J.*, **14**, 5179–90.
203. Woo, M., Hakem, R., Soengas, M. S., Duncan, G. S., Shahinian, A., Kagi, D., *et al.* (1998). Essential contribution of caspase 3/CPP32 to apoptosis and its associated nuclear changes. *Genes Dev.*, **12**, 806–19.

204. Li, F., Srinivasan, A., Wang, Y., Armstrong, R. C., Tomaselli, K. J., and Fritz, L. C. (1997). Cell-specific induction of apoptosis by microinjection of cytochrome c. Bcl-xL has activity independent of cytochrome c release. *J. Biol. Chem.*, **272**, 30299–305.
205. Yoshida, H., Kong, Y. Y., Yoshida, R., Elia, A. J., Hakem, A., Hakem, R., *et al.* (1998). Apaf1 is required for mitochondrial pathways of apoptosis and brain development. *Cell*, **94**, 739–50.
206. Varfolomeev, E. E., Schuchmann, M., Luria, V., Chiannilkulchai, N., Beckmann, J. S., Mett, I. L., *et al.* (1998). Targeted disruption of the mouse Caspase 8 gene ablates cell death induction by the TNF receptors, Fas/Apo1, and DR3 and is lethal prenatally. *Immunity*, **9**, 267–76.
207. Chandler, J. M., Cohen, G. M., and MacFarlane, M. (1998). Different subcellular distribution of caspase-3 and caspase-7 following Fas-induced apoptosis in mouse liver. *J. Biol. Chem.*, **273**, 10815–18.
208. Csordas, G., Thomas, A. P., and Hajnoczky, G. (1999). Quasi-synaptic calcium signal transmission between endoplasmic reticulum and mitochondria. *EMBO J.*, **18**, 96–108.
209. Kass, G. E. and Orrenius, S. (1999). Calcium signaling and cytotoxicity. *Environ. Health Perspect.*, **107 Suppl 1**, 25–35.
210. Nakagawa, T., Zhu, H., Monishima, N., Li, E., Xu, J., Yankner, B. A. (2000). Caspase-12 mediates endoplasmic-reticulum-specific apoptosis and cytotoxicity by amyloid-beta. *Nature*, **403**, 98–103.
211. Ng, F. W. and Shore, G. C. (1998). Bcl-XL cooperatively associates with the Bap31 complex in the endoplasmic reticulum, dependent on procaspase-8 and Ced-4 adaptor. *J. Biol. Chem.*, **273**, 3140–3.
212. Soengas, M. S., Alarcon, R. M., Yoshida, H., Giaccia, A. J., Hakem, R., Mak, T. W., *et al.* (1999). Apaf-1 and caspase-9 in p53-dependent apoptosis and tumor inhibition. *Science*, **284**, 156–9.
213. Willis, T. G., Jadayel, D. M., Du, M. Q., Peng, H., Perry, A. R., Abdul-Rauf, M., *et al.* (1999). Bcl10 is involved in t(1;14) (p22;q32) of MALT B cell lymphoma and mutated in multiple tumor types. *Cell*, **96**, 35–45.
214. Zhang, Q., Siebert, R., Yan, M., Hinzmann, B., Cui, X., Xue, L., *et al.* (1999). Inactivating mutations and overexpression of BCL10, a caspase recruitment domain-containing gene, in MALT lymphoma with t(1;14) (p22;q32). *Nature Genet.*, **22**, 63–8.
215. Haas, C., Hung, A. Y., Citron, M., Teplow, D. B., and Selkoe, D. J. (1995). beta-Amyloid, protein processing and Alzheimer's disease. *Arzneimittelforschung*, **45**, 398–402.
216. Weidemann, A., Paliga, K., U. D. r., Reinhard, F. B., Schuckert, O., Evin, G., *et al.* (1999). Proteolytic processing of the Alzheimer's disease amyloid precursor protein within its cytoplasmic domain by caspase-like proteases. *J. Biol. Chem.*, **274**, 5823–9.
217. Gervais, F. G., Xu, D., Robertson, G. S., Vaillancourt, J. P., Zhu, Y., Huang, J., *et al.* (1999). Involvement of caspases in proteolytic cleavage of Alzheimer's amyloid-beta precursor protein and amyloidogenic A beta peptide formation. *Cell*, **97**, 395–406.
218. Haass, C., Lemere, C. A., Cappell, A., Citron, M., Seubert, P., Schenk, D., *et al.* (1995). The Swedish mutation causes early-onset Alzheimer's disease by beta-secretase cleavage within the secretory pathway. *Nature Med.*, **1**, 1291–6.
219. Grunberg, J., Walter, J., Loetscher, H., Deuschle, U., Jacobsen, H., and Haass, C. (1998). Alzheimer's disease associated presenilin-1 holoprotein and its 18–20 kDa C-terminal fragment are death substrates for proteases of the caspase family. *Biochemistry*, **37**, 2263–70.
220. van de Craen, M., de Jonghe, C., van den Brande, I., Declercq, W., van Gassen, G., van

Criekinge, W., *et al.* (1999). Identification of caspases that cleave presenilin-1 and presenilin-2. Five presenilin-1 (PS1) mutations do not alter the sensitivity of PS1 to caspases. *FEBS Lett.*, **445**, 149–54.

221. Wang, J., Zheng, L., Lobito, A., Chan, F. K., Dale, J., Sneller, M., *et al.* (1999). Inherited human Caspase 10 mutations underlie defective lymphocyte and dendritic cell apoptosis in autoimmune lymphoproliferative syndrome type II. *Cell*, **98**, 47–58.
222. Ishizaki, Y., Jacobson, M. D., and Raff, M. C. (1998). The role for caspases in lens fiber differentiation. *J. Cell Biol.*, **140**, 153–8.
223. Watanabe, Y. and Akaike, T. (1999). Possible involvement of caspase-like family in maintenance of cytoskeletal integrity. *J. Cell. Physiol.*, **179**, 45–51.
224. Stoh C, and Schulze-Osthoff, K. (1998). Death by a thousand cuts: an ever increasing list of caspase substrates. *Cell Death Differ.*, **5**, 997.

5 | Regulation of apoptosis by the Bcl-2 family of proteins

YOSHIHIDE TSUJIMOTO

1. Introduction

Molecular biological studies of apoptosis, a cellular self-destruction mechanism, have made quite rapid progress in unveiling the apoptotic machinery over the last ten years. The framework of the apoptotic signal transduction pathway appears to be as follows: various pro-apoptotic signals initially activate stimulus-specific signalling pathways, which eventually converge into a common mechanism driven by a unique family of cysteine protease, called caspases (1, 2). The common mechanism of apoptosis is negatively regulated by several sets of genes, the best characterized of which is the still-growing *bcl-2* family consisting of both anti-apoptotic and pro-apoptotic members (3, 4). The early observations that *C. elegans* possesses a caspase (CED-3) (5) and a death suppressor (CED-9) (6) showing structural and functional homology to mammalian Bcl-2, initially led us to the notion that the basic mechanism of apoptosis appears to have been highly conserved during evolution. This notion is strengthened further by the subsequent finding that two other apoptosis-regulating factors are also shared by *C. elegans* and mammals, including the caspase activator Ced-4/Apaf-1 (7) and pro-apoptotic regulator Egl-1/BH3-only proteins, which represents a subfamily of Bcl-2 family of proteins (8).

The notion that the Bcl-2 family proteins are critical regulators of apoptosis is supported by the observation that many viruses produce Bcl-2-like proteins to prevent or delay viral infection-induced apoptosis (see Chapter 11). These findings suggest that Bcl-2 family proteins are suitable targets for manipulation in attempts to prevent diseases associated with the deregulation of apoptosis. This article will focus on how the Bcl-2 family proteins determine life or death of a cell.

2. Discovery of *bcl-2* gene and its anti-apoptotic activity

bcl-2 (9) was originally identified as an oncogene activated by a t(14;18) chromosomal translocation in human follicular B cell lymphoma (10–13). The *bcl-2* gene is also activated by variant translocations t(18;22) and t(2;18) associated with chronic lymphocytic leukaemia (14). Up-regulation of Bcl-2 expression without any apparent gene rearrangement has also been found in other solid tumours, suggesting an important role of the *bcl-2* gene in tumorigenesis.

In 1988, Vaux *et al.* (15) discovered the anti-apoptotic activity of Bcl-2 protein in a system in which apoptosis was induced by deprivation of IL-3. Soon after, Bcl-2 was shown to prevent apoptosis induced by several different factors, including serum deprivation, heat shock, chemotherapy agents, ethanol (16), and various other stimuli (17), suggesting that Bcl-2 prevents apoptosis via a common pathway. Bcl-2 inhibits most modes of apoptosis, with a few exceptions such as apoptosis induced by death receptors like Fas, although only in certain cells (18) (see Chapter 8). Bcl-2 is also able to inhibit necrosis induced by hypoxia and inhibition of mitochondrial respiration (19–22), suggesting that both apoptosis and some forms of necrosis share common mechanisms. Such a conclusion is also consistent with the inhibition of these types of necrosis by caspase inhibitors (21, 22).

Disruption of pathways that regulate apoptosis can give rise to various diseases, including cancer and neurodegenerative disorders. It has been generally predicted that deregulation of apoptosis resulting in an enhancement of cell death would occur in degenerative disorders such as Alzheimer's. However, inhibition of apoptosis, as occurs in the development of cancer, was unexpected and was only detected after discovery of the anti-apoptotic function of the *bcl-2* oncogene. Subsequently, apoptosis-related functions of several oncogenes and suppressor oncogenes were discovered, further supporting the important role of inhibition of apoptosis in tumorigenesis. Bax, a pro-apoptotic Bcl-2 family member, would be predicted to be a tumour suppressor gene, and indeed Bax mutations have been reported in human haematopoietic tumours (23) as well as in colon cancer (24).

3. Bcl-2 family proteins and regulation of apoptosis

Based upon their sequence similarity and ability to bind to other Bcl-2-related proteins, a large number of Bcl-2 protein family members have been isolated (3, 4) (Table 1). This family of proteins comprises both anti-apoptotic and pro-apoptotic molecules that form a critical intracellular decision point within the common cell death pathway. The Bcl-2 family is classified into the following three subfamilies, based on the sequence and function of each protein (see Table 1):

(a) A Bcl-2 subfamily, all members of which exert anti-cell death activity and share sequence homology particularly within four regions, called BH (*B*cl-2 *H*omology) 1 through 4 (see Fig. 1). Examples are Bcl-2, Bcl-x_L, and Bcl-w.

(b) A Bax subfamily, all members of which exert pro-apoptotic activity and share sequence homology within their BH1, BH2, and BH3, but not BH4 regions, with the exception of some members such as Mtd where there is sequence similarity in BH4 (25). Examples are Bax and Bak.

(c) A BH3-only subfamily, all members of which are pro-apoptotic and share primary sequence homology only within BH3. Examples are Bik and Bid.

Table 1 Bcl-2 family proteins

	Function	
	Anti-apoptotic	**Pro-apoptotic**
Bcl-2 subfamily		
Bcl-2	+	
Bcl-x Bcl-x_L	+	
Bcl-x_S		
Bcl-w	+	
Mcl-1	+	
A1 (Bfl-1)	+	
Boo	+	
Viral proteins		
BHRF-1 (Epstein–Barr virus)	+	
LMW-5-HL (African swine fever virus)	+	
E1B 19 kDa (adenovirus)	+	
Ksbcl-2 (HHV8)	+	
ORF16 (herpesvirus saimiri)	+	
Bax subfamily		
Bax		+
Bak		+
Mtd (Bok)		+
Diva		+
BH3-only subfamily		
Bik		+
Bid		+
Bad		+
Bim		+
Hrk (DP5)		+
Blk		+
Bnip3		+
Bnip3L/Nix/B5		+
Brag-1	?	?

One of the striking features of Bcl-2 family proteins is their ability to interact with each other to form homodimers and heterodimers (26). Heterodimerization between the anti-apoptotic and pro-apoptotic members of this family may provide a mechanism by which these proteins regulate apoptosis (26, 27). Heterodimerization of Bcl-2 family proteins is mediated by insertion of the BH3 region of a pro-apoptotic protein into the hydrophobic pocket (composed of BH1, BH2, and BH3) of an anti-apoptotic protein, as shown by crystallographic studies of Bcl-x_L and BH3 peptide (28). It has been suggested that pro-apoptotic Bcl-2 family members like Bax, Bak, and Bid need to activate (expose) the BH3 domain, which is considered to be normally inactive (possibly through being buried in the molecule), in order to heterodimerize with partner proteins. This can be achieved by a conformational change, as has been suggested for Bax (29), Bak (30), or by N-terminal truncation, which seems to be critical for the full pro-apoptotic activity of Bid (31, 32).

Overall, the BH4 domain in addition to BH1 and BH2 is required for anti-apoptotic

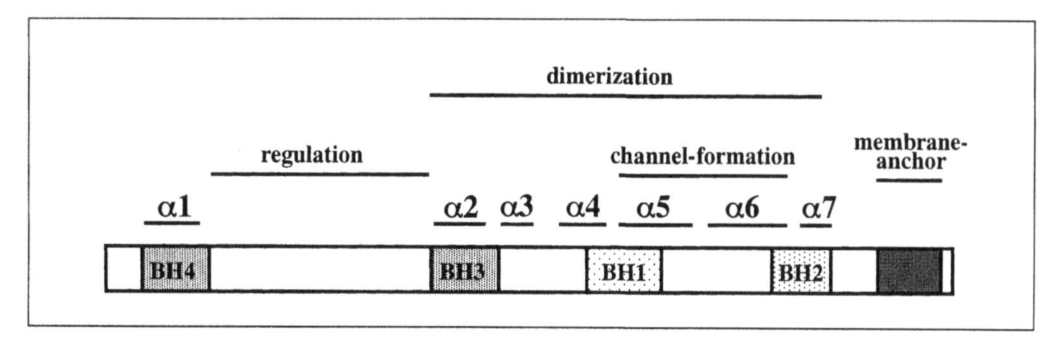

Fig. 1 Diagram of Bcl-2 protein. BH1 to BH4 are conserved sequence motifs. Several functional domains are shown (see text). α1 to α7 indicate helices identified in Bcl-x_L, in which a core of two hydrophobic helices α5 and α6 is surrounded by five amphipathic helices. The region between α1 and α2 (Bcl-2 and Bcl-x_L) is a loop that appears to have a regulatory role and contains several phosphorylation sites.

activity (33), while BH3 is essential for pro-apoptotic activity (34). Oligopeptides corresponding to slightly more than the BH3 region of Bax and Bak are able to exert pro-apoptotic activity in mammalian cells (34) and in a cell-free system (35, 36). Replacement of the BH3 region of Bcl-2 by the BH3 region of Bax converts Bcl-2 from a death antagonist to a death agonist (37), although some reports indicate that Bcl-2 carrying the BH3 region of Bax does not possess any activity (38). These results suggest that a critical sequence difference may distinguish the BH3 domain of the anti-apoptotic and pro-apoptotic members of the Bcl-2 family.

In addition to the regulation of apoptosis by heterodimerization of anti-apoptotic and pro-apoptotic members of this family, Bax and Bcl-2 have been suggested to regulate apoptosis independently from each other, based on inferences from studies of *bcl-2* and *bax* knockout and transgenic mice (39). This idea of independent function is supported by the existence of several mutant forms of Bcl-2 family proteins that fail to heterodimerize with their antagonistic partners but still exert activity (40–44). The independent functioning of some Bcl-2 family proteins might be attributed to a unique biochemical property that allows these proteins to form ion channels in synthetic lipid membranes (Section 4.2). Thus, Bcl-2 family proteins may determine the life or death of a cell in two ways: by heterodimerization between anti-apoptotic and pro-apoptotic members, or through the independent functions of these proteins. In either case, the ratio between anti-apoptotic and pro-apoptotic members of the Bcl-2 family may determine the susceptibility of a cell to apoptosis.

4. Biochemical functions

A variety of apoptotic stimuli ultimately converge on caspases, activating them and leading to the progression of apoptosis. Anti-apoptotic members of the Bcl-2 family shut off the apoptotic signal transduction pathway upstream of caspase activation (22, 45–47); they do not appear to act downstream of effector caspases (48). The biochemical mechanisms by which Bcl-2 family proteins regulate caspase activation, and thereby control cell death, are not fully understood. Nevertheless, specific biochemical and physiological activities of some Bcl-2 family proteins have been elucidated which may be important aspects of these proteins' function. These biochemical functions are described in this section. Ever since Bcl-2 was found localized to mitochondria (49, 50), attention has focused on the effect of Bcl-2 on mitochondrial functions and integrity. Three mitochondrial actions of Bcl-2-related proteins are discussed in Section 4.1. They include:

(a) Prevention of the release of apoptosis-inducing factors from the mitochondrial inner membrane.

(b) Regulation of mitochondrial membrane permeability through the permeability transition (PT) pore.

(c) Possible effects on the F_0F_1ATPase.

Section 4.2 discusses the formation of ion channels, which may occur in mitochondrial membranes as well as other intracellular membranes where Bcl-2 family proteins are localized. Section 4.3 concerns the sequestration of caspases via an adaptor molecule, and Section 4.4 addresses the possible role of Bcl-2-related proteins in the endoplasmic reticulum (ER).

4.1 Role in the mitochondria

4.1.1 Prevention of release of mitochondrial apoptogenic factors by Bcl-2 and Bcl-x_L

Bcl-2 and Bcl-x_L prevent a step leading to the activation of caspases (45–47). Several mechanisms have been proposed for this effect (summarized in Fig. 2). Bcl-2 and Bcl-x_L are localized in the mitochondrial outer membrane as well as in the endoplasmic reticulum and the nuclear envelope (49, 50). Mitochondrial Bcl-2 and Bcl-x_L can prevent release of the apoptogenic factor cytochrome *c* (51–53) from the mitochondrial intermembrane space into the cytoplasm. Once cytochrome *c* reaches the cytoplasm, it binds to a cytoplasmic protein known as Apaf-1 (7, 54). This complex recruits and activates procasapse-9 by inducing its oligomerization (see Fig. 2) (55).

Mitochondria also sequester other apoptogenic factors in addition to cytochrome *c*, such as AIF (apoptosis-inducing factor) (56) and reportedly procasapse-2, -3, and -9 (Fig. 2) (57, 58), although the significance of these mitochondrial caspases is not known (see Chapter 6). AIF is a flavoprotein that shares homology with bacterial oxidoreductases (56), and has the ability to induce large scale DNA fragmentation (~ 50 kb) and apoptotic morphological changes of the nucleus such as chromatin condensation after translocation into the nucleus. The release of AIF appears to be dependent upon the occurrence of mitochondrial dysfunction, including loss of mitochondrial membrane potential ($\Delta\psi_m$) and opening of membrane permeability

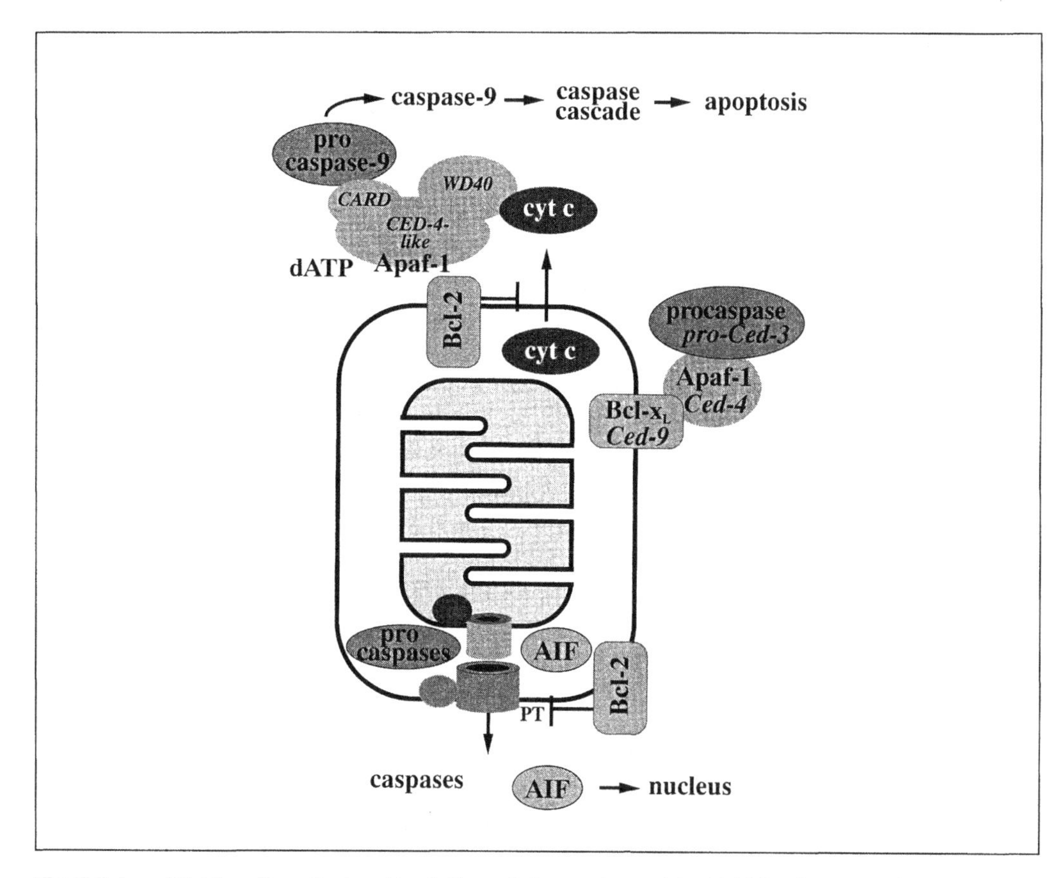

Fig. 2 Roles of Bcl-2 on the mitochondria. Anti-apoptotic members of the Bcl-2 family prevent the activation of caspases either by blocking the release of apoptogenic cytochrome *c* from the mitochondria into the cytoplasm or by sequestering the pro-forms of caspases. In mammals, caspase activation involves two independent steps, release of Apaf-1 from death antagonists of the Bcl-2 family and release of cytochrome *c* from the mitochondria, both of which are triggered by pro-apoptotic members of the Bcl-2 family. AIF is released through the PT pore which is inhibited by Bcl-2. The molecular architecture of the PT pore is not known but is thought to consist of several proteins located on the mitochondrial outer and inner membranes.

transition (PT) pores (59, 60), whereas cytochrome *c* release does not necessarily depend on $\Delta\psi_m$-loss/PT (52, 53). Bcl-2 and Bcl-x_L prevent all of these mitochondrial changes (59, 61).

4.1.2 Mitochondrial PT pores

Our understanding of how Bcl-2 family proteins regulate mitochondrial changes associated with apoptosis has largely developed from studies using isolated mitochondria and recombinant forms of Bcl-2 family proteins. Addition of recombinant pro-apoptotic Bcl-2 family members (such as Bax, Bak, and Bid), or BH3 oligopeptides to isolated mitochondria induces cytochrome *c* release (31, 32, 35, 36, 62, 63) and $\Delta\psi_m$-loss (36, 62, 64), resulting in AIF release. Recombinant Bcl-2 and

Bcl-x_L efficiently inhibit all of the mitochondrial changes induced by these pro-apoptotic proteins (31, 32, 35, 36, 62, 63) as well as by other PT inducers (36). Mitochondrial changes mediated by Bax, Bak, and BH3 peptides are dependent on Ca^{2+} and are inhibited by cyclosporin A (36, 62), a drug that targets mitochondrial cyclophilin (cyclosporin D), one of the components of the permeability transition (PT) pore. These findings suggest that Bax, Bak, and BH3 peptide-mediated cytochrome *c* release may require the PT pore, although some studies have found that cytochrome *c* release can occur independent of the PT pore (63, 65). Nevertheless, the involvement of the PT pore in Bax/Bak-mediated cytochrome *c* release *in vitro* has helped to identify a functional target of Bcl-2 family proteins.

The PT pore is a poly-protein channel, the molecular nature of which is still largely unknown. It is considered to consist of porin (also known as the voltage-dependent anion channel, VDAC) and peripheral benzodiazepine receptor (PBR), both on the outer membrane, plus adenine nucleotide translocator (ANT) on the inner membrane, cyclophilin D, a matrix protein, and other molecules (66, 67). Thus, the PT pore is probably localized at the contact site of the inner and outer mitochondrial membranes. An important role of the PT pore in cytochrome *c* release is supported by the observation that Bax and Bak interact with the pore (36, 68). It has been suggested that cytochrome *c* is released by outer membrane rupture resulting from mitochondrial swelling caused by PT pore opening (69). It has also been proposed that Bax might form a large pore (76) through which cytochrome *c* is able to pass, but the PT pore does not seem to be directly involved in this process. Consistent with the interaction of Bax and Bak with the PT pore, it has been shown, using a VDAC-liposome system, that recombinant Bax and Bak proteins interact with VDAC through which cytochrome *c* is released, and this is inhibited by recombinant Bcl-x_L (70). An essential role of the VDAC in cytochrome *c* release is further supported by the observation that Bax and Bak induce cytochrome *c* release from wild-type yeast mitochondria but not from the mitochondria of VDAC-deficient yeast cells (70). Although Bax has been shown to bind to ANT (68, 70) and to sensitize ANT to atractyloside, an ANT ligand (68), Bax and Bcl-x_L do not affect ANT channel activity in the isolated mitochondria (our unpublished observations). The ability of pharmacological inhibitors of the PT pore to inhibit cytochrome *c* release induced by Bax or Bak under some conditions *in vitro* implies that VDAC-mediated cytochrome *c* release is not only regulated by Bcl-2 family proteins but also by other components of the PT pore. This might also be true for certain circumstances *in vivo*. PT pore opening leads to mitochondrial swelling *in vitro*, but this is not necessarily the case in whole cells.

4.1.3 Bax and F_0F_1ATPase

Bax and Bak induce cell death in yeast, which is prevented by co-expression of Bcl-2 and Bcl-x_L (71). A genetic approach employed to study Bax-mediated death of yeast has lead to the novel finding that Bax-mediated cell death is dependent upon the F_0F_1ATPase (72). Oligomycin, an inhibitor of the F_0F_1ATPase, has also been shown to inhibit Bax-mediated apoptosis in mammalian cells. Most apoptosis in mammalian cells is not inhibited by oligomycin, however, suggesting that, if Bax functions through the F_0F_1ATPase, it may not play a major role in apoptosis. Nevertheless, it is possible that F_0F_1ATPase may indirectly regulate the PT pore by modulating the ATP/ADP ratio. Consistent with this concept, Bax does not seem to bind to the F_0F_1ATPase.

4.2 Ion channels formed by the Bcl-2 family proteins

Structural analysis of Bcl-x_L has revealed striking similarity to a pore-forming region of the bacterial proteins diphtheria toxin and colicins (73). This region includes a pair of hydrophobic core helices (α5 and α6 for Bcl-2/Bcl-x_L) that are long enough to penetrate the lipid bilayer. These core helices are surrounded by five amphipathic helices, their hydrophobic surfaces facing toward the core helices and their hydrophilic surfaces directed outward (see Fig. 1). Indeed, Bcl-x_L, Bcl-2, and Bax have all been shown to form ion channels in synthetic lipid membranes (74–77). Bcl-2 and Bcl-x_L form channels with a cation preference, whereas Bax tends to form larger ion channels with a Cl^- preference, although Bax reportedly shows a weak cation selectivity in some systems (76; our unpublished observations). It remains to be determined whether Bcl-2 family proteins form ion channels *in vivo* and whether these proteins regulate apoptosis via ion channels. It is conceivable that Bax or other pro-apoptotic members of the Bcl-2 family may form channels *in vivo* to destroy ion homeostasis, such as that maintaining the mitochondrial membrane potential.

BH3-only proteins share sequence similarity with Bcl-2 only within the BH3 domain, and they have been postulated to function as ligands, binding to Bcl-2 and Bax subfamily members and affecting their activity. In contrast, pro-apoptotic Bax subfamily members like Bax and Bak, which have greater similarity to Bcl-2, appear to behave both independently and by virtue of their interactions with other family members. However, BH3-only proteins may be structurally more similar to their Bcl-2 and Bax subfamily siblings than primary sequence comparisons imply. The three-dimensional structure of Bid, a BH3-only protein, has been shown to consist of eight α-helices, with two central hydrophobic helices surrounded by amphipathic helices on either side (78, 79), an overall tertiary structure similar to that of Bcl-x_L. Moreover, Bid has been shown to form an ion channel on synthetic lipid membranes, with tBid more active than Bid (115). A comparison of the structures of Bid and truncated Bid has led to the view that pro-apoptotic Bcl-2 family members can be categorized into two groups. One (including Bax, Bak, Bid) with the BH3 domain masked inside the molecule and the other (tBid, Bad) with the BH3 domain exposed (78). Thus, pro-apoptotic Bcl-2 family members such as Bax and Bak must require some conformational change to expose the BH3 domain before exerting their pro-apoptotic activity.

4.3 Sequestration of caspases by Bcl-2 and Bcl-x_L

4.3.1 Binding of Bcl-x_L to Apaf-1, a caspase activator

Cytochrome *c* binds to Apaf-1 via the C-terminal WD40 repeat domain of Apaf-1 in the presence of ATP or dATP, and thus exposes the N-terminal caspase recruitment domain (CARD), to which procaspase-9 binds through its N-terminal CARD domain (54) (see Fig. 2). Apaf-1 possesses sequences similar to Ced-4 (a protein required for death in *C. elegans*) and is considered to be a mammalian homologue of Ced-4 (7). Apaf-1 is able to oligomerize via its Ced-4-like domain, and the resulting oligomerization of procaspases-9 induces its autoactivation (55). Based upon the observation that Bcl-x_L binds indirectly to procaspase-8 (80) and directly to Apaf-1 (or an Apaf-1/procaspase-9 complex) via the Ced-4-like domain (81, 82), it was proposed that Bcl-x_L sequesters caspases or caspase activator(s) to prevent the activation of caspases. For example, Bcl-x_L might prevent Ced-4 domain-dependent oligomerization of Apaf-1 by direct binding to the overlapping region. Genetic studies in *C. elegans* suggest that a similar mechanism operates (see Fig. 2): Ced-9 (a Bcl-2 homologue) binds to pro-Ced-3 (*C. elegans* caspase) via Ced-4 (83–85). This complex is called the 'apoptosome'. However, whether Bcl-x_L in mammalian cells and Ced-9 in the nematode prevent apoptotic cell death by forming apoptosomes inside cells still needs to be formally proved. It has been reported that Bcl-2 does not seem to have the same ability to bind to Apaf-1 as does Bcl-x_L, suggesting that regulation of apoptosis through apoptosomes might not be essential, at least in mammalian cells. The physiological role of Bcl-2/Bcl-x_L to sequester Apaf-1 in preventing apoptosis has recently been questioned, based upon the failure to detect a stable interaction between Apaf-1 and Bcl-2 family proteins (116).

If this mechanism operates the release of Apaf-1 and Ced-4 from their inhibitory proteins Bcl-x_L and Ced-9, respectively, seems to be critical for determining the survival or death of a cell. In mammalian systems, the release of Apaf-1 from Bcl-x_L can be mediated by heterodimerization with Bax and other pro-apoptotic members of the Bcl-2 family (81). In *C. elegans*, Egl-1, a death protein that belongs to the BH3 family, plays a similar role in releasing Ced-4 from Ced-9 (8) (see Chapter 2). It has also been reported that a pro-apoptotic member of the Bcl-2 family, Diva, does not heterodimerize with Bcl-2 family members except for a viral Bcl-2 homologue, KsBcl-2 encoded by HHV-8, but rather heterodimerizes with and activates Apaf-1 (86).

There is no evidence for the involvement of cytochrome *c* in the programmed cell death of *C. elegans*. Consistently, Ced-3 activation by Ced-4 does not require cytochrome *c*, unlike in mammalian systems. This striking difference seems to be attributable to the WD40 repeat of Apaf-1, which is not present on Ced-4. The WD repeats are thought to suppress N-terminal CARD binding since Apaf-1 lacking the WD40 domain is active without cytochrome *c*, resembling Ced-4 (55). Therefore, it seems likely that the activation of caspase(s) is regulated by protein–protein interaction in *C. elegans*, whereas in mammalian cells, caspase activation appears to be regulated mainly through mitochondrial dysfunction.

4.4 Role of Bcl-2 localized on the endoplasmic reticulum and nuclear envelope

A significant amount of Bcl-2 and Bcl-x_L is localized on the endoplasmic reticulum (ER) and topologically equivalent outer nuclear membrane. Bcl-2 specifically targeted to the ER has been shown to exert anti-apoptotic activity in a limited number of systems (87). This might be mediated by heterodimerization with pro-apoptotic Bcl-2 family proteins or by sequestration of caspases through binding to an ER protein, p28Bap31, which associates with procaspase-8 (80), or possibly through binding to the Apaf-1/procaspase-9 complex. It has also been shown that Bcl-2 can reduce Ca^{2+} efflux from the ER induced by thapsigargin or H_2O_2 (88, 89), although it has not been determined whether this is mediated by any direct action of Bcl-2 on the ER.

Bcl-2 has been shown to affect the nuclear translocation of some transcription factors, including p53 (90, 91) and NF-AT (92). Bcl-2 binds to calcineurin to inhibit the dephosphorylation of NF-AT, keeping NF-AT in the cytoplasm and arresting T cells in the G1 phase of the cell cycle (92). Bcl-2 seems to have the ability to reduce Ca^{2+} uptake into isolated nuclei (93), and it has been proposed that Bcl-2 might be a gatekeeper for nuclear apoptosis. However, it is still unclear whether these activities are significant in the general and broad anti-apoptotic effect of Bcl-2.

5. Regulation of Bcl-2 family proteins

5.1 Regulation of pro-apoptotic members of the Bcl-2 family through post-translational modification

Given that pro-apoptotic proteins of the Bcl-2 family induce release cytochrome *c* from the mitochondria, one of the critical issues is how these pro-apoptotic proteins are activated or how anti-apoptotic proteins are inactivated. Compelling evidence for regulation by protein phosphorylation is available in the case of Bad (94, 95). In the presence of survival factors, Akt (95) or PKA (117) phosphorylates Bad. Phosphorylated Bad binds to the cytoplasmic protein 14-3-3, and is unable to bind to Bcl-x_L (94). In the absence of survival factors, Bad is dephosphorylated and binds to Bcl-x_L to antagonize its anti-apoptotic activity. In Ca^{2+}-induced apoptosis, Bad is dephosphorylated by calcineurin (118). Thus, the activity of Bad seems to be directly regulated by both phosphorylation and dephosphorylation.

Bax normally exists in both the cytoplasm and mitochondria via peripheral association, i.e. it shows no stable integration into membranes. Moreover, a substantial fraction of cytoplasmic monomeric Bax moves to the mitochondria during apoptosis, where it forms oligomers that are stably integrated into the membrane using its hydrophobic membrane-anchoring domain, suggesting that cytoplasmic Bax undergoes conformational modification during apoptosis (29, 96, 97). Conversely, it has also been reported that without translocation of cytosolic Bax to the mitochondria, mitochondrial Bax undergoes a similar conformational change during apoptosis, as detected using an antibody specific to the N-terminal epitope (98). This conformational change of Bax probably leads to exposure of the BH3 domain, and manifests as oligomerization, translocation to the mitochondrial membrane, and apoptotic mitochondrial dysfunction that results in cell death. Consistently, forced dimerization of Bax results in both its translocation from the cytosol to the mitochondria and in apoptosis (97). Translocation of Bax to the mitochondria has been reported using apoptotic cell lysates (29, 119), suggesting that some factor promoting translocation is present in lysates from apoptotic but not normal cells. Bcl-2 expression prevents all of the apoptosis-associated changes of Bax (97). Since N-terminal truncation of Bax facilitates its translocation to the mitochondria, the conformational change of Bax may suppress a negative regulatory function of the N-terminal region. This might be related to exposure of the N-terminal epitope of Bax and also to increased susceptibility of this region to proteolytic cleavage by exogenous proteases (29). Mitochondrial Bak also undergoes a conformational change during apoptosis, as shown by a change in the binding of an N-terminal epitope-specific antibody (30). How Bax and Bak undergo conformational change remains unclear, although it has been shown that Bid binds to Bax to induce conformational change, leading to exposure of the N-terminal epitope (98).

Another mechanism for the activation of pro-apoptotic members of the Bcl-2 family is proteolytic cleavage. Bid, a BH3-only protein, is cleaved at its N-terminal region by caspase-8 (directly activated via Fas stimulation). Truncated Bid then efficiently translocates to the mitochondria and induces cytochrome *c* release (31, 32). Several different mechanisms have been suggested to explain how caspase-dependent cleavage activates Bid, including exposure of the BH3 domain required for pro-apoptotic activity (78), removal of an inhibitory activity present in the N-terminal fragment, and exposure of central core helices that possibly form an ion channel (79). Bid has also been shown to translocate to the mitochondria during apoptosis independently of proteolytic cleavage by caspase(s) (98). It is possible that pro-apoptotic signals might be transmitted directly to pro-apoptotic members of the Bcl-2 family by inducing conformational changes or, alternatively, such signals might be transmitted indirectly by other factors which modify the conformation of pro-apoptotic family members.

Anti-apoptotic members of Bcl-2 family, namely Bcl-2 and Bcl-x_L, are also phosphorylated at several serine/threonine residues within the regulatory loop region (see Fig. 1) *in vivo*, although effects of phosphorylation are still controversial (99, 100). Hyperphosphorylation of Bcl-2 is induced by drugs affecting microtubule integrity, such as paclitaxel and vincristine, probably through activation of PKA, and seems to block the anti-apoptotic activity of this protein, partly due to loss of the ability to bind with Bax (101). In contrast, phosphorylation of Bcl-2 at Ser70 has been shown to be required for its anti-apoptotic activity (100). The anti-apoptotic activity of Bcl-2 and Bcl-x_L also seems to be regulated through proteolytic cleavage by caspases (102, 103), because cleavage at the loop region reduces anti-apoptotic activity and the cleaved products appear to become pro-apoptotic. Although it is unclear how this process contributes to regulation of apoptosis, the pro-apoptotic activity of truncated Bcl-2 and Bcl-x_L seems to be at least partly due to exposure of the BH3 domain (78).

5.2 Bcl-2 binding proteins

A number of Bcl-2 binding proteins, which do not belong to the Bcl-2 family, have been identified and are summarized in Table 2. Some of them are implicated in apoptosis, including, Bag-1 (104), Raf-1 (105), and Smn (106), all of which enhance the anti-apoptotic activity of Bcl-2. Synergistic anti-apoptotic activity between Smn and Bcl-2 is quite remarkable (106), and Smn seems to be the only binding protein with a demonstrated physiological role—lack of functional Smn is observed in all patients with spinal muscular atrophy, a disease characterized by degeneration of lower motor neurons (107). Interaction between Bcl-2 and Smn has been observed in various species, including mice (our unpublished results). Bag-1 has been shown to have a role in the chaperone system (108). BI-1 was recently identified as a repressor of Bax-induced apoptosis, and was shown to be another Bcl-2 but not Bax binding protein with anti-apoptotic activity (109). Bag-1 and Raf-1 bind to the N-terminal region of Bcl-2 that includes the BH4 domain. Three proteins known as BNip-1, -2, and -3 (previously called Nip's), originally identified as proteins with the ability to bind to the adenovirus-encoded anti-apoptotic protein E1B 19 kDa, also bind to Bcl-2 (110). A protein called Bnip-3L/Nix/B5, homologous to Bnip-3, has also been

Table 2 Bcl-2 binding proteins

Protein	Function	Remarks
Bnip1	?	Localized to nuclear envelope
Bnip2	?	Localized to nuclear envelope
Bag-1	Anti-apoptotic	Chaperone regulator
Raf-1	Anti-apoptotic	Serine/threonine kinase
R-ras p23	?	Small G protein
p53BP2	?	p53 binding protein
CPT-1	?	Mitochondrial carnitine palmitoyl transferase
Prion	?	
Calcineurin	Anti-apoptotic	Phosphatase
Smn	Anti-apoptotic	SMA-determining gene
BI-1	Anti-apoptotic	Multiple membrane spanning
Apaf-1[a]	Pro-apoptotic	Ced-4 homologue
p28Bap31	?	Localized to ER
Beclin	Anti-apoptotic	
PKC	Anti-apoptotic?	
ANT	Apoptosis regulation	Adenine nucleotide translocator
VDAC	Apoptosis regulation	Mitochondrial porin or voltage-dependent anion channel
SERCA2	?	Ca^{2+}-ATPase in ER
MRIT/c-FLIP[a]	?	DED-containing protein
FLASH	Pro-apoptotic	A component of DISC

[a] Interaction with Bcl-x_L has been reported.

described (111). The function of BNip-1 and BNip-2 is still unknown, whereas BNip-3 and Bnip3L/Nix/B5 have been shown to have a BH3 domain and to exert pro-apoptotic activity (112–114).

5.3 Bcl-2 family proteins as a convergence point for various life–death signals

The activities of Bcl-2 family proteins are regulated by multiple mechanisms, including binding to other proteins and post-translational modifications such as phosphorylation and proteolytic cleavage. This regulatory control suggests that the Bcl-2 family might be viewed as a convergence point for the integration of various apoptosis-regulating signals (4). Some apoptotic stimuli activate protein phosphorylation cascades that transmit death signals to specific members of the Bcl-2 family, subsequently activating death agonists or inactivating death antagonists. Survival signals also activate protein kinase signalling pathways that modulate Bcl-2 family proteins. Other apoptotic stimuli might signal to Bcl-2 family proteins by generating a small amount of active caspases, which in turn activate some pro-apoptotic Bcl-2 family members and inactivate anti-apoptotic members. Yet other death and survival signals might modulate the activities of Bcl-2 binding proteins, leading indirectly to changes in the activity of the Bcl-2 family proteins to which they bind. One important convergence point of these signals is on Bcl-2 family proteins located on the mitochondria, determining whether apoptogenic factors are released to the cytoplasm or not. This concept might explain why there are so many Bcl-2 family proteins and is consistent with the observation that each Bcl-2 family protein seems to be involved in specific molecular pathways of apoptosis.

6. Conclusion

Studies over the last ten years have made great progress in determining molecular basis of apoptosis, including the mechanism by which the Bcl-2 family proteins regulate apoptosis. The Bcl-2 family seems to play a major role in determining the life or death of cells primarily through regulation of the redistribution of mitochondrial apoptogenic factors. Elucidation of the functional targets of Bcl-2 family proteins will be important not only for understanding the molecular basis of cell death but also for developing therapeutic strategies for various diseases that involve aberrant cell death.

Acknowledgements

I wish to thank all the members of my laboratory at Osaka University Graduate School of Medicine. The work performed at this laboratory was supported in part by a grant for Scientific Research on Priority Areas, by a grant for Center of Excellence Research, and by grants for Scientific Research from the Ministry of Education, Science, Sports, and Culture of Japan.

References

1. Alnemri, E. S., Livingston, D. J., Nicholson, D. W., Salvesen, G., Thornberry, N. A., Wong, W. W., *et al.* (1996). Human ICE/CED-3 protease nomenclature. *Cell,* **87**, 171.
2. Thornberry, N. A. and Lazebnik, Y. (1998). Caspases: Enemies within. *Science,* **281**, 1312.
3. Adams, J. M. and Cory, S. (1998). The Bcl-2 protein family: arbiters of cell survival. *Science,* **281**, 1322.
4. Tsujimoto, Y. (1998). Role of Bcl-2 family proteins in apoptosis: apoptosomes or mitochondria? *Genes Cells,* **3**, 697.
5. Yuan, J., Shaham, S., Ledoux, S., Ellis, H. M., and Horvitz, H. R. (1993). The *C. elegans* cell death gene ced-3 encodes a protein similar to mammalian interleukin-1 beta-converting enzyme. *Cell,* **75**, 641.
6. Hengartner, M. O. and Horvitz, H. R. (1994). *C. elegans* cell survival gene ced-9 encodes a functional homolog of the mammalian proto-oncogene bcl-2. *Cell,* **76**, 665.
7. Zou, H., Henzel, W. J., Liu, X., Lutschg, A., and Wang, X. (1997). Apaf-1, a human protein homologous to *C. elegans* CED-4, participates in cytochrome c-dependent activation of caspase-3. *Cell,* **90**, 405.
8. Conradt, B. and Horvitz, H. R. (1998). The *C. elegans* protein EGL-1 is required for programmed cell death and interacts with the Bcl-2-like protein CED-9. *Cell,* **93**, 519.
9. Tsujimoto, Y., Finger, L. R., Yunis, J., Nowell, P. C., and Croce, C. M. (1984). Cloning of the chromosome breakpoint of neoplastic B cells with the t(14;18) chromosome translocation. *Science,* **226**, 1097.
10. Tsujimoto, Y., Cossman, J., Jaffe, E., and Croce, C. M. (1985). Involvement of the bcl-2 gene in human follicular lymphoma. *Science,* **228**, 440.
11. Bakhshi, A., Jensen, J. P., Goldman, P., Wright, J. J., McBride, O. W., Epstein, A. L., *et al.* (1985). Cloning the chromosomal breakpoint of t(14;18) human lymphomas: clustering around JH on chromosome 14 and near a transcriptional unit on 18. *Cell,* **41**, 899.
12. Cleary, M. L. and Sklar, J. (1985). Nucleotide sequence of a t(14;18) chromosomal breakpoint in follicular lymphoma and demonstration of a breakpoint-cluster region near a transcriptionally active locus on chromosome 18. *Proc. Natl. Acad. Sci. USA,* **82**, 7439.
13. Tsujimoto, Y. and Croce, C. M. (1986). Analysis of the structure transcripts and protein products of bcl-2, the gene involved in human follicular lymphoma. *Proc. Natl. Acad. Sci. USA,* **83**, 5214.
14. Tashiro, S., Takechi, M., Asou, H., Takauchi, K., Kyo, T., Dohy, H., *et al.* (1992). Cytogenetic 2;18 and 18;22 translocation in chronic lymphocytic leukemia with juxtaposition of bcl-2 and immunoglobulin light chain genes. *Oncogene,* **7**, 573.
15. Vaux, D. L., Cory, S., and Adams, J. M. (1988). bcl-2 gene promotes haematopoietic cell survival and cooperates with c-myc to immortalize pre-B cells. *Nature,* **335**, 440.
16. Tsujimoto, Y. (1989). Stress-resistance conferred by high level of bcl-2α protein in a human B lymphoblastoid cell. *Oncogene,* **4**, 1331.
17. Nunez, G., London, L., Hockenbery, D., Alexander, M., McKearn, J. P., and Korsmeyer, S. J. (1990). Deregulated Bcl-2 gene expression selectively prolongs survival of growth factor-deprived hemopoietic cell lines. *J. Immunol.,* **144**, 3602.
18. Scaffidi, C., Fulda, S., Srinivasan, A., Friesen, C., Li, F., Tomaselli, K. J., *et al.* (1998). Two CD95 (APO-1/Fas) signaling pathways. *EMBO J.,* **17**, 1675.

19. Strasser, A., Harris, A. W., and Cory, S. (1991). bcl-2 transgene inhibits T cell death and perturbs thymic self-censorship. *Cell*, **67**, 889.
20. Kane, D. J., Ord, T., Anton, R., and Bredesen, D. E. (1995). Expression of bcl-2 inhibits necrotic neural cell death. *J. Neurosci. Res.*, **40**, 1274.
21. Shimizu, S., Eguchi, Y., Kamiike, W., Waguri, S., Uchiyama, Y., Matsuda, H., *et al.* (1996). Retardation of chemical hypoxia-induced necrotic cell death by Bcl-2 and ICE inhibitors: Possible common steps in apoptosis and necrosis. *Oncogene*, **12**, 2045.
22. Shimizu, S., Eguchi, Y., Kamiike, W., Waguri, S., Uchiyama, Y., Matsuda, H., *et al.* (1996). Bcl-2 blocks loss of mitochondrial membrane potential while ICE inhibitors act at a different step during inhibition of death induced by respiratory chain inhibitors. *Oncogene*, **13**, 21.
23. Meijerink, J. P., Smetsers, T. F., Sloetjes, A. W., Linders, E. H., and Mensink, E. J. (1995). Bax mutations in cell lines derived from hematological malignancies. *Leukemia*, **9**, 1828.
24. Rampino, N., Yamamoto, H., Ionov, Y., Li, Y., Sawai, H., Reed, J. C., *et al.* (1997). Somatic frameshift mutations in the Bax gene in colon cancers of the microsatellite mutator phenotype. *Science*, **275**, 967.
25. Inohara, N., Ekhterae, D., Garcia, I., Carrio, R., Merino, J., Merry, A., *et al.* (1998). Mtd, a novel Bcl-2 family member activates apoptosis in the absence of heterodimerization with Bcl-2 and Bcl-xL. *J. Biol. Chem.*, **273**, 8705.
26. Oltvai, Z. N., Milliman, C. L., and Korsmeyer, S. J. (1993), Bcl-2 heterodimerizes in vivo with a conserved homolog, Bax, that accelerates programmed cell death. *Cell*, **74**, 609.
27. Yang, E., Zha, J., Jockel, J., Boise, L. H., Thompson, C. B., and Korsmeyer, S. J. (1995). Bad, a heterodimeric partner for Bcl-x_L and Bcl-2, displaces Bax and promotes cell death. *Cell*, **80**, 285.
28. Sattler, M., Liang, H., Nettesheim, D., Meadows, R. P., Harlan, J. E., Eberstadt, M., *et al.* (1997). Structure of Bcl-x_L-Bak peptide complex: recognition between regulators of apoptosis. *Science*, **275**, 983.
29. Goping, I. S., Gross, A., Lavoie, J. N., Nguyen, M., Jemmerson, R., Roth, K., *et al.* (1998). Regulated targeting of BAX to mitochondria. *J. Cell Biol.*, **143**, 207.
30. Griffiths, G. J., Dubrez, L., Morgan, C. P., Jones, N. A., Whitehouse, J., Corfe, B. M., *et al.* (1999). Cell damage-induced conformational changes of the pro-apoptotic Protein Bak *in vivo* precede the onset of apoptosis. *J. Cell Biol.*, **144**, 903.
31. Luo, X., Budihardjo, I., Zou, H., Slaughter, C., and Wang, X. (1998). Bid, a Bcl-2 interacting protein, mediates cytochrome c release from mitochondria in response to activation of cell surface death receptors. *Cell*, **94**, 471.
32. Li, H., Zhu, H., Xu, C., and Yuan, J. (1998). Cleavage of BID by caspase 8 mediates the mitochondrial damage in the Fas pathway of apoptosis. *Cell*, **94**, 481.
33. Huang, D. C., Adams, J. M., and Cory, S. (1998). The conserved N-terminal BH4 domain of Bcl-2 homologues is essential for inhibition of apoptosis and interaction with CED-4. *EMBO J.*, **17**, 1029.
34. Chittenden, T., Flemington, C., Houghton, A. B., Ebb, R. G., Gallo, G. J., Elangovan, B., *et al.* (1995). A conserved domain in Bak, distinct from BH1 and BH2, mediates cell death and protein binding functions. *EMBO J.*, **14**, 5589.
35. Cosulich, S., Worrall, V., Hedge, P., Green, S., and Clarke, P. (1997). Regulation of apoptosis by BH3 domains in a cell-free system. *Curr. Biol.*, **7**, 913.

36. Narita, M., Shimizu, S., Ito, T., Chittenden, T., Lutz, R. J., Matsuda, H., *et al.* (1998). Bax interacts with the permeability transition pore to induce permeability transition and cytochrome c release in isolated mitochondria. *Proc. Natl. Acad. Sci. USA*, **95**, 14681.
37. Hunter, J. J. and Parslow, T. G. (1996). A peptide sequence from Bax that converts Bcl-2 into an activator of apoptosis. *J. Biol. Chem.*, **271**, 8521.
38. Han, J., Modha, D., and White, E. (1998). Interaction of E1B 19K with Bax is required to block Bax-induced loss of mitochondrial membrane potential and apoptosis. *Oncogene*, **17**, 2993.
39. Knudson, C. M., Tung, K. S., Tourtellotte, W. G., Brown, G. A., and Korsmeyer, S. J. (1995). Bax-deficient mice with lymphoid hyperplasia and male germ cell death. *Science*, **270**, 96.
40. Cheng, E. H., Levine, B., Boise, L. H., Thompson, C. B., and Hardwick, J. M. (1996). Bax-independent inhibition of apoptosis by Bcl-xL. *Nature*, **379**, 554.
41. Minn, A. J., Kettlun, C. S., Liang, H., Kelekar, A., Vander Heiden, M. G., Chang, B. S., *et al.* (1999). Bcl-xL regulates apoptosis by heterodimerization-dependent and -independent mechanisms. *EMBO J.*, **18**, 632.
42. Simonian, P. L., Grillot, D. A., and Nunez, G. (1997). Bak can accelerate chemotherapy-induced cell death independently of its heterodimerization with Bcl-xL and Bcl-2. *Oncogene*, **15**, 1871.
43. Wang, K., Gross, A., Waksman, G., and Korsmeyer, S. J. (1998). Mutagenesis of the BH3 domain of Bax identifies residues critical for dimerization and killing. *Mol. Cell. Biol.*, **18**, 6083.
44. Zha, H. and Reed, J. C. (1997). Heterodimerization-independent functions of cell death regulatory proteins Bax and Bcl-2 in yeast and mammalian cells. *J. Biol. Chem.*, **272**, 31482.
45. Chinnaiyan, A. M., Orth, K., O'Rouke, K., Duan, H., Poirier, G. G., and Dixit, V. M. (1996). Molecular ordering of the cell death pathway: Bcl-2 and Bcl-xL function upstream of the Ced-3-like apoptotic proteases. *J. Biol. Chem.*, **271**, 4573.
46. Shimizu, S., Eguchi, Y., Kamiike, W., Matsuda, H., and Tsujimoto, Y. (1996). Bcl-2 expression prevents activation of the ICE protease cascade. *Oncogene*, **12**, 2251.
47. Boulakia, C. A., Chen, G., Ng, F. W., Teodoro, J. G., Branton, P. E., Nicholson, D. W., *et al.* (1996). Bcl-2 and adenovirus E1B 19 kDA protein prevent E1A-induced processing of CPP32 and cleavage of poly(ADP-ribose) polymerase. *Oncogene*, **12**, 529.
48. Yasuhara, N., Sahara, S., Kamada, S., Eguchi, Y., and Tsujimoto, Y. (1997). Evidence against a functional site for Bcl-2 downstream of caspase cascade in preventing apoptosis. *Oncogene*, **15**, 1921.
49. Monaghan, P., Robertson, D., Amos, T. A., Dyer, M. J., Mason, D. Y., and Greaves, M. F. (1992). Ultrastructural localization of bcl-2 protein. *J. Histochem. Cytochem.*, **40**, 1819.
50. Akao, Y., Otsuki, Y., Kataoka, S., Ito, Y., and Tsujimoto, Y. (1994). Multiple subcellular localization of bcl-2: Detection in nuclear outer membrane, endoplasmic reticulum membrane and mitochondrial membranes. *Cancer Res.*, **54**, 2468.
51. Liu, X., Kim, C. N., Yang, J., Jemmerson, R., and Wang, Z. (1996). Induction of apoptotic program in cell-free extracts: Requirement for dATP and cytochrome c. *Cell*, **86**, 147.
52. Yang, J., Liu, X., Bhalla, K., Kim, C. N., Ibrado, A. M., Cai, J., *et al.* (1997). Prevention of apoptosis by Bcl-2: Release of cytochrome c from mitochondria blocked. *Science*, **275**, 1129.

53. Kluck, R. M., Bossy-Wetzel, E., Green, D. R., and Newmeyer, D. D. (1997). The release of cytochrome c from mitochondria: A primary site for Bcl-2 regulation of apoptosis. *Science*, **275**, 1132.
54. Li, P., Nijhawan, D., Budihardjo, I., Srinivasula, S. M., Ahmad, M., Alnemri, E. S., *et al.* (1997). Cytochorme c and dATP-dependent formation of Apaf-1/caspase-9 complex initiates an apoptotic protease cascade. *Cell*, **91**, 479.
55. Srinivasula, S. M., Ahmad, M., Fernandes-Alnemri, T., and Alnemri, S. (1998). Autoactivation of procaspase-9 by Apaf-1-mediated oligomerization. *Mol. Cell*, **1**, 949.
56. Susin, S. A., Lorenzo, H. K., Zamzami, N., Marzo, I., Snow, B. E., Brothers, G. M., *et al.* (1999). Molecular characterization of mitochondrial apoptosis-inducing factor. *Nature*, **397**, 441.
57. Susin, S. A., Lorenzo, H. K., Zamzami, N., Marzo, I., Brenner, C., Larochette, N., *et al.* (1999). Mitochondrial release of caspase-2 and -9 during the apoptotic process. *J. Exp. Med.*, **189**, 381.
58. Mancini, M., Nicholson, D. W., Roy, S., Thornberry, N. A., Peterson, E. P., Casciola-Rosen, L. A., *et al.* (1998). The caspase-3 precursor has a cytosolic and mitochondrial distribution: implications for apoptotic signaling. *J. Cell Biol.*, **140**, 1485.
59. Susin, S. A., Zamzami, N., Castedo, M., Daugas, E., Wang, H. G., Geley, S., *et al.* (1997). The central executioner of apoptosis: Multiple connections between protease activation and mitochondria in Fas/Apo-1/CD95- and ceramide-induced apoptosis. *J. Exp. Med.*, **186**, 25.
60. Kroemer, G., Dallaporta, B., and Resche-Rigon, M. (1998). The mitochondrial death/life regulator in apoptosis and necrosis. *Annu. Rev. Physiol.*, **60**, 619.
61. Shimizu, S., Eguchi, Y., Kamiike, W., Funahashi, Y., Mignon, A., Lacronique, V., *et al.* (1998). Bcl-2 prevents apoptotic mitochondrial dysfunction by regulating proton flux. *Proc. Natl. Acad. Sci. USA*, **95**, 1455.
62. Jurgensmeier, J. M., Xie, Z., Deveraux, Q., Ellerby, L., Bredesen, D., and Reed, J. C. (1998). Bax directly induces release of cytochrome c from isolated mitochondria. *Proc. Natl. Acad. Sci. USA*, **95**, 4997.
63. Eskes, R., Antonsson, B., Osen-Sand, A., Montessuit, S., Richter, C., Sadoul, R., *et al.* (1998). Bax-induced cytochrome c release from mitochondria is independent of the permeability transition pore but highly dependent on Mg^{2+} ions. *J. Cell Biol.*, **143**, 217.
64. Zamzami, N., Brenner, C., Marzo, I., Susin, S. A., and Kroemer, G. (1998). Subcellular and submitochondrial mode of action of Bcl-2-like oncoproteins. *Oncogene*, **16**, 2265.
65. Bossy-Wetzel, E., Newmeyer, D. D., and Green, D. R. (1998). Mitochondrial cytochrome c release in apoptosis occurs upstream of DEVD-specific caspase activation and independently of mitochondrial transmembrane depolarization. *EMBO J.*, **17**, 37.
66. Bernardi., P., Broekemeier, K. M., and Pfeiffer, D. R. (1994). Recent progress on regulation of the mitochondrial permeability transition pore; a cyclosporin-sensitive pore in the inner mitochondrial membrane. *J. Bioenerg. Biomembr.*, **26**, 509.
67. Zoratti, M. and Szabo, I. (1995). The mitochondrial permeability transition. *Biochim. Biophys. Acta*, **1241**, 139.
68. Marzo, I., Brenner, C., Zamzami, N., Jurgensmeier, J. M., Susin, S. A., Vieira, H. L., *et al.* (1998). Bax and adenine nucleotide translocator cooperate in the mitochondrial control of apoptosis. *Science*, **281**, 2027.
69. Vander Heiden, M. G., Chandel, N. S., Williamson, E. K., Schumacker, P. T., and Thompson, C. B. (1997). Bcl-xL regulates the membrane potential and volume homeostasis of mitochondria. *Cell*, **91**, 627.

70. Shimizu, S., Narita, M., and Tsujimoto, Y. (1999). Bcl-2 family proteins target a mitochondrial channel VDAC to regulate apoptogenic cytochrome c release. *Nature*, **399**, 483.
71. Sato, T., Hanada, M., Bodrug, S., Irie, S., Iwama, N., Boise, L. H., *et al.* (1994). Interactions among members of the Bcl-2 protein family analyzed with a yeast two-hybrid system. *Proc. Natl. Acad. Sci. USA*, **91**, 9238.
72. Matsuyama, S., Xu, Q., Velours, J., and Reed, J. C. (1998). The mitochondrial F0F1-ATPase proton pump is required for function of the proapoptotic protein Bax in yeast and mammalian cells. *Mol. Cell*, **1**, 327.
73. Muchmore, S. W., Sattler, M., Liang, H., Meadows, R. P., Harlan, J. E., Yoon, H. S., *et al.* (1996). X-ray and NMR structure of human Bcl-x_L, an inhibitor of programmed cell death. *Nature*, **381**, 3351.
74. Minn, A. J., Velez, P., Schendel, S. L., Liang, H., Muchmore, S. W., Fesik, S. W., *et al.* (1997). Bcl-xL forms as ion channel in synthetic lipid membranes. *Nature*, **385**, 353.
75. Schendel, S. L., Xie, Z., Montal, M. O., Matsuyama, S., Montal, M., and Reed, J. C. (1997). Channel formation by antiapoptotic protein Bcl-2. *Proc. Natl. Acad. Sci. USA*, **94**, 5113.
76. Antonsson, B., Conti, F., Ciavatta, A., Montessuit, S., Lewis, S., Martinou, I., *et al.* (1997). Inhibition of Bax channel-forming activity by Bcl-2. *Science*, **277**, 370.
77. Schlesinger, P. H., Gross, A., Yin, X. M., Yamamoto, K., Saito, M., Waksman, G., *et al.* (1997). Comparison of the ion channel characteristics of proapoptotic Bax and anti-apoptotic Bcl-2. *Proc. Natl. Acad. Sci. USA*, **94**, 11357.
78. McDonnell, J. M., Fushman, D., Milliman, C. L., Korsmeyer, S. J., and Cowburn, D. (1999). Solution structure of the proapoptotic molecule Bid: A structural basis for apoptotic agonists and antagonists. *Cell*, **96**, 625.
79. Chou, J. J., Li, H., Salvesen, G. S., Yuan, J., and Wagner, G. (1999). Solution structure of BID, an intracellular amplifier of apoptotic signaling. *Cell*, **96**, 615.
80. Ng, F. W., Nguyen, M., Kwan, T., Branton, P. E., Nicholson, D. W., Cromlish, J. A., *et al.* (1997). p28 Bap31, a Bcl-2/Bcl-x_L- and procaspase-8-associated protein in the endoplasmic reticulum. *J. Cell Biol.*, **139**, 327.
81. Pan, G., O'Rourke, K., and Dixit, V. M. (1998). Caspase-9, Bcl-xL, and Apaf-1 form a ternary complex. *J. Biol. Chem.*, **273**, 5841.
82. Hu, Y., Benedict, M. A., Wu, D., Inohara, N., and Nunez, G. (1998). Bcl-xL interacts with Apaf-1 and inhibits Apaf-1-dependent caspase-9 activation. *Proc. Natl. Acad. Sci. USA*, **95**, 4386.
83. Chinnaiyan, A. M., O'Rourke, K., Lane, B. R., and Dixit, V. M. (1997). Interaction of CED-4 with CED-3 and CED-9: A molecular framework for cell death. *Science*, **275**, 1122.
84. Wu, D., Wallen, H. D., and Nunez, G. (1997). Interaction and regulation of subcellular localization of CED-4 by CED-9. *Science*, **275**, 1126.
85. Spector, M. S., Desnoyers, S., Hoeppner, D. J., and Hengartner, M. O. (1997). Interaction between the *C. elegans* cell-death regulators CED-9 and CED-4. *Nature*, **385**, 653.
86. Inohara, N., Gourley, T. S., Carrio, R., Muniz, M., Merino, J., Garcia I., *et al.* (1998). Diva, a Bcl-2 homologue that binds directly to Apaf-1 and induces BH3-independent cell death. *J. Biol. Chem.*, **273**, 32479.
87. Zhu, W., Cowie, A., Wasfy, G. W., Penn, L. Z., Leber, B., and Andrews, D. W. (1996). Bcl-2 mutants with restricted subcellular location reveal spatially distinct pathways for apoptosis in different cell types. *EMBO J.*, **15**, 4130.

88. Distelhorst, C. W., Lam, M., and McCormick, T. S. (1996). Bcl-2 inhibits hydrogen peroxide-induced ER Ca^{2+} pool depletion. *Oncogene*, **12**, 2051.
89. He, H., Lam, M., McCormick, T. S., and Distelhorst, C. W. (997). Maintenance of calcium homeostasis in the endoplasmic reticulum by Bcl-2. *J. Cell Biol.*, **138**, 1219.
90. Beham, A., Marin, M. C., Fernandez, A., Herrmann, J., Brisbay, S., Tari, A. M., *et al.* (1997). Bcl-2 inhibits p53 nuclear import following DNA damage. *Oncogene*, **4**, 2767.
91. Ryan, J. J., Prochownik, E., Gottlieb, C. A., Apel, I. J., Merino, R., Nunez, G., *et al.* (1994). c-myc and bcl-2 modulate p53 function by altering p53 subcellular trafficking during the cell cycle. *Proc. Natl. Acad. Sci. USA*, **91**, 5878.
92. Shibasaki, F., Kondo, E., Akagi, T., and McKeon, F. (1997). Suppression of signalling through transcription factor NF-AT by interactions between calcineurin and Bcl-2. *Nature*, **386**, 728.
93. Marin, M. C., Fernandez, A., Bick, R. J., Brisbay, S., Buja, L. M., Snuggs, M., *et al.* (1996). Apoptosis suppression by bcl-2 is correlated with the regulation of nuclear and cytosolic Ca^{2+}. *Oncogene*, **12**, 2259.
94. Zha, J., Harada, H., Yang, E., Jockel, J., and Korsmeyer, S. J. (1996). Serine phosphorylation of death agonist Bad in response to survival factor results in binding to 14-3-3 not Bcl-x_L. *Cell*, **15**, 619.
95. Datta, S. R., Dudek, H., Tao, X., Masters, S., Fu, H., Gotoh, Y., *et al.* (1997). Akt phosphorylation of BAD couples survival signals to the cell-intrinsic death machinery. *Cell*, **91**, 231.
96. Wolter, K. G., Hsu, Y. T., Smith, C. L., Nechushtan, A., Xi, X. G., and Youle, R. J. (1997). Movement of Bax from the cytosol to mitochondria during apoptosis. *J. Cell Biol.*, **139**, 1281.
97. Gross, A., Jockel, J., Wei, M. C., and Korsmeyer, S. J. (1998). Enforced dimerization of Bax results in its translocation, mitochondrial dysfunction and apoptosis. *EMBO J.*, **17**, 3878.
98. Desagher, S., Osen-Sand, A., Nichols, A., Eskes, R., Montessuit, S., Lauper, S., *et al.* (1999). Bid-induced conformational change of Bax is responsible for mitochondrial cytochrome c release during apoptosis. *J. Cell Biol.*, **144**, 891.
99. Haldar, S., Jena, N., and Croce, C. M. (1995). Inactivation of Bcl-2 by phosphorylation. *Proc. Natl. Acad. Sci. USA*, **92**, 4507.
100. Ito, T., Deng, X., Carr, B., and May, W. S. (1997). Bcl-2 phosphorylation required for anti-apoptosis function. *J. Biol. Chem.*, **272**, 11671.
101. Srivastava, R. K., Srivastava, A. R., Korsmeyer, S. J., Nesterova, M., Cho-Chung, Y. S., and Longo, D. L. (1998). Involvement of microtubules in the regulation of Bcl2 phosphorylation and apoptosis through cyclic AMP-dependent protein kinase. *Mol. Cell. Biol.*, **18**, 3509.
102. Cheng, E. H., Kirsch, D. G., Clem, R. J., Ravi, R., Kastan, M. B., Bedi, A., *et al.* (1997). Conversion of Bcl-2 to a Bax-like death effector by caspases. *Science*, **278**, 1966.
103. Clem, R. J., Cheng, E. H., Karp, C. L., Kirsch, D. G., Ueno, K., Takahashi, A., *et al.* (1998). Modulation of cell death by Bcl-xL through caspase interaction. *Proc. Natl. Acad. Sci. USA*, **95**, 554.
104. Takayama, S., Sato, T., Krajewski, S., Kochel, K., Irie, S., Millan, J. A., *et al.* (1995). Cloning and functional analysis of BAG-1: a novel Bcl-2-binding protein with anti-cell death activity. *Cell*, **80**, 279.
105. Wang, H.-G., Rapp, U., and Reed, J. C. (1996). Bcl-2 targets the protein kinase Raf-1to mitochondria. *Cell*, **87**, 629.

106. Iwahashi, H., Eguchi, Y., Yasuhara, N., Hanafusa, T., Matsuzawa, Y., and Tsujimoto, Y. (1997). Synergy between Bcl-2 and SMN, a protein implicated in spinal muscular atrophy. *Nature*, **390**, 413.
107. Lefebvre, S., Burglen, L., Reboullet, S., Clermont, O., Burlet, P., Viollet, L., *et al.* (1995). Identification and characterization of a spinal muscular atrophy-determining gene. *Cell*, **80**, 155.
108. Takayama, S., Bimston, D. N., Matsuzawa, S., Freeman, B. C., Aime-Sempe, C., Xie, Z., *et al.* (1997). BAG-1 modulates the chaperone activity of Hsp70/Hsc70. *EMBO J.*, **16**, 4887.
109. Xu, Q. and Reed, J. C. (1998). Bax inhibitor-1, a mammalian apoptosis suppressor identified by functional screening in yeast. *Mol. Cell*, **1**, 337.
110. Boyd, J. M., Malstrom, S., Subramanian, T., Venkatesh, L. K., Schaeper, U., Elangovan, B., *et al.* (1994). Adenovirus E1B 19 kDa and Bcl-2 proteins interact with a common set of cellular proteins. *Cell*, **79**, 341.
111. Matsushima, M., Fujiwara, T., Takahashi, E., Minaguchi, T., Eguchi, Y., Tsujimoto, Y., *et al.* (1998). Isolation, mapping, and functional analysis of a novel human cDNA (BNIP3L) encoding a protein homologous to human NIP3. *Genes Chromosomes Cancer*, **21**, 230.
112. Yasuda, M., Theodorakis, P., Subramanian, T., and Chinnadurai, G. (1998). Adenovirus E1B-19K/BCL-2 interacting protein BNIP3 contains a BH3 domain and a mitochondrial targeting sequence. *J. Biol. Chem.*, **273**, 12415.
113. Chen, G., Cizeau, J., Vande Velde, C., Park, J. H., Bozek, G., Bolton, J., *et al.* (1999). Nix and Nip3 form a subfamily of pro-apoptotic mitochondrial proteins. *J. Biol. Chem.*, **274**, 7.
114. Imazu, T., Shimizu, S., Tagami, S., Matsushima, M., Nakamura, Y., Miki, T., *et al.* (1999). Bcl-2/E1B 19 kDa-interacting protein 3-like protein (Bnip3L) interacts with Bcl-2/Bcl-xL and induces apoptosis by altering mitochondrial membrane permeability. *Oncogene*, **18**, 4523.
115. Schendel, S. L., Azimov, R., Pawlowski, K., Godzik, A., Kagan, B. L., and Reed, J. C. (1999). Ion channel activity of the BH3 only Bcl-2 family member, BID. *J. Biol. Chem.*, **274**, 21932.
116. Moriishi, K., Huang, D. C., Cory, S., and Adams, J. M. (1999). Bcl-2 family members do not inhibit apoptosis by binding the caspase activator Apaf-1. *Proc. Natl. Acad. Sci. USA*, **96**, 9683.
117. Harada, H., Becknell, B., Wilm, M., Mann, M., Huang, L. J., Taylor, S. S., *et al.* (1999). Phosphorylation and inactivation of BAD by mitochondria-anchored protein kinase A. *Mol. Cell*, **3**, 413.
118. Wang, H. G., Pathan, N., Ethell, I. M., Krajewski, S., Yamaguchi, Y., Shibasaki, F., *et al.* (1999). Ca^{2+}-induced apoptosis through calcineurin dephosphorylation of BAD. *Science*, **284**, 339.
119. Nomura, M., Shimizu, S., Ito, T., Narita, M., Matsuda, H., and Tsujimoto, Y. (1999). Apoptotic cytosol facilitates Bax translocation to mitochondria that involves cytosolic factor regulated by Bcl-2. *Cancer Res.*, **59**, 5542.

Addendum: recent discoveries on Bcl-2 family proteins

This addendum covers recent progress in the field of Bcl-2 family proteins that has occurred since this chapter was written.

1. Mechanisms for breaking outer mitochondrial membrane permeability

An increase in outer mitochondrial membrane permeability is central to apoptotic signaling pathways regulated by Bcl-2 family proteins. The best-characterized apoptogenic factors known to be released from mitochondria during apoptosis are cytochrome c, AIF and Smac/Diablo (an inhibitor of IAP family proteins that inhibits caspases via direct binding) (1, 2). All of these proteins are localized in the mitochondrial intermembrane space, suggesting that their release occurs via openings in the outer mitochondrial membrane (reviewed in 3, 4). One model (discussed below) suggests that apoptogenic proteins are released through specific pores in the outer membrane. The alternative model, physical tear of the outer mitochondrial membrane, involves mitochondrial swelling or membrane instability that might be directly induced by Bax and Bid. Previous studies have suggested, however, that mitochondrial swelling and membrane rupture rarely accompany apoptosis, suggesting that this form of mitochondrial change may be more relevant to necrotic cell death.

2. Protein conducting pores for release of mitochondrial apoptogenic factors

There are two main models for the formation of protein-conducting pores. First, that they are formed directly by oligomerization of Bax and Bak and second, that Bax and Bak induce the formation of a pore consisting of the mitochondrial protein VDAC (voltage-dependent anion channel, also known as mitochondrial porin). Evidence for the Bax/Bak oligomerization model is supported by observations of Bax and Bak oligomers within cells that have been stimulated to undergo apoptosis (5, 6). Bax oligomers have also been found on synthetic lipid membranes, where they form tetrameric cytochrome c-conducting pores (7). Bax and Bak oligomers have also been found when isolated mitochondria were incubated with tBid (8, 9). However, it is not yet clear whether tBid directly induces oligomerization of Bax and Bak. Since little stable complex of tBid and Bax is seen, it has been proposed that tBid only transiently interacts with Bax and induces a conformational change, leading to Bax oligomerization (9).

The VDAC pore model is also supported by recent findings. VDAC contributes to the formation of a cytochrome c-conducting pore and is one of the functional targets for Bcl-2 family members (10, 11). In the presence of Bax/Bak, VDAC forms a large pore on liposomes that is cytochrome c permeable and is inhibited by a VDAC inhibitor (10, 12). Anti-apoptotic members Bcl-x_L and Bcl-2 directly bind to and inhibit the activity of VDAC (10, 13), thereby preventing the formation of a cytochrome c-conducting pore. Furthermore, anti-human VDAC antibodies inhibit the apoptotic changes of isolated mitochondria *in vitro* induced by apoptogenic agents such as recombinant Bax and calcium, as well as the apoptotic deaths of cells in culture that are induced by microinjection of Bax, etoposide or staurosporine (11).

It is possible that both of these mechanisms may be correct. For example, Bax or Bak oligomers may serve as a pore for the initial release of cytochrome c, which is then followed by a larger efflux of cytochrome c through VDAC-containing pores. Alternatively, these two kinds of pores might be generated simultaneously. In either case, Bcl-2 and Bcl-x_L appear to inhibit the formation of both types of protein-conducting pores.

3. BH3-only proteins serve as sensors for various death signals

BH3-only proteins seem to have distinct roles in different forms of apoptosis, by localizing different subcellular compartments or by being activated by different signals. For example, Bax, Noxa (14), and Puma (15, 16) are transcriptionally activated upon DNA damage, whereas cytoplasmic Bid is cleaved and the truncated form translocates to the mitochondria during death receptor-mediated apoptosis. Bim and Bmf respond to apoptotic signals that are transmitted via proteins that associate with the cytoskeleton. Bim is normally sequestered to the microtubular dynein motor complex through dynein light chain 1 (or LC8), and Bmf localizes to myosin V actin motor complex through dynein light chain 2. Upon apoptotic stimuli, these proteins translocate to the mitochondria to induce release of apoptogenic factors (17, 18). Bim is released from the microtubular complex by apoptosis induced by paclitaxel, which polymerizes microtubules. Similarly Bmf, but not Bim, is released from the actin-myosin complex by apoptotic stimuli, such as Anoikis, that affect the actin cytoskeleton. Both Bim and Bmf appear to be released by UV irradiation. Thus, various BH3-only proteins appear to serve as intracellular death signal sensors that transmit various signals to the mitochondria through multi-domain pro-apoptotic Bcl-2 family members (Fig. 1).

4. Convergence point for life-or-death signals

Cells from Bax/Bak-double knockout mice have been shown to be resistant to a variety of apoptotic stimuli including the BH3-only proteins tBid, Bad, Noxa and Bim (19, 20), suggesting that Bax and Bak function as a convergence point for multiple apoptotic signalling pathways (Fig. 1). VDAC may also function as a second convergence point, directly downstream of Bax and Bak as well as through interactions with Bcl-2 and Bcl-x_L. Activated BH-3-only proteins appear to function through both Bax/Bak and Bcl-2/Bcl-xL pathways by interacting with and inactivating the anti-apoptotic members, and by activating multi-domain pro-apoptotic members such as Bax and Bak by a still unidentified mechanism. Although Bcl-2 and Bcl-x_L function through heterodimerization with various pro-apoptotic members, Bcl-x_L can also function independently of other pro-apoptotic members (21). It is

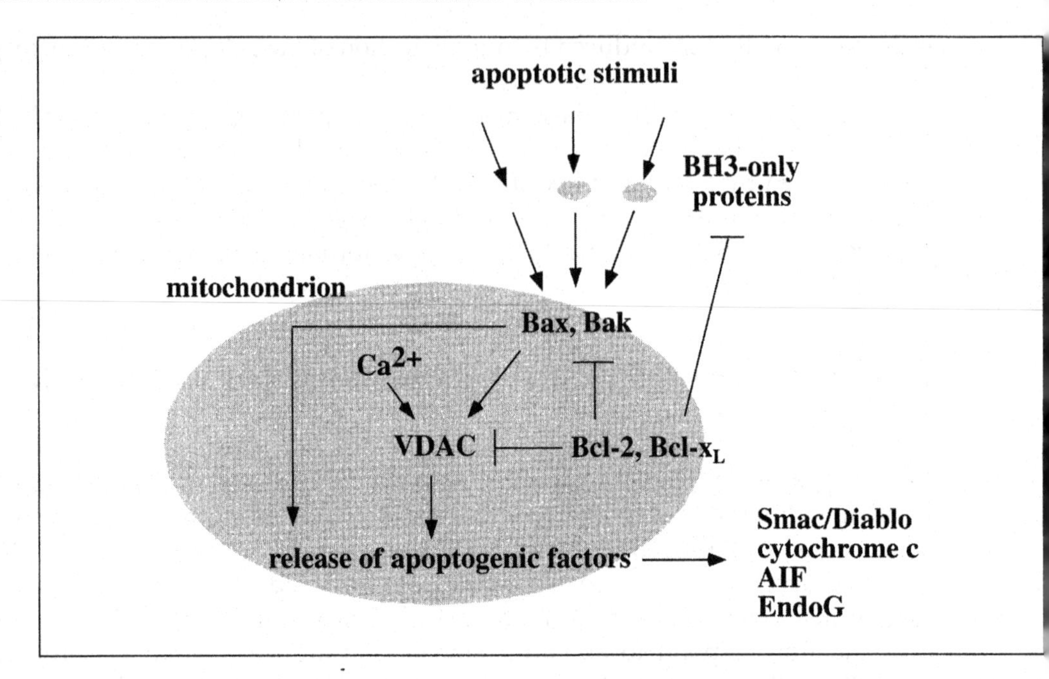

Fig. 1 Convergence point for various apoptotic signals. Apoptotic signals transmitted via BH-3-only proteins converge on pro-apoptotic proteins Bax and Bak. VDAC also functions as a convergence point for signals conveyed to Bax and Bak, as well as the anti-apoptotic proteins Bcl-2 and Bcl-x_L.

possible that this activity might be exhibited by functioning with VDAC (Fig. 1). The notion that VDAC constitutes a convergence point for various death signals is also supported by observations that Bcl-2 is capable of preventing apoptotic changes of yeast mitochondria that do not possess Bcl-2 family members but do possess VDAC (22). Moreover, VDAC is directly involved in the mitochondrial permeability transition (11), suggesting that the inhibition of VDAC by Bcl-2 and Bcl-x_L underlies the ability of Bcl-2 and Bcl-x_L to prevent permeability transition.

References

1. Verhagen, A. M., Ekert, P. G., Pakusch, M., Silke, J., Connolly, L. M., Reid, G. E., *et al.* (2000). Identification of DIABLO, a mammalian protein that promotes apoptosis by binding to and antagonizing IAP proteins. *Cell*, **102**, 43-53.
2. Du, C., Fang, M., Li, Y., Li, L., and Wang, X. (2000). Smac, a mitochondrial protein that promotes cytochrome c-dependent caspase activation by eliminating IAP inhibition. *Cell*, **102**, 33-42.
3. Tsujimoto, Y. and Shimizu, S. (2000). VDAC regulation by the Bcl-2 family of proteins. *Cell Death Differ.*, **7**, 1174-1181.
4. Tsujimoto, Y. and Shimizu, S. (2000). Bcl-2 family: life-or-death switch. *FEBS Lett*, **466**, 6-10.

5. Wei, M. C., Zong, W. X., Cheng, E. H., Lindsten, T., Panoutsakopoulou, V., Ross, A. J., *et al.* (2001). Proapoptotic BAX and BAK: a requisite gateway to mitochondrial dysfunction and death. *Science,* **292,** 727-730.
6. Antonsson, B., Montessuit, S., Sanchez, B., and Martinou, J. C. (2001). Bax is present as a high molecular weight oligomer/complex in the mitochondrial membrane of apoptotic cells. *J. Biol. Chem.,* **276,** 11615-11623.
7. Saito, M., Korsmeyer, S. J., and Schlesinger, P. H. (2000). BAX-dependent transport of cytochrome c reconstituted in pure liposomes. *Nat. Cell. Biol.,* **2,** 553-555.
8. Eskes, R., Desagher, S., Antonsson, B., and Martinou, J. C. (2000). Bid induces the oligomerization and insertion of Bax into the outer mitochondrial membrane. *Mol. Cell. Biol.,* **20,** 929-935.
9. Wei, M. C., Lindsten, T., Mootha, V. K., Weiler, S., Gross, A., Ashiya, M., *et al.* (2000). tBID, a membrane-targeted death ligand, oligomerizes BAK to release cytochrome c. *Genes Dev.,* **14,** 2060-2071.
10. Shimizu, S., Narita, M., and Tsujimoto, Y. (1999). Bcl-2 family proteins regulate the release of apoptogenic cytochrome c by the mitochondrial channel VDAC. *Nature,* **399,** 483-487.
11. Shimizu, S., Matsuoka, Y., Shinohara, Y., Yoneda, Y., and Tsujimoto, Y. (2001). Essential role of voltage-dependent anion channel in various forms of apoptosis in mammalian cells. *J. Cell. Biol.,* **152,** 237-250.
12. Shimizu, S., Ide, T., Yanagida, T., and Tsujimoto, Y. (2000). Electrophysiological study of a novel large pore formed by Bax and the voltage-dependent anion channel that is permeable to cytochrome c. *J. Biol. Chem.,* **275,** 12321-12325.
13. Shimizu, S., Konishi, A., Kodama, T., and Tsujimoto, Y. (2000). BH4 domain of antiapoptotic Bcl-2 family members closes voltage-dependent anion channel and inhibits apoptotic mitochondrial changes and cell death. *Proc. Natl. Acad. Sci. USA,* **97,** 3100-3105.
14. Oda, E., Ohki, R., Murasawa, H., Nemoto, J., Shibue, T., Yamashita, T., *et al.* (2000). Noxa, a BH3-only member of the Bcl-2 family and candidate mediator of p53-induced apoptosis. *Science,* **288,** 1053-1058.
15. Yu, J., Zhang, L., Hwang, P. M., Kinzler, K. W., and Vogelstein, B. (2001). PUMA induces the rapid apoptosis of colorectal cancer cells. *Mol. Cell.,* **7,** 673-682.
16. Nakano, K. and Vousden, K. H. (2001). PUMA, a novel proapoptotic gene, is induced by p53. *Mol. Cell.,* **7,** 683-694.
17. Puthalakath, H., Huang, D. C., O'Reilly, L. A., King, S. M., and Strasser, A. (1999). The proapoptotic activity of the Bcl-2 family member Bim is regulated by interaction with the dynein motor complex. *Mol. Cell.,* **3,** 287-296.
18. Puthalakath, H., Villunger, A., O'Reilly, L. A., Beaumont, J. G., Coultas, L., Cheney, R. E., *et al.* (2001). Bmf: a proapoptotic BH3-only protein regulated by interaction with the myosin V actin motor complex, activated by anoikis. *Science,* **293,** 1829-1832.
19. Lindsten, T., Ross, A. J., King, A., Zong, W. X., Rathmell, J. C., Shiels, H. A., *et al.* (2000). The combined functions of proapoptotic Bcl-2 family members bak and bax are essential for normal development of multiple tissues. *Mol. Cell.,* **6,** 1389-1399.
20. Cheng, E. H., Wei, M. C., Weiler, S., Flavell, R. A., Mak, T. W., Lindsten, T., *et al.* (2001). BCL-2, BCL-X(L) sequester BH3 domain-only molecules preventing BAX- and BAK-mediated mitochondrial apoptosis. *Mol. Cell.,* **8,** 705-711.

21. Minn, A. J., Kettlun, C. S., Liang, H., Kelekar, A., Vander Heiden, M. G., Chang, B. S., *et al.* (1999). Bcl-xL regulates apoptosis by heterodimerization-dependent and -independent mechanisms. *Embo. J.*, **18**, 632-643.
22. Shimizu, S., Shinohara, Y., and Tsujimoto, Y. (2000). Bax and Bcl-xL independently regulate apoptotic changes of yeast mitochondria that require VDAC but not adenine nucleotide translocator. *Oncogene*, **19**, 4309-4318.

6 | Mitochondria in apoptosis: Pandora's Box

NAOUFAL ZAMZAMI, SANTOS A. SUSIN, and GUIDO KROEMER

1. Introduction

Permeabilization of mitochondrial membranes is an early rate-limiting event of the apoptotic process. At least five different classes of effector molecules are released from the intermembrane space, via the outer mitochondrial membrane:

(a) Cytochrome *c*, which participates in the activation of caspases.

(b) Procaspases, in particular procaspase-2, -3, and -9, which become activated upon release from the intermembrane space.

(c) Heat shock proteins hsp10 and hsp60 which may facilitate the activation of procaspase-3.

(d) Apoptosis-inducing factor (AIF), a flavoprotein which induces large scale chromatin fragmentation (~ 50 kb) when added to isolated nuclei *in vitro*.

(e) A DNase which is different from the above activities.

Among these proteins, cytochrome *c*, caspases, as well as AIF can each induce nuclear apoptosis when microinjected into the cytoplasm of live cells. Thus, several redundant pathways link mitochondrial membrane permeabilization to the induction of apoptotic cell death. In addition, the bioenergetic and redox catastrophe that results from mitochondrial membrane permeabilization is itself sufficient to cause cell death.

2. Mitochondrial contributions to cell death

The division of apoptosis into three functionally distinct phases (1–5), though artificial, provides an attractive theoretical framework for the molecular dissection of apoptotic pathways. According to this triphasic scheme (Fig. 1), the pre-mitochondrial initiation phase would include diverse signal transduction and damage pathways which are 'private' in the sense that they are not activated in a universal fashion and rather depend on the death-inducing primary stimulus and/or cell type.

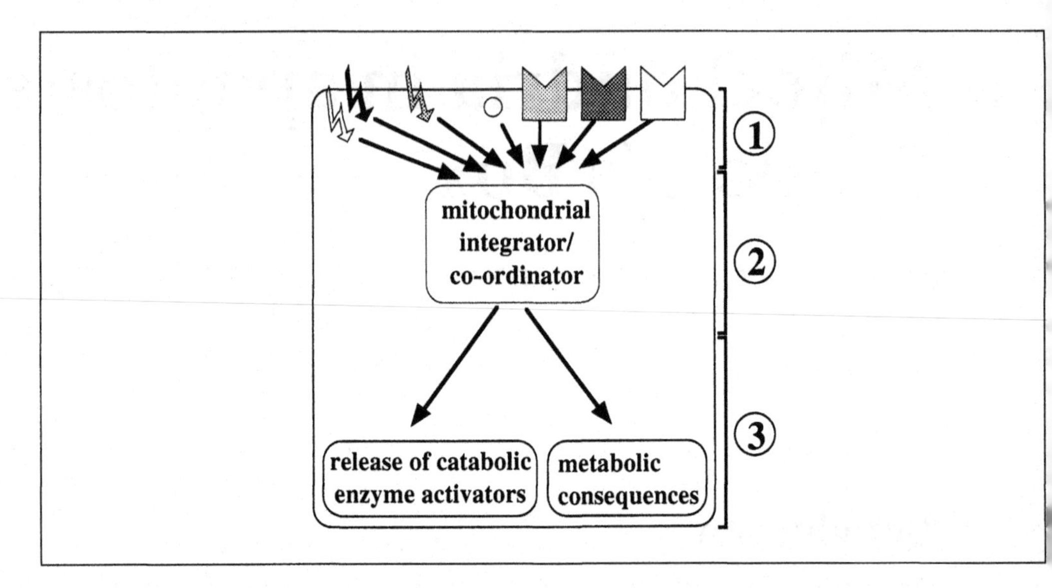

Fig. 1 Triphasic model of the apoptotic process. (1) During the pre-mitochondria initiation phase, multiple different damage or signal transduction pathways converge on the mitochondrion. (2) Mitochondrial membrane permeabilization occurs during the decision/effector phase of apoptosis. (3) As a consequence of outer membrane permeabilization, intermembrane proteins are released into the cytosol and participate in catabolic reaction and/or in the activation of catabolic enzymes (nucleases and proteases) which participate in the post-mitochondrial degradation phase. Moreover, the irreversible loss of mitochondrial function is lethal for the cell.

These pathways would converge on the mitochondrion, which integrates them into a common pathway. The common mitochondrial phase is characterized by an irreversible loss of mitochondrial membrane barrier function (1–5). Thus, the inner mitochondrial transmembrane potential ($\Delta\Psi_m$) is frequently lost during the early phase of apoptosis (6–9), and the outer mitochondrial membrane is permeabilized, leading to the release of soluble intermembrane proteins into the cytosol (10–13). It is only after this process, beyond the point-of-no-return, that downstream caspases (e.g. caspase-3 and -6) are activated and endonucleases come into action, leading to the acquisition of the biochemical and morphological hallmarks of apoptosis (10, 14–19). According to this model, the mitochondrion would function as a sort of Pandora's Box, in which the release of proteins that are usually well secluded in the mitochondrial intermembrane compartment determines the cell's fate.

Most authors agree upon the fact that mitochondrial intermembrane proteins such as cytochrome *c* are generally released during apoptosis, at least in mammalian cells, although some rare exceptions have been reported (20–22) (and still need to be confirmed). However, the importance of this release is subject to rather divergent interpretations. Some data may be interpreted to mean that the release of mitochondrial intermembrane proteins would be a decisive, perhaps obligatory step of the apoptotic cascade (1–5, 13, 23–25), or at least an amplifying mechanism accelerating cell death (26). The release of intermembrane protein, in particular cytochrome *c*, has also been viewed by a few authors as a reversible event (27, 28).

Some investigators suggest that mitochondrial membrane changes are irrelevant to the apoptotic process, at least in those cases in which caspases become activated in a direct fashion. This is sometimes the case when the Fas/CD95 receptor is crosslinked (29, 30) or when granzyme B translocates from cytotoxic T cells into the target cell (31). Yet another group of researchers simply ignore the contribution of mitochondria to apoptosis.

A further issue of polemics concerns the mechanism leading to the release of intermembrane proteins through the outer mitochondrial membrane. Numerous publications have implied tacitly that cytochrome *c* (14.5 kDa) would be the only apoptosis-relevant protein released from mitochondria and that it would be liberated through a 'specific' mechanism (10–12, 16, 17). However, at present, it appears that the mechanism of release is not specific for one determined class of intermembrane proteins and that larger molecules such as adenylate kinase (50 kDa) and apoptosis-inducing factor (AIF; 57 kDa) are also released from mitochondria (13–15, 32). Moreover, some authors have described a physical rupture of the outer mitochondrial membrane (33, 34) that would be responsible for the release of these proteins, whereas others postulate that physical rupture would be a downstream event (35). The putative involvement of the permeability transition (PT) pore, a composite protein channel which is inhibited by cyclosporin A (36) and by Bcl-2 (14, 15, 37, 38), is also a subject of debate. According to some authors, cyclosporin A prevents the release of mitochondrial intermembrane proteins induced by agents such as TNF (39–41), or Bax (9, 42, 43), or reactive oxygen species (44). However, some results suggest that other, PT pore-independent mechanisms might account for outer mitochondrial membrane permeabilization (12, 45).

The above controversies have been reviewed recently (5, 25, 46) and will not be touched on in this review. Rather, we will concentrate on the proteins leaking out from mitochondria during apoptosis (Fig. 2), and we will discuss their relative contribution to the death process. Although much emphasis has been laid on the role of cytochrome *c* as a molecule linking mitochondrial membrane permeabilization to the activation of cytosolic caspases (10–12, 16, 17) (see Chapter 4), it is clear that additional mitochondrial intermembrane proteins including AIF (13–15) and mitochondrial caspases (19, 47) can participate in the apoptotic process. Moreover, the metabolic consequences of mitochondrial membrane permeabilization are likely to have a major impact on the death process.

3. Cytochrome *c*

Cytochrome *c* was the first mitochondrial intermembrane protein to be molecularly identified as a potentially apoptogenic protein. When searching for caspase-3-activating factors in cell extracts, Wang and collaborators discovered that both dATP (or ATP) and cytochrome *c* were required for the proteolytic activation of procaspase-3 (10). In normal conditions, cytochrome *c* is involved in electron transfer between complexes III and IV of the respiratory chain. Cytochrome *c* is a haem-containing protein that is assembled in the mitochondrial intermembrane space (48).

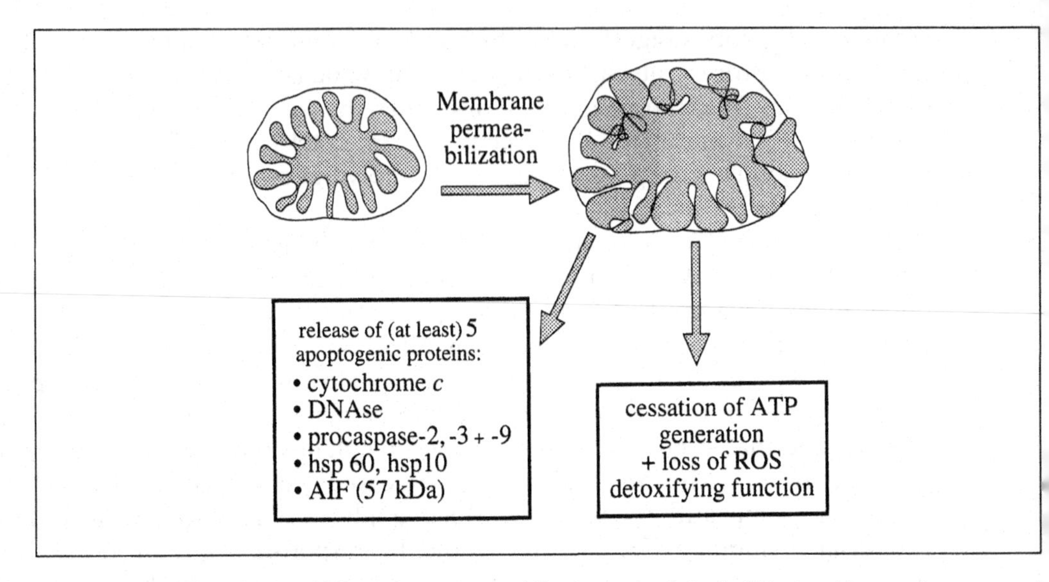

Fig. 2 Links between mitochondrial membrane permeabilization and cell death. Mitochondria can release several intermembrane proteins with potential apoptogenic effect: cytochrome *c*, caspases, heat shock proteins (hsp), AIF, and a DNase. Moreover, the bioenergetic and redox state of the cells is compromised as a direct consequence of mitochondrial disruption.

Apocytochrome *c* (which lacks haem) is translated from a nuclear DNA-encoded mRNA in the cytosol and is imported into mitochondria via a mechanism which is different from the general mitochondrial import pathway (49). Upon transport into the mitochondrial intermembrane space, haem lyase catalyses the attachment of haem, to yield holocytochrome *c* (or cytochrome *c tout court*). Because the outer membrane is normally impermeable for holocytochrome *c*, this reaction is irreversible. Thus, cytochrome *c* is normally confined to the intermembrane space unless the outer membrane is permeabilized.

Only holocytochrome *c*, not apocytochrome *c* or haem alone, can activate caspase-3 when added to cell-free extracts containing procaspase-3 and other cytoplasmic components (50). However, the oxidoreductase function of cytochrome *c* is not required for *in vitro* caspase activation since substitution of the Fe^{2+} ion of haem by a Cu^{2+} atom (which abrogates its oxidoreductase function) does not affect its caspase-3 activating potential (50). Moreover, yeast (*Saccharomyces cerevsiae*) cytochrome *c*, which has a normal oxidoreductase function, cannot activate caspase-3, indicating that the overall tertiary conformation of (animal) cytochrome *c* determines its apoptogenic capacity (50). Cytochrome *c* binds to the Apaf-1 protein, an ATP binding protein bearing a region homologous to the *Caenorrhabditis elegans* protein CED-4, as well as several WD repeats (16, 51–53). Cytochrome *c* binding probably concerns this latter region of the Apaf-1 protein. In the presence of cytochrome *c*, Apaf-1, and ATP (or dATP), Apaf-1 promotes the autoactivation of procaspase-9 (16, 51), which becomes proteolytically cleaved to yield active caspase-9. Caspase-9 then acts on procaspase-3 to initiate a caspase activation cascade (16, 51). While this is the case in mammalian and *Xenopus laevis* cell extracts, it appears that in *C. elegans* the activation of CED-3 (which is a caspase resembling caspase-9) by CED-4 and ATP (or dATP) does not require cytochrome *c*, in line with the absence of a cytochrome *c* binding domain in CED-4 (53). Based on this latter observation, it may be speculated that the phylogenetically conserved cell death pathway represented by *C. elegans* does not require cytochrome *c*. Alternatively, it is possible that *C. elegans* has lost the 'original' death pathway which involved mitochondrial intermembrane proteins including cytochrome *c*.

The ectopic (that is extra-mitochondrial) introduction of cytochrome *c* into cells by microinjection or electroporation generally suffices to induce apoptosis (54–57), although in some cell types this manipulation does not induce apoptosis—for example MCF7 cells (54), CEM cells (56), and sympathetic neurons cultured in the presence of nerve cell growth factor (57). This may be explained in several alternative but non-exclusive ways. First, it is highly likely that endogenous inhibitors, such as proteins that bind to or degrade cytochrome *c*, may determine the threshold that cytochrome *c* must attain to induce apoptosis. On teleological grounds, it appears tempting to postulate such a mechanism that would prevent accidental cytochrome *c* release from few mitochondria to cause cell death. Moreover, we have observed that the total amount of immunodetectable cytochrome *c* diminishes during late apoptosis (unpublished observation). Secondly, the downstream targets of cytochrome *c* may be missing, as this is the case for MCF7 cells, which lack caspase-3 expression. Transfection of these cells with an intact procaspase-3 gene restores the response to cytochrome *c* microinjection (54). Similarly, cytosol from skeletal muscle cells, which lack Apaf-1, fail to activate caspases in the presence of cytochrome *c* (58). Thirdly, cytochrome *c*-mediated caspase activation may be subjected to inhibitory regulation. Thus, activation of caspase-9 by cytochrome *c* is inhibited by a small family of proteins known as the inhibitors of apoptosis (IAP), probably due to direct interaction with caspase-9 (59) and other caspases (60). Caspase-9 is also subject to phosphorylation by the serine/threonine protein kinase Akt (61). Binding of Apaf-1 to members of the Bcl-2 family (62) could also result in the inhibition of caspase activation. Thus, the mere ectopic presence of cytochrome *c* may be insufficient to trigger caspase activation and subsequent cell death in the presence of caspase-inhibitory molecules. Fourthly, since cytochrome *c* is not the only mitochondrial apoptogenic factor (see below), these results may imply that, at least in some cases, mitochondrial factors other than cytochrome *c* are required for the induction of cell death.

4. Mitochondrial caspases

Caspases have conventionally been thought of as being cytosolic. However, in some cell types, the mitochondrial intermembrane space contains procaspase-3, as shown by Rosen and co-workers (47). After induction of apoptosis, the procaspase is released into the cytosol and becomes activated (47). We have recently performed a different approach to search for mitochondrial caspases, working with liver as starting material which contains little (47) or no (our observation) procaspase-3. Mitochondrial intermembrane protein preparations were found to contain an enzymatic activity that cleaves Z-VAD.afc, a synthetic caspase substrate. Moreover, in this preparation at least five distinct proteins with apparent molecular weight between 18 and 33 kDa bind Z-VAD.biotin (19), a pseudo-substrate which covalently reacts with the large subunit of caspases (63). When we purified the Z-VAD.afc-cleaving activity, only one Z-VAD.biotin binding protein of ~ 33 kDa was found (19). Since the large subunit of caspases is heterogeneous in length, this observation allowed us to predict that, among the 14 different known caspases, only caspase-2 and -9 were candidates to account for the mitochondrial Z-VAD.afc-cleaving activity. Western blot analysis revealed indeed that mitochondria from liver, kidney, spleen, brain, and heart, as well as several cell lines contain procaspase-2 and -9 (19). This result has been confirmed for caspase-9 in Jurkat cells (64).

Immunodepletion of caspase-2 and -9 from the intermembrane fraction of mouse liver mitochondria removed most of the Z-VAD.afc-cleaving activity, indicating that these two caspases are responsible for this enzymatic activity (19). At present, we do not know whether the other Z-VAD.biotin binding proteins (< 33 kDa) found in the intermembrane fraction of mitochondria represent other caspases with preference for other substrates than Z-VAD.afc or whether they are artefacts. Irrespective of these considerations, subcellular fractionation and *in situ* immunofluorescence has confirmed that caspase-2 and -9 are translocated from the mitochondrion to the cytosol after induction of apoptosis (19), in a manner akin to that described for cytochrome *c*. Similar results have been reported for mitochondrial caspase-3, which is liberated into the cytosol when apoptosis is induced (47).

After osmotic lysis or opening of the PT pore, procaspase-2 and -9 become activated in a Z-VAD.fmk-inhibitable fashion (19). Thus, during or after the release from the mitochondrial intermembrane space, the caspase-2 and -9 zymogens become proteolytically processed and enzymatically active (19). At present, the mechanisms responsible for this phenomenon are not clear. One possibility is that the release of caspase zymogens from mitochondria alters the equilibrium between caspase activators and local caspase inhibitors, thereby favouring caspase activation. Alternatively, mitochondrial surface proteins with which they normally cannot interact could activate procaspases. Irrespective of the exact mode of activation, cytochrome *c* is not a rate-limiting factor for caspase-2 and -9 activation upon mitochondrial release, as suggested by experiments in which cytochrome *c* has been immunodepleted from mitochondrial supernatant (19). These results however do not rule out the possibility that a rather low level of cytochrome *c* and/or early interactions between caspase zymogens and cytochrome *c*, within the intermembrane space, may participate in caspase activation.

The finding that mitochondria contain procaspases, as well as the molecules required to activate them, is highly intriguing. In the nematode C. *elegans*, the death regulatory machine is composed of three core interacting proteins: CED-4 (a homologue of mammalian Apaf-1), the caspase CED-3, and the Bcl-2 homologue CED-9 (see Chapter 2). Similarly, in mammalian cells Bcl-x_L, Apaf-1, and caspase-9 have been shown to interact (65, 66) (see Chapter 5). It is tempting to speculate that these complexes or 'apoptosomes' are formed at the mitochondrial outer membrane/intermembrane interface, where at least two of the three molecules reside. However, at present it is not clear whether the Bcl-x_L/Apaf-1/caspase-9 complex participates in the Bcl-2/Bcl-x_L-mediated regulation of mitochondrial membrane permeability or whether it is only formed after permeabilization of the outer mitochondrial membrane. If Apaf-1 is truly a cytosolic protein, as previously suggested (16, 51), then this latter possibility would apply.

5. Heat shock proteins

Caspase-3 can interact with hsp60 and hsp10, two mitochondrial heat shock proteins (64, 67). These proteins facilitate the activation of caspase-3 by other upstream caspases. The ATP-dependent 'foldase' activity of hsp60 may improve the vulnerability of procaspase-3 to proteolytic maturation by other caspases (67). Hsp60 is mainly a mitochondrial matrix protein. Immunostaining of normal and apoptotic cells yields a punctate cytoplasmic staining, even in conditions in which cytochrome *c* is released into the cytosol, indicating that the inner mitochondrial membrane maintains the capacity to retain matrix proteins in conditions in which the outer membrane is permeable. However, upon apoptosis induction a minor fraction of hsp60 redistributes from the mitochondrion to the cytosol, based on subcellular fractionation studies (64). At present it is not known whether this redistribution is due to a partial destruction of the inner membrane or whether it selectively affects a minor pool of hsp60 present in the mitochondrial intermembrane space. Irrespective of these possibilities, it appears that mitochondrial hsp60 and hsp10 can participate in the acceleration of caspase activation.

It should be noted that other heat shock proteins such as hsp70 and hsp27 have been reported to have anti-apoptotic effects, hsp27 presumably acting at a pre-mitochondrial level (68) and hsp70 at the post-mitochondrial level (69). Thus, different classes of heat shock proteins can have pro-apoptotic and anti-apoptotic activities.

6. Apoptosis-inducing factor (AIF)

The mitochondrial intermembrane protein fraction, which is released during apoptosis, contains an activity that suffices to force isolated HeLa nuclei to adopt apoptotic morphology and to lose their chromatin (14, 15). We have named this activity 'apoptosis-inducing factor' (AIF). Based on a cytofluorometric assay designed to measure the frequency of subdiploid nuclei exposed to mitochondrial protein (70), we have purified a protein that maintains its bioactivity in the presence of the caspase inhibitor Z-VAD.fmk (19) and thus cannot be a caspase. This protein was found to be a FAD binding flavoprotein (13). We have recently succeeded in cloning the corresponding full-length cDNAs from mouse (612 AA) and human (613 AA) (13). AIF is strongly conserved between the two mammalian species (92% AA identity) and bears a highly significant homology with several eubacterial and archaebacterial ferredoxin or NADH oxidoreductases in its C-terminal portion (aa 128–612 for mAIF; 95% AA identity between mouse and human). Its N-terminal portion has no such homology to oxidoreductases and instead bears a mitochondrial targeting sequence (aa 1–101 for mAIF; 84% AA identity between mouse and human), as well as 'spacer' region (aa 101–127 for mAIF; 60% AA identity between mouse and human) without any obvious function (Fig. 3). Based on Northern blot analysis, one 2.4 kb AIF mRNA species is expressed ubiquitously in human tissue. This finding was confirmed at the protein level for mouse tissues using an antibody raised against residues 151–200 of AIF, which recognizes a single ~ 57 kDa protein. The primary transcription/translation product of mAIF cDNA obtained *in vitro* has an apparent molecular weight close to the expected 66.8 kDa. When imported into mitochondria *in vitro*, it gives rise to a shorter protein (57 kDa) due to the removal of the mitochondrial targeting sequence. A recombinant protein corresponding to the mAIF precursor does not bind FAD, whereas a shorter protein lacking the mitochondrial targeting sequence and part of the 'spacer' region (Δ1–120) does bind FAD. These data suggest that the FAD prosthetic group is attached to the AIF protein within the mitochondrion, after removal of the targeting sequence, as has been described for other mitochondrial flavoproteins (71, 72).

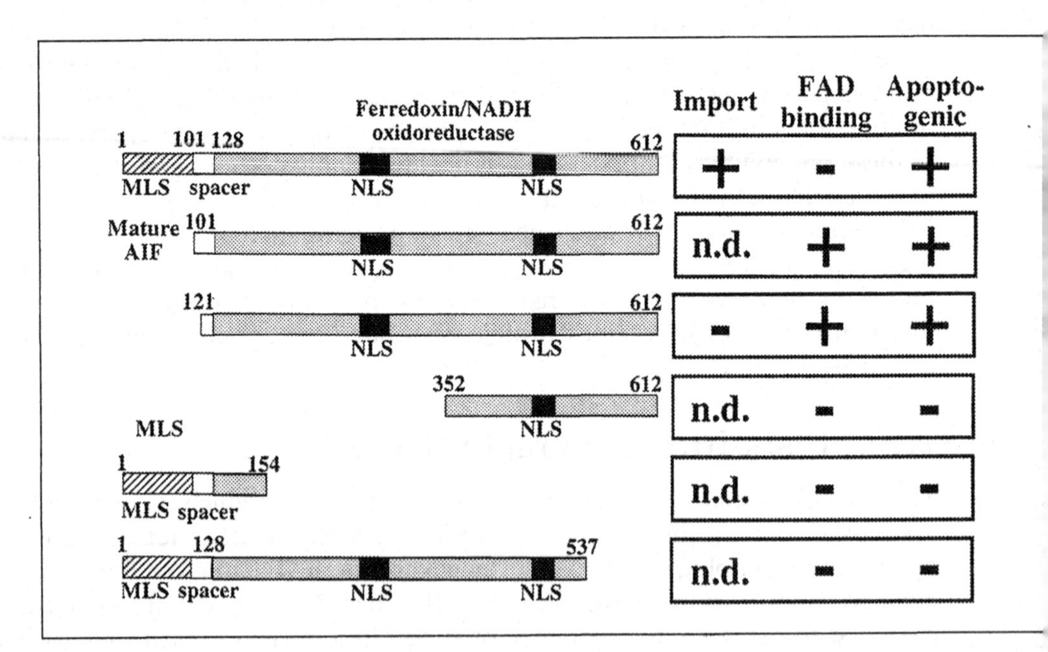

Fig. 3 Analysis of the structure–function relationship of mouse AIF. The AIF precursor carries an N-terminal mitochondrion targeting sequence which is cleaved off after import into the intermembrane space. The oxidoreductase domain of AIF contains several putative nuclear localization sequences (NLS). The following parameters were assessed for mature AIF (purified from liver mitochondria) or for several different recombinant proteins: mitochondrial import, binding of flavin adenine nucleotide (FAD), or apoptogenic effect on isolated nuclei *in vitro*.

The AIF bioactivity and immunoreactivity is exclusively found within the mitochondrial intermembrane space (13). Immunodepletion of AIF from the entire pool of mitochondrial intermembrane proteins also removes most of the biological activity inducing nuclear apoptosis *in vitro*, indicating that AIF is the principal mitochondrial factor causing nuclear apoptosis. Subcellular fractionation, immunofluorescence analysis, and immunoelectron microscopy confirmed that AIF is normally confined to mitochondria, yet subject to translocation from mitochondria to the nucleus upon induction of apoptosis by diverse agents such as ceramide, staurosporine, etoposide, or glucocorticoids. Thus, in contrast to cytochrome *c* (which stays cytosolic), AIF moves to the nucleus, concomitant to the initial phase of chromatin condensation. This nuclear localization of AIF is compatible with the presence of several putative nuclear localization signals (74). It is likely to require active import into the nucleus because microinjection of wheat germ agglutinin (WGA), an inhibitor of nuclear pores, prevents the translocation of AIF (13).

When added to purified nuclei from HeLa cells, recombinant AIF protein induces DNA loss, peripheral chromatin condensation, and digestion of chromatin into ~ 50 kb fragments (but no oligonucleosomal fragmentation) (13). Thus, AIF causes nuclear changes resembling those found during early apoptosis. These *in vitro* AIF effects are observed for the entire protein (aa 1–612), for the mature protein (Δ1–101), but not for several deletion mutations (Δ1–351, Δ155–612, or Δ538–612). They are likely to be independent of its putative oxidoreductase function, because the recombinant AIF precursor lacking the prosthetic FAD group does have an apoptogenic effect in this *in vitro* assay (Fig. 3). The effect of AIF on isolated nuclei does not require additional cytoplasmic factors. Recombinant AIF does not cleave purified plasmid DNA. The fact that it does cause large scale DNA fragmentation in isolated nuclei thus suggests that it activates a sessile nuclear DNase.

In addition to its nuclear effects, recombinant AIF acts on mitochondria. In the presence of cytosol, AIF causes purified mitochondria to dissipate their $\Delta\Psi_m$ and to release cytochrome *c* and caspase-9. None of these AIF effects, either on isolated nuclei or on mitochondria, is prevented by the broad-spectrum caspase inhibitor Z-VAD.fmk, suggesting that they are caspase independent. However, the supernatant from mitochondria treated with AIF plus cytosol contains a Z-VAD.fmk-inhibitable enzymatic activity which cleaves the caspase substrate Z-VAD.afc. This activity is at least in part due to the presence of activated caspase-9 (19). Thus, AIF may activate caspase-9 (and presumably other caspases) via an indirect, mitochondrion-dependent mechanism.

The ectopic (extra-mitochondrial) localization of AIF induces apoptosis *in vivo*. Thus, microinjection of recombinant AIF into the cytoplasm of live cells induces several hallmarks of apoptosis: nuclear chromatin condensation and DNA loss, dissipation of the $\Delta\Psi_m$, and exposure of phosphatidylserine on the outer leaflet of the plasma membrane (13). None of the effects mediated by microinjected AIF is inhibited by Z-VAD.fmk (13), although Z-VAD.fmk succeeds in preventing cytochrome *c*-induced, caspase-dependent (12, 51) apoptosis. Microinjection of an AIF-specific antiserum abolishes the nuclear (but not the cytoplasmic) signs of apoptosis induced by the PT pore-opening agent atractylocide or the protein kinase inhibitor staurosporine. In contrast, it failed to prevent nuclear apoptosis induced by microinjection of active caspase-8 (13), thus confirming the existence of caspase-dependent, presumably mitochondrion-independent pathway leading to nuclear apoptosis (29, 30).

7. Additional factors and metabolic changes due to mitochondrial membrane permeabilization

In addition to the above factors (cytochrome *c*, caspases, AIF), the mitochondrial intermembrane fraction may contain further potentially apoptogenic factors (19). We have recently described the presence of a DNase which cleaves purified plasmid DNA *in vitro*. The mitochondrial DNase activity is not affected by immunodepletion of cytochrome *c* nor by caspase inhibitors. It is not inhibited by aurintricarboxylic acid (ATA) (19), indicating that it is distinct from DFF/CAD, a caspase-activated DNase which is inhibited by ATA (75, 76). It also differs from a previously described mitochondrial DNase, endonuclease G, which acts in a sequence-specific fashion (77). At present, we have not yet performed the molecular characterization of this mitochondrial DNase.

In general terms, it remains elusive whether additional catabolic enzymes or enzyme activators are present in the mitochondrial intermembrane space and are released during the apoptotic process. We are currently designing experiments to actively search for such activities, and it may be a mere question of time until new potentially pro-apoptotic proteins will be identified.

Permeabilization of the outer mitochondrial membrane with consequent loss of cytochrome *c* interrupts electron flow between respiratory chain complexes III and IV. This antimycin-like effect causes an increase in the generation of reactive oxygen species (ROS), mainly superoxide anion (78) and this occurs in apoptosis (7). This effect, together with the uncoupling of the respiratory chain due to opening of the PT pore can be expected to have several lethal consequences: oxidation of NAD(P)H, loss of ATP, and oxidation/depletion of glutathione (GSH). All these changes have been described to occur during apoptosis (1) and may be expected to contribute to cell death.

8. Open questions

In our view, two major questions remain open. First, are mitochondria required for the process of apoptosis? This is a complex issue because mitochondrial membrane permeabilization is such a general phenomenon in apoptosis, yet it is not yet clear whether mitochondria are always an active participant in apoptotic cascade or whether they are sometimes just a victim of the process. This question requires further exploration in cell-free systems (see Chapter 7).

The second major question concerns the functional relationship between the intermembrane proteins released from mitochondria during the apoptotic process. Is their relationship to be explained in terms of hierarchy, complementarity, or redundancy? Microinjection experiments clearly show that cytochrome *c* (54–57), recombinant active caspase-2 and -9 (19), as well as recombinant AIF (13) suffice to induce cell death with prominent features of apoptosis, suggesting the existence of redundant pathways. This interpretation is complicated by the fact that both AIF (13) and caspases (18, 37) can act on purified mitochondria to permeabilize their outer membrane and to release cytochrome *c*. Recombinant caspases release AIF from mitochondria (18, 37), whereas recombinant AIF releases caspase-9 from the intermembrane space (13). In the same way, it appears plausible that caspases activated by cytochrome *c* can have an effect on mitochondria. These data underline the function of mitochondria as a crossroad between different pro-apoptotic molecules and pathways. However, the molecular identity of several of the players in the lethal gamble—namely the cytosolic and mitochondrial AIF and caspase targets—requires urgent clarification.

References

1. Kroemer, G., Petit, P. X., Zamzami, N., Vayssière, J.-L., and Mignotte, B. (1995). The biochemistry of apoptosis. *FASEB J.*, **9**, 1277–87.
2. Kroemer, G. (1997). The proto-oncogene Bcl-2 and its role in regulating apoptosis. *Nature Med.*, **3**, 614–20.
3. Kroemer, G., Zamzami, N., and Susin, S. A. (1997). Mitochondrial control of apoptosis. *Immunol. Today*, **18**, 44–51.
4. Kroemer, G., Dallaporta, B., and Resche-Rigon, M. (1998). The mitochondrial death/life regulator in apoptosis and necrosis. *Annu. Rev. Physiol.*, **60**, 619–42.
5. Susin, S. A., Zamzami, N., and Kroemer, G. (1998). Mitochondrial regulation of apoptosis. Doubt no more. *Biochim. Biophys. Acta (Bioenergetics)*, **1366**, 151–65.
6. Zamzami, N., Marchetti, P., Castedo, M., Zanin, C., Vayssière, J.-L., Petit, P. X., *et al.* (1995). Reduction in mitochondrial potential constitutes an early irreversible step of programmed lymphocyte death *in vivo*. *J. Exp. Med.*, **181**, 1661–72.
7. Zamzami, N., Marchetti, P., Castedo, M., Decaudin, D., Macho, A., Hirsch, T., *et al.* (1995). Sequential reduction of mitochondrial transmembrane potential and generation of reactive oxygen species in early programmed cell death. *J. Exp. Med.*, **182**, 367–77.
8. Castedo, M., Hirsch, T., Susin, S. A., Zamzami, N., Marchetti, P., Macho, A., *et al.* (1996). Sequential acquisition of mitochondrial and plasma membrane alterations during early lymphocyte apoptosis. *J. Immunol.*, **157**, 512–21.
9. Marzo, I., Brenner, C., Zamzami, N., Jürgensmeier, J., Susin, S. A., Vieira, H. L. A., Prévost, M.-C., *et al.* (1998). Bax and adenine nucleotide translocator cooperate in the mitochondrial control of apoptosis. *Science*, **281**, 2027–31.
10. Liu, X. S., Kim, C. N., Yang, J., Jemmerson, R., and Wang, X. (1996). Induction of apoptotic program in cell-free extracts: requirement for dATP and cytochrome c. *Cell*, **86**, 147–57.
11. Yang, J., Liu, X., Bhalla, K., Kim, C. N., Ibrado, A. M., Cai, J., *et al.* (1997). Prevention of apoptosis by Bcl-2: release of cytochrome c from mitochondria blocked. *Science*, **275**, 1129–32.
12. Kluck, R. M., Bossy-Wetzel, E., Green, D. R., and Newmeyer, D. D. (1997). The release of cytochrome c from mitochondria: a primary site for Bcl-2 regulation of apoptosis. *Science*, **275**, 1132–6.

13. Susin, S. A., Lorenzo, H. K., Zamzami, N., Marzo, I., Snow, B. E., Brothers, G. M., *et al.* (1999). Molecular characterization of mitochondrial apoptosis-inducing factor (AIF). *Nature*, **397**, 441–6.
14. Zamzami, N., Susin, S. A., Marchetti, P., Hirsch, T., Gómez-Monterrey, I., Castedo, M., *et al.* (1996). Mitochondrial control of nuclear apoptosis. *J. Exp. Med.*, **183**, 1533–44.
15. Susin, S. A., Zamzami, N., Castedo, M., Hirsch, T., Marchetti, P., Macho, A., *et al.* (1996). Bcl-2 inhibits the mitochondrial release of an apoptogenic protease. *J. Exp. Med.*, **184**, 1331–42.
16. Zhou, H., Henzel, W. J., Liu, X., Lutschg, A., and Wang, X. D. (1997). Apaf-1, a human protein homologous to *C. elegans* Ced-4, participates in cytochrome c-dependent activation of caspase-3. *Cell*, **90**, 405–13.
17. Liu, X., Zou, H., Slaughter, C., and Wang, X. (1997). DFF, a heterodimeric protein that functions downstream of caspase 3 to trigger DNA fragmentation during apoptosis. *Cell*, **89**, 175–84.
18. Susin, S. A., Zamzami, N., Castedo, M., Daugas, E., Wang, H.-G., Geley, S., *et al.* (1997). The central executioner of apoptosis. Multiple links between protease activation and mitochondria in Fas/Apo-1/CD95- and ceramide-induced apoptosis. *J. Exp. Med.*, **186**, 25–37.
19. Susin, S. A., Lorenzo, H. K., Zamzami, N., Marzo, I., Larochette, N., Alzari, P. M., *et al.* (1999). Mitochondrial release of caspases-2 and -9 during the apoptotic process. *J. Exp. Med.*, **189**, 381–94.
20. Chauhan, D., Pandey, P., Ogata, A., Teoh, G., Krett, N., Halgren, R., *et al.* (1997). Cytochrome c-dependent and -independent induction of apoptosis in multiple myeloma cells. *J. Biol. Chem.*, **272**, 29995–7.
21. Tang, D. G., Li, L., Zhu, Z. Y., and Joshi, B. (1998). Apoptosis in the absence of cytochrome c accumulation in the cytosol. *Biochem. Biophys. Res. Commun.*, **242**, 380–4.
22. Adachi, S., Gottlieb, R. A., and Babior, B. M. (1998). Lack of release of cytochrome c from mitochondria into cytosol early in the course of Fas-mediated apoptosis of Jurkat cells. *J. Biol. Chem.*, **273**, 19892–4.
23. Penninger, J. M. and Kroemer, G. (1998). Molecular and cellular mechanisms of T lymphocyte apoptosis. *Adv. Immunol.*, **68**, 51–144.
24. Green, D. R. and Kroemer, G. (1998). The central execution of apoptosis: mitochondria or caspases? *Trends Cell Biol.*, **8**, 267–71.
25. Green, D. R. and Reed, J. C. (1998). Mitochondria and apoptosis. *Science*, **281**, 1309–12.
26. Kuwana, T., Smith, J. J., Muzio, M., Dixit, V., Newmeyer, D. D., and Kornbluth, S. (1998). Apoptosis induction by caspase-8 is amplified through the mitochondrial release of cytochrome c. *J. Biol. Chem.*, **273**, 16589–94.
27. Chen, Q., Takeyama, N., Brady, G., Watson, A. J. M., and Dive, C. (1998). Blood cells with reduced mitochondrial membrane potential and cytosolic cytochrome c can survive and maintain clonogenicity given appropriate signals to suppress apoptosis. *Blood*, **92**, 4545–53.
28. Martinou, I., Desagher, S., Eskes, R., Antonsson, B., Andre, E., Fakan, S., *et al.* (1999). The release of cytochrome c from mitochondria during apoptosis of NGF-deprived sympathetic neurons is a reversible event. *J. Cell Biol.*, **144**, 883–9.
29. Stennicke, H. R., Jurgensmeier, J. M., Shin, H., Deveraux, Q., Wolf, B. B., Yang, X. H., *et al.* (1998). Pro-caspase-3 is a major physiological target of caspase-8. *J. Biol. Chem.*, **273**, 27084–90.
30. Scaffidi, C., Fulda, S., Srinivasan, A., Friesen, C., Li, F., Tomaselli, K. J., *et al.* (1998). Two CD95 (APO-1/Fas) signaling pathways. *EMBO J.*, **17**, 1675–87.

31. Quan, L. T., Tewari, M., Orourke, K., Dixit, V., Snipas, S. J., Poirier, G. G., *et al.* (1996). Proteolytic activation of the cell death protease Yama / CPP32 by granzyme B. *Proc. Natl. Acad. Sci. USA*, **93**, 1972–6.
32. Single, B., Leist, M., and Nicotera, P. (1998). Simultaneous release of adenylate kinase and cytochrome c in cell death. *Cell Death Differ.*, **5**, 1001–3.
33. vander Heiden, M. G., Chandal, N. S., Williamson, E. K., Schumacker, P. T., and Thompson, C. B. (1997). Bcl-XL regulates the membrane potential and volume homeostasis of mitochondria. *Cell*, **91**, 627–37.
34. Angermuller, S., Kunstle, G., and Tiegs, G. (1998). Pre-apoptotic alterations in hepatocytes of TNF alpha-treated galactosamine-sensitized mice. *J. Histochem. Cytochem.*, **46**, 1175–83.
35. Zhuang, J., Dinsdale, D., and Cohen, G. M. (1998). Apoptosis, in human monocytic THP.1 cells, results in the release of cytochrome c from mitochondria prior to their ultracondensation, formation of outer membrane discontinuities and reduction in inner membrane potential. *Cell Death Differ.*, **5**, 953–62.
36. Zoratti, M. and Szabò, I. (1995). The mitochondrial permeability transition. *Biochim. Biophys. Acta—Rev. Biomembranes*, **1241**, 139–76.
37. Marzo, I., Brenner, C., Zamzami, N., Susin, S. A., Beutner, G., Brdiczka, D., *et al.* (1998). The permeability transition pore complex: a target for apoptosis regulation by caspases and Bcl-2 related proteins. *J. Exp. Med.*, **187**, 1261–71.
38. Zamzami, N., Brenner, C., Marzo, I., Susin, S. A., and Kroemer, G. (1998). Subcellular and submitochondrial mechanisms of apoptosis inhibition by Bcl-2-related proteins. *Oncogene*, **16**, 2265–82.
39. Pastorino, J. G., Simbula, G., Yamamoto, K., Glascott, P. A. J., Rothman, R. J., and Farber, J. L. (1996). The cytotoxicity of tumor necrosis factor depends on induction of the mitochondrial permeability transition. *J. Biol. Chem.*, **271**, 29792–9.
40. Bradham, C. A., Quian, T., Streetz, K., Trautwein, C., Brenner, D. A., and Lemasters, J. J. (1998). The mitochondrial permeability transition is required for tumor necrosis factor alpha-mediated apoptosis and cytochrome *c* release. *Mol. Cell. Biol.*, **18**, 6353–64.
41. Walter, D. H., Haendeler, J., Galle, J., Zeiher, A. M., and Dimmeler, S. (1998). Cyclosporin A inhibits apoptosis of human endothelial cells by preventing release of cytochrome c from mitochondria. *Circulation*, **98**, 1153–7.
42. Jürgensmeier, J. M., Xie, Z., Deveraux, Q., Ellerby, L., Bredesen, D., and Reed, J. C. (1998). Bax directly induces release of cytochrome c from isolated mitochondria. *Proc. Natl. Acad. Sci. USA*, **95**, 4997–5002.
43. Pastorino, J. G., Chen, S.-T., Tafani, M., Snyder, J. W., and Farber, J. L. (1998). The overexpression of Bax produces cell death upon induction of the mitochondrial permeability transition. *J. Biol. Chem.*, **273**, 7770–7.
44. Sugano, N., Ito, K., and Murai, S. (1999). Cyclosporin A inhibits H202-induced apoptosis of human fibroblasts. *FEBS Lett.*, **447**, 274–6.
45. Eskes, R., Antonsson, B., Osen Sand, A., Montessuit, S., Richter, C., Sadoul, R., *et al.* (1998). Bax-induced cytochrome c release from mitochondria is independent of the permeability transition pore but highly dependent on Mg^{2+} ions. *J. Cell Biol.*, **143**, 217–24.
46. Jacotot, E., Costantini, P., Labourau, E., Zamzami, N., Susin, S. A., and Kroemer, G. (1999). Mitochondrial membrane permeabilization in the apoptotic process. *Ann. NY Acad. Sci.*, **887**, 18–30.
47. Mancini, M., Nicholson, D. W., Roy, S., Thornberry, N. A., Peterson, E. P., Casciola-Rosen, L. A., *et al.* (1998). The caspase-3 precursor has a cytosolic and mitochondrial distribution: implications for apoptotic signaling. *J. Cell Biol.*, **140**, 1485–95.

48. Neupert, W. (1997). Protein import into mitochondria. *Annu. Rev. Biochem.*, **66**, 863–917
49. Mayer, A., Neupert, W., and Lill, R. (1995). Translocation of apocytochrome c across the outer membrane of mitochondria. *J. Biol. Chem.*, **270**, 12390–7.
50. Kluck, R. M., Martin, S. J., Hoffman, B. M., Zhou, J. S., Green, D. R., and Newmeyer, D. D. (1997). Cytochrome c activation of CPP32-like proteolysis plays a critical role in a *Xenopus* cell-free apoptosis system. *EMBO J.*, **16**, 4639–49.
51. Li, P., Nijhawan, D., Budihardjo, I., Srinivasula, S. M., Ahmad, M., Alnemri, E. S., *et al.* (1997). Cytochrome c and dATP-dependent formation of Apaf-1/caspase-9 complex initiates an apoptotic protease cascade. *Cell*, **91**, 479–89.
52. Pan, G. H., O'Rourke, K., and Dixit, V. M. (1998). Caspase-9, BclXL, and Apaf-1 form a ternary complex. *J. Biol. Chem.*, **273**, 5841–5.
53. Hu, Y. M., Benedict, M. A., Wu, D. Y., Inohara, N., and Nunez, G. (1998). Bcl-XL interacts with Apaf-1 and inhibits Apaf-1-dependent caspase-9 activation. *Proc. Natl. Acad. Sci. USA*, **95**, 4386–91.
54. Li, F., Srinivasan, A., Wang, Y., Armstrong, R. C., Tomaselli, K. J., and Fritz, L. C. (1997). Cell-specific induction of apoptosis by microinjection of cytochrome c—Bcl-XL has activity independent of cytochrome c release. *J. Biol. Chem.*, **272**, 30299–305.
55. Brustugun, O. T., Fladmark, K. E., Doskeland, S. O., Orrenius, S., and Zhivotovsky, B. (1998). Apoptosis induced by microinjection of cytochrome c is caspase-dependent and is inhibited by Bcl-2. *Cell Death Differ.*, **5**, 660–8.
56. Garland, J. M. and Rudin, C. (1998). Cytochrome c induces caspase-dependent apoptosis in intact hematopoietic cells and overrides apoptosis suppression mediated by Bcl-2, growth factor signaling, MAP-kinase-kinase, and malignant change. *Blood*, **92**, 1235–46.
57. Deshmukh, M. and Johnson, E. M. J. (1998). Evidence of a novel event during neuronal death: Development of competence-o-die in response to cytoplasmic cytochrome *c*. *Neuron*, **21**, 695–705.
58. Burgess, D. H., Svensson, M., Dandrea, T., Gronlund, K., Hammarquist, F., Orrenius, S., *et al.* (1999). Human skeletal muscle cytosols are refractory to cytochrome c-dependent activation of type-II caspases and lack Apaf-1. *Cell Death Differ.*, **6**, 256–61.
59. Deveraux, Q. L., Takahashi, R., Salvesen, G. S., and Reed, J. C. (1997). X-linked IAP is a direct inhibitor of cell-death proteases. *Nature*, **388**, 300–4.
60. Duckett, C. S., Li, F., Wang, Y., Tomaselli, K. J., Thompson, C. B., and Armstrong, R. C. (1998). Human IAP-like protein regulates programmed cell death downstream of Bcl-XL and cytochrome c. *Mol. Cell. Biol.*, **18**, 608–15.
61. Cardone, M. H., Roy, N., Stennicke, H., Salvesen, G. S., Franke, T. F., Stanbridge, E., *et al.* (1998). Regulation of cell death protease caspase-9 by phosphorylation. *Science*, **282**, 1318–21.
62. Song, Q. Z., Kuang, Y. P., Dixit, V. M., and Vincenz, C. (1999). Boo, a novel negative regulator of cell death, interacts with Apaf-1. *EMBO J.*, **18**, 167–78.
63. Faleiro, L., Kobayashi, R., Fearnhead, H., and Lazebnik, Y. (1997). Multiple species of CPP32 and Mch2 are the major active caspases present in apoptotic cells. *EMBO J.*, **16**, 2271–81.
64. Samali, A., Cai, J., Zhivotovsky, B., Jones, D. P., and Orrenius, S. (1999). Presence of a pre-apoptotic complex of pro-caspase-3, hsp60, and hsp10 in the mitochondrial fraction of Jurkat cells. *EMBO J.*, **18**, 2040–8.
65. Pan, G. H., Humke, E. W., and Dixit, V. M. (1998). Activation of caspases triggered by cytochrome c *in vitro*. *FEBS Lett.*, **426**, 151–4.

66. Srinivasula, S. M., Ahmad, M., Ferndes-Alnemri, T., and Alnemri, E. S. (1998). Autoactivation of procaspase-9 by Apaf-1-mediated oligomerization. *Mol. Cell*, **1**, 949–57.
67. Xanthoudakis, S., Roy, S., Rasper, D., Hennessey, T., Aubin, Y., Cassady, R., *et al.* (1999). Hsp60 accelerates the maturation of pro-caspase-3 by upstream activator proteases during apoptosis. *EMBO J.*, **18**, 2049–56.
68. Preville, X., Salvemi, F., Giraud, S., Chaufour, S., Paul, C., Stepien, G., *et al.* (1999). Mammalian small stress proteins protect against oxidative stress through their availability to increase glucose-6-phosphate dehydrogenase activity and by maintaining optimal cellular detoxifying machinery. *Exp. Cell Res.*, **247**, 61–78.
69. Jaattela, M., Wissing, D., Kokholm, K., Kallunki, T., and Egeblad, M. (1998). Hsp70 exerts its anti-apoptotic function downstream of caspase-3-like proteases. *EMBO J.*, **17**, 6124–34.
70. Susin, S. A., Zamzami, N., Larochette, N., Dallaporta, B., Marzo, I., Brenner, C., *et al.* (1997). A cytofluorometric assay of nuclear apoptosis induced in a cell-free system. Application to ceramide-induced apoptosis. *Exp. Cell Res.*, **236**, 397–403.
71. Saijo, T. and Tanaka, K. (1995). Isoalloxazine ring of FAD is required for the formation of the core in the hsp60-assisted folding of medium-chain acyl-CoA dehydrogenase subunit into the assembly competent conformation in mitochondria. *J. Biol. Chem.*, **270**, 1899–907.
72. Robinson, K. M. and Lemire, B. D. (1996). A requirement for matrix processing peptidase but not for mitochondrial chaperonin in the covalent attachment of FAD to yeast succinate dehydrogenase flavoprotein. *J. Biol. Chem.*, **271**, 4061–7.
73. Cedano, J., Aloy, P., Pérez-Pons, J. A., and Quero, E. (1997). Relation between amino acid composition and cellular location of proteins. *J. Mol. Biol.*, **266**, 231–4
74. Boulikas, T. (1993). Nuclear localization signals (NLS). *Crit. Rev. Euk. Gene Express.*, **3**, 193–227.
75. Enari, M., Sakahira, H., Yokoyoma, H., Okawa, K., Iwamtsu, A., and Nagata, S. (1998). A caspase-activated DNase that degrades DNA during apoptosis, and its inhibitor ICAD. *Nature*, **391**, 43–50.
76. Sakahira, H., Enari, M., and Nagata, S. (1998). Cleavage of CAD inhibitor in CAD activation and DNA degradation during apoptosis. *Nature*, **391**, 96–9.
77. Côté, J. and Ruiz-Carrillo, A. (1993). Primers for mitochondrial DNA replication generated by endonuclease G. *Science*, **261**, 765–9.
78. Cai, J. and Jones, D. P. (1998). Superoxide in apoptosis. Mitochondrial generation triggered by cytochrome c loss. *J. Biol. Chem.*, **273**, 11401–4.

7 Apoptosis: lessons from cell-free systems

PAUL R. CLARKE

1. Introduction

In the age of genome sequencing, the major challenge facing molecular and cellular biologists is to determine the functions of individual gene products and understand how higher orders of complexity are created by their interactions in living systems. Model genetic organisms (yeasts, nematodes, fruit flies, the mouse) provide invaluable systems to delineate the pathways in which genes are involved. Yet, biochemical models are also required to understand the roles of individual gene products, their post-translational regulation, enzymatic activities, and molecular interactions. The challenge is to be able to study how molecules interact and function in a cellular context, to bridge the gap between detailed biochemical analysis of structure and function in solution with what happens in intact cells and tissues.

Concentrated cell extracts that attempt to maintain the cellular environment provide a valuable resource to address this problem, readily permitting dissection and manipulation of pathways, analysis of the activities of individual gene products, and study of their interactions. Such cell-free systems, particularly those derived from early embryos of marine invertebrates and amphibians, but also in some instances those derived from mammalian somatic cells, have been used very successfully to reproduce in the test-tube complicated cellular processes such as DNA replication (1), nuclear assembly and disassembly (2), membrane dynamics (3), and microtubule dynamics (4, 5). More recently, cell-free systems have also been developed to study the molecular processes involved in apoptosis (6, 7). These systems promise to be particularly useful for identifying novel factors, determining mechanisms of post-translational control, and studying the execution phase when the characteristic morphological and biochemical changes of apoptosis occur. Here, I will discuss the development of cell-free systems for apoptosis, with some examples of their use. Particular emphasis is given to the *Xenopus laevis* egg extract system (8, 9) which has certain advantages and is the one that we have used most extensively.

2. Studying the biochemistry of apoptosis

Apoptosis, as described by Kerr, Wyllie, and Currie in 1972 (10), involves the condensation of chromatin, restructuring of the cytoplasm, blebbing of cytoplasmic membranes, and finally fragmentation of the cell into apoptotic bodies that are phagocytosed by neighbouring cells, features that distinguish the process from necrotic cell death. These morphological characteristics, and the cleavage of DNA at inter-nucleosomal sites to produce a series of discrete fragments that run as a ladder on a gel (11), are found in many cells undergoing cell death during development and in response to a wide variety of insults (12). However, some dying cells exhibit only some of these features, and apoptotic and necrotic forms of cell death are often not easily separated. Indeed, what are the differences between these processes at the molecular level and do they share some events? While different cells undergoing apoptosis may exhibit similar morphological symptoms, is there common biochemical pathway controlling and executing these events, or are there different ways of producing similar end results in different cells? Most importantly, from a medical point of view, we must understand the process at the molecular level in order to understand how apoptosis functions normally and how it may be aberrant in major human diseases such as cancer and neurodegeneration. Such molecular characterization also has great potential to identify novel targets for pharmacological intervention in these diseases.

A model genetic system, developmental cell death in the nematode worm *Caenorhabditis elegans*, has proved invaluable in identifying the genes involved in apoptosis. During development of the worm, the fate of each cell is known precisely, and 131 cells die by apoptosis. *ced-3* and *ced-4* are required for these cell deaths, while loss of *ced-9* induces widespread cell death acting via *ced-3* and *ced-4* (13). We now know that *ced-3*, *ced-4*, and *ced-9* encode homologues of mammalian caspases (14), the caspase-activating factor Apaf-1 (15), and the anti-apoptotic protein Bcl-2 (16), respectively. Other members of the Bcl-2 family are either anti-apoptotic, such as Bcl-x_L, or pro-apoptotic, such as Bax, Bak, and Bid. In *C. elegans*, *egl-1* encodes an activator of apoptosis containing a Bcl-2 homology domain, BH3, shown to be critical for the activity of vertebrate pro-apoptotic Bcl-2 family proteins (16). Thus, this central pathway is conserved, probably in all metazoans (Fig. 1). Less is known about how these gene products interact, how they cause the morphological and biochemical events of apoptosis, and in particular how the pathway is controlled by signals from within and without the cell.

The major difficulty in studying the biochemistry of apoptosis in intact cells and tissues lies in the nature of the process itself (7). An apoptotic signal may come either from within the cell, for instance in response to DNA damage or other cellular stresses, or from without the cell, for instance a pro-apoptotic signal acting through a cell surface receptor or the removal of anti-apoptotic signals that otherwise maintain cell survival. There then follows a period in which the cell responds to the signal, often by activating or suppressing protein synthesis and changing the balance of pro- and anti-apoptotic regulators in the cell. Some signals, however, may act more

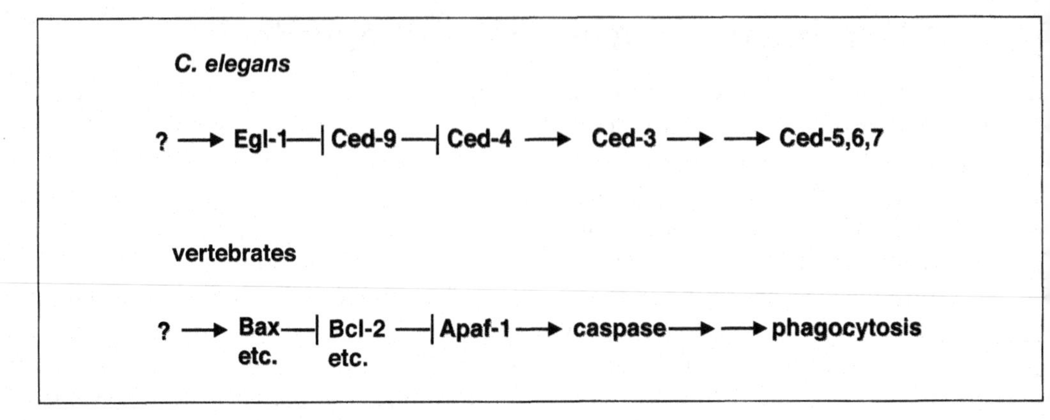

Fig. 1 A common pathway for apoptosis in *C. elegans* and vertebrates. In *C. elegans*, developmental cell death requires a caspase, ced-3, which requires ced-4 for its activation. ced-3 regulation is negatively regulated by ced-9, a Bcl-2 family homologue. ced-3 is activated by a BH3 containing pro-apoptotic protein encoded by *egl-1*. Downstream of ced-3, the products of *ced-5*, *ced-6*, and *ced-7* play roles in phagocytosis. In vertebrates, there are multiple pro- and anti-apoptotic Bcl-2 family proteins which may respond to different signals controlling apoptosis. Their activity is controlled at the level of expression or post-translationally by caspase-catalysed cleavage or phosphorylation. The apoptotic programme in vertebrates is executed by multiple caspases. Activation of an initiator caspase requires Apaf-1, a large protein that contains a domain homologous to ced-4.

directly through post-transcriptional mechanisms, modulating signalling pathways and changing the activities of target proteins that are apoptotic regulators. During this period, there are no obvious morphological changes in cells viewed under the microscope—it is therefore often referred to as a latent stage. At some point during the temporal progression of this process, a cell becomes 'committed' to subsequent death. Commitment is unlikely to be a single molecular event, but rather the summation of multiple signals. It may occur at a different stage of the apoptotic pathway in different cells or indeed in the same cell type in response to different stimuli. At commitment, a cell enters a condemned phase in which it can proceed to undergo apoptosis even without further apoptotic stimulus. This does not necessarily mean that subsequent stages in the process are not controlled and anti-apoptotic signals or pharmacological intervention may still inhibit some or all of the subsequent events of apoptosis and may even rescue the cell from death. Thus, the commitment point may be analogous to the restriction point of cells entering the cell cycle, after which point growth factor stimulation does not need to be maintained to proceed to the initiation of DNA replication. Nevertheless, many subsequent biochemical events are required and the process can be arrested at any one of several checkpoints. Cells then enter the phase of execution, when specific proteolysis and nuclease activation brings about the classical changes associated with apoptosis. Finally, fragmented cells are phagocytosed by neighbouring cells.

During the latent and committed phases, cells may continue to carry out other cellular processes, even continuing through the cell cycle. This makes it difficult to distinguish the critical molecular events leading to apoptosis. Furthermore, during this period, changes in protein expression may be part of an attempt to recover from the apoptotic stimulus—thus increased expression of a protein does not necessarily mean that it is pro-apoptotic or that it functions as part of the apoptotic machinery. The pre-execution period may vary in duration from hours to days, depending on the stimulus. Even for a given stimulus acting on a clonal population of cells, the mechanism is stochastic, often leading to considerable variability in duration. As a result, sampling for biochemical analysis during the process will mix cells at different stages. Analysis of the execution phase is even more problematic, since it may be very rapid, being completed in as little as 15 minutes and involving a rapid series of minute-by-minute changes that are very difficult to dissect. Finally, in culture, cells may not be phagocytosed but start to undergo processes more akin to necrosis, with loss of mitochondrial function, acidification, and generalized degradation of proteins and nucleic acids which are difficult to distinguish from the controlled events of apoptosis proper.

3. Development of cell-free systems for apoptosis

One successful approach to studying the biochemistry of apoptosis has been the development of cell-free systems to study the process *in vitro*. These systems have the advantages of allowing experiments to be performed quickly, they are generally synchronous in biochemical changes and sufficient material can be prepared to allow purification of factors. Since none of the apoptotic cell-free systems developed to date carry out transcription, regulation by changing gene expression cannot be studied directly, although these systems may be useful for studying the effects of transcriptional and post-transcriptional changes in the intact cells prior to the preparation of extracts. However, post-translational control mechanisms may still be functional in the extracts. Cell-free systems for apoptosis have already been used to great effect to study the activities of known components of the apoptotic machinery and to assay and identify novel components. Such systems are likely to play an important role in future discoveries about the central machinery of apoptosis, how it brings about the cellular events of execution, and how the initiation of the process is regulated.

The cell-free systems for apoptosis have been developed fall into three types:

(a) Extracts from somatic cells undergoing apoptosis that have caspases activated. These extracts are useful for investigation of the execution phase of apoptosis, e.g. DNA fragmentation in nuclei added to the extracts. They can also be used as a way of measuring apoptosis *in vitro* after manipulation of protein expression or signalling pathways in the cells *before* their homogenization and fractionation.

(b) Extracts made from non-apoptotic somatic cells that can be induced to undergo aspects of apoptosis *in vitro*. Extracts from non-apoptotic cells have been spectacularly useful in identifying cytochrome *c* and Apaf-1 as a major cofactors for caspase activation.

(c) *Xenopus* egg extracts, which have advantages for the study of signals controlling the initiation of apoptosis.

4. Extracts of apoptotic cells

4.1 Investigation of nuclear apoptosis

Extracts made from cells given a pro-apoptotic stimulus before preparation of the extracts provide a source of apoptotic factors, particularly active caspases. Such extracts have proved useful in studying the cleavage of specific proteins, either endogenous proteins detected on Western blots with antibodies or recombinant proteins synthesized and radioactively labelled in reticulocyte lysates before addition to the extracts. Apoptotic extracts have also proved useful for studying the nuclear events of apoptosis by adding nuclei isolated from non-apoptotic cells to them (17). The first such system described, by Lazebnik *et al.* (18), was produced using a protocol originally designed to yield mitotic extracts. Proliferating chicken hepatoma DU249 cells were treated first with the DNA polymerase inhibitor aphidicolin, to synchronize them in S phase, released from the block, then synchronized in mitosis with the spindle poison, nocadazole. Concentrated homogenates of these cells were freeze-thawed, ground up, and centrifuged at 150 000 *g* to prepare a cytosolic fraction (S/M extracts). Nuclei added to these extracts underwent very rapid chromatin condensation, DNA fragmentation, and lamina breakdown, showing that active apoptotic factors were present in the extracts. Using this cell-free system, Lazebnik *et al.* (19) identified the first caspase substrate (other than IL-1 beta), poly (ADP-ribose) polymerase (PARP), an enzyme involved in DNA repair, and showed that PARP was cleaved by an enzyme other than caspase-1 (ICE), which they named prICE. Importantly, they identified the site of cleavage on PARP as C-terminal to an aspartate residue, a specificity related to that of caspase-1 and, as shown subsequently, for all other caspases. Later, the major PARP cleaving activity in apoptotic cell extracts was identified as caspase-3 (20), a major workhorse caspase that also cleaves substrates such as the catalytic subunit of DNA-dependent protein kinase (DNA-PKcs) (21–23) and the proenzyme forms of other caspases (14). A tetrapeptide inhibitor of prICE/ caspase-3 derived from the recognition site in PARP (DEVD coupled to reactive group such as chloromethyl ketone) completely inhibits PARP cleavage, DNA fragmentation, and chromatin condensation in vertebrate cell-free systems (20, 24). Thus, caspase-independent factors, if present in these cell-free systems, are insufficient to bring about nuclear events of apoptosis.

Structural proteins cleaved in nuclei in cell-free systems include lamins (25, 26) and the nuclear matrix antigen, NuMA (our unpublished data). Lazebnik *et al.* (25) also used the DU249 cell extract system to examine the cleavage of lamin A. They showed that lamin A is cleaved by a different caspase from PARP. Selective inhibition by the poxvirus caspase inhibitor, CrmA, indicated that lamin A cleavage plays a role in the disintegration of the nucleus after chromatin condensation. Using purified caspases, lamin A is preferentially cleaved by caspase-6 (Mch2), and the cleavage site VEID↓NG fits the consensus of caspase-6, although it remains to be established which caspase(s) are responsible in cells.

More recently, Samejima *et al.* (27) have modified the DU249 cell-free system by preparing extracts at different times during the apoptotic process. They managed to separate a phase when cells made extracts in which nuclear disintegration was dependent upon caspase activity, whereas extracts made from cells late in the apoptotic pathway no longer required active caspases, indicating that downstream effectors which induce nuclear disintegration were already activated. This suggests that, during apoptotic progression, it is the activation of downstream effectors such as the nuclease and chromatin condensation factor that signals the transition from the latent (condemned) phase to the execution phase, rather than the activation of caspases which proceed this transition. This is consistent with the *Xenopus* egg extract model system in which DNA fragmentation follows caspase activation in the extracts after ~ one hour and can still be blocked by caspase inhibitors during this time (24).

Other cell-free systems have been described that are made from cells given a direct apoptotic stimulus. Enari *et al.* (28) described a cell extract prepared from Fas-expressing cells treated with anti-Fas antibody which triggers the apoptotic response. Concentrated cell lysates prepared as a 10 000 *g* supernatant from the cells after 90 minutes treatment with the antibody underwent caspase-dependent DNA fragmentation. Similarly, Martin *et al.* (29) described cell extracts prepared as 14 000 *g* supernatants from UV irradiated CEM T lymphoblastoid cells and anti-Fas antibody treated Jurkat cells that also reproduced caspase-dependent nuclear fragmentation. Both groups found that addition of Bcl-2 to the lysates at the start of incubations partially blocked DNA fragmentation, although the stage at which Bcl-2 was acting was unclear in these experiments (28, 29). Importantly, Martin *et al.* (29) also produced extracts from untreated cells in which DNA fragmentation did not occur. Using these non-apoptotic extracts they found that C2-ceramide could produce nuclear disintegration, suggesting that the apoptotic machinery could be controlled *in the extract*.

4.2 Identification of apoptotic DNA fragmentation and chromatin condensation factors

The biochemical approach to studying apoptosis has proved very successful in the identification of the endonuclease that cleaves DNA into discrete fragments during the later stages of execution. Using non-apoptotic S_{100} extracts of HeLa cells, Liu *et al.* (30) looked for factors that were activated by caspase-3 and induced DNA cleavage, purifying a heterodimer of 40 and 44 kDa that they named DFF, or DNA fragmentation factor. They showed that caspase-3 cleaved the 44 kDa subunit at two sites (30). Nagata and colleagues also purified the 40 kDa subunit as caspase-activated deoxyribonuclease (CAD) in a dimer with the 44 kDa inhibitor, ICAD (inhibitor of CAD) (31). Treatment with caspase-3 cleaved ICAD, releasing the active nuclease (32). Later work has shown that ICAD exists in two forms $ICAD_S$ and $ICAD_L$, the long form being active as a chaperone, probably ensuring the correct folding of CAD during synthesis (33).

Biochemical fractionation is likely to be a fruitful approach to identify other cytoplasmic factors promoting nuclear events of apoptosis, although it may be more difficult to identify components that are attached to structures in the nucleus by this approach. There is evidence that a factor distinct from CAD/DFF40 is responsible for apoptotic chromatin condensation (27, 34) and in cells, DNA is first cleaved into very large fragments before inter-nucleosomal cleavage (35), suggesting that another DNA cleavage factor may exist. Recently, Sahara and colleagues (36) have identified a novel protein, acinus, that is activated after cleavage by caspase-3 and induces chromatin condensation in permeabilized HeLa cells, an *in vitro* system similar to that often used to study nuclear protein import. Jurkat cell lysates treated with caspase-3 were fractionated to purify the factor responsible for inducing chromatin condensation. A 17 kDa active fragment was identified, which was generated by caspase-3-dependent cleavage of three isoforms of the acinus protein expressed as 220, 98, and 94 kDa polypeptides (denoted acinus L, S, and S′ respectively), probably generated by alternative splicing of transcripts. One important question in the mechanism of nuclear apoptosis is whether factors are activated by caspase cleavage in the cytosol, then move to the nucleus, or if the caspases themselves translocate. In the case of acinus, the endogenous protein is present in the nucleus, suggesting that caspase-3 moves to the nucleus and carries out the cleavage there, although it is possible that a small amount of active acinus is generated in the cytoplasm before being transported to the nucleus (36).

Another pathway for nuclear disintegration has been identified by Kroemer and colleagues (37, 38) who used a system in which factors are released from isolated mitochondria following induction of the permeability transition when a large mitochondrial pore is thought to be opened and the electrochemical potential across the inner membrane is dissipated (see Chapter 6). One such factor, known as apoptosis-inducing factor (AIF), caused large scale DNA fragmentation and chromatin condensation in isolated nuclei. AIF was purified and cloned, and is an oxidoreductase, judging from its primary sequence (37). AIF translocates from the mitochondria to the nucleus after apoptotic stimulus, and is presumably actively imported since it has putative nuclear localization signals (NLS) in its primary sequence. The release of AIF from mitochondria is prevented by overexpression of Bcl-2. In the analysis of such apoptogenic factors, critical questions are: is the factor activated by the cytochrome *c*/caspase-9/caspase-3 pathway (Fig. 2), does it act through this pathway to trigger downstream events, or does it act via a separate mechanism? AIF causes the release of cytochrome *c* and caspase-9 from mitochondria, suggesting that it does activate this pathway. However, its ability to induce chromatin condensation when microinjected into cells was not blocked by the general caspase inhibitor, zVAD.fmk, indicating that caspases are not required for its action. Nevertheless, in many cells and cell-free systems, it is clear that polynucleosomal DNA fragmentation and nuclear morphological changes are caspase-3 dependent (20, 24), indicating that AIF is not a dominant factor in these systems. At present it is unclear whether the release

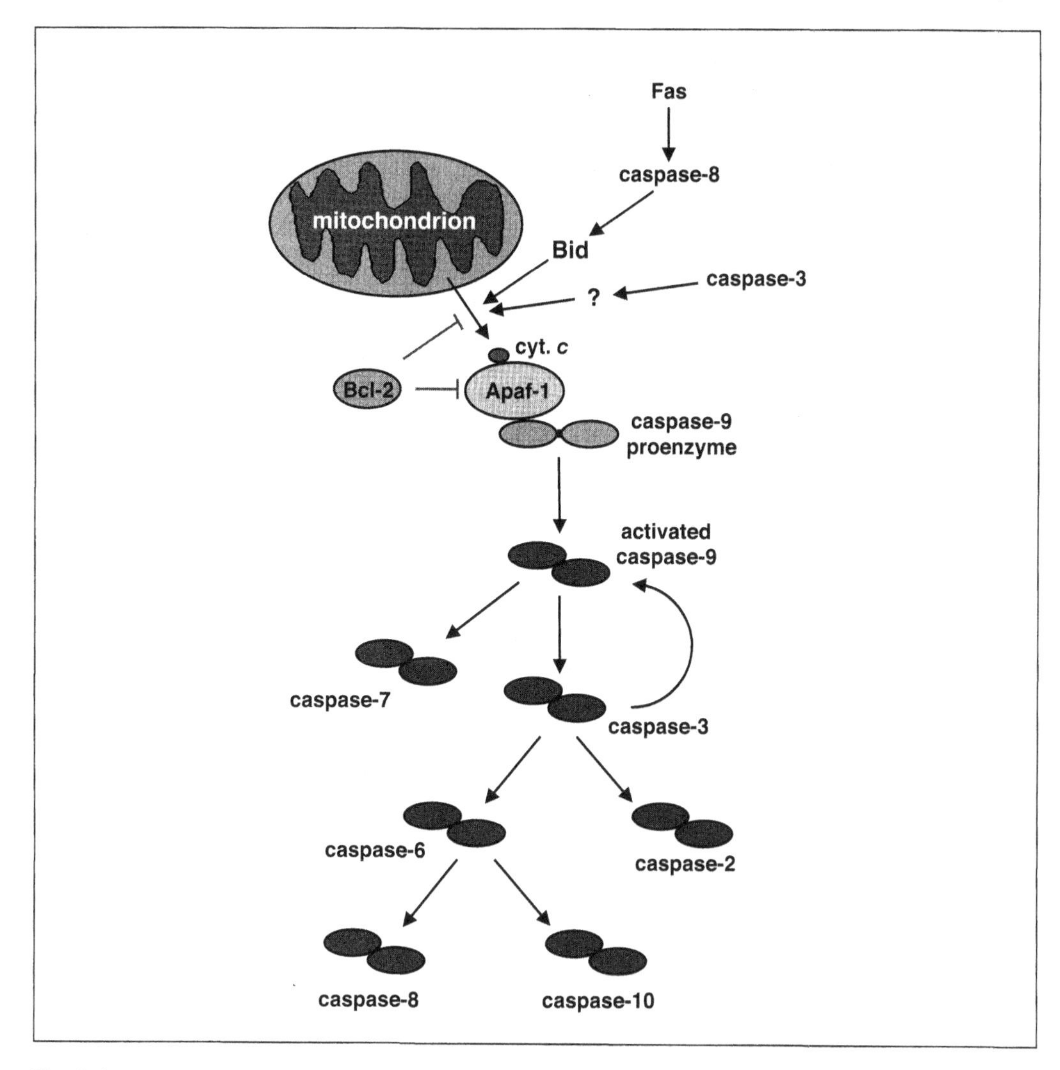

Fig. 2 A caspase cascade is activated by cytochrome *c* release from mitochondria. Cytochrome *c* activates caspase-9 through Apaf-1. In turn caspase-9 cleaves and activates the proenzymes of caspase-3 and -7. Active caspase-3 directly cleaves structural and regulatory factors as well as other caspase proenzymes as part of a cascade that may amplify and diversify the signal. Cytochrome *c* release is promoted by some caspases as part of a feedback amplification that also allows signals that activate caspases independently of cytochrome *c* to act through the mitochondria to amplify the apoptotic signal. For instance, Fas signalling directly activates caspase-8 which cleaves and activates the BH3-containing protein Bid. Cleaved Bid promotes cytochrome *c* release, whereas this process is suppressed by anti-apoptotic proteins such as Bcl-2 which also acts downstream of cytochrome *c* release to inhibit caspase activation.

of proteins in response to the permeability transition plays a general role in apoptosis, since mitochondrial membrane potential is maintained at least during the early stages of apoptotic execution in response to variety of signals in intact cells and in cell extracts, even after cytochrome *c* release (40, 41).

4.3 Oncogene-regulated apoptosis

The cell-free systems that have been developed to date do not reproduce any of the transcriptional events which play an important role in the response of cells to many apoptotic stimuli. Therefore they cannot be used to study regulation at this level directly. However, cell extracts have been used to analyse the later stages of the apoptotic pathway induced by expression of oncogenes and other regulators in the cells prior to extract preparation. Fearnhead *et al.* (42) found that expression of the adenoviral oncogene E1A results in extracts that undergo caspase activation and disintegrate added nuclei. E1A induces cytochrome *c* release from mitochondria and subsequent activation of the Apaf-1/caspase-9/caspase-3 pathway (43). Recently, Ding *et al.* (44) have described a cell-free extract (100 000 *g* supernatant) derived from $p53^{+/+}$ MEFs (mouse embryo fibroblasts) transformed with E1A/Ras or $p53^{+/+}$ REFs (rat embryo fibroblasts) transformed with Myc/Ras. Extracts prepared after the cells were gamma-irradiated for one hour showed PARP cleavage and nuclear DNA fragmentation after 10–12 hours incubation at 32 °C. However, extracts from transformed $p53^{-/-}$ MEFs or untransformed REFs were non-apoptotic for at least 24 hours, but could nevertheless respond to addition of cytochrome *c*, showing that they contain latent caspases. This suggests that the transformed cells were sensitized to irradiation in a p53-dependent manner and that this sensitivity carried through to the extracts. Perhaps surprisingly, immunodepletion of p53 from the transformed and irradiated cell extracts or addition of SV40 T antigen blocked subsequent PARP cleavage, whereas removal of Bax or cytochrome *c* had no effect. One explanation for this result is that the caspase activation pathway is sensitized in these extracts by a mechanism working in a parallel or downstream of Bax and cytochrome *c* which is removed by depletion of p53 from the extracts. Perhaps p53 suppresses caspase activation in the extracts by an unknown post-translational mechanism that may involve cAMP-dependent protein kinase (44).

5. Triggering apoptosis in non-apoptotic extracts

Extracts from some cells such as HeLa (45) or Jurkat (29) can be prepared without active caspases and can therefore be used to study caspase activation *in vitro*. It is likely that these cells are resistant to apoptosis since they are transformed and immortalized—this may permit extract preparation, but could mean that their apoptotic pathways are controlled differently from non-transformed cells. Wang and colleagues (45) described a cell-free system derived from 100 000 *g* supernatants (S_{100}) of HeLa cells. They made the discovery that caspase activation and apoptosis was initiated in these extracts by addition of dATP (or dADP), but not other nucleotides. Caspase substrates such as PARP and the precursor of caspase-3 were cleaved in the extracts, and DNA was fragmented in added nuclei. While the physiological relevance of this stimulation by dATP remains unclear (it may have a pathological role) (45), this procedure allowed them to fractionate extracts and purify pro-apoptotic factors that caused the cleavage and activation of caspase-3 proenzyme.

Remarkably, they identified an apoptotic protease activating factor (Apaf-2) as cytochrome *c* (45). This startling and unexpected result showed the power of a biochemical approach to studying apoptosis. Since cytochrome *c* also plays an essential role in respiratory electron transport it is unlikely that it would have been identified by a genetic approach. With hindsight, it was apparent that the cytochrome *c* present in the extracts and required for caspase activation stimulated by dATP had been fortuitously released from mitochondria during extract preparation. Wang and colleagues also used the caspase-3 activation assay to identify two other factors required for cytochrome *c* and dATP-induced caspase-3 cleavage—Apaf-1, a protein containing a Ced-4 related domain (15), and caspase-9 (46). This *tour de force* of biochemistry suddenly opened up research on the mechanism controlling caspase activation and the induction of apoptotic execution. Subsequent work has shown that caspase-9 acts as an initiator caspase that is activated by oligomerization induced by Apaf-1. This activity of Apaf-1 is stimulated by dATP and requires the interaction of cytochrome *c*. Caspase-9, once processed and activated cleaves caspase-3 proenzyme and thereby induces the morphological and biochemical changes associated with late stages of apoptotic execution (47–49). This pathway represents the major mechanism in p53- and oncogene-mediated apoptosis (43, 50–52).

Cytochrome *c*-induced caspase activation can be reproduced in extracts from a variety of immortalized cultured cells (45, 53), some primary cells (44), *Xenopus* egg extracts (26), and in reticulocyte lysates (54). This effect of cytochrome *c* indicated a possible explanation for the role of mitochondria in apoptosis, previously indicated by a number of approaches (38), including the cell-free system of *Xenopus* egg extracts developed by Newmeyer (8). It was also known that Bcl-2 family proteins were localized to the outer mitochondrial membrane (16). Wang and colleagues (45) showed that apoptotic stimuli such as the protein kinase inhibitor staurosporine triggered the release of cytochrome *c* from mitochondria into the cytosol of intact cells. Subsequently, they showed that when Bcl-2 was overexpressed in cells, mitochondria prepared from them were resistant to induction of cytochrome *c* release (55). In a complementary approach, Kluck *et al.* (41) showed that recombinant Bcl-2 blocked cytochrome *c* release from mitochondria in *Xenopus* egg extracts. Many further experiments have confirmed the central role of cytochrome *c* release from mitochondria both in cell-free systems and intact cells in response to a wide variety of pro-apoptotic stimuli (56). In the case of most stimuli, the mechanism triggering cytochrome *c* release remains unclear, but one pathway has been determined, in part using cell-free systems. Activation of caspase-8 in response to ligation of the Fas cell surface receptor results in cleavage of Bid, generating a pro-apoptotic fragment that translocates to the mitochondria and triggers cytochrome *c* release and activation of the caspase-9/caspase-3 pathway (57–59). Cytochrome *c* release may play an important role in amplification of caspase activation in response to many apoptotic signals, since it is induced by several caspases (60–62), acting either through Bid or other cytosolic factors (62). However, direct activation of the caspase cascade by some signals may bypass the cytochrome *c*/Apaf-1/caspase-9 step and still be sufficient to induce full apoptosis, for instance during granzyme B-induced killing by cytotoxic T cells (63, 64) or during TNF-mediated cell killing (51).

The ability of cytochrome *c* to initiate caspase activation in cell-free systems derived from human cells has been used to investigate the relationship between different caspases, delineating the cascade of activities. In HeLa cell S_{100} extracts, the activation of caspase-2 is dependent upon caspase-3 (65). In post-nuclear (15 000 *g*) supernatants of Jurkat T lymphoblastoid cell homogenates, cytochrome *c* activates caspase-2, -3, -6, -7, -8, and -10 (53). Immunodepletion experiments, readily carried out in a cell-free system, showed that caspase-9 is required for activation of all of these caspases, whereas caspase-3 is required for the activation of caspase-2, -6, -8, and 10 (Fig. 2), consistent with studies using caspase-$9^{-/-}$ and Apaf-$1^{-/-}$ cells derived from mouse knockouts (51, 52).

6. *Xenopus* egg extracts

6.1 Development of the *Xenopus* egg extract system

Extracts of early embryos, particularly those of the South African clawed frog *Xenopus laevis*, have played a critical role in determining the molecular mechanisms underlying cell cycle control (66). In *Xenopus*, the first 12 cell cycles following fertilization of the egg are each about 30 minutes long and consist of alternating S and M phases without a requirement for transcription and without growth. The egg is a very large cell, approximately 1 mm in diameter, and contains stores of all the proteins required for the first 12 cell cycles, with the exception of mitotic cyclins, which are stored as mRNA. There are also abundant stores of mitochondria and the lipid vesicles that form intracellular membranes. This means that concentrated extracts prepared from *Xenopus* eggs have all the components required to reproduce complex processes *in vitro*, including nuclear assembly and nucleocytoplasmic transport (2), DNA replication (1), mitotic spindle assembly (67), and regulation of cell cycle kinases (68), without input from transcription. They can also be used to reproduce cell cycle checkpoints activated by DNA replication arrest (69) or mitotic spindle disruption (70) by adding nuclei or forming spindles in the extracts, respectively.

Xenopus egg extracts are prepared by crushing unfertilized eggs in a centrifuge tube with the minimum volume of buffer, so that a concentrated cell-free system is produced that behaves much like the egg cytoplasm (Fig. 3). Unless steps are taken to preserve the M phase arrested state of the unfertilized egg, the extracts tend to be released into interphase during preparation. This release can be ensured by activating the eggs by treatment with a Ca^{2+}-ionophore or an electric pulse. Low speed extracts containing ribosomes continue to synthesize cyclins and may cycle once or more through mitosis and back into interphase. In the presence of the protein synthesis inhibitor, cycloheximide, the extracts are arrested in an interphase state (9, 66).

A common experience when preparing and utilizing *Xenopus* egg extracts for cell cycle studies is that some batches are 'bad', failing to carry out nuclear assembly

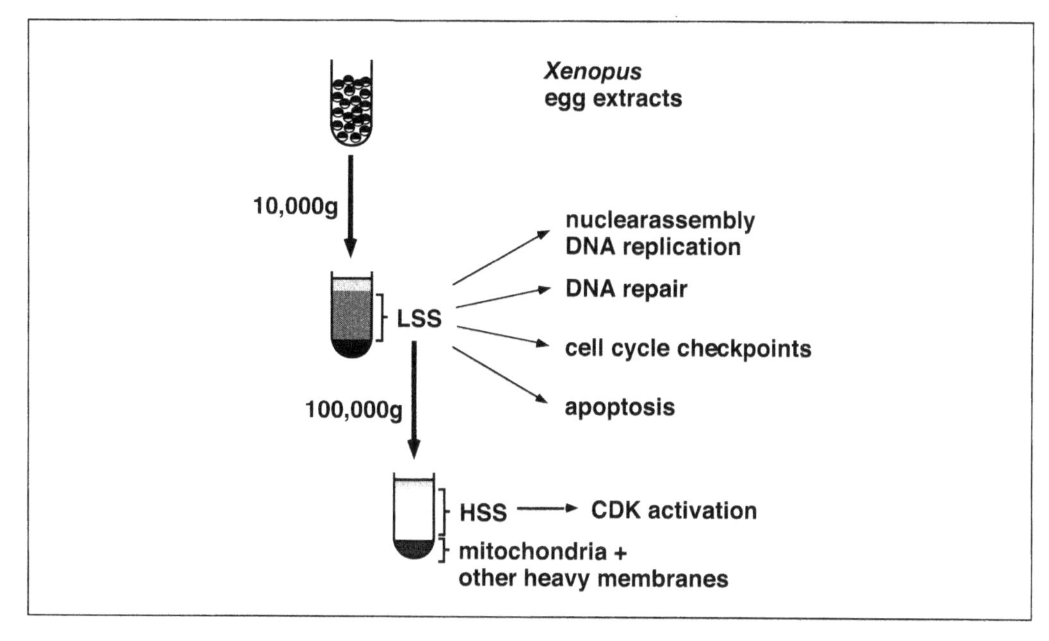

Fig. 3 Preparation of *Xenopus* egg extracts. Unfertilized eggs are packed in a centrifuge tube and crushed by centrifugation at about 10 000 *g*. The central fraction is removed, avoiding lipids at the top of the tube and cellular debris at the bottom. This low speed extract contains mitochondria and membrane vesicles, allowing reconstitution of complex processes *in vitro*, including apoptosis. Centrifugation at 100 000 *g* removes mitochondria and dense vesicles, making a soluble extract that is stable to incubation.

reactions and indeed causing nuclei to disintegrate. Newmeyer *et al.* (8) first recognized that this process resembled apoptosis, since nuclei from a variety of sources added to extracts could undergo chromatin condensation very reminiscent of apoptotic nuclei and indeed DNA was fragmented into the classical poly-nucleosomal fragments. Critically, they showed that this process was completely blocked by addition of cell lysates containing human Bcl-2 protein, confirming that it resembled at least some aspects of apoptosis. Induction of apoptosis was prevented by the removal of a dense membrane fraction that contained mitochondria. With hindsight, it can be seen that the procedures originally developed for cell cycle extracts, which add sucrose or glycerol to buffers during extract preparation, protect mitochondrial integrity (66). Low speed extracts containing mitochondria are more refractory to freezing, whereas clarified high speed supernatants lacking store well and are stable for many hours incubation.

The physiological mechanisms controlling apoptosis during *Xenopus* and early development are poorly understood, although it is clear that the machinery to carry out the process is present in the oocyte and during early embryonic development. In hydroxyurea- or cycloheximide-treated embryos, apoptosis is triggered at around the early gastrulation transition (EGT) (71). At this time degradation of maternal transcripts is compensated for by new zygotic transcription that begins prior to EGT at the mid-blastula transition (MBT). Stack and Newport (71) suggest that prior to

EGT, apoptosis may be suppressed by maternally encoded inhibitors. During oogenesis, apoptosis may be under hormonal control in the ovary, since extending the period between the hormonal injections used to first promote oocyte development, then to induce maturation and ovulation may promote apoptosis in extracts made from the resulting eggs (8). It has been suggested that this is result of atresia, a process whereby unused oocytes die and are reabsorbed by the ovary. However, our experience is that eggs from different animals treated with the same hormonal regime can produce extracts with quite different propensities to apoptosis. Furthermore, during preparation of egg extracts, the buffers used and degree of separation of subcellular fractions have a critical influence on whether or not the extracts become apoptotic. Indeed, increasing the proportion of a heavy membrane fraction containing mitochondria is sufficient to induce apoptosis in otherwise stable extracts (72; our unpublished data). Inhibition of protein synthesis and arrest of the cell cycle with cycloheximide, routinely added to the extracts, could also influence the propensity to apoptosis. Fortunately, although there is variability in the time course of apoptosis between batches of extract, aliquots of a given batch show very reproducible kinetics of apoptosis, permitting many directly comparable results to be obtained.

6.2 Control of apoptosis in *Xenopus* egg extracts

The most important difference between the *Xenopus* egg extract system and those prepared from somatic cells is that the extracts are initially not apoptotic, but become apoptotic upon incubation (8, 24, 72). No transcription or protein synthesis is required for this transition, although loss of suppressors might be involved. When low speed (10 000 *g*) *Xenopus* egg extracts containing mitochondria are incubated at room temperature, there is a latent period of typically two to three hours before a caspase-3-like activity is initiated. DNA fragmentation and chromatin condensation occur about one hour afterwards (24). Caspase activation can be followed by the cleavage of fluorogenic substrates or specific nuclear and cytoplasmic proteins. Nuclear apoptosis is followed by DNA fragmentation or chromatin condensation in nuclei added to the extracts (Fig. 4). One of the advantages of such a synchronous cell-free system is that the temporal progression of the process is readily tested by the addition of regulators during the time course. Caspase activation is blocked by recombinant Bcl-2 (or Bcl-x_L) proteins added to the extracts at the start of the incubation, but these regulators fail to prevent subsequent nuclear apoptosis after caspases are activated, showing that they regulate the execution phase by acting upstream of the caspases. However, addition of the caspase-3 inhibitor, Ac-DEVD-CHO, completely blocks nuclear apoptosis even when added after caspase activation but before DNA fragmentation becomes apparent (24). Thus, a caspase-dependent pathway is essential for nuclear apoptosis in this system and Bcl-2 regulates caspase activation.

The transition from latent to execution phases involves the mitochondria present in the extracts (8). Kluck *et al.* (41) showed that this critical step corresponds to the

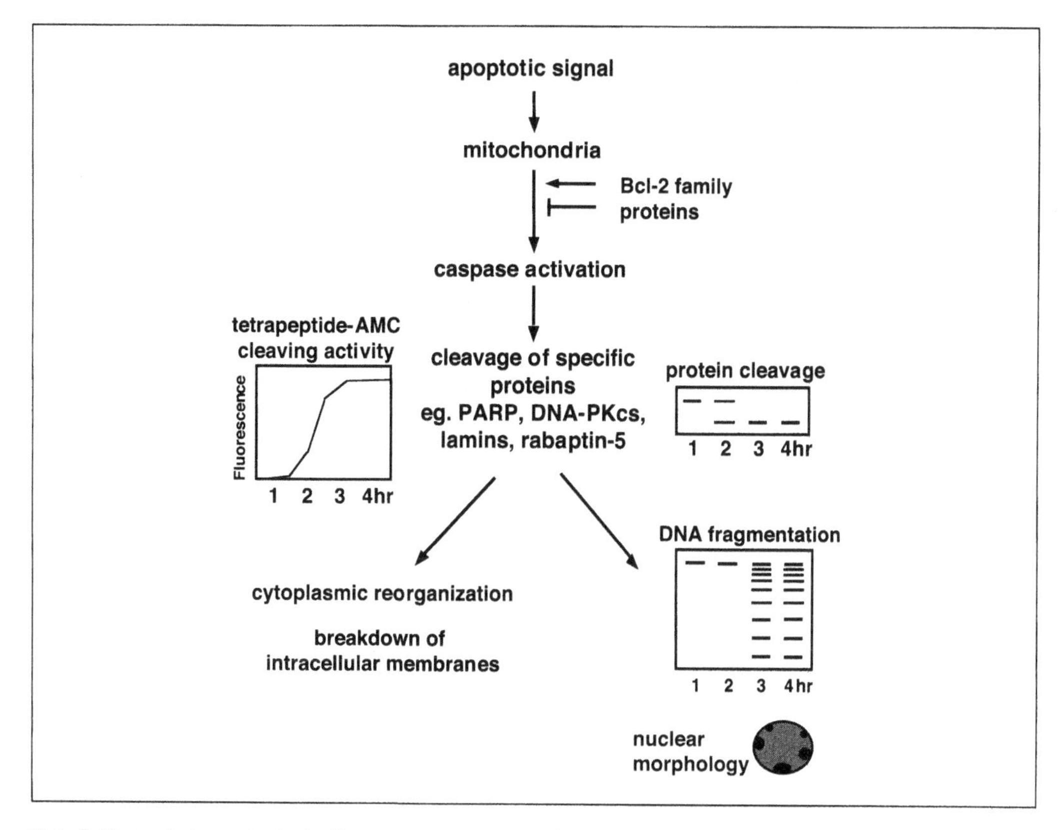

Fig. 4 Measuring apoptosis in *Xenopus* egg extracts. Caspase activation can be measured directly using tetrapeptide substrates that release a fluorogenic group (AMC) that is detected in a fluorimeter. The substrate DEVD-AMC measures a caspase-3-like activity. The temporal regulation of caspase activation can also be followed by the cleavage of specific substrates separated by SDS–PAGE, detected with antibodies or by autoradiography of radioactively-labelled recombinant proteins. Nuclear events can be followed by chromatin condensation or more quantitatively by analysis of DNA fragmentation on agarose gels. Cytoplasmic processes such as membrane vesicle fusion can also be assayed.

release of cytochrome *c* from the mitochondria into the soluble fraction of the extracts (Fig. 5). Cytochrome *c* is sufficient to trigger activation of a caspase-3-like activity that is essential for the downstream cleavage of DNA and the morphological characteristics of apoptosis in nuclei added to the extracts (26). Most importantly, the release of cytochrome *c* is completely blocked by anti-apoptotic Bcl-2 protein (41).

Cytochrome *c* release in this system precedes loss of mitochondrial membrane potential and is therefore not the result of crude disruption of mitochondrial function or the permeability transition (26). Rather it seems that the outer membrane is selectively permeabilized to release cytochrome *c* and possibly other factors. Although cytosolic cytochrome *c* is sufficient to cause nuclear disintegration, it remains possible that additional apoptogenic factors are released from the mitochondria. Nevertheless, caspase-3-like activity is essential for downstream events leading to DNA fragmentation and chromatin condensation (Fig. 5) (24, 26, 73).

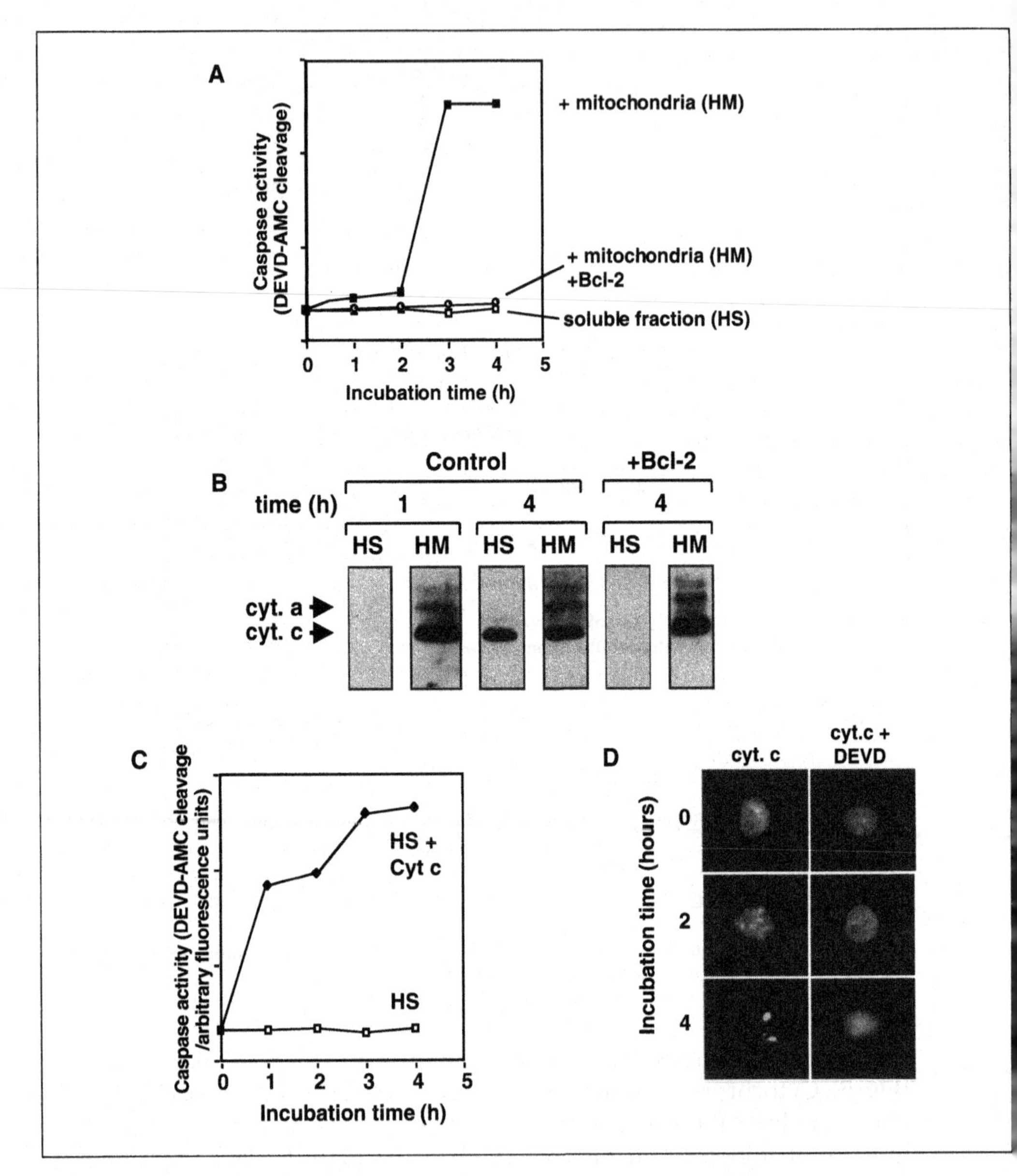

Fig. 5 Cytochrome *c* plays a critical role in caspase activation in *Xenopus* egg extracts. (A) A membrane fraction containing mitochondria is required for the spontaneous activation of caspase during incubation. Low speed supernatants (10 000 *g*; LS) of *Xenopus* egg extracts were fractionated by centrifugation at 100 000 *g* into soluble (HS) and heavy membrane (mitochondrial) fractions (HM). Caspase activation when mitochondria are added back to the soluble fraction is blocked by recombinant Bcl-2 protein (24). (B) Bcl-2 inhibits the release of cytochrome *c* (cyt. c) from mitochondria (HM) into the soluble fraction (HS) of the extracts after four hours incubation. Cytochrome *a* (cyt. a) is used as a control. (C) Cytochrome *c* triggers rapid caspase activation in soluble extracts prepared as 100 000 *g* supernatants (HS). (D) Cytochrome *c* initiates chromatin condensation and nuclear disintegration in HeLa nuclei added to soluble extracts. The effect of cytochrome *c* is blocked by the caspase-3 inhibitor AcDEVD-CHO, showing that the nuclear apoptotic effects of cytochrome *c* are mediated through a caspase-3-like enzyme.

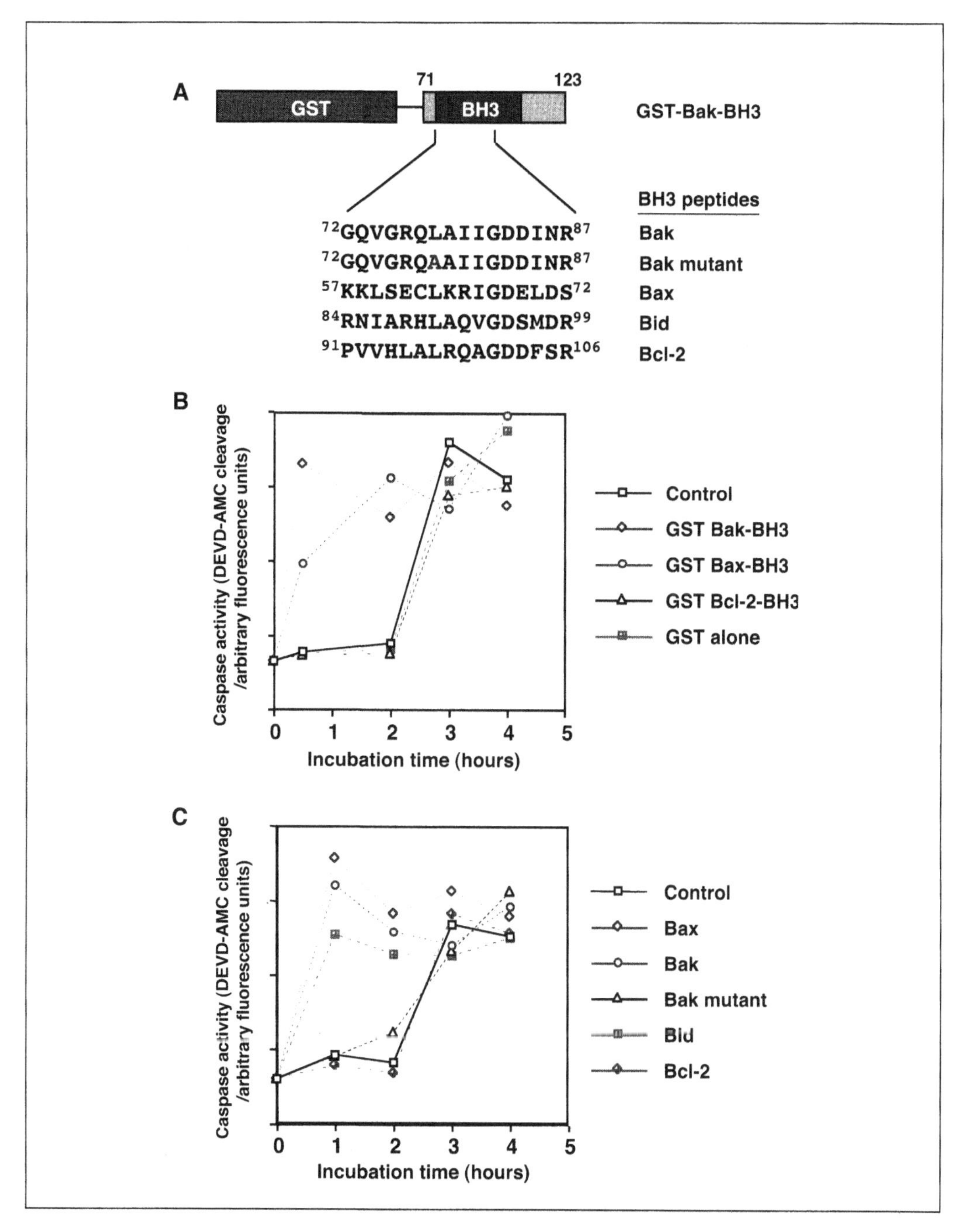

Fig. 6 Activation of caspases by pro-apoptotic BH3 domains in *Xenopus* egg extracts (73). (A) BH3 domains were constructed as fusion proteins with glutathione-*S*-transferase (GST) and expressed in bacteria. Peptides derived from these domains were synthesized chemically. (B) Bax and Bak BH3 domains accelerate caspase activation in extracts. They promote cytochrome *c* release from mitochondria (73). Bcl-2-BH3 is inactive. (C) Peptides derived from pro-apoptotic BH3 domains trigger caspase activation. A peptide derived from Bak with a single leucine to alanine change (Bak mutant) abolishes its activity.

In contrast to Bcl-2, pro-apoptotic family members can promote cytochrome *c* release and caspase activation in *Xenopus* egg extracts (73). Isolated BH3 domains of proteins such as Bax and Bak, produced as recombinant fusion proteins and added to the extracts trigger apoptosis. BH3 domains require cytosolic fraction of the extracts as well as the mitochondrial fraction to trigger caspase activation. The effects of BH3 domains can be mimicked by short 16 amino acid peptides (Fig. 6) (73). These BH3 peptides could serve as the basis of peptide derivatives or mimetic small molecules to trigger apoptosis. The *Xenopus* egg extract system could prove a useful system in which to assay the biological activity of such compounds.

In addition to the control of cytochrome *c* release from mitochondria, Bcl-2 and Bcl-x_L inhibit the activation of caspases downstream of cytochrome *c* in a soluble system of *Xenopus* egg extracts prepared as 100 000 *g* supernatants that lack mitochondria (61), albeit using relatively higher (1 μM) concentrations of Bcl-2 and Bcl-x_L. This result has also been confirmed in soluble human (HeLa S_{100}) cell extracts (65). In these cell-free systems, there is a threshold in the response to cytochrome *c*, since very low levels fail to activate caspases. Bcl-2 and Bcl-x_L act in these soluble extracts by increasing the concentration of cytochrome *c* required to induce caspase activation (61). This threshold is reduced by dATP or high concentrations of ATP, which presumably stimulate an Apaf-1-like factor. Thus, addition of dATP or ATP to the extracts can overcome the inhibitory effects of Bcl-2 and Bcl-x_L, an important technical point when studying regulation of this stage of the apoptotic pathway by Bcl-2 family proteins. The stimulation of caspase activation by dATP requires the presence of some cytochrome *c* in soluble *Xenopus* egg extracts to accelerate caspase activation (61), as in HeLa cell extracts (45), indicating that cytochrome *c* is an essential cofactor of Apaf-1. Precisely how Bcl-2 and Bcl-x_L increase the threshold requirement for cytochrome *c* remains unclear at present. One possibility is that Bcl-2 binds to and inhibits Apaf-1, analogous to the direct inhibition by Ced-9 of Ced-4-catalysed processing of Ced-3 (74, 75). However, the interaction between Bcl-2 (or Bcl-x_L) and Apaf-1 remains controversial. The *Xenopus* system could potentially be very useful for the dissection of this control point, although one disadvantage of this system is that the components of the apoptotic machinery have not been fully described. Another disadvantage is that commercial reagents such as antibodies are not usually available.

One unique and very useful feature of the *Xenopus* egg cell-free system is that transition to apoptosis is under the control of signalling pathways that operate in the extract (72, 76, 77). Remarkably, Kornbluth and colleagues showed that a pathway activated by the pro-apoptotic *Drosophila* protein Reaper is operational in *Xenopus* egg extracts. Reaper triggers apoptosis in this system by inducing cytochrome *c* release from mitochondria (78). Subsequently, they purified a novel reaper-interacting protein named scythe that is required for reaper-induced apoptosis (79). Reaper binding to Scythe releases a sequestered apoptotic inducer (80). Application of the *Xenopus* egg extract system is likely to prove useful for further dissection of this pathway which is in all probability conserved in humans.

Caspase activation in *Xenopus* egg extracts is also under the control of protein kinase signalling pathways. The anti-apoptotic oncogenic tyrosine kinases Bcr-Abl and v-Abl block caspase activation (73; our unpublished results), whereas the depletion of the adaptor molecule Crk, a substrate for Abl kinases, prevents initiation of apoptosis (72). Isolated SH2 domains of Crk and other proteins inhibit caspase activation (72, 76), also suggesting that tyrosine phosphorylation is involved. *Xenopus* egg extracts can reproduce *in vitro* some of the signalling pathways that may regulate apoptosis and are therefore likely to be useful in dissecting these pathways and understanding how they control apoptosis. A *Xenopus* protein serine/threonine kinase of the Ste20/PAK family has been shown to inhibit caspase activation and apoptosis in egg extracts. This kinase or a closely related enzyme also plays a role in maintaining the G2/prophase arrest of oocytes and may be important for the suppression of apoptosis during the prolonged cell cycle arrest of these cells (77).

6.3 Analysis of specific processes in *Xenopus* egg extracts

Xenopus egg extracts faithfully reproduce many cellular processes *in vitro*, including cytoplasmic processes that have been relatively understudied in apoptosis. They offer an attractive system for studying the effects of caspase cleavage of specific substrates in such processes. One example of such a use has been the discovery of the role of cleavage of rabaptin-5, a protein interacting with the small GTPase rab5 that plays a critical role in endosome vesicle fusion and hence the structure and function of the endocytic pathway. Cleavage of rabaptin-5 prevents vesicle fusion in a cell-free model system derived from *Xenopus* eggs (81). In cells, rabaptin-5 cleavage may play a role in the fragmentation of intracellular membranes that occurs during apoptosis.

7. Conclusions

Cell-free systems can offer important advantages for the study of apoptosis:

(a) They reproduce the process synchronously and large amounts of material can be obtained to study apoptosis biochemically. The temporal regulation of the process, particularly the execution phase, can be dissected *in vitro*.

(b) They facilitate the assay and purification of novel factors that promote or inhibit apoptosis.

(c) They can be readily manipulated by direct addition of recombinant proteins, antibodies, and small molecule reagents. Some cell-free systems can be used to assay activities of regulators such as Bcl-2 family proteins and could be utilized in toxicological and pharmaceutical studies.

(d) Some cell-free systems are useful for dissecting signalling pathways controlling the initiation of apoptosis.

Cell-free systems have played a critical role in the identification of factors involved in nuclear apoptosis and in determining the central function of caspases in apoptosis. Cell-free systems have also been instrumental in the identification of cytochrome *c* and Apaf-1 as major factors in the initiation of caspase activation and in demonstrating the role of Bcl-2 family proteins in controlling the release of cytochrome *c* into the cytosol. This is clearly a critical control point in the progression of apoptosis, but it is also emerging that the apoptotic mechanism is controlled at multiple steps, both upstream and downstream of this point. In the future, cell-free systems are likely to be particularly useful for studying these controls: for determining the regulation of the apoptosis downstream of cytochrome *c* release by Bcl-2 family proteins (61), regulation by reversible protein phosphorylation, e.g. of the caspases (82), and in understanding the functions of endogenous caspase inhibitors such as the IAPs (83). Furthermore, the apoptotic mechanism is not a simple linear pathway, but involves feedback mechanisms that amplify the signal and thresholds that determine the response (61). Cell-free systems offer a unique opportunity to study these mechanisms biochemically. Another development may be the use of cell-free systems derived from differentiated cells (84) to study the apoptotic controls particularly important to a particular cell type. Furthermore, cell-free systems derived from cells with different genetic backgrounds, e.g. cells generated from mouse knockouts, will facilitate investigation of the role of specific gene products in the control of apoptosis. *Xenopus* egg extracts (9) provide a particularly useful *in vitro* model for studying the execution phase and for investigating the effects of apoptosis on the complex nuclear and cytoplasmic processes that are readily reproduced in this system. Uniquely, it can be used to dissect intracellular signalling pathways controlling the process, including those acting upstream of mitochondria.

Acknowledgements

Drs Lindsey Allan, Sabina Cosulich, Shalini Pathak, Cathy Sampson, Peter Savory, and Chuanmao Zhang have carried out the work on cell-free systems for apoptosis in my laboratory. Our work is supported by the Association for International Cancer Research (AICR), the Biotechnology and Biological Sciences Research Council (BBSRC), and The Cancer Research Campaign (CRC).

References

1. Hutchison, C. J. (1993). The use of cell-free extracts of *Xenopus* eggs for studying DNA replication *in vitro*. In *The cell cycle: a practical approach* (ed. P. Fantes and R. Brooks), p. 177. IRL Press, Oxford.
2. Newmeyer, D. D. and Wilson, K. L. (1991). Egg extracts for nuclear import and nuclear assembly reactions. In *Xenopus laevis: Practical uses in cell and molecular biology* (ed. B. K. Kay and H. B. Peng), Vol. 36, p. 607. Academic Press, San Diego.
3. Woodman, P. G. and Warren, G. (1989). Fusion of endocytic vesicles in a cell-free system. *Methods Cell Biol.*, **31**, 197.
4. Belmont, L. D., Hyman, A. A., Sawin, K. E., and Mitchison, T. J. (1990). Real-time visualization of cell cycle-dependent changes in microtubule dynamics in cytoplasmic extracts. *Cell*, **62**, 579.
5. Verde, F., Labbe, J. C., Doree, M., and Karsenti, E. (1990). Regulation of microtubule dynamics by cdc2 protein kinase in cell-free extracts of *Xenopus* eggs. *Nature*, **343**, 233.

6. Hengartner, M. O. (1995). Out-of body experiences: cell-free cell death. *Bioessays*, **17**, 549.
7. Earnshaw, W. C. (1995). Apoptosis: lessons from *in vitro* systems. *Trends Cell Biol.*, **5**, 217.
8. Newmeyer, D. D., Farschon, D. M., and Reed, J. C. (1994). Cell-free apoptosis in *Xenopus* egg extracts: inhibition by Bcl-2 and requirement for an organelle fraction enriched in mitochondria. *Cell*, **79**, 353.
9. Kornbluth, S. (1997). Apoptosis in *Xenopus* egg extracts. In *Cell Cycle Control* (ed. W. G. Dunphy), *Methods in enzymology*, Vol. 283, p. 600. Academic Press, San Diego.
10. Kerr, J. F. R., Wyllie, A. H., and Currie, A. R. (1972). Apoptosis: a basic biological phenomenon with wide-ranging implications in tissue kinetics. *Br. J. Cancer*, **26**, 239.
11. Wyllie, A. H. (1980). Glucocorticoid-induced thymocyte apoptosis is associated with endogenous endonuclease activation. *Nature*, **284**, 555.
12. Raff, M. (1998). Cell suicide for beginners. *Nature*, **396**, 119.
13. Hengartner, M. O. and Horvitz, H. R. (1994). Programmed cell death in *Caenorhabditis elegans*. *Curr. Opin. Genet. Dev.*, **4**, 581.
14. Thornbury, N. A. and Lazebnik, Y. (1998). Caspases: Enemies within. *Science*, **281**, 1312.
15. Zou, H., Henzel, W. J., Liu, X., Lutschg, A., and Wang, X. (1997). Apaf-1, a human protein homologous to *C. elegans* CED-4, participates in cytochrome c-dependent activation of caspase-3. *Cell*, **90**, 405.
16. Adams, J. M. and Cory, S. (1998). The Bcl-2 family: arbiters of cell survival. *Science*, **281**, 1322.
17. Earnshaw, W. C. (1995). Nuclear changes in apoptosis. *Curr. Opin. Cell Biol.*, **7**, 337.
18. Lazebnik, Y. A., Cole, S., Cooke, C. A., Nelson, W. G., and Earnshaw, W. C. (1993). Nuclear events of apoptosis *in vitro* in cell-free mitotic extracts: a model system for analysis of the active phase of apoptosis. *J. Cell Biol.*, **123**, 7.
19. Lazebnik, Y. A., Kaufmann, S. H., Desnoyers, S., Poirier, G. G., and Earnshaw, W. C. (1994). Cleavage of poly(ADP-ribose) polymerase by a proteinase with properties like ICE. *Nature*, **371**, 346.
20. Nicholson, D. W., Ali, A., Thornberry, N. A., Vaillancourt, J. P., Ding, C. K., Gallant, M., *et al.* (1995). Identification and inhibition of the ICE/CED-3 protease necessary for mammalian apoptosis. *Nature*, **376**, 37.
21. Le Romancer, M., Cosulich, S. C., Jackson, S. P., and Clarke, P. R. (1996). Cleavage and inactivation of DNA-dependent protein kinase catalytic subunit during apoptosis in *Xenopus* egg extracts. *J. Cell Sci.*, **109**, 3121.
22. Casciola-Rosen, L., Nicholson, D. W., Chong, T., Rowan, K. R., Thornberry, N. A., Miller, D. K., *et al.* (1996). Apopain/Cpp32 cleaves proteins that are essential for cellular repair—a fundamental principle of apoptotic death. *J. Exp. Med.*, **183**, 1957.
23. Song, Q., Lees-Miller, S. P., Kumar, S., Zhang, N., Chan, D. W., Smith, G. C. M., *et al.* (1996). DNA-dependent protein kinase catalytic subunit: a target for an ICE-like protease in apoptosis. *EMBO J.*, **15**, 3238.
24. Cosulich, S. C., Green, S., and Clarke, P. R. (1996). Bcl-2 regulates activation of apoptotic proteases in a cell-free system. *Curr. Biol.*, **6**, 997.
25. Lazebnik, Y. A., Takahashi, A., Moir, R. D., Goldman, R. D., Poirier, G. G., Kaufmann, S. H., *et al.* (1995). Studies of the lamin proteinase reveal multiple parallel biochemical pathways during apoptotic execution. *Proc. Natl. Acad. Sci. USA*, **92**, 9042.
26. Kluck, R. M., Martin, S. J., Hoffman, B. M., Zhou, J. S., Green, D. R., and Newmeyer, D. D. (1997). Cytochrome c activation of CPP32-like proteolysis plays a critical role in a *Xenopus* cell-free apoptosis system. *EMBO J.*, **16**, 4639.

27. Samejima, K., Toné, S., Kottke, T. J., Enari, M., Sakahira, H., Cooke, C. A., *et al.* (1998). Transition from caspase-dependent to caspase-independent mechanisms at the onset of apoptotic execution. *J. Cell Biol.*, **143**, 225.
28. Enari, M., Hase, A., and Nagata, S. (1995). Apoptosis by a cytosolic extract from Fas-activated cells. *EMBO J.*, **14**, 5201.
29. Martin, S. J., Newmeyer, D. D., Mathias, S., Farschon, D. M., Wang, H. G., Reed, J. C., *et al.* (1995). Cell-free reconstitution of Fas-, UV radiation- and ceramide-induced apoptosis. *EMBO J.*, **14**, 5191.
30. Liu, X. S., Zou, H., Slaughter, C., and Wang, X. D. (1997). DFF, a heterodimeric protein that functions downstream of caspase-3 to trigger DNA fragmentation during apoptosis. *Cell*, **89**, 175.
31. Enari, M., Sakahira, H., Yokoyama, H., Okawa, K., Iwamatsu, A., and Nagata, S. (1998). A caspase-activated DNase that degrades DNA during apoptosis, and its inhibitor ICAD. *Nature*, **391**, 43.
32. Sakahira, H., Enari, M., and Nagata, S. (1998). Cleavage of CAD inhibitor in CAD activation and DNA degradation during apoptosis. *Nature*, **391**, 96.
33. Sakahira, H., Enari, M., and Nagata, S. (1999). Functional differences of two forms of the inhibitor of caspase-activated DNase, ICAD-L, and ICAD-S. *J. Biol. Chem.*, **274**, 15740.
34. Sakahira, H., Enari, M., Ohsawa, Y., Uchiyama, Y., and Nagata, S. (1999). Apoptotic nuclear morphological change without DNA fragmentation. *Curr. Biol.*, **9**, 543.
35. Oberhammer, F., Wilson, J. W., Dive, C., Morris, I. D., Hickman, J. A., Wakeling, A. E., *et al.* (1993). Apoptotic death in epithelial cells: cleavage of DNA to 300 and/or 50 kb fragments prior to or in the absence of internucleosomal fragmentation. *EMBO J.*, **12**, 3679.
36. Sahara, S., Aoto, M., Eguchi, Y., Imamoto, N., Yoneda, Y., and Tsujimoto, Y. (1999). Acinus is a caspase-3-activated protein required for apoptotic chromatin condensation. *Nature*, **401**, 168.
37. Susin, S. A., Lorenzo, H. K., Zamzami, N., Marzo, I., Snow, B. E., Brothers, G. M., *et al.* (1999). Molecular characterization of mitochondrial apoptosis-inducing factor. *Nature*, **397**, 441.
38. Susin, S. A., Zamzami, N., Castedo, M., Hirsch, T., Marchetti, P., Macho, A., *et al.* (1996). Bcl-2 inhibits the mitochondrial release of an apoptogenic protease. *J. Exp. Med.*, **184**, 1331.
39. Janicke, R. U., Sprengart, M. L., Wati, M. R., and Porter, A. G. (1998). Caspase-3 is required for DNA fragmentation and morphological changes associated with apoptosis. *J. Biol. Chem.*, **273**, 9357.
40. Bossy-Wetzel, E., Newmeyer, D. D., and Green, D. R. (1998). Mitochondrial cytochrome c release in apoptosis occurs upstream of DEVD-specific caspase activation and independently of mitochondrial transmembrane depolarization. *EMBO J.*, **17**, 37.
41. Kluck, R. M., Bossy-Wetzel, E., Green, D. R., and Newmeyer, D. D. (1997). The release of cytochrome c from mitochondria: a primary site for Bcl-2 regulation of apoptosis. *Science*, **275**, 1132.
42. Fearnhead, H. O., McCurrach, M. E., O'Neill, J., Zhang, K., Lowe, S. W., and Lazebnik, Y. A. (1997). Oncogene-dependent apoptosis in extracts from drug-resistant cells. *Genes Dev.*, **11**, 1266.
43. Fearnhead, H. O., Rodriguez, J., Govek, E. E., Guo, W., Kobayashi, R., Hannon, G., *et al.* (1998). Oncogene-dependent apoptosis is mediated by caspase-9. *Proc. Natl. Acad. Sci. USA*, **95**, 13664.

44. Ding, H. F., McGill, G., Rowan, S., Schmaltz, C., Shimamura, A., and Fisher, D. E. (1998). Oncogene-dependent regulation of caspase activation by p53 protein in a cell-free system. *J. Biol. Chem.*, **273**, 28378.
45. Liu, X., Kim, C. N., Yang, J., Jemmerson, R., and Wang, X. (1996). Induction of apoptotic program in cell-free extracts: requirement for dATP and cytochrome c. *Cell*, **86**, 147.
46. Li, P., Nijhawan, D., Budihardjo, I., Srinivasula, S. M., Ahmad, M., Alnemri, E. S., *et al.* (1997). Cytochrome c and dATP-dependent formation of Apaf-1/caspase-9 complex initiates an apoptotic protease cascade. *Cell*, **91**, 479.
47. Zou, H., Li, Y., Liu, X., and Wang, X. (1999). An APAF-1.cytochrome c multimeric complex is a functional apoptosome that activates procaspase-9. *J. Biol. Chem.*, **274**, 11549.
48. Srinivasula, S. M., Ahmad, M., Fernandes-Alnemri, T., and Alnemri, E. S. (1998). Autoactivation of procaspase-9 by Apaf-1-mediated oligomerization. *Mol. Cell*, **1**, 949.
49. Yang, X., Chang, H. Y., and Baltimore, D. (1998). Autoproteolytic activation of procaspases by oligomerization. *Mol. Cell*, **1**, 319.
50. Soengas, M. S., Alarcón, R. M., Yoshida, H., Giaccia, A. J., Hakem, R., Mak, T. W., *et al.* (1999). Apaf-1 and caspase-9 in p53-dependent apoptosis and tumor inhibition. *Science*, **284**, 156.
51. Hakem, R., Hakem, A., Duncan, G. S., Henderson, J. T., Woo, M., Soengas, M. S., *et al.* (1998). Differential requirement for caspase-9 in apoptotic pathways *in vivo*. *Cell*, **94**, 339.
52. Yoshida, H., Kong, Y. Y., Yoshida, R., Elia, A. J., Hakem, A., Hakem, R., *et al.* (1998). Apaf-1 is required for mitochondrial pathways of apoptosis and brain development. *Cell*, **94**, 739.
53. Slee, E. A., Harte, M. T., Kluck, R. M., Wolf, B. B., Casiano, C. A., Newmeyer, D. D., *et al.* (1999). Ordering the cytochrome c-initiated caspase cascade: hierarchical activation of caspases-2, -3, -6, -7, -8, and -10 in a caspase-9-dependent manner. *J. Cell Biol.*, **144**, 281.
54. Pan, G., Humke, E. W., and Dixit, V. M. (1998). Activation of caspases triggered by cytochrome c *in vitro*. *FEBS Lett.*, **426**, 151.
55. Yang, J., Liu, X., Bhalla, K., Kim, C. N., Ibrado, A. M., Cai, J., *et al.* (1997). Prevention of apoptosis by Bcl-2: release of cytochrome c from mitochondria blocked. *Science*, **275**, 1129.
56. Reed, J. C. (1997). Cytochrome c: Can't live with it—Can't live without it. *Cell*, **91**, 559.
57. Luo, X., Budihardjo, I., Zou, H., Slaughter, C., and Wang, X. (1998). Bid, a Bcl2 interacting protein, mediates cytochrome c release from mitochondria in response to activation of cell surface death receptors. *Cell*, **94**, 481.
58. Han, Z., Bhalla, K., Pantazis, P., Hendrickson, E. A., and Wyche, J. H. (1999). Cif (Cytochrome c efflux-inducing factor) activity is regulated by Bcl-2 and caspases and correlates with the activation of Bid. *Mol. Cell. Biol.*, **19**, 1381.
59. Gross, A., Yin, X. M., Wang, K., Wei, M. C., Jockel, J., Milliman, C., *et al.* (1999). Caspase cleaved Bid targets mitochondria and is required for cytochrome c release, while Bcl-xL prevents this release but not tumor necrosis factor-R1/Fas death. *J. Biol. Chem.*, **274**, 1156.
60. Kuwana, T., Smith, J. J., Muzio, M., Dixit, V., Newmeyer, D. D., and Kornbluth, S. (1998). Apoptosis induction by caspase-8 is amplified through the mitochondrial release of cytochrome c. *J. Biol. Chem.*, **273**, 16589.
61. Cosulich, S. C., Savory, P. J., and Clarke, P. R. (1999). Bcl-2 regulates amplification of caspase activation by cytochrome c. *Curr. Biol.*, **9**, 147.
62. Bossy-Wetzel, E. and Green, D. R. (1999). Caspases induce cytochrome c release from mitochondria by activating cytosolic factors. *J. Biol. Chem.*, **274**, 17484.
63. Darmon, A. J., Nicholson, D. W., and Bleackley, R. C. (1995). Activation of the apoptotic protease Cpp32 by cytotoxic T-cell-derived granzyme-B. *Nature*, **377**, 446.

64. Martin, S. J., Amarante-Mendes, G. P., Shi, L., Chuang, T. H., Casiano, C. A., O'Brien, G. A., *et al.* (1996). The cytotoxic cell protease granzyme B initiates apoptosis in a cell-free system by proteolytic processing and activation of the ICE/CED-3 family protease, CPP32, via a novel two-step mechanism. *EMBO J.*, **15**, 2407.
65. Swanton, E., Savory, P., Cosulich, S., Clarke, P., and Woodman, P. (1999). Bcl-2 regulates a caspase-3/caspase-2 apoptotic cascade in cytosolic extracts. *Oncogene*, **18**, 1781.
66. Murray, A. W. (1991). Cell cycle extracts. In *Xenopus laevis: Practical uses in cell and molecular biology* (ed. B. K. Kay and H. B. Peng), *Methods in Cell Biology*, Vol. 36, p. 581. Academic Press, San Diego.
67. Sawin, K. E. and Mitchison, T. J. (1991). Mitotic spindle assembly by two different pathways *in vitro*. *J. Cell Biol.*, **112**, 925.
68. Dunphy, W. G. (1994). The decision to enter mitosis. *Trends Cell Biol.*, **4**, 202.
69. Dasso, M. and Newport, J. W. (1990). Completion of DNA replication is monitored by a feedback system that controls the initiation of mitosis *in vitro*: studies in *Xenopus*. *Cell*, **61**, 811.
70. Minshull, J., Sun, H., Tonks, N. K., and Murray, A. W. (1994). A MAP kinase-dependent spindle assembly checkpoint in *Xenopus* egg extracts. *Cell*, **79**, 475.
71. Stack, J. H. and Newport, J. W. (1997). Developmentally regulated activation of apoptosis early in *Xenopus* gastrulation results in cyclin A degradation during interphase of the cell cycle. *Development*, **124**, 3185.
72. Evans, E. K., Lu, W., Strum, S. L., Mayer, B. J., and Kornbluth, S. (1997). Crk is required for apoptosis in *Xenopus* egg extracts. *EMBO J.*, **16**, 230.
73. Cosulich, S. C., Worrall, V., Hedge, P. J., Green, S., and Clarke, P. R. (1997). Regulation of apoptosis by BH3 domains in a cell-free system. *Curr. Biol.*, **7**, 913.
74. Reed, J. C. (1997). Double identity for proteins of the Bcl-2 family. *Nature*, **387**, 773.
75. Chinnaiyan, A. M., O'Rouke, K., Lane, B., and Dixit, V. M. (1997). Interaction of CED-4 with CED-3 and CED-9: a molecular framework for cell death. *Science*, **275**, 1122.
76. Farschon, D. M., Couture, C., Mustelin, T., and Newmeyer, D. D. (1997). Temporal phases in apoptosis defined by the actions of Src homology 2 domains, ceramide, Bcl-2, interleukin-1beta converting enzyme family proteases, and a dense membrane fraction. *J. Cell Biol.*, **137**, 1117.
77. Faure, S., Vigneron, S., Doree, M., and Morin, N. (1997). A member of the Ste20/PAK family of protein kinases is involved in both arrest of *Xenopus* oocytes at G2/prophase of the first meiotic cell cycle and in prevention of apoptosis. *EMBO J.*, **16**, 5550.
78. Evans, E. K., Kuwana, T., Strum, S. L., Smith, J. J., Newmeyer, D. D., and Kornbluth, S. (1997). Reaper-induced apoptosis in a vertebrate system. *EMBO J.*, **16**, 7372.
79. Thress, K., Henzel, W., Shillinglaw, W., and Kornbluth, S. (1998). Scythe: a novel reaper-binding apoptotic regulator. *EMBO J.*, **17**, 6135.
80. Thress, K., Evans, E. K., and Kornbluth, S. (1998). Reaper-induced dissociation of a Scythe-sequestered cytochrome c-releasing activity. *EMBO J.*, **20**, 5486.
81. Cosulich, S. C., Horiuchi, H., Zerial, M., Clarke, P. R., and Woodman, P. G. (1997). Cleavage of rabaptin-5 blocks endosome fusion during apoptosis. *EMBO J.*, **16**, 6182.
82. Martins, L. M., Kottke, T. J., Kaufmann, S. H., and Earnshaw, W. C. (1998). Phosphorylated forms of activated caspases are present in cytosol from HL-60 cells during etoposide-induced apoptosis. *Blood*, **92**, 3042.
83. Tamm, I., Wang, Y., Sausville, E., Scudiero, D. A., Vigna, N., Oltersdorf, T., *et al.* (1998). IAP-family protein survivin inhibits caspase activity and apoptosis induced by Fas (CD95), bax, caspases and anticancer drugs. *Cancer Res.*, **58**, 5315.

84. Ellerby, H. M., Martin, S. J., Ellerby, L. M., Naiem, S. S., Rabizadeh, S., Salvesen, G. S., *et al.* (1997). Establishment of a cell-free system of neuronal apoptosis: comparison of premitochondrial, mitochondrial, and postmitochondrial phases. *J. Neurosci.*, **17**, 6165.

8 | Death signalling by the CD95/TNFR family of death domain containing receptors

NICOLA J. McCARTHY and MARTIN R. BENNETT

1. The death receptor superfamily

1.1 Introduction

The discovery that receptors expressed on the cell surface could induce apoptosis when triggered, opened up a whole new area of cell death research. From the initial discovery of the Apo-1/Fas receptor in 1989 (1, 2) has come the description of an ever increasing number of death domain bearing receptors which form part of the tumour necrosis receptor 1 (TNF-R1) superfamily. All members of the TNF-R1 family are type I transmembrane receptors. However, not all are able to induce cell death and are homologous only within their cysteine-rich extracellular domains. A subfamily of this superfamily also exhibits a high degree of homology within their intracellular domains. This region of additional homology is termed the death domain and represents the minimal region required to induce cell death (3). Like the initial member of this family, TNF-R1, each receptor binds one or more ligands and like TNFα, these ligands are type II transmembrane receptors (4) which can be either soluble or membrane bound. Triggering of the receptor by the ligand requires trimerization of the ligand. This subsequent activation of the receptor allows other cytoplasmic factors to bind, which can elicit several responses within the cell including death, proliferation, or differentiation.

Binding of an individual ligand to receptor is not monogamous (see Fig. 1). For example, one TNF family member, TRAIL, is able to bind to five receptors, so signal specificity may not always be exhibited at the level of ligand/receptor binding (5). Components of the downstream signalling machinery bind to many receptor family members suggesting that specificity of any resulting signal is also not determined at this level. Instead, the outcome of whether a cell lives, dies, or differentiates, may well be conferred by a combination of all of these binding events, as well as other signalling pathways occurring in the cell at the same time.

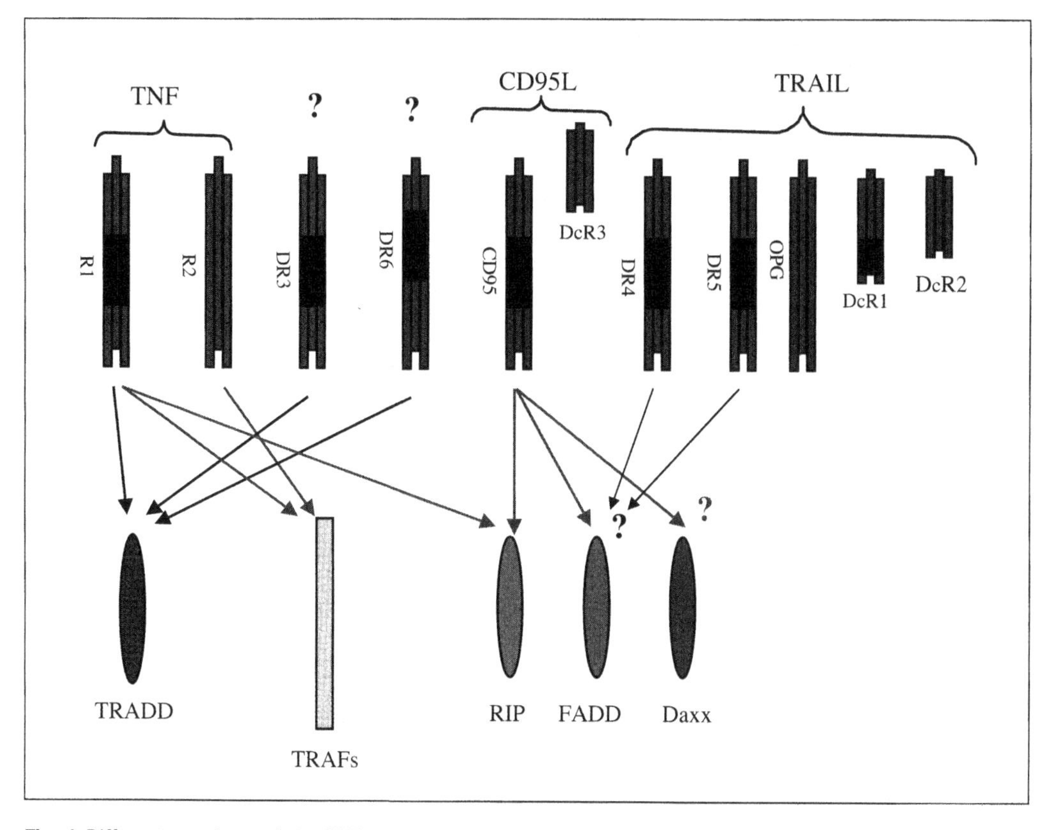

Fig. 1 Different members of the TNF superfamily of ligands bind members of the TNF receptor family. Several members of the TNF receptor superfamily share regions of homology within their intracellular domains, known as the death domain (shown in black). This domain is a protein–protein binding domain that enables interaction with a host of other downstream proteins. Question marks define possible or unknown interactions between death receptors and proteins or ligands.

The TNF receptor family plays an important role in the homeostasis of many, if not all tissues within the body (4, 6, 7). Hence, any disruption of either the receptors themselves or their downstream signalling pathways can result in several pathological outcomes. This chapter seeks to review the main functions of the TNF-R1 family of receptors and to discuss their relevance in a selected number of human diseases.

1.2 TNF receptor superfamily

Tumour necrosis factor was first described in 1984 after its isolation from activated macrophages and T cells. TNFα is part of the army of molecules involved in the immune response; however if given to tumour-bearing mice, it is also able to cause profound tumour 'necrosis' (4, 7). TNF binds to two receptors TNF-R1 (TNF-R55) and TNF-R2 (TNF-R75) of which TNF-R1 is the main receptor mediating TNFα-induced cytotoxicity. TNF-R1 and R2 are type 1 transmembrane proteins that have conserved cysteine-rich domains (CRD) within their extracellular domains. Receptors bearing similar extracellular domains have also been identified and define the TNF-R1 superfamily (see Table 1). All members have varying numbers of cysteine-rich domains, each of which is characterized by the presence of approximately six cysteine residues in a stretch of 40 amino acids. Overall however, family members have a low degree of homology at the protein level (20–25%). It is the CRDs that stabilize the trimeric or multimeric structures that the receptors form (4, 6, 7).

Although, despite its name, TNFα induces apoptotic cell death, interest in this family of receptors did not escalate until the discovery of Fas/Apo-1 (1, 2). The Apo-1 (for apoptosis-1) receptor was identified in a search for monoclonal antibodies that reacted with cell surface markers on tumour cells. Apo-1 is capable of inducing apoptosis when crosslinked by anti-Apo-1 antibody (1). A receptor with similar properties was also identified on fibroblasts called Fas for FS-7 associated cell surface protein. Engagement of the Fas antigen, which is expressed on many cells including myeloid cells, T lymphoblastoid cells, and diploid fibroblasts (2), also leads to cell death. Moreover, expression of fas cDNA in normally non-expressing cells also results in cell death in the presence of anti-fas antibodies (8). These two receptors have now been shown to be one and the same (9) and to belong to the TNF receptor

Table 1 Interactions between TNF receptor and TNF ligand superfamilies

Receptor	Interacting ligand	Signalling complex
TNF-R1 (*CD120a*)	TNFα, LTα	TRADD, RAIDD, TRAFs, RIP, cIAP1, 2, SODD
TNF-R2 (*CD120b*)	TNFα, LTα	TRAFs, cIAP1, 2
CD95 (*Fas/Apo-1*)	CD95L	FADD, RIP, Daxx, UBC-FAP, FAF-1, FAP-1, sentrin
DR3 (*wsl-1, APO-3, TRAMP, LARD*)	Apo3L/Tweak	TRADD
DR4 (*TRAIL-R1*)	TRAIL	TRADD, FADD, RIP
DR5 (*TRAIL-R2, TRICK2, KILLER*)	TRAIL	TRADD, FADD, RIP
DR6		TRADD, TRAFs?
Osteoprotegerin (*OPG, OCIF*)	TRAIL, TRANCE	Unknown
RANK	TRANCE (*ODF, RANKL, OPGL*)	TRAF1, 2, 3, 5, 6
CAR1	Unknown	Unknown
LTβ-R	LTβ	Unknown
GITR	Unknown	Unknown
CD27	CD70	TRAF2, 5
CD30	CD30L	TRAF1, 2
CD40	CD40L	TRAF2, 3, 5, 6
ATAR	Unknown	TRAF2, 5
4-IBB	4-IBBL	TRAF1, 2, 3
OX-40	OX-40L	TRAF2, 3, 5
$p75^{NTR}$	NGF	Unknown
DcR1 (*LIT, TRIDD, TRAIL-R3*)	TRAIL	Unknown
DcR2 (*TRUNDD, TRAIL-R4*)	TRAIL	Unknown
DcR3	CD95L	Unknown

superfamily. Apo-1/Fas, united by the term CD95 (10), is widely expressed and is bound by a specific ligand, CD95L. Interaction of CD95L with CD95 induces rapid apoptosis and the mapping of this cell death signalling pathway has lead to a greater understanding of the function of all members of the TNF receptor superfamily.

Biologically, the actions of members of this family are many and varied (4, 7). Many of this family are expressed on the surface of activated T cells and macrophages, making them important mediators of the immune response. This blanket coverage by members of the same family allows a very complex regulation of cellular responses through ligand and cell to cell interactions. Initially, only a subfamily of these receptors was thought to induce cell death since they possessed a death domain (DD) sequence within their C-terminus. This intracellular DD, required for the binding of 'adaptor' protein molecules, is conserved in TNF-R1, CD95, death receptor 3 (DR3), DR4, DR5, and DR6 (see Fig. 1). Mutations within the DD abolish the receptors' ability to induce cell death. Indeed, mutation of valine 238, which naturally occurs in mice, prevents all CD95-induced cell deaths leading to severe perturbations within the immune system (11).

Recently, other members of the TNF receptor family that do not contain a death domain, such as CD30, CD40, CD27, TNF-R2, and the low affinity nerve growth factor receptor ($p75^{NTR}$) have also been shown to trigger cell death (12–16). There is little doubt now that this family of receptors is able to trigger both proliferation and apoptosis in cells and in some instances may also regulate the differentiation pathways within specific cells. In order to explore more comprehensively the functions of this family, we will discuss each receptor individually.

1.2.1 The TNF receptors

TNFα is a highly cytotoxic molecule, indeed when initially discovered it was felt to be a potentially potent tumour treatment. Unfortunately, the TNF-R1 receptor is widely expressed giving TNFα a very low maximum tolerated dose. Most of the biological affects of TNFα such as apoptosis, antiviral activity, and activation of NF-κB are mediated via TNF-R1 (4). There are differences though in the receptor's interaction with ligand depending on whether TNFα is membrane-bound or soluble (7, 17). TNF-R2 has a higher affinity for membrane-bound TNFα, although TNF-R2 will transiently bind soluble TNFα. This has prompted the suggestion that TNF R1 and R2 act co-operatively at the cell surface, with TNF-R2 initially binding soluble ligand that it then passes to TNF-R1 which binds the ligand for a more prolonged period giving a much stronger signal.

TNF-R1 has four extracellular cysteine-rich domains and is most homologous to DR3 (18) (see Section 1.2.2). TNF-R1 invokes both pro and anti-apoptotic responses mainly due to the downstream proteins that associate with the activated receptor, primarily through the DD (4) (see Fig. 2A). From these downstream factors TNFα is able to activate pro-apoptotic proteases, termed caspases, resulting in apoptosis, or to activate NF-κB and initiate proliferation or differentiation. It is the nature of these downstream interactions that are responsible for the many effects of TNFα on tissues throughout the body.

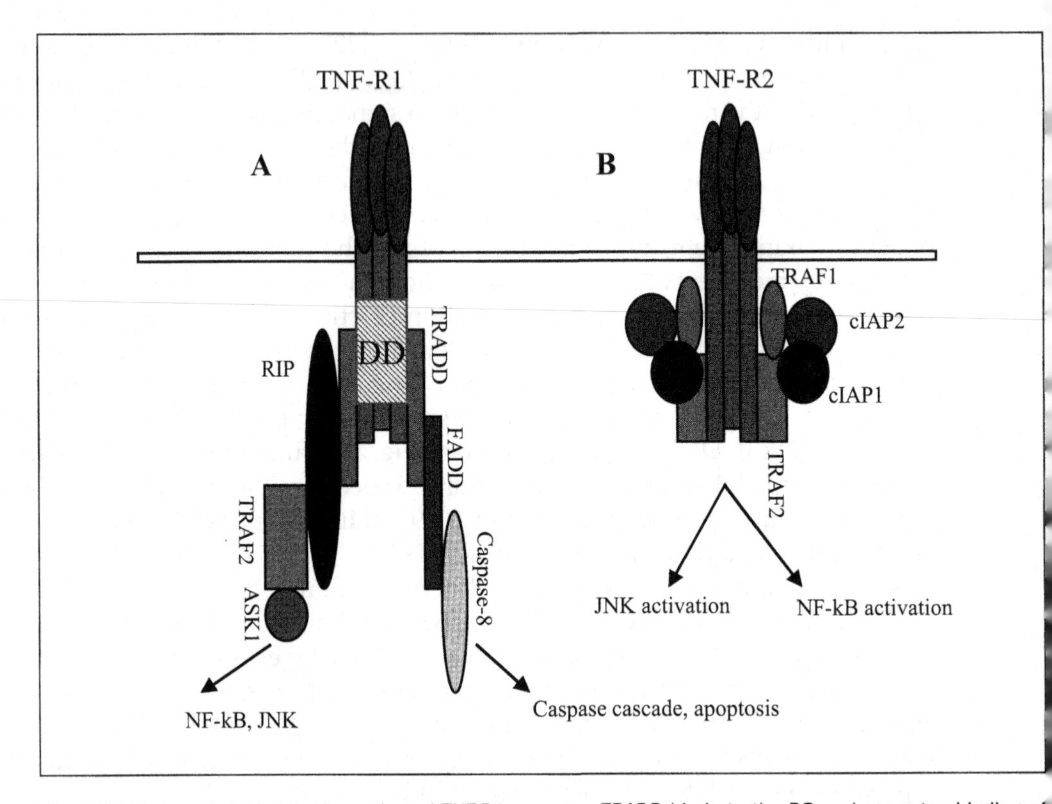

Fig. 2 (A) Factors that bind to the activated TNF-R1 receptor. TRADD binds to the DD and promotes binding of FADD or RIP also through a DD interaction. FADD facilitates, through its DED, the binding of procapsase-8 and activation of the caspase-8 molecule. This in turn leads to cleavage of other downstream caspases, activation of the caspase cascade, and death. RIP mediates binding of TRAF2 which in turn binds ASK1 and activates both NF-κB (survival signals) and JNK (possible death signals). (B) Factors that bind to the activated TNF-R2 receptor. Both TRAF1 and TRAF2 bind TNF-R2 and enable the binding of cellular IAP1 and 2 in mammalian cells. This triggers both JNK and NF-κB signalling pathways. A positive feed-forward loop is suggested given that NF-κB activates transcription of cIAP1 and 2 leading to an increase in the abundance of these proteins, all of which have been implicated in suppressing TNFα-induced apoptosis.

Like TNF-R1, TNF-R2 is also able to signal both cell death and proliferation. TNF-R2 has three extracellular cysteine-rich domains and is most similar to the osteoprotegerin receptor (see Section 1.2.5). Although it does not possess a death domain, TNF-R2 can induce apoptosis in specific cell lines (16) and in some cells is required in tandem with TNF-R1 to induce apoptosis following triggering of either receptor (17).

1.2.2 CD95, the death receptors, and CAR1

CD95 is a glycosylated cell surface molecule of 335 amino acids in length. The human CD95 gene is located on chromosome 10q23 and the mouse gene on chromosome 19 (19, 20). Although CD95 is very widely expressed, much of its biological function has been determined in the context of the immune response in which it plays a significant role (21). Self-tolerance (elimination of cells that recognize 'self' tissues as foreign) within the immune system is determined at several stages. Mature T and B cells are subject to tolerance in the periphery and it is at this stage that the CD95 signalling pathway is critical for the deletion of harmful self-recognizing cells and for the down-regulation of the immune response after challenge with foreign antigen. In both humans and mice that have non-functional CD95 responses, an accumulation of aberrant T cells is seen, resulting in non-malignant lymphadenopathy (21, 22). These mice also develop autoimmunity due to the aberrant accumulation of B cells, hence both B and T cells are affected by the absence of CD95 function (23).

Another, more recent role for CD95 has been in the regulation of tumour cell death (6). Expression of the CD95 receptor in some cancer cells is down-regulated, thereby protecting the tumour cells from killing by CD95 ligand bearing T cells and natural killer cells. Some tumour cells have also been shown to express soluble forms of the receptor in order to 'mop up' soluble CD95L (24). Moreover, it now appears that CD95 expression enhances a cells susceptibility to various triggers of cell death including p53, c-Myc, and cancer chemotherapeutic drugs (see Section 5.3).

The description of the death domain region in both CD95 and TNF-R1 prompted a search for other such receptors and resulted in the discovery of the so-called death receptors DR3 (18, 25–28), DR4 (29, 30), DR5 (29, 31–33), and DR6 (34). The first of these receptors, DR3, is most similar to TNF-R1 and contains four extracellular cysteine-rich domains and a death domain. DR3 was identified by various groups and is also known by the following names: WSL-1, Apo3, LARD, and TRAMP (25–28). All these groups demonstrated that DR3 induces apoptosis when over-expressed and is also able to activate NF-κB, much like TNF-R1. DR3 is highly expressed on thymocytes and lymphocytes implying that it too plays an important role in immune homeostasis. The reputed ligand for DR3, identified via an expressed sequence tag database search, is called Apo3 ligand or TWEAK. TWEAK is a type II transmembrane receptor that has greatest homology with TNFα and appears to mediate apoptosis in association with TNF-R1:TNFα (see Section 2.1.2 and references therein).

DR4 and DR5 bind the same ligand termed TRAIL. DR4 (29, 30) (also called TRAIL-R1) (30) contains two cysteine-rich domains and overexpression results in apoptosis that requires the presence of a functional DD. A soluble form of DR4 is able to inhibit TRAIL-induced cell death. DR4 mRNA is found in many tissues including thymus, activated T cells, small intestine, and some tumour cell lines. DR5 (29) was identified through searching an expressed sequence tag (EST) database with the DR4 death domain sequence and through ligand-based affinity purification using TRAIL. DR5 (also known as TRAIL-R2 (31), Trick2 (32) and KILLER/DR5 (33)) is very similar to DR4 in that it has two extracellular cysteine-rich repeats and a DD that is 65% homologous to the DD of DR4. Like CD95, DR5 will induce apoptosis when bound by agnostic antibody and is again widely expressed in human tissues.

DR6, the most recently identified death receptor, has four extracellular cysteine repeats that have highest homology to the osteoprotegerin receptor and TNF-R2 (34). Like DR3, the DD of DR6 is most like that of TNF-R1 (27.2%) and least like DR5 (19.7%). Unusually, the DD of DR6 is located very close to the transmembrane sequence and has a longer than normal tail sequence that is reminiscent of an SH3 binding domain. The relevance of this has yet to be investigated. DR6 is expressed in most human tissues and is highly expressed in heart, brain, placenta, pancreas, thymus, lymph node, and non-lymphoid cancer cell lines. DR6 overexpression triggers apoptosis and is able to activate NF-κB and c-Jun N-terminal kinase (JNK) (34).

Another member of the TNFR superfamily able to induce cell death is the chicken cell surface receptor, termed CAR1 (35). This receptor has two extracellular cysteine repeats and a death domain. CAR1 is a receptor for cytopathic avain leukosis and sarcoma viruses. Specifically it is bound by the envelope protein of the virus. Interestingly, the binding of the virus to CAR1 can mediate the death of the target cell, which appears somewhat paradoxical for the survival of the virus. The physiological ligand of CAR1 is not yet known.

The biological function of DRs 3–6 is as yet unclear, but their possession of a death domain implicates their involvement in regulating cell death. However, other members of the TNFR superfamily that do not possess a death domain can also induce cell death.

1.2.3 CD40, CD30, CD27, and RANK

CD40, CD30, and CD27, along with TNF-R2 and p75^{NTR}, are members of the TNFR family that induce cell death without possession of a death domain. All are expressed on the surface of lymphocytes and have been shown to regulate both cell death and cell proliferation/differentiation.

CD30 is a lymphocyte surface molecule that can augment activation and survival (36) through its ability to induce the transcription factor NF-κB (37–39). In addition, CD30 also induces cell death in lymphocytes (40) and is implicated in the process of negative selection in the thymus (41). Recent evidence has shown that CD30 is able to sensitize cells to TNFα/TNF-R1-mediated cell death. This co-operation does not occur at the level of the receptors on the cell surface, but through the binding to the intracellular C-terminus of downstream proteins known as TNF receptor-associated factors (TRAFs) (15) (see Section 3.4.1). TRAFs bind to many of the TNF receptor family and act to integrate many cellular pathways giving rise to a complex framework of signalling cross-talk.

CD40 is another lymphocyte cell surface receptor that can modulate cellular survival, differentiation, and apoptosis (reviewed in refs 7 and 12). CD40 is expressed on antigen-presenting cells (APC) and its cognate ligand CD40L/CD154, is expressed on T cells. The CD40/CD40L interaction is required for the generation of humoral immunity since blocking this interaction results in the loss of B cell responses to thymus-dependent antigens. CD40-mediated signals can also affect dendritic cell function leading to the up-regulation in expression of adhesion molecules, increased peptide–MHC complex expression, and anti-apoptotic signals that prolong the lifespan of APCs. Recently another member of the TNFR family has been cloned and called RANK for receptor activator of NF-κB (42). RANK is the receptor for TRANCE, a member of the TNFα superfamily. Given that RANK is also expressed on mature dendritic cells and has considerable homology to CD40, as

TRANCE has to CD40L, it was hypothesized that these receptor/ligand pairs had similar functions. Although CD40L and TRANCE are functionally similar, one cannot substitute for the other in animals that are deficient in either molecule (43). Moreover, evidence from TRANCE knockout animals suggests that the RANK/TRANCE signalling pathway has other functions within different cell types. TRANCE-deficient mice have defects in their lymphoid compartments as well as other defects (44) (see Section 2.1.3). Thymi from TRANCE null mice are much smaller than normal wild-type littermates due to a selective block in early thymocyte development prior to expression of the mature T cell receptor (TCR). A similar retardation in pre-B cell development in the bone marrow was also noted. Despite their involvement in dendritic cell function, TRANCE/RANK interactions are not required for dendritic cell development, but are required for the efficient stimulation of T cells upon binding dendritic cells (42).

CD27 is expressed specifically on lymphocytes. It acts as a co-stimulatory molecule, enhancing the TCR-induced proliferative response of T cells and enhancing both proliferation and immunoglobulin production in B cells (45). Like CD40 and CD30, CD27 has also been shown to induce apoptosis when bound by its ligand CD70 (a TNF type II transmembrane protein) (14). Moreover, crosslinking of surface immunoglobulin receptors in activated B cells augments CD27-induced apoptosis. Since CD27 does not possess a DD within its C-terminus, a yeast two-hybrid system was employed to determine which molecules might bind to this receptor. A novel protein was identified and called Siva (after the Hindu god of death) that is 189 amino acids long and has a conserved amino terminal region which bears some homology to the death domains of FADD and RIP (14) (see Section 3.1). Siva is thought to mediate the pro-apoptotic signals generated upon CD27 ligation. CD27 also interacts with TRAF2 and TRAF5 (46, 47) (see Section 3.4.1) through which it activates JNK, enhancing the proliferative signal generated in primary lymph node T cells by TCR engagement.

Two other novel TNFR family homologues have been described, ATAR and GITR (48, 49). ATAR is shorter than all other family members so far identified and is expressed mainly in spleen, thymus, bone marrow, small intestine, and lung (48). ATAR interacts with both TRAF2 and 5 and activates NF-κB. Its ligand and biological function are unknown. GITR (glucocorticoid-induced tumour necrosis factor receptor family related gene) is expressed only in lymphoid tissues and prevents activation-induced cell death in T cells (49). GITR shows greatest homology to 4-IBB, another member of the TNFR superfamily. The physiological ligand of GITR is unknown.

Thus, although the above proteins do not possess DDs, they are able to induce apoptosis as well as proliferation and differentiation. These different cellular outcomes are dependent upon which other signals are occurring concurrently in the cell and which downstream factors are binding to the receptor C-terminus. The restricted expression of CD40, CD30, and CD27 on lymphocytes and their ability to regulate very similar responses exemplifies the complex regulation of immune cell activation and homeostasis.

1.2.4 The low affinity nerve growth factor receptor

Nerve growth factor (NGF) is generally thought of as a survival factor; however it has been shown to induce cell death in specific cells during development or terminal differentiation (13). NGF binds to two different receptors Trk-A and p75 low affinity nerve growth factor receptor (p75 neurotrophin, $p75^{NTR}$) which is a member of the TNF receptor superfamily. Several lines of evidence now suggest that $p75^{NTR}$ affects different cell signalling pathways depending upon which NGF receptors are expressed in the cell lineage concerned. Binding of ligand to $p75^{NTR}$ induces activation of NF-κB and JNK. If however, the Trk-A receptor is also activated then the pro-apoptotic JNK signal from $p75^{NTR}$ is inhibited and Trk-A generates a mitogen-activated protein kinase cascade mediating cell survival (50). Moreover, *in vivo* models have shown that neurons within the developing spinal cord and retina of the mouse express only $p75^{NTR}$ and this correlates with high levels of cell death (51). Deletions within the *ngf* gene or $p75^{NTR}$ gene reduce the amount of cell death occurring in these tissues, suggesting that binding of NGF to $p75^{NTR}$ is lethal prior to the expression of Trk-A.

Thus, the $p75^{NTR}$ receptor is able to mediate cell death, without a death domain, in specific cell types in absence of Trk-A signalling. How $p75^{NTR}$ affects this pro-apoptotic response is not yet understood, but given the suggested requirement for JNK activation it may well involve one of the TRAF proteins (see Section 3.4.1).

1.2.5 Osteoprotegerin

Osteoprotegerin (OPG) was identified as a novel secreted member of the TNFR superfamily through an EST sequencing project (52). OPG (also known as osteoclastogenesis inhibitory factor, OCIF) is involved in the regulation of bone formation and inhibits the formation of osteoclasts (bone reabsorbing cells). Transgenic OPG mice and mice injected with OPG have dense bones due to the inhibition of osteoclast formation. OPG may also have a role in lymphocyte activation given that it is up-regulated in lymphoid cells upon CD40 stimulation (43, 53). OPG appears to bind to two ligands of the type II TNF transmembrane family, TRAIL (52) and TRANCE (54, 55) (also known as RANK ligand, osteoclast differentiation factor [ODF]). TRAIL binds to OPG and inhibits its function, but how this affects the regulation of osteoclastogenesis *in vivo* is unclear (52). OPG also prevents the interaction between the putative ODF receptor (RANK) and ODF ligand (TRANCE/RANKL). It is this mechanism in part that appears to regulate the ability of bone marrow stromal cells to differentiate into osteoblasts and osteoclasts (54) (see Section 2.1.3).

It has also been suggested that OPG acts as a decoy receptor for TRAIL thereby inhibiting the pro-apoptotic activity of this ligand (52). Other decoy receptors for TRAIL and CD95 have recently been described.

1.2.6 Decoy receptors

TRAIL, as discussed above, binds a multitude of receptors, which include decoy receptors (DcR) 1 and 2. DcR1 and DcR2 are members of the TNFR superfamily and have two conserved cysteine-rich domains and a transmembrane region, however they both have truncated cytoplasmic tails (reviewed in ref. 5). This lack of a cytoplasmic tail suggested that these receptors would limit the effect of TRAIL since they lack the C-terminal binding sites to which downstream pro-apoptotic signalling molecules bind.

DcR1 (also called LIT, TRID, or TRAIL-R3) (33, 56–58) is anchored at the membrane by glycophosphatidylinositol (GPI) and does not possess a death domain region. Overexpression of DcR1 does not induce cell death; instead it protects cells from TRAIL-induced cell death. DcR1 transcripts are found in some human tissues, but perhaps surprisingly not in most cancer cell lines.

DcR2 (also called TRUNND, TRAIL-R4) (59–61) is not a GPI-anchored receptor. Instead DcR2 possesses a truncated death domain within its C-terminus. The truncated DD is not able to transduce a cell death signal and thus DcR2 behaves as a decoy for TRAIL. DcR2 is also expressed in many human tissues, but not in most human tumour lines. Interestingly, the chromosomal localization of DcR1 and 2, 8q21, is the same as DR4 and DR5 suggesting these receptors arose from a common precursor (5).

A third decoy receptor, termed DcR3, has also been described (62). This receptor does not bind TRAIL, but binds CD95L. DcR3 is expressed in lung and colon cancer cell lines and has highest homology to OPG (31%) and TNF-R2 (29%). It has three extracellular cysteine-rich domains and like osteoprotegerin has no transmembrane domain suggesting that it is secreted from the cell. The DcR3 receptor has a considerably shorter C-terminus and lacks a death domain. Secreted DcR3 binds to CD95L and inhibits its function. DcR3 is expressed at high levels in some tumour cell lines suggesting this may be the mechanism by which some tumour cells evade the immune response. For example, DcR3 expression could prevent CD95L on natural killer cells or cytotoxic T lymphocytes binding to CD95 on the tumour cell.

The discovery of decoy receptors increases the complexity of cell death regulation still further, giving regulation at the stage of ligand/receptor binding as well as through downstream intracellular signalling pathways.

2. Ligands of the TNF receptor superfamily

2.1 Introduction

The TNF family of ligands (see Table 1) are type II transmembrane proteins (with the exception of LTαII, and are expressed as membrane-bound forms on the surface of cells that can be cleaved by metalloproteinases which release the extracellular domain from the cell surface (reviewed in refs 4, 6, and 7). The structures of TNFα and CD95L have shown that these ligands exist as trimeric molecules and it is this ligand formation that causes receptor activation upon binding (63, 64). Normally a ligand will bind specifically to an individual receptor; however one ligand, TRAIL, binds to two death receptors DR4 and DR5, two decoy receptors DcR1 and DcR2, and to OPG (5). Moreover, the soluble forms of TNF; TNFα and LTα (also known as lymphotoxin or TNFβ) bind both TNF receptors. In addition, LTα in combination with transmembrane LTβ, binds as a heterotrimer to the LTβ receptor (6). Hence these ligands can effect many tissues that express a variety of receptors.

2.1.1 TNF and CD95L

TNFα has many diverse actions and exerts pro-inflammatory responses in almost all cell types (reviewed in refs 4 and 7). Moreover, its inappropriate expression is crucial in the development of various acute and chronic inflammatory disorders. Major cellular sources of TNFα are activated macrophages, lymphoid cells, neutrophils, keratinocytes, NK cells, smooth muscle cells, and fibroblasts (4). TNFα is required for B cell and dendritic cell development (65–67). TNF null mice lack primary B cell follicles, follicular dendritic cell networks, and germinal centres. Unsurprisingly, macrophage function in these mice is also disrupted with clearance of intracellular pathogens such as Listeria being significantly impaired.

CD95L was initially characterized using activated T lymphocytes and was thought to function solely within the immune system due to its limited expression (6). However, other non-lymphoid cells do express CD95L, especially tissues that maintain immune privilege, such as sertoli cells of the testis and epithelial cells of the anterior eye chamber (68–70). Inoculation of virus into the anterior eye chamber results in an infiltrate of CD95-expressing lymphocytes which are killed, presumably due to the high expression of CD95L (68, 70). This role of CD95 in immune privilege has identified CD95L as a potential target in the search for molecules that may aid the success of organ transplantation (69). High levels of CD95L may protect some tissues from activated lymphocytes and may exert the same effect when overexpressed on tumour cells. Indeed, high levels of CD95L expression have been found in tumours of the colon, lung, kidney, skin, and liver, however the precise role of CD95L in tumour biology has yet to be clarified (see Section 5.3).

2.1.2 TRAIL and TWEAK

TRAIL (for TNF-related apoptosis-inducing ligand) shares homology with other TNF ligands in its extracellular carboxy terminal domain (71, 72). TRAIL is most similar to CD95L (28%) closely followed by TNF (23%). The amino terminal intracellular domain of TRAIL is short and not conserved between mouse and man, suggesting that this domain in TRAIL is of little functional significance. The TRAIL gene is located on chromosome 3 at 3q26 and is not close to any other TNF family members, nor its cluster of receptors. TRAIL mRNA is expressed in many tissues; however not all these tissues are sensitive to TRAIL-mediated death. Indeed, only tumour cells have been shown to be responsive to cell death mediated by this ligand, consistent with the suggestion that this reflects the absence of decoy receptors in these tumour cell lines (71, 72). The full biological function of TRAIL is at present unclear although some results have suggested a functional role in T cell activation-induced cell death (72).

TWEAK or Apo3 ligand appears to bind to DR3 and weakly induces apoptosis and NF-κB activation (73, 74). The extracellular domain of TWEAK has greatest homology to TNFα and is widely expressed in human tissues. TWEAK, like TNFα, may induce several different biological outcomes in its target cells. It has been implicated in angiogenesis, since it induces proliferation in both endothelial cells and vascular smooth muscle cells (75). Moreover, TWEAK may also be a cofactor in the TNFα/TNFR signalling pathway and it is this interaction that maybe responsible for the weak induction of apoptosis by TWEAK in the absence of TNFα (76). Evidence suggests that TWEAK may also interact with another, as yet unidentified, receptor that may not necessarily be involved in initiating a death signal (76).

2.1.3 TRANCE

TRANCE (TNF-related activation-induced cytokine, also known as RANKL, OPGL, and ODF) was identified during a search for novel apoptotic genes and is a type II trans-membrane protein of 316 amino acids (42, 43, 54, 55). It shares reasonable homology with TRAIL (20%), CD95L (19%), and TNFα (17%), and is located on human chromosome 13q14 and mouse chromosome 14. Northern analysis shows that expression of TRANCE, unlike other family members, is restricted to thymus, lymph node, and osteoblast/stromal cells of the bone marrow. A soluble form of TRANCE can activate JNK like other members of the TNF family, although the lack of information on TRANCE at present means its biochemical signalling pathways are not characterized (77). Initial co-immunoprecipitation studies demonstrated that TRANCE does not bind to any of the receptors associated with inducing apoptosis. Instead TRANCE has been recently shown to bind both the OPG receptor preventing osteoclast formation in bone marrow stromal cells and RANK (receptor activator of NF-κB). The interaction between RANK and TRANCE is, in concert with CD40 and CD40L, important for the regulation of the immune response (42). Moreover, TRANCE/ RANK has also been shown to play a critical part in osteoclast regulation. The generation of TRANCE-deficient mice has shown that, not only do they show disruptions in immune cell development and function (see Section 1.2.3), but that they also develop osteopetrosis (very dense bones) (54, 55). Furthermore, these animals have a defect in tooth eruption since no bone re-absorption occurs to enable the teeth to pass through openings in the developing jaw bone. TRANCE null mice do possess osteoblast progenitor cells, but these cells are unable to develop into mature osteoclasts due to the lack of TRANCE presentation by accessory (osteoblast/stromal) cells. Hence, TRANCE appears to be a novel TNF family ligand in that has no clear role in initiating a cell death signal. However, it appears to be critical in the development and function of both the bone and immune systems.

2.2 Summary

Overall, the TNF family of ligands regulate a large and diverse number of cellular processes, many of which have been studied within the context of the immune system. Their effects in other tissues require more detailed investigation, but as exemplified by TNFα and TRANCE null animals, these ligands do not simply regulate cell survival pathways. The wide ranging tissue effects generated by these ligands are in part dependent upon which receptors they bind, but are also influenced by the binding of downstream adaptor proteins to the activated receptors.

3. Adaptor proteins

3.1 FADD, TRADD, and RIP and RIP2

The DD found in several of the TNF family of receptors is a protein–protein interaction domain that facilitates the binding of a number of different adaptor proteins involved in generating downstream signalling events (3, 78). The DD is similar to a protein found in *Drosophila* called 'Reaper' (79) that is required for programmed cell death (80). Using the cytoplasmic domains of CD95 or TNF-R1 in the yeast two-hybrid system initially identified three proteins that bound to the DD (see Figs 2A and 3A). FADD (also known as Mort1) is a 208 amino acid protein that binds the cytoplasmic domain of CD95 (81, 82). Residues 111–170 of FADD match

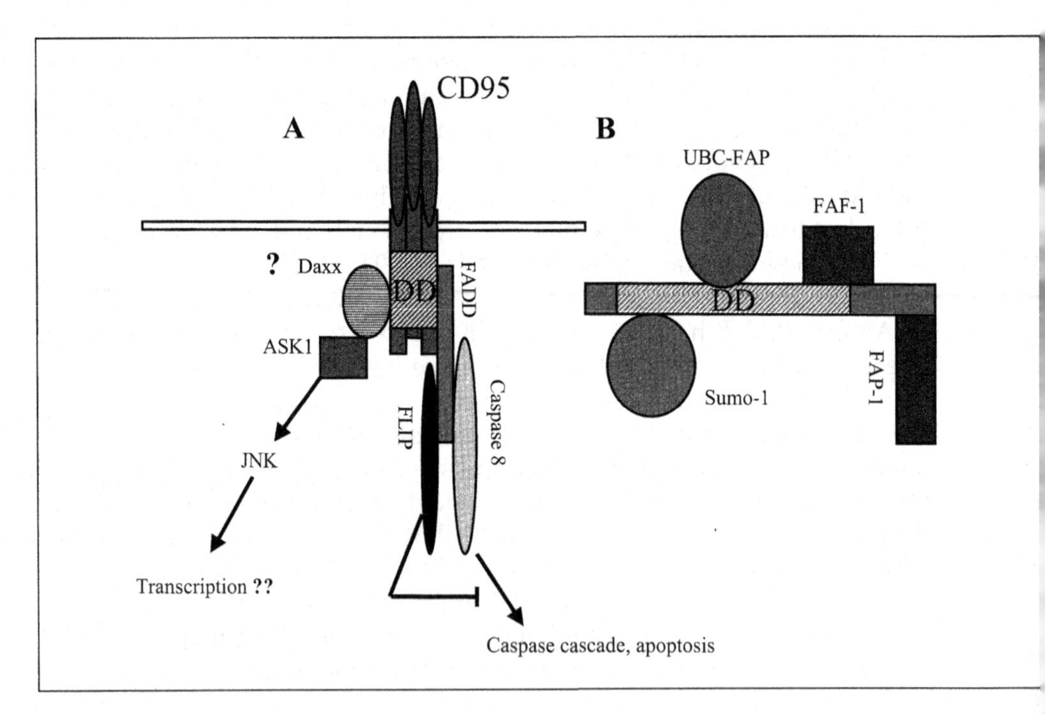

Fig. 3 (A) Downstream proteins that bind to the activated CD95 receptor. FADD interacts with the DD of CD95 and allows binding of procaspase-8 as described in Fig. 2A. Activation of caspase-8 can be inhibited by the binding of FLIP to FADD, also through its DED (see Section 4.1). Mouse Daxx also appears to interact with the CD95 receptor and mediates JNK activation via activation of ASK1. This pathway may synergize with the FADD-mediated pathway to increase sensitivity to CD95-induced cell death. However, human Daxx does not appear to physically interact with the CD95 receptor, although it does still sensitize to CD95-induced cell death, but this seems likely to require the nuclear localization of Daxx (see Section 3.2 for discussion). (B) Proteins identified using the yeast two-hybrid assay system as potential CD95 interacting proteins. The function of these proteins within the CD95 pathway are not yet fully understood (see Section 3.3).

residues 233–292 of rat CD95, share 27% homology, and define the DD. Valine 121 of FADD is aligned and conserved with valine 238 of CD95, the amino acid that is mutated in *lpr* mice (83 and see below). Mutation of V121 in FADD destroys its ability to bind CD95 suggesting that these proteins associate via their DDs. Indeed, FADD does not associate with the *lpr* CD95 mutant, or a CD95 mutant lacking the death domain.

FADD mediates the CD95 signal for apoptosis since overexpression of this protein in MCF7 (a breast carcinoma line sensitive to CD95-induced apoptosis) initiates cell death. Moreover, cells from FADD-deficient animals are completely insensitive to CD95-mediated death. Interestingly, the DD of FADD is not required to initiate cell death. Instead the N-terminal region of FADD is critical (82). This region has a second protein–protein interaction motif called the death effector domain (DED). Deletions within the DED abolish FADDs death-inducing ability. The DED of FADD binds to a family of proteases involved in cell death known as the caspases (see Chapter 4). Caspases, once activated, facilitate the rapid degradation of the apoptotic cell. FADD binds directly to one caspase, caspase-8 (also known as FLICE, Mach1) and it is this binding that is involved in activation of the caspase cascade (84) (see Fig. 3A and Chapter 4). Binding is achieved through the DEDs located in the C-terminus of caspase-8. Caspase-8, once activated, cleaves other 'downstream' caspases leading to cell death. Superficially, the CD95 pathway is simple, requiring the interaction of just five molecules to produce cell death. However, as one might predict, death is not so simple and many other proteins interact at the level of FADD and caspase-8 binding and can change a pro-apoptotic signal into an anti-apoptotic one. FADD not only interacts with the DD of CD95, it also binds DR3 (18, 25–28), DR4, DR5 (85), and TNF-R1 (86) (see Fig. 1), although DR3 and TNF-R1 binding is via a second DD-containing protein TRADD (TNF receptor-associated protein with death domain) (87).

Although TRADD has homology to CD95 in the death domain, it has no other homology to CD95, TNF-R1, or FADD. TRADD is a 34 kDa protein that, like FADD, induces apoptosis when ectopically expressed in a human carcinoma cell line. Unlike FADD, however, it does not bind CD95, but instead binds TNF-R1 (87), DR3 (18, 25–28), and DR6 (35) (see Figs 1 and 2A). Its role in DR3- and DR6-mediated cell death is at present unclear, however TRADDs function is better understood in the TNF-R1 signalling pathway. Binding of TRADD to TNF-R1 allows the binding of two other proteins FADD and RIP (88,89). Binding of FADD to TRADD enables a pro-apoptotic signal to be produced due to the recruitment of caspase-8 by FADD (84). Binding of RIP however, can produce two different outcomes, activation of NF-κB or programmed cell death.

Once again RIP was identified by use of the yeast two-hybrid system and binds both TNF-R1 and CD95 in the C-terminal death domain (see Fig. 2A) (89). Northern blot analysis of human RIP reveals a 4.2 kb transcript expressed in the majority of tissues with high expression in the lungs. The N-terminal of RIP has homology to serine/threonine kinases, and the 98 C-terminal amino acids of RIP share 87% homology between murine and human forms. Expression of RIP mutants shows that the absence of the C-terminus prevents induction of cell death, whereas a mutant missing the N-terminal kinase domain induces death, indicating that the kinase domain is not essential for the induction of apoptosis. The intermediate amino acids that lie between the DD and kinase domains are required for binding of TNF receptor-associated factors (TRAFs) (90). There are six TRAFs identified to date and they play a critical role in activation of non-apoptotic signals generated by many members of the TNF-R1 family (reviewed in ref. 91). Primarily TRAFs mediate the activation of NF-κB and JNK. The role of RIP in CD95-induced cell death remains controversial given that mutated Jurkat cell lines that do not express RIP are sensitive to CD95-induced death, but not TNFα-induced death (92). Thus, the binding of RIP to CD95 may not occur at physiological levels of protein expression *in vivo*.

A second RIP kinase (RIP2, RICK, CARDIAK) has been identified that binds to the TRAFs and mediates both cell death and NF-κB activation (93–95). Unlike RIP, RIP2 does not possess a DD. Instead, it has a caspase activation and recruitment domain (CARD) (96), which like the DEDs of FADD and caspase-8, interacts with apical caspases such as caspase-9, that also possess a CARD sequence. Caspase-9 is involved in mediating apoptosis via cleavage of downstream caspases (see Chapter 4). Another protein that binds to RIP is RAIDD (CRADD). RAIDD (97) has a DD through which it interacts with RIP and a CARD motif that binds to and activates CARD-containing caspases.

3.2 Daxx

Daxx has proved to be a particularly interesting putative CD95 interacting protein. Originally identified by binding to the TNF-R1 and CD95 receptors in a yeast two-hybrid assay (98), Daxx has recently been shown to reside primarily in the nucleus and influence gene transcription. Mouse (m) Daxx can act to augment CD95-induced apoptosis and activate JNK, as well as trigger death in the absence of CD95, albeit at a much reduced potency. Construction of a dominant-negative mDaxx that apparently binds only to the CD95 DD blocks apoptosis and JNK activation and demonstrates that the C-terminal region of mDaxx is required for its functionality. mDaxx is able to activate the JNK kinase kinase ASK1 through releasing an intramolecular interaction between the C- and N-termini of ASK1, producing an open, active ASK1 protein (99). Moreover, a dominant-negative form of ASK1 inhibits both mDaxx- and CD95-induced apoptosis suggesting that cell death signals on the surface can modulate nuclear targets of the JNK cascade.

Recent work however, adds further complexity to this story. First, the CD95 death signal can be mediated in the absence of functional Daxx (100, 101). Moreover, Daxx-deficient mice are embryonic lethal and exhibit profound tissue abnormalities and massive apoptosis, suggesting that Daxx maybe involved in early tissue organization. Daxx-deficient cells are also more sensitive to apoptosis. These results have been suggested to show that Daxx is not involved in CD95-mediated death since one would hypothesize that loss of Daxx should prohibit CD95 death leading to cellular accumulation, not increased cell death. Secondly, however, is evidence that Daxx binds to a number of other cellular proteins that may explain its requirement during embryogenesis. Moreover, human (h) Daxx resides primarily in the nucleus in PML oncogenic domains or PODs (102). PML is a tumour suppressor gene involved in the regulation of gene transcription whose deletion results in resistance to CD95-induced and X-ray-induced apoptosis (103). Indeed, recent evidence suggests that hDaxx may act as a transcriptional repressor and may augment CD95 killing by preventing the transcription of genes involved in survival pathways (102). Hence it is likely that Daxx can be involved, albeit indirectly, in CD95-induced cell death as well as a number of other critical cellular processes.

3.3 Other CD95/TNF-R1 receptor interacting proteins

A host of other proteins have been found to bind to the CD95 receptor. Many of these have been demonstrated using the yeast two-hybrid assay system, and have little *in vitro* significance at the present time. Further *in vitro* and *in vivo* analyses are required on all of the proteins discussed in the next section.

UBC-FAP is the human homologue of ubiquitin conjugating enzyme 9 found in yeast (104). UBC-FAP binds to CD95 within the DD and its overexpression has been shown to augment CD95-induced apoptosis (see Fig. 3B). FAF-1 (Fas associated factor-1) is another protein that enhances CD95-induced cell death (105). The avian homologue of FAF (qFAF) has been identified. Its nuclear localization is mediated by phosphorylation at a serine residue within a short 35 amino acid alpha-helical nuclear localization signal located within the N-terminus (106). Deletion mapping within the C-terminus has identified a domain that is anti-apoptotic, its deletion results in a protein that constitutively induces apoptosis. This C-terminal region has homology to a ubiquitin conjugating protein in *Caenorhabditis elegans*. These results suggest that the FAF protein may have complex pleiotropic effects within the CD95 signalling pathway.

FAP-1 and Sentrin/Sumo1 both mitigate the CD95-induced cell death pathway (see Fig. 3B). FAP-1 (Fas-associated phosphatase) binds to the terminal 15 amino acids of CD95 (a region also thought to act as a negative regulatory domain) (107). Mutants lacking this region were initially shown to have enhanced sensitivity to CD95-induced cell death. However, in some cell lines, the inability of FAP-1 to bind due to the absence of these amino acids does not seem to affect cell death. Thus, the role of FAP-1 in CD95-mediated signalling pathways is still unclear.

Sentrin/Sumo1 is a homologue of ubiquitin-conjugating enzymes (E2s) and binds to CD95 and TNF-R1 (108, 109). Sumo1 inhibits death induced by both TNFα and CD95L when overexpressed, but the mechanism via which it achieves this is not clear. Interestingly, however, PML is a protein that undergoes sumoylation and this increases its localization within PODs, again suggesting a link between PML and Daxx in the nucleus and CD95 (see Section 3.2 and ref. 110). Sumo1 is not directly involved with the addition of multiple ubiquitin moieties to proteins resulting in their degradation via the proteosome. Instead, Sumo1 is one of a new class of ubiquitin-like molecules that are added singularly to proteins and modify protein function (110). Many proteins have been shown to be sumoylated including p53 and MDM2, but precisely what sumoylation of CD95 achieves is as yet unclear.

3.4 TRAFs: TNF receptor-associated factors

Further studies on the cytoplasmic domains of the TNF receptors identified a new family of TNF receptor-associated factors or TRAFs (reviewed in ref. 91). TRAFs have been shown to interact with many members of the TNF receptor superfamily and are involved in mediating downstream signalling events, especially those involving activation of either NF-κB or JNK. There are six TRAFs so far identified in mammalian cells and all share a common set of domains (see Fig. 4A). The C-terminus is split into two domains, the TRAF-C and the TRAF-N domains and at the N-terminus are a RING finger domain and a variable number of zinc finger structures. The TRAF-N domain is critical for binding of the TRAFs to members of the TNFR family. Activation of NF-κB requires the N-terminal RING finger domain.

TRAFs have been shown to bind TNF-R1 and R2 (88), Ox-40 (111, 112), CD40 (113–116), CD30 (15), CD27 (47), RANK (117–119), and 4-IBB (113, 120) as well as other TNFR interacting proteins such as Sumo1, RIP, and RIP2 (89, 91, 93, 108). However, not all the TRAFs bind to these proteins, but instead bind selectively to specific sites on receptors. For example, CD40 has two C-terminal TRAF binding

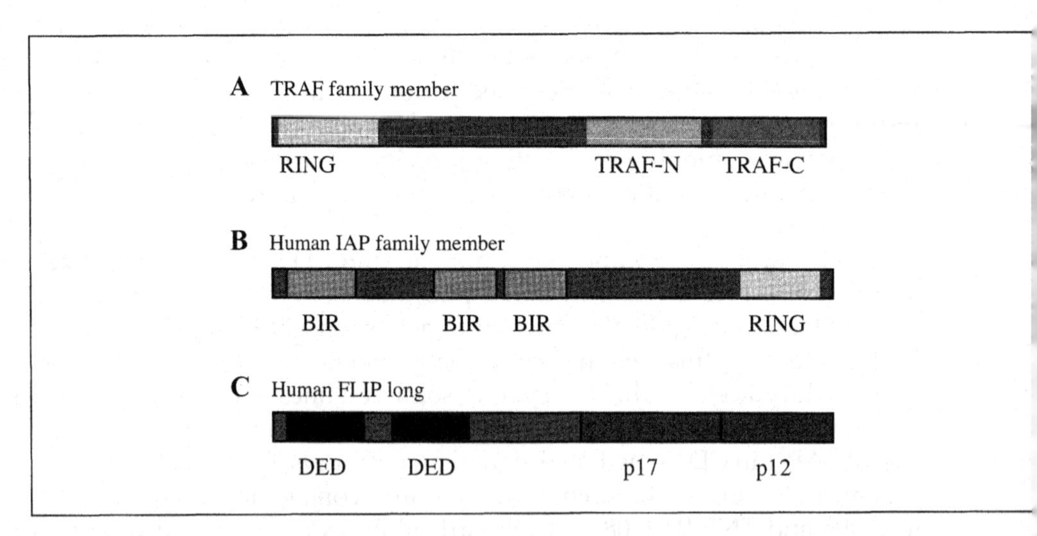

Fig. 4 (A) A representative TRAF family member. The N-terminal RING finger domain is required for activation of the NF-κB pathway. The C-terminal has both TRAF-C and TRAF-N domains which were first identified in this family of proteins. The TRAF-N domain is required for interactions with members of the TNF receptor superfamily. (B) IAP (inhibitor of apoptosis) family member. The structure of IAPs in humans varies with regard to the number of BIR (baculovirus IAP repeat) motifs and the presence of a RING finger domain (see ref. 125 for a review). Binding of IAPs to TNF-R2 activates NF-κB, preventing apoptosis induced by TNFα. (C) Human FLIP long has homology to caspase-8. It possesses two DED motifs within its prodomain and has a similar C-terminal structure. However, FLIP does not contain a caspase active site (QACQ/RG), unlike caspase-8. FLIP is able to bind efficiently to FADD to prevent activation of caspase-8.

domains termed cyt-C and cyt-N (113, 116). Deletion analysis reveals that TRAFs 2, 3, 5 bind to cyt-C and TRAF6 to cyt-N. These alternative binding sites allow signals to be transmitted to different transcription factors. The cyt-C site appears to regulate activation of p38 kinase and JNK activation, whereas TRAF6 binding is required for ERK and NF-κB activation. RANK, like CD40 also has two TRAF-interacting domains. However, the order in which TRAF members bind these sites appears to be in dispute (117–119). At present the full significance of this differential binding is unclear. However, it is clear that TRAFs regulate signalling to other pathways within the cell and for the most part the signals generated are anti-apoptotic.

4. Regulation of death receptor-induced killing

Many proteins interact with the intracellular tails of the TNF receptor superfamily. The signalling permutations generated by the number of proteins that can interact exemplifies the complex signalling pathways triggered by ligands such as TNFα or CD95L. The CD95 signalling pathway is perhaps still the best understood of the TNF receptor signalling pathways, mainly as a result of the identification of which proteins bind immediately after receptor activation.

4.1 The death-inducing signalling complex (DISC)

The death-inducing signalling complex or DISC describes the proteins that bind immediately to the intracellular domain of the CD95 receptor upon ligand binding (see Fig. 3A) (reviewed in ref. 6). These proteins, originally called CAP proteins for cytotoxicity-dependent Apo-1-associated proteins, are essential for the rapid induction of apoptosis induced upon CD95 activation. CAP1 and CAP2 are two different serine phosphorylated forms of FADD, although the significance of FADD phosphorylation for CD95 signalling remains enigmatic. CAP3 and CAP4 are different forms of caspase-8, an apical caspase involved in activating the caspase cascade. Thus, the binding of relatively few proteins to the activated CD95 receptor is enough to trigger cell death. However, binding of FADD and caspase-8 to the CD95 receptor does not ensure the death of the cell. In some cells there are abundant levels of caspase-8 molecules within the cytoplasm and these bind in large numbers to the activated receptors. In these cells, known as type 1 cells, anti-apoptotic proteins, such as Bcl-2, cannot inhibit death (121, 122). Bcl-2 is the archetypal member of a family of pro- and anti-apoptotic proteins that can induce or prevent cell death by influencing the release of cytochrome *c* from mitochondria (see Chapters 5 and 6). Release of cytochrome *c* from mitochondria, in concert with Apaf-1 and ATP, is required to initiate caspase-9 activation and activate downstream caspase cascades (reviewed in Chapter 4). In a type 1 cell where caspase-8 is abundant, the caspase-9 mitochondrial pathway is probably not required to activate cell death. Thus, Bcl-2-like proteins cannot inhibit CD95-induced cell death in type 1 cells. However, in some cells caspase-8 appears not to be abundantly expressed in the cytoplasm and only a minimal amount binds to the CD95 receptor upon activation. In type 2 cells, the minimal caspase-8 activation occurring at the DISC cannot activate downstream caspases directly. Instead caspase-8 activation indirectly promotes cytochrome *c* release from the mitochondria, enabling activation of caspase-9. It is the release of cytochrome *c* from mitochondria in type 2 cells that Bcl-2 is able to inhibit, thereby preventing type 2 CD95-induced cell death.

A recently identified 'new' member of the DISC is FIST/HIPK3, a 130 kDa Fas interacting serine/threonine kinase/homeodomain interacting protein kinase (123). FIST/HIPK3 is able to bind CD95 via a specific sequence within its C-terminus and is widely expressed in mammalian tissues. In cells transfected with a FIST/HIPK3 cDNA, FADD is phosphorylated promoting the interaction between CD95-FIST/HIPK3 and FADD. This overexpression of FIST/HIPK3 appears to impair activation of JNK by CD95 (implicating a role for Daxx), but does not inhibit CD95-mediated death. These results suggest that FIST/HIPK3 may act to alter one signalling pathway mediated via CD95 ligand/receptor interactions.

CD95-induced cell death can also be modulated by a caspase-8-like molecule that binds to the DISC. FLIP, for FLICE inhibitory protein, was initially identified as a viral gene product of γ-herpesviruses with homology to caspase-8 (124, 125, and references therein). Both viral and cellular FLIP have an identical structure with two caspase-8 DED motifs in the N-terminus that enable FLIP to bind to FADD (see Fig. 4C). However, FLIPs C-terminus does not contain a caspase active site (QACXG); instead FLIP has a mutated caspase site. Upon receptor activation caspase-8 undergoes one of two cleavages that activate the enzyme. First is the cleavage of the small p12 subunit from the p17 subunit and DED containing prodomain and this must occur prior to the second cleavage of the inhibitory DED prodomain, releasing active caspase-8. c-FLIP undergoes the first cleavage reaction in an identical fashion, but this does not result in the release of FLIP; instead FLIP is able to bind much more tightly to the DISC preventing release of caspase-8 and induction of apoptosis (126). Thus, CD95 triggered apoptosis can be regulated at several stages by different anti-apoptotic proteins, although the efficacy of inhibition appears to be cell type-specific.

Preventing the activity of FADD can also mitigate the CD95-mediated death signal. Embryonic fibroblasts from FADD-deficient mice are resistant to CD95-, TNF-R1-, and DR3-induced cell death (127). Apart from this predictable resistance, the mice exhibit profound abnormalities and die *in utero* at day 11.5. They show cardiac failure and abdominal haemorrhage, phenotypes not seen in TNF-R1 or CD95 knockout animals, suggesting that additional receptors or proteins required for development signal through FADD. It is interesting to note that caspase-8-deficient mice exhibit a similar phenotype including cardiac failure and haemorrhage. The reasons for this are still unknown, but it maybe that caspase-8 also functions in other pathways that do not necessarily involve apoptosis (see ref. 128 for review).

A role for FADD in proliferation is implicated in two transgenic dominant-negative (DN) FADD mouse models. Thymocytes that constitutively express DN-FADD are resistant to CD95-induced cell death and are also profoundly growth arrested (129, 130). Moreover, this growth arrest appears to be p53-dependent since DN-FADD/p53 null cells do not show this growth arrest phenotype, even though they remain insensitive to CD95-induced cell death. The connection between FADD and p53 is not clear, but may become more evident once the proliferative capacity of FADD is more fully understood.

4.1.1 Summary

Overall, CD95 signalling pathways are influenced through the binding of number of adaptor proteins. These proteins interact with a wide variety of intracellular signalling pathways, from activating JNK through RIP binding, to influencing the degradation of proteins through the binding of several ubiquitin homologues. More detailed analysis is now required to determine precisely how each of these pathways serves to regulate CD95-mediated death. Moreover, the ability of FADD to potentially regulate both proliferation and death suggests that other, as yet unknown, proteins may bind and influence these signalling pathways.

4.2 TNFR signalling pathways

A somewhat bewildering array of proteins can intact with the activated TNF receptors (see Fig. 2). As discussed above TRADD and FADD can interact with TNF-R1 and TRAF2 can bind directly to TNF-R2 and recruit TRAF1 (4, 86). TRADD can mediate apoptosis via FADD binding and caspase activation, or it can bind RIP or TRAF2 and activate NF-κB or JNK/SAPK activation. RIP and RIP2 also recruit the TRAFs to mediate the activation of kinase pathways (4, 93). The binding of TRADD, FADD, RIP, and TRAF2 form the TNF-R1 signalling complex (see Fig. 2A). The TRAFs in turn bind another cell death regulating family of proteins, the inhibitors of apoptosis or IAPs (131–133). IAPs were first identified in baculovirus (134), but since then IAPs have been found in many different species, including nematode, *Drosophila*, and man. The IAPs, like the TRAFs, have several conserved domains (see Fig. 4B). In their N-terminus they contain a variable number of baculovirus IAP repeat (BIR) motifs and in their C-terminus is a RING finger motif. Between these two domains human IAP1 and 2 possess a stretch of some 200 amino acids that contains a CARD motif, facilitating interaction with CARD-containing caspases (96). Human IAP1 and 2 are recruited to TNF-R2 by a TRAF1 and TRAF2 heterocomplex and are involved in activating NF-κB (see Fig. 2B) (131). NF-κB can also up-regulate the expression of human IAP1 and human IAP2, suggesting a positive feedback loop where IAPs activate NF-κB which in turn induces more IAP expression (133). In some cell models NF-κB-induced expression of IAPs protects against TNFα-induced apoptosis. The human IAPs (IAP1, IAP2, survivin, and x-IAP) have all been shown to inhibit TNFα- and CD95L-mediated apoptosis; however since IAPs have also been shown to inhibit caspase activation, it is at present unclear at what stage in the signalling pathway these inhibitory effects occur (reviewed in ref. 132). In addition, work in *C. elegans* demonstrates that the nematode IAPs BIR-1 and BIR-2 have other functions. BIR-1 disruption does not affect nematode apoptosis, but instead inhibits cytokinesis (135). Moreover, the human IAP survivin, is required for G2/M transition and binds to the mitotic spindle (136). More detailed analysis is therefore required to determine fully the contribution of IAP recruitment to TNFR superfamily signalling pathways and the role of IAPs in apoptosis.

4.2.1 Death domain self-association

Self-association between the death domains resulting in receptor oligomerization was initially detected in yeast two-hybrid screens (82). Thus, expression of TNF-R1 or just its DD in cells causes cell death, presumably due to the multimerization of the DDs and recruitment of TRADD and FADD. How these domains are prevented from self-association in normal cells has recently been elucidated (137). SODD (silencer of death domains) is a widely expressed protein that interacts with the intracellular DD of TNF-R1 and DR3, but not TNF-R2, CD95, DR4, or DR5. Moreover, SODD does not interact with TRADD, FADD, or RIP DDs. TNF-R1 and SODD are pre-associated in cells that have not encountered TNFα. However, upon TNFα binding, SODD dissociates from TNF-R1 and is replaced by TRADD, consistent with *in vitro* data demonstrating that TRADD and SODD do not bind TNF-R1 simultaneously. SODD begins to reassociate with TNF-R1 about 10 minutes after ligand binding. Overexpression of SODD negates both NF-κB and JNK activation demonstrating that SODDs function is to inhibit any TNF-R1 signalling in the absence of ligand. SODD represents yet another level of regulation controlling the activation of TNF-R1. Several other SODD-like proteins have been found in the GenBank sequence database, implying that keeping DDs in a monomeric form is a generic regulation mechanism.

4.3 Summary

The recent rapid progress on the nature of proteins that interact with the intracellular domains of TNF-R1 and CD95 shows that these signalling pathways are very tightly regulated. Many members of the TNFR superfamily can induce both proliferation and death making them fundamental for cellular regulation. The regulation of the signalling pathways emanating from the activated receptors occurs at multiple levels, from decoy receptors that prevent ligand binding, to proteins that inhibit DD association or promote survival signalling pathways (such as those mediated by the activation of JNK or NF-κB). It is not surprising then that perturbations in the signalling of both the TNFα and CD95L pathways have severe pathological consequences.

5. Disease states with which death receptor signalling pathways are associated

5.1 Introduction

TNFα is involved in many different diseases associated with an inflammatory response. There is extensive literature on this subject and it will not be reviewed here. Instead, the role of CD95 in several diseases will be discussed along with other defects within the downstream pathways that also affect disease progression.

5.1.1 CD95 and immune disorders

CD95 is important in the regulation of the immune system and mutations affecting the expression of this receptor and its ligand lead to the development of autoimmune diseases that have been extensively characterized in two mouse models. The *lpr* mutation (lymphoproliferative disease) is a recessive mutation causing lymphoproliferation and autoimmune disease similar to human systemic lupus erythromatosus (SLE). The *gld* mutation (generalized lymphoproliferative disease) results in a very similar disease to that described for *lpr* (6). Both cases show an accumulation of T cells with unusual phenotypes (138). CD95 is the gene mutated in *lpr* due to the insertion of a retroviral early transposable element into intron 2 resulting in a splicing defect and low abundance of receptor protein (22, 138, 139). *gld* mice have a point mutation in the C-terminus of the CD95 ligand, impairing its ability to interact with the CD95 receptor and trigger apoptosis (138, 140).

Mature T cells from *lpr* and *gld* mice are resistant to activation-induced apoptosis *in vitro* (141) suggesting that the lymphadenopathy seen in these mice is a direct result of the accumulation of aberrant, surviving T cells. So what is the evidence that CD95 expression regulates T cell number? Activation of T cells with PMA and ionomycin induces expression of CD95 *in vitro*. Although the receptor is expressed immediately after stimulation, susceptibility to CD95-induced apoptosis only occurs after a delay of several days (142). This implies that CD95 expression is involved in the termination of the immune response with the apoptotic machinery becoming coupled to the CD95 receptor after a period of time that allows for the clearance of the invading pathogen.

A similar response is evident in activated B cells. In *gld* and *lpr* mice the absence of CD95-mediated down-regulation of the immune response also results in an accumulation of activated B cells that probably contributes to autoantibody production (21).

To investigate further the role of CD95 in termination of the immune response, anti-CD95 F(ab)2 antibody (binds but does not trigger CD95) or CD95-Fc fusion proteins (inhibits CD95L binding) were added to human Jurkat T cells stimulated with anti-CD3 antibodies. Only in the absence of either of these agents did apoptosis occur. Although evidence suggests that both the CD95 receptor and ligand are crucial for activation-induced cell death in T cells, precisely how CD95 and its ligand trigger the death of T cells is still unclear. Cells expressing both receptor and ligand may kill one another by interaction between ligand on one cell and the receptor on the other, or cells may secrete ligand inducing death in an adjacent CD95-expressing cell. Isolated cells also undergo CD95-induced death, possibly due to shedding of their own CD95L, or through cell surface interactions between CD95 and the TCR (reviewed in refs 6 and 21). Identification of the primary mechanism will enable us to understand better the method of peripheral T cell tolerance and down-regulation of the immune response. Although CD95 signalling pathways are important in peripheral T cell deletion, TNF-R2 and TNF are also able to mediate T cell death late on after TCR triggering (143). Moreover, it is highly likely that TNFR family members such as DRs 3–6 play very similar roles which should be elucidated through the generation of deficient and transgenic animals.

5.1.2 Canale–Smith Syndrome

Perturbations of the CD95 pathway in humans also cause immunological diseases. One disease where CD95 mutations are apparent is in Canale–Smith syndrome (144). Heterozygous mutations with the CD95 receptor are associated with defective lymphocyte apoptosis resulting in lymphadenopathy, splenomegaly, and systemic autoimmunity, comparable to the *lpr* and *gld* mouse phenotypes. Mutations within CD95 have been characterized in these patients and affect both the intra- and extracellular domains. Mutations within the intracellular domain either mutate the DD, preventing binding of adaptor proteins, or cause frameshift mutations leading to premature termination and a truncated or absent DD. The extracellular domain mutants have either mutated cysteine-rich binding domains, preventing ligand interaction, or have very early frameshift terminations resulting in almost no CD95 protein being made. Similar mutations have been documented in other patients (145). Thus, impairment of the CD95 pathway in humans leads to a complex clinical disease.

CD95 can also have a reverse role in immunological diseases. That is, if sensitivity to CD95 is somehow increased then cells can be induced to die in response to signals that would normally mediate proliferation or differentiation. Sensitization to CD95-induced apoptosis is proposed to be one mechanism for $CD4^+$ T cell loss in HIV infection.

5.2 CD95 and AIDS

Apoptosis as a mechanism for the loss of $CD4^+$ T cells in AIDS was initially proposed a decade ago (146, 147). Since this time the number of papers on this subject has grown steadily, although the potential role of apoptosis in the aetiology of AIDS remains unclear.

5.2.1 HIV sensitizes T cells to TCR-mediated apoptosis

HIV infects $CD4^+$ T cells by binding to the CD4 receptor on the T cell via the viral envelope protein gp120. This complex is then internalized by the cell and HIV begins to replicate itself inside the newly infected cell. The demonstration that infected T cells stimulated via their CD4 receptors are primed for apoptosis suggests that binding of gp120 to CD4 primes the cells for death upon subsequent TCR stimulation (148–152). Crosslinking of the CD4 receptor by gp120 and anti-gp120 antibodies leads to cell death *in vitro* in normal human purified T cells upon stimulation of the TCR (149). Thus, uninfected T cells maybe primed for death upon contact with HIV envelope expressing infected T cells. The recent discovery of the second cell surface receptor that HIV binds to on the cell surface, a chemokine receptor known as CXCR4 (fusin/LESTR), has not altered the significance of the above findings (6, 153). In fact, CXCR4 and CD4 can lead to extremely rapid cell death in T cell lines, peripheral blood lymphocytes, and CD4/CXCR4 transfectants, and that death is specific to CD4 and not CD8 T cells. Thus, inappropriate signalling caused by HIV infection can lead to the rapid deletion of T helper cells as is seen in AIDS patients.

5.2.2 Tat and CD95

Activated, but non-infected $CD4^+$ and $CD8^+$ positive T cells from HIV-infected patients express higher levels of CD95 receptor than cells from uninfected individuals and these cells are more sensitive to apoptosis. Krammer and colleagues (154, 155) have suggested a role for HIV Tat protein in sensitizing uninfected $CD4^+$ cells to CD95-induced apoptosis. Tat is an HIV protein involved in regulating HIV gene transcription and is imported across the cell membrane of uninfected cells. *In vitro* T cells cultured in the presence of synthetic Tat protein or Tat protein and gp120 and anti-gp120 antibodies show CD95-dependent apoptosis at a higher rate compared to normal controls. Apoptosis induced by these proteins can be significantly inhibited by the presence of F(ab)2 anti-CD95 or soluble CD95 Fc receptor that prevent ligand–receptor interaction. Hence, the hypothesis is that loss of $CD4^+$ non-infected cells is a result of TCR triggered CD95-mediated suicide hypersensitized by the presence of HIV Tat and gp120 and anti-gp120 antibodies in the patients sera.

So far it seems likely that CD95 does have some role in mediating the death of $CD4^+$ T cells seen in AIDS. However, the *in vitro* data has yet to be fully tested in patients with HIV. If Tat and CD95 have significant roles *in vivo*, then they are possible candidates for therapeutic intervention in this disease.

5.3 Cancer

CD95-induced cell death is an efficient mechanism for mediating the death of cells, whether this is via cytotoxic T cell-induced death, or through CD95 ligand–receptor interactions within the same cell. As discussed in Section 4 regulation of the CD95 pathway occurs at many different levels. Each of the mechanisms that can suppress CD95-induced apoptosis is a candidate for mutation during the evolution of diseases that evade the immune response. In theory, this would be true for all cancers since all cancerous cells are targeted by the immune response, which uses in part, along with perforin and granzyme B, CD95-mediated T cell-induced cell death. Thus, cells that overexpress anti-apoptotic proteins such as IAPs, FLIP, and Bcl-2 family members can be more resistant to cytotoxic T cells (126, 132, 156). Moreover, in theory, overexpression of either decoy receptors or SODD-like proteins could also inhibit CD95-mediated killing. Tumour cells have been shown to secrete CD95, preventing T cells binding to the tumour cell, or to express high levels of CD95L on their cell surface that kills any immune cell that attempts to interact with the tumour cell. The effectiveness of these evasive mechanisms is not well documented and may not always lead to an advantage *in vivo* (21, 157). Overall, tumour cells can exhibit resistance to CD95-mediated cell death at a number of levels. Which of these are important for tumour progression is only now being investigated.

Another more surprising role for CD95 in cancer has emerged recently. CD95, in some cell systems is involved in the death of cells in response to oncogene activation or chemotherapy. For example, Myc-induced apoptosis in fibroblasts is mediated by an increased sensitization to pro-apoptotic stimuli, one of which is CD95 (158). If the interaction between CD95 and CD95L is blocked then Myc-induction of cell death is significantly delayed, although not inhibited. c-Myc does not appear to increase either the expression of CD95 or CD95L. Instead, c-Myc expression leads to the release of cytochrome *c* which makes cells more sensitive to apoptotic stimuli, an effect that can be blocked by the presence of survival cytokines or anti-apoptotic proteins such as Bcl-2 (159). This finding suggests that nuclear and cytoplasmic signals generated within the cell can mediate cell death via receptors expressed on the cell surface. A related phenomenon has been shown for chemotherapy. Tumour cells that express low levels of CD95 are more resistant to chemotherapy (160). Moreover, several groups have shown that chemotherapy induces up-regulation of CD95 and CD95L, aiding both sensitization to apoptosis and providing targets for the immune response (161). In this model CD95 up-regulation requires functional p53, the protein responsible for the detection of DNA damage within the cell (162). Thus, in tumour cells that have mutant p53, the response to chemotherapy would be lessened by both the inability to detect DNA damage and the inability to increase expression of CD95. The up-regulation of CD95L expression and the requirement of CD95 for chemotherapeutic-induced cell death is not evident in all cell types (163, 164). It is possible that these differences reflect the fact that downstream elements of the CD95 pathway are active in response to other stimuli. Thus, no one trigger is required absolutely for cell death.

Although not yet proven, the link between oncogene- and chemotherapeutic-induced cell death and the death receptors is intriguing and may yet prove useful in the development of novel cancer therapies for specific cell types.

6. Conclusion

The TNF family of receptors and ligands encompasses molecules that are able to induce pleiotropic effects in many cell lineages. Primarily, the activated receptors are able to regulate cell death and proliferation and the downstream pathways that enable these different outcomes to be manifest have been partially mapped. Future research should establish how many of the known proteins that bind both TNF-R1 and R2 and CD95 also interact with other TNFR family members. Moreover, as this information becomes available it will enable further understanding of diseases in which this family of receptors and ligands are implicated.

References

1. Trauth, B. C., Klas, C., Peters, A. M., Matzku, S., Moller, P., Falk, W., *et al.* (1989). Monoclonal antibody-mediated tumor regression by induction of apoptosis. *Science*, **245**, 301.
2. Yonehara, S., Ishii, A., and Yonehara, M. (1989). A cell-killing monoclonal antibody (anti-Fas) to a cell surface antigen co-downregulated with the receptor of tumor necrosis factor. *J. Exp. Med.*, **169**, 1747.

3. Itoh, N. and Nagata, S. (1993). A novel protein domain required for apoptosis. Mutational analysis of human Fas antigen. *J. Biol. Chem.*, **268**, 10932.
4. Schulze-Osthoff, C., Ferrari, D., Los, M., Wesselborg, S., and Peter, M. E. (1998). Apoptosis signaling by death receptors. *Eur. J. Biochem.*, **254**, 439.
5. Golstein, P. (1997). Cell death: TRAIL and its receptors. *Curr. Biol.*, **7**, R750.
6. Krammer, P. H. (1999). CD95 (Apo-1/Fas)-mediated apoptosis: Live and let die. *Adv. Immunol.*, **71**, 163.
7. Baker, S. J. and Reddy, E. P. (1998). Modulation of life and death by the TNF receptor superfamily. *Oncogene*, **17**, 3261.
8. Itoh, N., Yonehara, S., Ishii, A., Yonehara, M., Mizushima, S-I., Sameshima, M., *et al.* (1991). The polypeptide encoded by the cDNA for human cell surface antigen Fas can mediate apoptosis. *Cell*, **66**, 233.
9. Oehm, A., Behrmann, I., Falk, W., Pawlita, M., Maier, G., Klas, C., *et al.* (1992). Purification and molecular cloning of the APO-1 cell surface antigen, a member of the tumor necrosis factor/nerve growth factor receptor superfamily. Sequence identity with the Fas antigen. *J. Biol. Chem.*, **267**, 10709.
10. Schlossman, S. F., Boumsell, L., Gilks, W., Harlan, J. M., Kishimoto, T., Morimoto, C., *et al.* (ed.) (1995). Leukocyte typing V: White cell differentiating antigens. *Proceedings of the fifth international workshop and conference*, Vol. 1. Oxford University, Press, London.
11. Watanabe Fukunaga, R., Brannan, C. I., Copeland, N. G., Jenkins, N. A., and Nagata, S. (1992). Lymphoproliferation disorder in mice explained by defects in Fas antigen that mediates apoptosis. *Nature*, **356**, 314.
12. Tewari, M. and Dixit, T. M. (1996). Recent advances in tumour necrosis factor and CD40 signaling. *Curr. Opin. Genet. Dev.*, **6**, 39.
13. Casaccia-Bonnefil, P., Kong, H., and Chao, M. V. (1998). Neurotrophins: the biological paradox of survival factors eliciting apoptosis. *Cell Death Differ.*, **5**, 357.
14. Prasad, K. V. S., Ao, Z., Yoon, Y., Wu, M., Rizik, M., Jacquot, S., *et al.* (1997). CD27, a member of the tumour necrosis factor receptor family, induces apoptosis and binds to Siva, a proapoptotic protein. *Proc. Natl. Acad. Sci. USA*, **94**, 6346.
15. Duckett, C. S. and Thompson, C. B. (1997). CD30-dependent degradation of TRAF2: implications for negative regulation of TRAF signaling and the control of cell survival. *Cell Death Differ.*, **11**, 2810.
16. Grell, M., Scheurich, P., Meager, A., and Pfizenmaier, K. (1993). TR60 and TR80 tumour necrosis factor (TNF)-receptors can independently mediate cytolysis. *Lymphokine Cytokine Res.*, **12**, 143.
17. Declercq, W., Denecker, G., Fiers, W., and Vandenabeele, P. (1998). Co operation of both TNF receptors in inducing apoptosis: Involvement of the TNF receptor-associated factor binding domain of the TNF receptor 75. *J. Immunol.*, **161**, 390.
18. Chinnaiyan, A. M., O'Rouke, K., Yu, G. L., Lyons, R. H., Grag, M., Duan, D. R., *et al.* (1996). Signal transduction by DR3, a death -domain containing receptor related to TNF-R1 and CD95. *Science*, **274**, 990.
19. Inazawa, J., Itoh, N., Abe, T., and Nagata, S. (1992). Assignment of the human Fas antigen gene (Fas) to 10q24.1.*Genomics*, **14**, 821.
20. Watanabe Fukunaga, R., Brannan, C. I., Itoh, N., Yonehara, S., Copeland, N. G., Jenkins, N. A., *et al.* (1992). The cDNA structure, expression, and chromosomal assignment of the mouse Fas antigen. *J. Immunol.*, **148**, 1274.
21. Krammer, P., Behrmann, I., Daniel, P., Dhein, J., and Debatin, K.-M. (1994). Regulation of apoptosis in the immune system. *Curr. Opin. Immunol.*, **6**, 279.

22. Nagata, S. (1994). Mutations in the Fas antigen gene in lpr mice. *Semin. Immunol.*, **6**, 3.
23. Sobel, E. S., Katagiri, T., Katagiri, K., Morris, S. C., Cohen, P. L., and Eisenberg, R. A. (1991). An intrinsic B cell defect is required for the production of autoantibodies in the lpr model of murine systemic autoimmunity. *J. Exp. Med.*, **173**, 1441.
24. Cheng, J., Zhou, T., Liu, C., Shapiro, J. P., Brauer, M. J., Kiefer, M. C., *et al.* (1994). Protection from Fas mediated apoptosis by a soluble form of the Fas molecule. *Science*, **263**, 1759.
25. Kitson, J., Raven, T., Jiang, Y. P., Goeddel, D. V., Giles, K. M., Pun, K. T., *et al.* (1996). A death domain containing receptor that mediates apoptosis. *Nature*, **384**, 372.
26. Marsters, S. A., Sheridan, J. P., Donahue, C. J., Pitti, R. M., Gray, C. L., Goddard, A. D., *et al.* (1996). Apo-3, a new member of the tumour necrosis factor receptor family contains a death domain and activates apoptosis and NF-kB. *Curr. Biol.*, **6**, 1669.
27. Bodmer, J. L., Burns, K., Schneider, P., Hofmann, K., Steiner, V., Thome, M., *et al.* (1997). TRAMP, a novel apoptosis-mediating receptor with sequence homology to tumour necrosis factor receptor 1 and Fas (Apo-1/CD95). *Immunity*, **6**, 79.
28. Screaton, G. R., Xu, X. N., Olsen, A. L., Cowper, A. E., Tan, R., McMicheal, A. J., *et al.* (1997). LARD: a new lymphoid specific death domain containing receptor regulated by alternate pre-mRNA splicing. *Proc. Natl. Acad. Sci. USA*, **94**, 4615.
29. Chaudhary, P. M., Eby, M., Jasmin, A., Bookwalter, A., Murray, J., and Hood, L. (1997). Death receptor 5, a new member of the TNFR family and DR4 induce FADD-dependent apoptosis and activate the NF-kB pathway. *Immunity*, **7**, 821.
30. Pan, G., O'Rouke, K., Chinniayan, A. M., Gentz, R., Ebner, R., Ni, J., *et al.* (1997). The receptor for the cytotoxic ligand TRAIL. *Science*, **276**, 111.
31. Walczak, H., Degli-Eposti, M. A., Johnson, R. S., Smolak, P. J., Waugh, J. Y., Boiani, N., *et al.* (1997). TRAIL-R2: A novel apoptosis-mediating receptor for TRAIL. *EMBO J.*, **16**, 5386.
32. Screaton, G. R., Mongkolsapaya, J., Xu, X.-N., Cowper, A. E., McMicheal, A. J., and Bell, J. I. (1997). TRICK2, a new alternatively spliced receptor that transduces the cytotoxic signal from TRAIL. *Curr. Biol.*, **7**, 693.
33. Pan, G., Ni, J., Wei, Y. F., Yu, G., Gentz, R., and Dixit, V. M. (1997). An agonist decoy receptor and a death domain containing receptor for TRAIL. *Science*, **277**, 815.
34. Pan, G., Bauer, J. H., Haridas, V., Wang, S., Liu, D., Yu, G., *et al.* (1998). Identification and functional characterisation of DR6, a novel death domain-containing TNF receptor. *FEBS Lett.*, **431**, 351.
35. Brojatch, J., Naughton, J., Rolls, M. M., Zingler, K., and Young, J. A. (1996). CAR1, a TNF-related protein is a cellular receptor for cytopathic avian leukosis-sarcoma viruses and mediates apoptosis. *Cell*, **87**, 845.
36. Smith, C. A., Grus, H. J., Davis, T., Anderson, D., Farrah, T., Baker, E., *et al.* (1993). CD30 antigen, a marker for Hodgkin's lymphoma, is a receptor whose ligand defines an emerging family of cytokines with homology to TNF. *Cell*, **73**, 1349.
37. Biswas, P., Smith, C. A., Goletti, D., Hardy, E. C., Jackson, R. W., and Fauci, A. S. (1995). Cross-linking of CD30 induces HIV expression in chronically infected T-cells. *Immunity*, **2**, 587.
38. Lee, S. Y., Park, C. G., and Choi, Y. (1996). T-cell receptor dependent cell death of T-cell hybridomas mediated by the CD30 cytoplasmic domain in association with tumour necrosis factor associated receptors. *J. Exp. Med.*, **183**, 669.
39. Schwarb, U., Stein, H., Gerdes, J., Lemke, H., Kirchner, H., Schaadt, M., *et al.* (1982). Production of a monoclonal antibody specific for Hodgkin and Sternberg-Reed cells of Hodgkin's disease and a subset of normal lymphoid cells. *Nature*, **299**, 65.

40. Falini, B., Pileri, S., Pizzolo, G., Durkop, H., Flenghi, L., Stripe, F., *et al.* (1995). CD30 (Ki-1) molecule: A new cytokine receptor of the tumour necrosis receptor superfamily as a tool for diagnosis and immunotherapy. *Blood*, **85**, 1.
41. Amakawa, R., Hakem, A., Kundig, T. M., Matsuyama, T., Simard, J. J. L., Timms, E., *et al.* (1996). Impaired negative selection of T-cells in Hodgkin's disease antigen CD30 deficient mice. *Cell*, **84**, 551.
42. Green, E. A. and Flavell, R. A. (1999). TRANCE-RANK, a new signal pathway involved in lymphocyte development and T-cell activation. *J. Exp. Med.*, **189**, 1017.
43. Bachmann, M. F., Wong, B. R., Josien, R., Steinman, R. M., Oxenius, A., and Choi, Y. (1999). TRANCE, a tumour necrosis family member critical for CD40-ligand-independent T-helper cell activation. *J. Exp. Med.*, **189** 1025.
44. Kong, Y-Y., Yoshida, H., Sarosi, I., Tan, H-L., Timms, E., Capparelli, C., *et al.* (1999). OPGL is a key regulator of osteoclastogenesis, lymphocyte development and lymph-node organogenesis. *Nature*, **397**, 315.
45. Hintzen, R. Q., Lens, S. M. A., Lammers, K., Kuper, H., Beckmann, P., and Van-Lier, R. A. W. (1995). Engagement of CD27 with its ligand CD70 provides a second signal for T cell activation. *J. Immunol.*, **154**, 2612.
46. Gravestein, L. A., Amsen, D., Boes, M., Calvo, C. R., Kruisbeek, A. M., and Borst, J. (1998). The TNF receptor family member CD27 signals to Jun N-terminal kinase via Traf2. *Eur. J. Immunol.*, **28**, 2208.
47. Akiba, H., Nakano, H., Nishinaka, S., Shindo, M., Kobata, T., Atsuta, M., *et al.* (1998). CD27, a member of the tumour necrosis receptor superfamily, activates NF-kB and stress-activated protein kinase/c-Jun N-terminal kinase via TRAF2, TRAF5 and NF-kB-inducing kinase. *J. Biol. Chem.*, **273**, 13353.
48. Hsu, H., Solovyev, I., Colombero, A., Elliot, R., Kelley, M., and Boyle, W. J. (1997). ATAR, a novel tumor necrosis factor receptor family member, signals through TRAF2 and TRAF5. *J. Biol. Chem.*, **272**, 13471.
49. Nocentini, G., Giunchi, L., Ronchetti, S., Krausz, L. T., Bartoli, A., Moraca, R., *et al.* (1997). A new member of the tumour necrosis/nerve growth factor receptor family inhibits T-cell receptor-induced apoptosis. *Cell Biol.*, **94**, 6216.
50. Yoon, S. O., Casaccai-Bonnefil, P., Carter, B., and Chao, M. V. (1998). Competitive signalling between TrkA and p75 nerve growth factor receptors determines cell survival. *J. Neurosci.*, **18**, 3273.
51. Frade, J. M. and Barde, Y. A. (1999). Genetic evidence for cell death mediated by nerve growth factor and the neurotrophin receptor p75 in the developing mouse retina and spinal cord. *Development*, **126**, 683.
52. Emery, J. G., McDonnell, P., Burke, M., Deen, K. C., Lyn, S., Silverman, C., *et al.* (1998). Osteoprotegerin is a receptor for the cytotoxic ligand TRAIL. *J. Biol. Chem.*, **273**, 14363.
53. Yun, T. J., Chaushary, P. M., Shu, G. L., Frazer, J. K., Ewings, M. K., Schwartz, S. M., *et al.* (1998). OPG/FDCR-1, a TNF receptor family member, is expressed in lymphoid cells and is upregulated by ligating CD40. *J. Immunol.*, **161**, 6113.
54. Takahashi., N., Udagawa, N., and Suda, T. (1999). A new member of tumour necrosis factor ligand family, ODF/OPGL/TRANCE/RANKL, regulates osteoclast differentiation and function. *Biochem. Biophys. Res. Commun.*, **256**, 449.
55. Yasuda, H., Shima, N., Nakagawa, N., Yamaguchi, K., Kinosaki, M., Mochizuki, S-I., *et al.* (1998). Osteoclast differentiation factor is as ligand for osteoprotegerin/osteoclastogenesis-inhibitory factor and is identical to TRANCE/RANKL. *Proc. Natl. Acad. Sci. USA*, **95**, 3597.

56. Sheridan, J. P., Marsters, S. A., Pitti, R. M., Gurney, A., Skubatch, M., Baldwin, D., *et al.* (1997). Control of TRAIL induced apoptosis by a family of signaling and decoy receptors. *Science*, **277**, 818.
57. Mongkolsapaya, J., Cowper, A. E., Xu, X. N., McMicheal, A. J., Bell, J. I., and Screaton, J. R. (1998). LIT a new receptor protecting lymphocytes from the death ligand TRAIL. *J. Immunol.*, **160**, 3.
58. Degli-Esposti, M. A., Smolak, P. J., Walczak, H., Waugh, J., Huang, C. P., DuBose, R. F., *et al.* (1997). Cloning and characterisation of TRAIL-R3 a novel member of the emerging TRAIL receptor family. *J. Exp. Med.*, **186**, 1165.
59. Masters, S. A., Sheridan, J. P., Pitti, R. M., Huang, A., Skubatch, M., Baldwin, D., *et al.* (1997). A novel receptor for Apo2L/TRAIL contains a truncated death domain. *Curr. Biol.*, **7**, 1003.
60. Pan, G., Ni, J., Yu, G., Wei, Y. F., and Dixit, V. M. (1998). TRUNDD, a new member of the TRAIL receptor family that antagonises TRAIL signalling. *FEBS Lett.*, **424**, 41.
61. Degli-Eposti, M. A., Dougall, W., Smolak, P. J., Waugh, J. Y., Smith, C. A., and Goodwin, R. G. (1997). The novel receptor TRAIL-R4 induces NF-kB and protects against TRAIL-mediated apoptosis, yet retains and incomplete death domain. *Immunity*, **7**, 813.
62. Pitti, R. M., Marsters, S. A., Lawrence, D. A., Roy, M., Kischkel, F. C., Dowd, P., *et al.* (1998). Genomic amplification of a decoy receptor for Fas ligand in lung and colon cancer. *Nature*, **396**, 699.
63. Eck, M. J. and Sprang, S. R. (1989). The structure of tumour necrosis factor-alpha at 2.6A resolution: implications for receptor binding. *J. Biol. Chem.*, **264**, 17595.
64. Bajorath, J. and Aruffo, A. (1997). Prediction of the three dimensional structure of the human Fas receptor by comparative molecular modelling. *J. Comupt. Aided Mol. Des.*, **11**, 3.
65. Pasparakis, M., Alexopoulou, C., Episkopou, V., and Kollias, G. (1996). Immune and inflammatory responses in TNF-alpha deficient mice: a critical requirement for TNF alpha in the formation of primary B-cell follicles, follicular dendritic cell networks and germinal centers, and the maturation of the humoral immune response. *J. Exp. Med.*, **184**, 1397.
66. Marino, M. W., Dunn, A., Grail, D., Inglese, M., Nuguchi, Y., Richards, E., *et al.* (1997). Characterisation of tumour necrosis factor deficient mice. *Proc. Natl. Acad. Sci. USA*, **94**, 8093.
67. Paspara, M., Alexoulou, L., Grell, M., Pfizenmaier, K., Bluethmann, H., and Kollias, G. (1997). Peyers patch organogenesis is defective in peripheral lymphoid organs of mice deficient for tumour necrosis factor and its 55KDa receptor. *Proc. Natl. Acad. Sci. USA*, **94**, 6319.
68. Griffith, T. S., Brunner, T., Fletcher, S. M., Green, D. R., and Ferguson, T. A. (1995). Fas ligand induced apoptosis as a mechanism of immune privilege. *Science*, **270**, 1189.
69. Bellgrau, D., Gold, D., Selawry, H., Moore, J., Franzusoff, A., and Duke, R. C. (1995). Role for CD95 ligand in preventing graft rejection. *Nature*, **377**, 630.
70. Griffith, T. S., Yu, X., Herndon, J. M., Green, D. R., and Ferguson, T. A. (1996). CD95-induced apoptosis of lymphocytes in an immune privilege site induces immunological tolerance. *Immunity*, **5**, 7.
71. Pitti, R. M., Masters, S. A., Ruppert, S., Donahue, C. J., Moore, A., and Ashkenazi, A. (1996). Induction of apoptosis by Apo2 ligand, a new member of the tumour necrosis factor cytokine family. *J. Biol. Chem.*, **271**, 12687.
72. Wiley, S. R., Schooley, K., Smolak, P. J., Din, W. S., Huang, C. P., Nicholl, J. K., *et al.* (1995). Identification and characterisation of a new member of the TNF family that induces apoptosis. *Immunity*, **3**, 673.

73. Masters, S. A., Sheridan, J. P., Pitti, R. M., Brush, J., Goddard, A., and Ashkenarzi, A. (1998). Identification of a ligand for the death-domain-containing receptor Apo3. *Curr. Biol.*, **8**, 525.
74. Chicheportiche, Y., Bourdon, P. R., Xu, H., Hsu, Y. M., Scott, H., Hession, C., *et al.* (1997). TWEAK, a new secreted ligand in the tumour necrosis family that weakly induces apoptosis. *J. Biol. Chem.*, **272**, 32401.
75. Lynch, C. N., Wang, Y. C., Lund, J. K., Chen, Y. W., Leal, J. A., and Wiley, S. R. (1999). TWEAK induces angiogenesis and proliferation of endothelial cells. *J. Biol. Chem.*, **274**, 8455.
76. Schneider, P., Schwenzer, R., Haas, E., Muhlenbeck, F., Schubert, G., Scheurich, P., *et al.* (1999). TWEAK can induce cell death via endogenous TNF and TNF receptor 1. *Eur. J. Immunol.*, **29**, 1785.
77. Wong, B. R., Rho, J., Arron, J., Robinson, E., Orlinink, J., Chao, M., *et al.* (1997). TRANCE is a novel ligand of the tumour necrosis factor receptor family that activates c-Jun N-terminal kinase in T-cells. *J. Biol. Chem.*, **272**, 25190.
78. Tartaglia, L. A., Ayers, T. M., Wong, G. H., and Goddel, D. V. (1993). A novel domain within the 55kD receptor signals cell death. *Cell*, **74**, 845.
79. Golstein, P., Marguet, D., and Depraetere, V. (1995). Homology between reaper and the cell death domains of FADD and TNFR1. *Cell*, **81**, 185.
80. White, K., Tahaoglu, E., and Steller, H. (1996). Cell killing by the *Drosophila* gene reaper. *Science*, **271**, 805.
81. Chinnaiyan, A. M., Orourke, K., Tewari, M., and Dixit, V. M. (1995). Fadd, a novel death domain-containing protein, interacts with the death domain of fas and initiates apoptosis. *Cell*, **81**, 505.
82. Boldin, M. P., Varfolomeev, E. E., Pancer, Z., Mett, I. L., Camonis, J. H., and Wallach, D. (1995). A novel protein that interacts with the death domain of Fas/Apo-1 contains a sequence motif related to the death domain. *J. Biol. Chem.*, **270**, 7795.
83. Chu, J. L., Drappa, J., Parnassa, A., and Elkon, K. B. (1993). The defect in Fas mRNA expression in MRL/lpr mice is associated with insertion of the retrotransposon, ETn. *J. Exp. Med.*, **178**, 723.
84. Muzio, M., Chinnaiyan, A. M., Kischkel, F. C., O'Rouke, K., Shevchenko, A., Ni, J., *et al.* (1996). FLICE, a novel FADD-homologous ICE/CED-3 like protease is recruited to the CD95 (Fas/Apo-1) death-inducing signalling complex. *Cell*, **85**, 817.
85. Schnider, P., Thome, M., Burns, K., Bodmer, J-L., Hofmann, K., Kataoka, T., *et al.* (1997). TRAIL receptors 1 (DR4) and 2 (DR5) signal FADD-dependent apoptosis and activate NF-kB. *Immunity*, **7**, 831.
86. Wallach, D. (1997). Cell death induction by TNF: a matter of self control. *FEBS Lett.*, **410**, 96.
87. Hsu, H. L., Xiong, J., and Goeddel, D. V. (1995). The TNF receptor 1-associated protein Tradd signals cell-death and NF-kappa-b activation. *Cell*, **81**, 495.
88. Varfolomeev, E. E., Boldin, M. P., Goncharov, T. M., and Wallach, D. (1996). A potential mechanism of 'cross-talk' between the p55 tumour necrosis factor receptor and Fas/APO-1: proteins binding to the death domains of the two receptors also bind to each other. *J. Exp. Med.*, **183**, 1271.
89. Stanger, B. Z., Leder, P., Lee, T. H., Kim, E., and Seed, B. (1995). RIP: a novel protein containing a death domain that interacts with Fas/Apo-1 (CD95) in yeast and causes death. *Cell*, **81**, 513.
90. Keilliher, A., Grimm, S., Ishida, Y., Kuo, F., Stranger, B. Z., and Leder, P. (1998). The death domain kinase RIP mediates the TNF-induced NF-kB signal. *Immunity*, **8**, 297.

91. Van-Antwerp, D. J., Martin, S. J., Verma, I. M., and Green, D. R. (1998). Inhibition of TNF-induced apoptosis by NF-kB. *Cell Biol.*, **8**, 107.
92 Ting, A. T., Pimentel-Muinos, F. X., and Seed, B. (1996). RIP mediates tumour necrosis factor receptor 1 activation of NF-kappaB, but not Fas/Apo-1-initiated apoptosis. *EMBO J.*, **15**, 6189.
93. McCarthy, J. V., Ni, J., and Dixit, V. M. (1998). RIP2 is a novel NF-kB-activating and cell death inducing kinase. *J. Biol. Chem.*, **273**, 16975.
94. Inohara, N., Peso, L., Koseki, T., Chen, S., and Nunez, G. (1998). RICK, a novel protein kinase containing a caspase recruitment domain, interacts with CLARP and regulates CD95-mediated apoptosis. *J. Biol. Chem.*, **273**, 12296.
95. Thome, M., Hofmann, K., Burns, K., Martinon, F., Bodmer, J. L., Mattmann, C., *et al.* (1998). Identification of CARDIAK, a RIP-like kinase that associates with caspase-1. *Curr. Biol.*, **8**, 885.
96. Hofmann, K., Bucher, P., and Tschopp, J. (1997). The CARD domain: a new apoptotic signalling motif. *Trends Biochem. Sci.*, **22**, 155.
97. Duan, H. and Dixit, V. M. (1997). RAIDD is a new 'death' adaptor molecule. *Nature*, **385**, 86.
98. Yang, X., Khosravi-Far, R., Chang, H., and Baltimore, D. (1997). Daxx, a novel Fas-binding protein that activates JNK and apoptosis. *Cell*, **89**, 106.
99. Chang, H. Y., Nishitoh, H., Yang, X., Ichijo, H., and Baltimore, D. (1998). Activation of apoptosis signal-regulating kinase 1 (ASK-1) by the adaptor protein Daxx. *Science*, **281**, 1860.
100. Chang, H. Y., Yang, X., and Baltimore, D. (1999). Dissecting Fas signaling with an altered specificity death-domain mutant: requirement of FADD binding for apoptosis, but not for Jun N-terminal kinase activation. *Proc. Natl. Acad. Sci. USA*, **96**, 1252.
101. Michaelson, J. S., Bader, D., Kuo, F., Kozak, C., and Leder, P. (1999). Loss of Daxx, a promiscuously interacting protein, results in extensive apoptosis in early mouse development. *Genes Dev.*, **13**, 1918.
102. Torii, S., Egan, D. A., Evans, R. A., and Reed, J. C. (1999). Human Daxx regulates Fas-induced apoptosis from mulear PML oncogenic domains (PODs). *EMBO J.*, **18**, 6037.
103. Wand, Z-G., Ruggero, D., Ronchetti, S., Zhong, S., Gaboli, M., Rivi, R., *et al.* (1998). Pml is essential for multiple apoptotic pathways. *Nature Genet.*, **20**, 266.
104. Wright, D. A., Fukher, B., Gosh, P., and Geha, R. S. (1996). Association of human Fas (CD95) with a ubiquitin conjugating enzyme (UBC-FAP). *J. Biol. Chem.*, **271**, 31037.
105. Ryu, S. W., Chae, S. K., Lee, K. J., and Kim, E. (1999). Identification and characterization of human Fas associated factor 1 hFAF1. *Biochem. Biophys. Res. Commun.*, **262**, 388.
106. Frohlich, T., Riasau, W., and Flamme, I. (1998). Characterisation of novel nuclear targeting and apoptosis inducing domains in Fas associated factor 1. *J. Cell Sci.*, **111**, 2353.
107. Sato, T., Irie, S., Kihada, S., and Reed, J. C. (1995). FAP-1 a protein tyrosine phosphatase that associates with Fas. *Science*, **268**, 411.
108. Okura, T., Gong, L., Kamitani, T., Wada, T., Okura, I., Wei, C. F., *et al.* (1996). Protection against Fas/Apo-1 and tumour necrosis factor mediated cell death by a novel protein Sentrin. *J. Immunol.*, **157**, 4277.
109. Mahajan, R., Delphin, C., Guan, T., Gerace, L., and Melchior, F. (1997). A small ubiquitin-related polypeptide involved in targeting RanGAP1 to nuclear pore complex protein RanBP2. *Cell*, **88**, 97.
110. Jentsch, S. and Pyrowolakis, G. (2000). Ubiquitin and its kin: how close are the family ties? *Trends Cell Biol.*, **10**, 335.

111. Arch, R. H. and Thompson, C. B. (1998). 4-IBB and Ox-40 are members of a tumour necrosis factor (TNF)-nerve growth factor receptor subfamily that bind TNF receptor-associated factors and activate nuclear factor kappaB. *Mol. Cell. Biol.*, **18**, 558.
112. Kawamata, S., Hori, T., Imura, A., Takaori-Kondo, A., and Uchiyama, T. (1998). Activation of OX-40 signal transduction pathways leads to tumour necrosis factor receptor associated factor (TRAF)-2 and TRAF-5 mediated NF-kB activation. *J. Biol. Chem.*, **273**, 5808.
113. Kashiwada, M., Shirakata, Y., Inoue, J. I., Nakano, H., Okazaki, K., Okumura, K., *et al.* (1998). Tumour necrosis factor receptor-associated factor 6 (TRAF6) stimulates extra-cellular signal-related kinase (ERK) activity in CD40 signalling in a ras-dependant pathyway. *J. Exp. Med.*, **187**, 237.
114. Sutherland, C. L., Krebs, D. L., and Gold, M. R. (1999). An 11-amino acid sequence in the cytoplasmic domain of CD40 is sufficient for activation of c-Jun N-terminal kinase, activation of MAPK kinase-2, phosphorylation of IkappaBalpha and protection of WEHI-231 cells from anti-IgM induced growth arrest. *J. Immunol.*, **162**, 4720.
115. Grammer, A. C., Swantek, J. L., McFarland, R. D., Miura, Y., Geppert, T., and Lipsky, P. L. (1998). TNF-receptor associated factor 3 signalling mediates activation of p38 and Jun N-terminal kinase, cytokine secretion and Ig production following ligation of CD40 on human B-cells. *J. Immunol.*, **161**, 11831.
116. Tsukamoto, N., Kobayashi, N., Azuma, S., Yamamoto, T., and Inoue, J-I. (1999). Two differentially regulated nuclear factor κB activation pathways triggered by the cytoplasmic tail of CD40. *Proc. Natl. Acad. Sci. USA*, **96**, 1234.
117. Darnay, B. G., Ni, J., Moore, P. A., and Aggawal, B. B. (1999). Activation of NF-kB by RANK requires tumour necrosis receptor associated factor (TRAF)6 and NF-κB inducing kinase. *J. Biol. Chem.*, **274**, 7724.
118. Kim, H. H., Lee, D. E., Shin, J. N., Lee, Y. S., Jeon, Y. M., Chung, C. H., *et al.* (1999). Receptor activator of NF-kappaB recruits multiple TRAF family adaptors and activates c-Jun N-termianl kinase. *FEBS Lett.*, **443**, 297.
119. Galibert, L., Tometsko, M. E., Anderson, D. M., Cosman, D., and Dougall, W. C. (1998). The involvement of multiple tumour necrosis factor receptor (TNFR)-associated factors in the signalling mechanisms of receptor activator of NF-kB, a member of the TNFR superfamily. *J. Biol. Chem.*, **273**, 34120.
120. Jang, I. K., Lee, Z. H., Kim, Y. J., Kim, S. H., and Kwon, B. S. (1998). Human 4-IBB (CD137) signals are mediated by TRAF2 and activate nuclear factor-kappa B. *Biochem. Biophys. Res. Commun.*, **26**, 613.
121. Scaffidi, C., Fulda, S., Srinivasan, A., Friesen, C., Li, F., Tomaselli, K., *et al.* (1998). Two CD95 (APO-1/Fas) signaling pathways. *EMBO J.*, **17**, 1675.
122. Eguchi, Y., Srinivasan, A., Tomaselli, K. J., Shimizu, S., and Tsujimoto, Y. (1999). ATP-dependent steps in apoptotic signal transduction. *Cancer Res.*, **59**, 2174.
123. Rochat-Steiner, V., Becker, K., Micheau, O., Schneider, P., Burns, K., and Tschopp, J. (2000). FIST/HIPK3. A Fas/FADD-interacting serine/threonine kinase that induces FADD phosphorylation and inhibits Fas-mediated jun nh(2)-terminal kinase activation. *J. Exp. Med.*, **192**, 1165.
124. Thome, M., Schneider, P., Hofmann, K., Fickenscher, H., Meinl, E., Neipel, F., *et al.* (1997). Viral-FLICE-Inhibitory proteins (FLIPs) prevent apoptosis induced by death receptors. *Nature*, **386**, 517.
125. Tschop, J., Irmler, M., and Thome, M. (1998). Inhibition of Fas death signals by FLIPs. *Curr. Opin. Immunol.*, **10**, 552.

126. Scaffidi, C., Schmitz, I., Krammer, P. H., and Peter, M. E. (1999). The role of c-FLIP in modulation of CD95-induced apoptosis. *J. Biol. Chem.*, **274**, 1541.
127. Yeh, W. C., Pompa, J. L., McCurrach, M. E., Shu, H. B., Elia, A. J., Shahinian, A., *et al.* (1998). FADD: essential for embryo development and signalling from some, but not all, inducers of apoptosis. *Science*, **279**, 1954.
128. Los, M., Wesselborg, S., and Schulze-Osthoff, K. (1999). The role of caspases in development, immunity and apoptotic signal transduction: Lessons from knockout mice. *Immunity*, **10**, 629.
129. Zornig, M., Hueber, A-O., and Evan, G. (1998). P53-dependent impairment fo T-cell proliferation in FADD dominant-negative transgenic mice. *Curr. Biol.*, **8**, 467.
130. Newton, K., Harris, A. W., Bath, M. L., Smith, K. G. C., and Strasser, A. (1998). A dominant interfering mutant of FADD/MORT1 enhances deletion of autoreactive thymocytes and inhibits proliferation of mature T-lymphocytes. *EMBO J.*, **17**, 706.
131. Rothe, M., Pan, M-G., Henzel, W. J., Ayres, T. M., and Goeddel, D. V. (1995). The TNFR2-TRAF signalling complex contains two novel proteins related to baculoviral inhibitor of apoptosis proteins. *Cell*, **83**, 1243.
132. LaCasse, E. C., Baird, S., Korneluk, R. G., and MacKenzie, A. E. (1998). The inhibitors of apoptosis (IAPs) and their emerging role in cancer. *Oncogene*, **17**, 3247.
133. Wang, C. Y., Mayo, M. W., Korneluk, R. G., Goeddel, D. V., and Baldwin, A. S. (1998). NF-kappaB antiapoptosis: induction of TRAF1 and TRAF2 and cIAP1 and cIAP2 to suppress caspase 8 activation. *Science*, **281**, 1680.
134. Clem, R. J. and Miller, L. K. (1994). Control of programmed cell death by the baculovirus genes p35 and iap. *Mol. Cell. Biol.*, **14**, 5212.
135. Fraser, A. G., James, C., Evan, G. I., and Hengartner, M. O. (1999). *Caenorhabditis elegans* inhibitor of apoptosis protein (IAP) homologue BIR-1 plays a conserved role in cytokinesis. *Curr. Biol.*, **9**, 292.
136. Fengzhi, L., Ambrosini, G., Chu, E. Y., Plescia, J., Tognin, S., Marchisio, P. C., *et al.* (1998). Control of apoptosis and mitotic spindle checkpoint by survivin. *Nature*, **396**, 580.
137. Jiang, Y., Woronicz, J. D., Liu, W., and Goeddel, D. V. (1999). Prevention of constitutive TNF receptor 1 signaling by silencer of death domains. *Science*, **283**, 543.
138. Cohen, P. L. and Eisenberg, R. A. (1992). The lpr and gld genes in systemic autoimmunity: life and death in the Fas lane. *Immunol. Today*, **13**, 427.
139. Adachi, M., Watanabe Fukunaga, R., and Nagata, S. (1993). Aberrant transcription caused by the insertion of an early transposable element in an intron of the Fas antigen gene of lpr mice. *Proc. Natl. Acad. Sci. USA*, **90**, 1756.
140. Takahashi, T., Tanaka, M., Brannan, C. I., Jenkins, N. A., Copeland, N. G., Suda, T., *et al.* (1994). Generalized lymphoproliferative disease in mice, caused by a point mutation in the Fas ligand. *Cell*, **76**, 969.
141. Daniel, P. T. and Krammer, P. H. (1994). Activation induces sensitivity toward APO-1 (CD95)-mediated apoptosis in human B-cells. *J. Immunol.*, **15**, 25624.
142. Vignaux, F. and Golstein, P. (1994). Fas-based lymphocyte-mediated cytotoxicity against syngeneic activated lymphocytes: a regulatory pathway? *Eur. J. Immunol.*, **24**, 923.
143. Zheng, L., Fisher, G., Miller, R. E., Peschon, J., Lynch, D. H., and Lenardo, M. J. (1995). Induction of apoptosis in mature T-cells by tumour necrosis factor. *Nature*, **377**, 348.
144. Vaishnaw, A. K., Ornlinick, J. R., Chu, J-L., Krammer, P. H., Chao, M. V., and Elkon, K. B. (1999). The molecular basis for apoptotic defects in patients with CD95 (Fas/Apo-1) mutations. *J. Clin. Invest.*, **103**, 355.

145. Fischer, G. H., Rosenberg, F. J., Straus, S. E., Dale, J. K., Middleton, L. A., Lin, A. Y. *et al.* (1995). Dominant interfering Fas gene mutations impair apoptosis in a human lymphoproliferative syndrome. *Cell*, **81**, 935.
146. Ameison, J. C. and Capron, A. (1991). Cell dysfunction and depletion in AIDS: the programmed cell death hypothesis. *Immunol. Today*, **12**, 102.
147. Laurent Crawford, A. G., Krust, B., Muller, S., Riviere, Y., Rey Cuille, M. A., Bechet, J. M., *et al.* (1991). The cytopathic effect of HIV is associated with apoptosis. *Virology*, **185**, 829.
148. Martin, S. J., Matear, P. M., and Vyakarnam, A. (1994). HIV-1 infection of human CD4+ T cells *in vitro*. Differential induction of apoptosis in these cells. *J. Immunol.*, **152**, 330.
149. Cameron, P. U., Pope, M., Gezelter, S., and Steinman, R. M. (1994). Infection and apoptotic cell death of CD4+ T cells during an immune response to HIV-1-pulsed dendritic cells. *AIDS Res. Hum. Retroviruses*, **10**, 61.
150. Lu, Y. Y., Koga, Y., Tanaka, K., Sasaki, M., Kimura, G., and Nomoto, K. (1994). Apoptosis induced in CD4+ cells expressing gp160 of human immunodeficiency virus type 1. *J. Virol.*, **68**, 390.
151. Tani, Y., Tian, H., Lane, H. C., and Cohen, D. I. (1993). Normal T cell receptor-mediated signaling in T cell lines stably expressing HIV-1 envelope glycoproteins. *J. Immunol.*, **151**, 7337.
152. Banda, N. K., Bernier, J., Kurahara, D. K., Kurrle, R., Haigwood, N., Sekaly, R. P., *et al.* (1992). Crosslinking CD4 by human immunodeficiency virus gp120 primes T cells for activation-induced apoptosis. *J. Exp. Med.*, **176**, 1099.
153. Berndt, C., Mopps, B., Angermuller, S., Gierschik, P., and Krammer, P. H. (1999). CXCR4 and CD4 mediate a novel type of apoptosis in CD4 T-cells. *Proc. Natl. Acad. Sci. USA* **95**, 12556.
154. Westendorp, M. O., Frank, R., Ochsenbauer, C., Sticker, K., Dhein, J., Walczk, H., *et al.* (1995). Sensitisation of T-cells to CD95-mediated apoptosis by HIV-1 Tat and gp120. *Nature*, **375**, 497.
155. Baumler, C., Hus, I., Bohler, T., Krammer, P. H., and Debatin, K. M. (1996). Increased *ex vivo* expression of the CD95 ligand in T-cells of HIV infected children. *Blood*, **88**, 1741.
156. Falk, M. H., Trauth, B. C., Debatin, K. M., Klas, C., Gregory, C. D., Rickinson, A. B., *et al.* (1992). Expression of the APO-1 antigen in Burkitt lymphoma cell lines correlates with a shift towards a lymphoblastoid phenotype. *Blood*, **79**, 3300.
157. Walker, P. R., Saas, P., and Dietrich, P-Y. (1998). Tumour expression of Fas ligand (CD95L) and the consequences. *Curr. Opin. Immunol.*, **10**, 564.
158. Hueber, A. O., Zornig, M., Lyon, D., Suda, T., Nagata, S., and Evan, G. I. (1997). Requirement for the CD95 receptor-ligand pathway in c-Myc-induced apoptisis. *Science*, **278**, 1305.
159. Juin, P., Hueber, A-O., Littlewood, T., and Evan, G. I. (1999). c-Myc-induced sensitisation to apoptosis is mediated through cytochrome c release. *Genes Dev.*, **13**, 1367.
160. Muller, M. S., Strnad, S., Hug, H., Heinemann, M., Walczak, H., Hofmann, W. J., *et al.* (1997). Drug-induced apoptosis in hepatoma cells is mediated by the CD95 (Apo-1/Fas) receptor ligand system and involves activation of wild type p53. *J. Clin. Invest.*, **99**, 403.
161. Debatin, K. M. (1997). Cytotxic drugs, programmed cell death and the immune system: defining new roles in an old play. *J. Natl. Cancer Inst.*, **89**, 750.
162. Muller, M., Wilder, S., Bannasch, D., Israeli, D., Lehlbach, K., Li-Weber, M., *et al.* (1998). p53 activates the CD95 (APO-1/Fas) gene in response to DNA damage by anticancer drugs. *J. Exp. Med.*, **188**, 2033.

163. Wesselborg, S., Engels, I. H., Rossmann, E., Los, M., and Schulze-Osthoff, K. (1999). Anticancer drugs induced caspase-8/FLICE activation and apoptosis in the absence of CD95receptor/ligand interaction. *Blood*, **93**, 3053.
164. Micheau, O., Solary, E., Hammann, A., and Dimanche-Boitrel, M-T. (1998). Fas-ligand-independent, FADD-mediated activation of the Fas death pathway by anticancer drugs. *J. Biol. Chem.*, **274**, 7987.

9 | Survival signalling by phosphorylation: PI3K/Akt sets the stage

THOMAS F. FRANKE

1. Introduction

Recent research has examined Akt and Akt-related serine/threonine kinases in signalling cascades that regulate cell survival and are important in the pathogenesis of degenerative diseases and in cancer. We will recapitulate the research that has helped to define the current understanding of the role of the Akt pathway under normal and pathological conditions. In particular, we will evaluate the mechanisms of Akt regulation and the role of Akt substrates in Akt-dependent biological responses in the decisions of cell death and cell survival. Here, we hope to establish mechanisms of apoptosis suppression by Akt kinase as a framework for a more general understanding of growth factor-dependent regulation of cell survival.

2. Structure of Akt and Akt-related kinases

Akt is the gene product of the cellular homologue of the *v-akt* oncogene transduced by AKT8, an acute transforming retrovirus in mice that was originally described in 1977 (1). Research over the past decade has not only elucidated the regulation of Akt kinase by upstream signalling events, mainly as a consequence of activation of the second messenger phospholipid kinase phosphatidylinositol 3-kinase (PI3K), but also defined a role for Akt in promoting cell survival in a variety of apoptotic paradigms (for review see refs 2–4). In 1991, three independent research groups cloned and characterized Akt. The group of Philip Tsichlis identified *v-akt* as the gene transduced by rodent retrovirus AKT8 (5) and subsequently that its cellular homologue, then named *c-akt*, encoded the cytoplasmic serine/threonine protein kinase Akt (6). Also in 1991, two European research groups in Switzerland and England identified Akt and related kinases when searching for novel kinases related to protein kinases A and C (7, 8). As a consequence, Akt is today frequently referred to as protein kinase B (PKB, as a kinase similar to protein kinases A and C) or in older literature sometimes described as RAC-PK (a protein kinase related to protein kinases A and C).

The Akt serine/threonine kinase consists of an N-terminal regulatory domain resembling a pleckstrin homology (PH) domain (9), a hinge region connecting the PH domain to a kinase domain with serine/threonine specificity (6), and a C-terminal region required for the induction and maintenance of its kinase activity (for review see ref. 4). In mammals, three closely related isoforms of Akt are encoded by distinct genetic loci: Akt (that in fact is Akt1), Akt2 and its differently spliced variants, and Akt3 (for review see ref. 3). Whereas Akt (Akt1) is ubiquitously expressed at high levels with the exception of kidney, liver, and spleen (6–8), Akt2 expression varies between different tissues, with higher expression levels in muscle, intestinal organs, and reproductive tissues (10, 11). Akt3 is expressed the highest in brain and testis and lower expression levels are found in intestinal organs and muscle tissue (12). It is conceivable that the different Akt isoforms are functionally redundant since they are activated similarly and phosphorylate downstream substrates with equal specificity and efficiency (13). Of the different isoforms of human AKT, human AKT2 and AKT3 are of particular clinical interest since gene amplifications and increases in basal activity are found in several human cancer diseases, suggesting an involvement of these isoforms in their pathogenesis (14, 15).

3. Rediscovery of a proto-oncogene

Akt kinase was discovered as the cellular homologue of a viral oncogene, implicating a pivotal role for Akt in cell growth and survival that has been verified in many different cell systems. Oncogenic transformation that involves Akt has not been linked to any singular event, but rather to multiple causes that coincide with the gain-of-function of the Akt signalling cascade. These changes lead to increased activity of Akt kinase, suppression of apoptosis and differentiation, and to increased cell cycle progression. The pathological conversion of the physiological pathway may be caused by several factors (see Fig. 1) that have the potential to co-operate. First, the extracellular signals that trigger the intracellular activation of Akt might be increased, or the receptors that receive the signal might be altered or present in increased numbers. Second, the activity of intracellular mediators of Akt activation that include G proteins including Ras and its regulators, second messenger-generating phosphatidylinositol-3-kinases (PI3K), and other molecules are increased either by oncogenic mutation or increased expression. Third, control mechanisms that regulate the signal leading to Akt activity are diminished by mutation or deletion. In particular, the cancer-associated finding of frequent mutations and deletions in the tumour suppressor PTEN/MMAC1, a phosphatase that dephosphorylates and inactivates the second messenger molecules leading to Akt activation, has further supported the critical role of PI3K/Akt signal transduction in the control of cell growth and proliferation (for review see ref. 16). Finally, in mammary, ovarian, pancreatic, and prostate cancer, the expression of Akt kinases is enhanced or altered (14, 15, 17–19). Taken together, Akt dysregulation—both indirect and direct—that

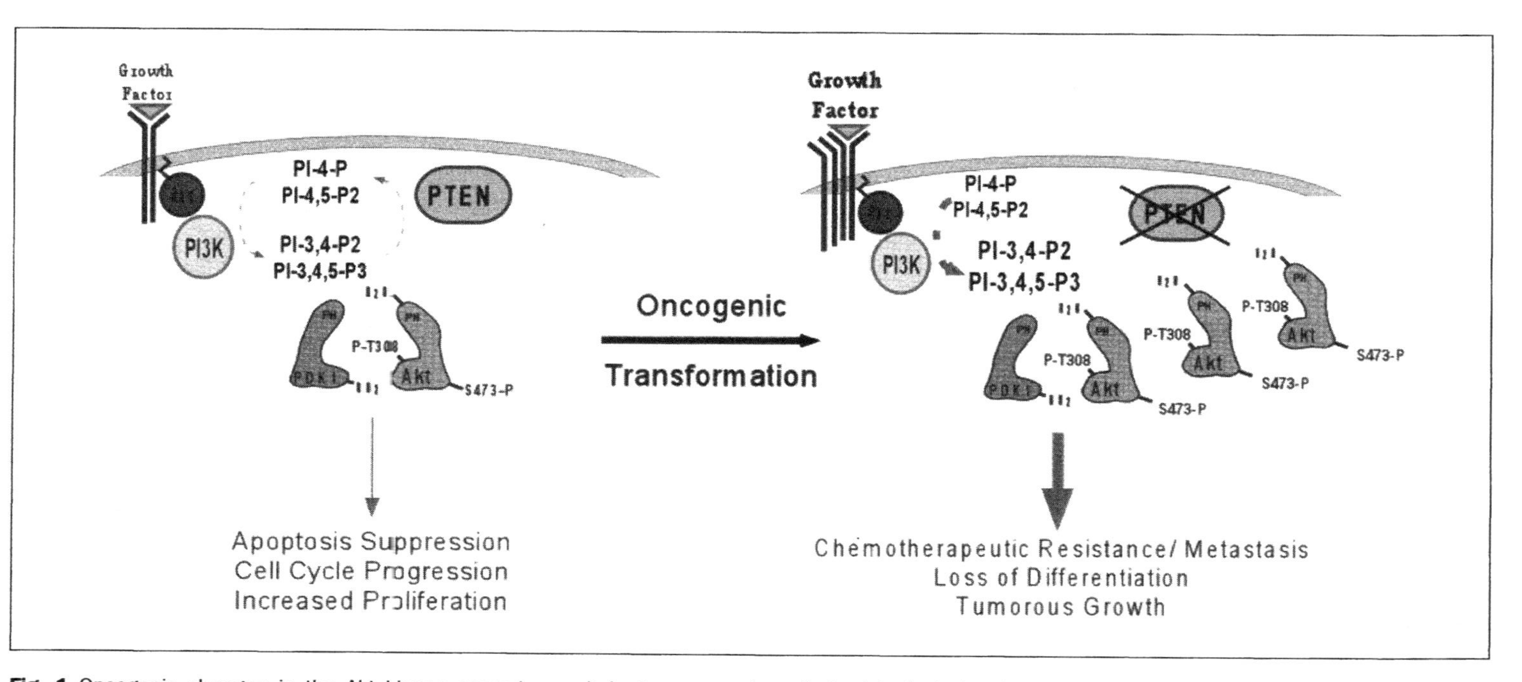

Fig. 1 Oncogenic changes in the Akt kinase cascade result in the conversion of physiological signal transduction processes (shown on the left) into pathologically altered growth, differentiation, and apoptosis regulation (shown on the right). Factors that result in oncogenic changes include increased growth factor production, receptor expression or expression of mutant receptor species, increased PI3K activation, loss of PTEN function, and amplified expression of Akt kinases. In some types of cancer, PTEN expression is frequently affected and the lack of PTEN activity results in constitutive Akt activity and dysregulation of downstream biological responses.

results in an increased signalling strength has been found in a wide variety of human tumours. It should, however, be remembered that except for direct amplification, none of the oncogenic factors involved in Akt activation are truly specific, and it is likely that some or all of the above changes in upstream signalling molecules will affect both Akt signalling and that of other parallel pathways.

A second argument for the biological relevance of the Akt signalling cascade is derived from the extent of evolutionary conservation of functional elements and their interrelation within the Akt pathway. Structural homologues of Akt kinases exist in all vertebrate and invertebrate species examined to date and they include Akt-related kinases in *Drosophila melanogaster*, *Caenorhabditis elegans*, and *Dictyostelium discoides* (9, 20, 21). Genetic analysis in *D. melanogaster* supports the relevance of the Akt signalling cascade as a critical regulatory mechanism of cell growth, and experiments using genetic mutants suggest a conserved role for Akt in apoptosis regulation (23). These findings are contrasted by studies of Akt-related protein kinases in the slime mould *D. discoides*, where Akt-homologue kinases are primarily involved in the regulation of chemotaxis and cell motility (21). The mechanisms of Akt activation in *D. discoides* are somewhat different from the mechanisms of Akt activation that prevail in animal cells, however, because in *D. discoides* Akt activity is primarily activated by upstream cAMP-dependent kinases (21), whereas in animal cells it is activated by PI3k (for review see ref. 22).

In *C. elegans*, genetic complementation groups have contributed significantly to the analysis and understanding of the Akt signal transduction pathway. A comparison of the Akt signalling pathways in *C. elegans* and mammals indicates strong evolutionary conservation of the pathway (see Fig. 2) (20). For example, the upstream regulatory molecules involved in activation of Akt are highly similar, including DAF-18 (24), a structural homologue of PTEN that inhibits Akt signalling downstream of activated PI3K. Furthermore, specific transcription factor substrates downstream of Akt are conserved to the extent that their phosphorylation site motifs are similar between nematodes and humans. These studies have not only provided a genetic model to complement biochemical studies of Akt signal transduction in vertebrate cells, but also defined a functional end-point for Akt signalling in the regulation of animal lifespan, suggesting some yet to be further clarified role for PI3K/Akt signal transduction in animal ageing (also see below). It appears, however, that the transcriptional consequences of activated Akt in *C. elegans* and mammalian cells and their biological consequences are distinct. In *C. elegans*, Akt activation is likely to result in the down-regulation of the cytoplasmic catalase Ctl-1 and in the acceleration of the exit from the *dauer* stage, both reducing the overall lifespan of animals (25). In mammalian cells, increased Akt activity leads to down-regulation of pro-apoptotic factors and an increased apoptosis resistance of cells (for review see ref. 22). It remains to be determined if increased Akt activity in mammalian cells also impinges on animal lifespan. At least in animals, where insulin-dependent metabolic responses are increased as a consequence of nutritional overload, overall animal life expectancy is decreased, supporting the genetic evidence in *C. elegans* (26), but in contrast to the protective role of Akt at the cellular level.

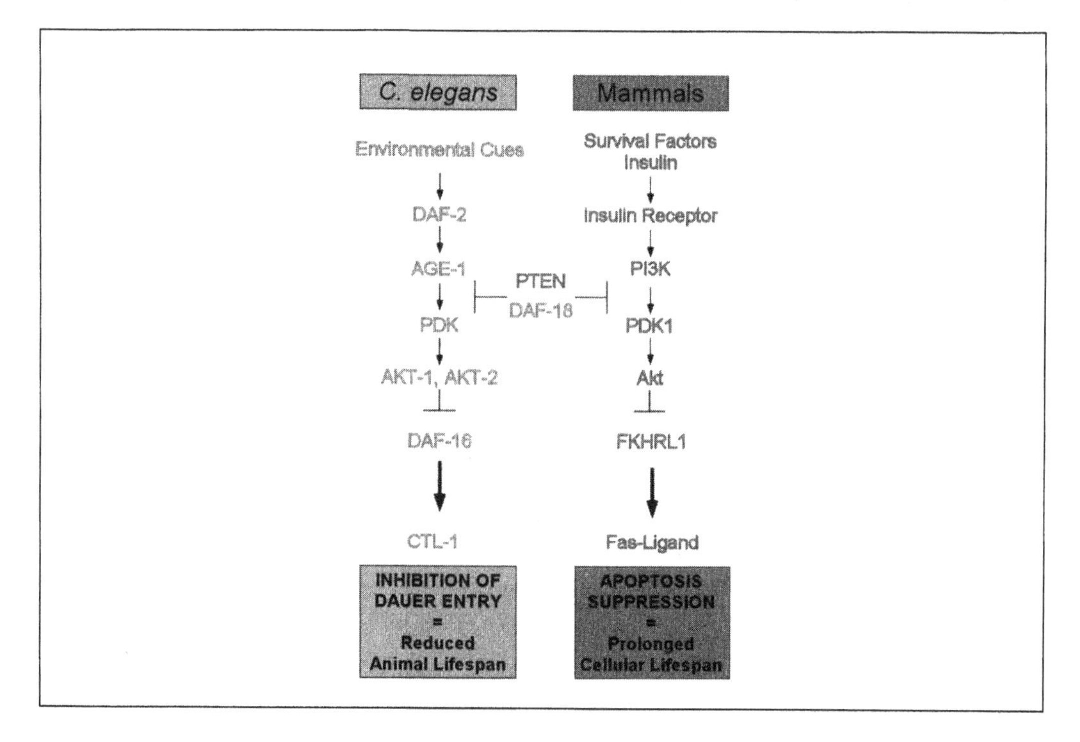

Fig. 2 The relevance of the Akt signalling cascade to eukaryotic signal transduction is underscored by a direct comparison of PI3K/Akt signalling cascades in *C. elegans* and in mammals. All steps leading from the activation of insulin receptor homologues (DAF-2) to the regulation of specific transcription factors as direct substrates of Akt (DAF-16) are preserved. PI3K (AGE-1), PTEN (DAF-18), and PDK1 (PDK) homologues contribute similarly activating or inhibiting events. In contrast to mammalian Akt-dependent signal transduction, Akt signal transduction in *C. elegans* is primarily involved in lifespan determination. In mammals Akt regulates several DAF-16 homologues. FKHR and FKHRL1 are directly involved in apoptosis and act as inducers of cell death when not phosphorylated by Akt. It is not known whether any of the wing-helix transcription factors in mammals have functions that resemble the lifespan-regulating function of DAF-16.

4. Akt as a direct downstream target of PI3K

Studies examining the role of PI3K, mainly by the use of inhibitors, have underscored the role of signals originating from activated PI3K in biological responses that encompass a wide diversity of functions ranging from the regulation of mitogenesis, cell growth and proliferation, cell survival, and cell motility, to the oxidative burst in neutrophils (for review see refs 27 and 28). Even though p70 ribosomal S6-kinase (p70S6K) was the first downstream effector of PI3K to be identified, Akt is the best understood. Critical to the understanding of the regulation of Akt in cells was the finding that Akt kinase activity is induced following PI3K activation in various growth factor receptor-mediated signalling cascades (29–31). Although PI3K has been recognized as an important signal transducer, the direct molecular consequences of PI3K activation have only recently been elucidated. The reason was most likely related to technical difficulties in examining the kinase activity of PI3K that is a phospholipid kinase and primarily phosphorylates plasma membrane-bound phospholipids rather than protein substrates. PI3K phosphorylates phosphoinositides on the 3′-OH position of the inositol ring, and second messenger products of the kinase reaction in growth factor-stimulated animal cells are phosphatidylinositol 3,4′-bisphosphate (PI-3,4-P2) and phosphatidylinositol 3,4,5′-trisphosphate (PIP3) (for review see ref. 32). It is likely that other cellular mechanisms of phosphoinositide generation exist as PI-3,4-P2 has been found elevated independently of PI3K activity. Recent findings of phosphoinositide-specific phospholipid phosphatases further increase our understanding of second messenger signalling by phospholipids and the regulation of phosphoinositide metabolism in cells (for review see ref. 16).

The exact mechanism of Akt activation by PI3K in cells has been studied thoroughly and it is initiated by the binding of specific 3′-phosphorylated phosphoinositides to the Akt PH domain (33, 34). In contrast to PH domains in other molecules, which specifically bind the PI3K product PIP3, the Akt PH domain binds to both PI-3,4-P2 and PIP3, and shows a relatively higher affinity for the binding of PI-3,4-P2. PI-3-P is another product of PI3K *in vitro*, but it can bind only poorly to the Akt PH domain—suggesting that the function of PI-3-P in cells is distinct from that of growth factor-induced second messengers such as PI-3,4-P2 and PIP3 (for review also see ref. 32). The binding specificities of different PH domains for different phosphoinositides have been confirmed by structural studies, and the specificity of the Akt PH domain for both PI-3,4-P2 and PIP3 could be relevant in achieving an extended signal of Akt activity in cells (for review see ref. 32).

An important consequence of the initial binding of Akt to the plasma membrane-bound phosphoinositides is the relocalization of the cytoplasmic Akt protein to signalling complexes at the plasma membrane. Deletion and mutation studies have found that the PH domain is critical in mediating this relocalization, in agreement with its abilities for binding phospholipid second messenger molecules. The finding of increased activity of an N-terminal deleted Akt mutant has been interpreted as an indication of intramolecular inhibition, but no evidence has been found that suggests an interaction of N- and C-terminal regions of the Akt molecule. Thus, it remains possible that phosphoinositide binding causes an initial conformational change in Akt to establish a basic kinase activity (33). The significance of phosphoinositide binding for Akt activation is underscored by findings that those mutations in the Akt PH domain that interfere with its phosphoinositide binding also abolish Akt activation by growth factor receptor signalling and activated PI3K. Consequently, oncogenic mutations of Akt that result in a constitutive, factor-independent activation include constitutive plasma membrane binding by N-terminal myristoylation and also explain the constitutive activity of v-akt, presumably by circumventing the initial translocation step (29, 35).

At the plasma membrane, phosphorylation of a threonine residue (residue threonine-308 in mouse Akt) in the catalytic loop of Akt potently activates the enzyme further. This phosphorylation event resembles the phosphorylation that is found in many AGC kinases [protein kinase A (PKA), G (PKG), and C (PKC)] and in all AGC kinases this is conferred by phosphoinositide-dependent kinase 1 (PDK1) (36, 37). PDK1 activity depends upon phospholipid products of PI3K and the binding of 3′-phosphorylated phosphoinositides to the PDK1 PH domain has been shown to determine its localization to the plasma membrane (38, 39). Thus, the two-step mechanism of Akt activation requires first phosphoinositide binding and subsequent translocation to the plasma membrane, followed by phosphorylation of Akt by an upstream kinase (PDK1) that it is common to the activation of Akt and other kinases in PI3K-dependent signalling systems (for review see ref. 3). Consistent with the role of PI3K products and the PH domain in facilitating the access of PDK1 to a threonine residue in the catalytic loop, PH domain-deleted Akt is phosphorylated by PDK1 even in the absence of PI3K products. Full activation of Akt requires the phosphorylation of a second serine residue in the C-terminus of the Akt enzyme (residue serine 473 in mouse Akt), and the second phosphorylation event also depends upon the activity of a kinase that is phosphoinositide-dependent (for review see ref. 40). The phosphorylation motif surrounding the catalytic site residue that is phosphorylated by PDK1 and conserved between different AGC kinases that are substrates of PDK1 is sufficiently different from that surrounding the C-terminal serine residue in Akt. Thus, in keeping with the established nomenclature, a distinct phosphoinositide-dependent kinase 'PDK2' activity was originally proposed to explain the phosphorylation of the C-terminal residue (40).

No additional PDK-type kinase has been isolated, however, that can explain the PDK2-like activity in cells. Experimental evidence suggests that PDK1 can phosphorylate the C-terminal residue *in vitro* in the presence of PIF peptide, raising the possibility that PDK1, under certain conditions, might pose as a PDK2 kinase by undergoing an unprecedented specificity switch after the initial phosphorylation of the catalytic site residue in Akt (41, 42). Experiments using PDK1-deficient embryonic fibroblasts cells strongly support the critical role of PDK1 for the phosphorylation of both the catalytic loop threonine 308 and of the C-terminal residue in Akt (43). Recent data, however, have suggested that the C-terminal residue is autophosphorylated by Akt itself, either in *cis* or in *trans*, in a mechanism that requires pre-phosphorylation of Akt by PDK1 (44), similar to the mechanism required for the maturation of PKC isoforms (45). Even though these data need to be expanded further toward a quantitative and structural analysis of the sequence of phosphorylation events, they present a simpler model for activation of Akt kinases. Studies of PDK1-deficient mouse embryonic fibroblast strongly support the importance of PI3K-dependent PDK1 activation for the induction of Akt activity in cells (43). Genetic models in *C. elegans* and *D. melanogaster* also underscore the importance of these upstream kinases for Akt signalling *in vivo*, suggesting that our current model of a primarily PI3K/PDK1-dependent activation of Akt is accurate (for review see ref. 16). Nevertheless, PI3K/PDK1-independent pathways of Akt activation may be important in mammalian cells under certain physiological conditions.

5. PDK1-independent mechanisms of Akt activation

Alternative mechanisms of Akt activation exist that are independent of PDK1 and sometimes even do not involve PI3K. One mechanism involves integrin-linked kinase (ILK1) (for review see ref. 46). Co-expression of Akt with ILK1 in cells induces phosphorylation of Akt on serine 473, and expression of an ATP binding site mutant form of ILK1 together with Akt interferes with cell survival signalling and with Akt-dependent oncogenesis in prostate cancer cells (47). The assumption that ILK1 phosphorylates Akt on the C-terminal residue serine 473 is an attractive explanation to link Akt signal transduction to those cell–matrix interactions that have significant consequences for cell survival (47–49). However, the ability of ILK1 to function not only as an adaptor molecule, but also as a protein kinase, is still controversial after alignments of the sequence of ILK1 with that of other kinases (50). In particular, ILK1 is missing residues that are found to be critical in other kinases to catalyse phosphotransfer reactions (50).

Other Akt-interacting proteins include proto-oncogenic proteins, such as Tcl-1, which have recently been shown to bind to the Akt PH domain, and to amplify PI3K-dependent signalling through complex formation with Akt or by altering Akt localization in cells (51, 52). These molecules are particularly interesting since their interaction with Akt is isoform-specific: both Tcl-1 and the related MTCP1 bind primarily AKT and not AKT2 (51, 52). The exact contribution and relative importance of this novel pathway of Akt regulation, however, has yet to be determined.

A growing number of studies have reported mechanisms of PI3K-independent induction of Akt activity. Ca^{2+}/calmodulin-dependent kinase kinase (CaM-KK) phosphorylates Akt on threonine 308, independently of PI3K, to trigger Akt-dependent survival responses (53). Protein kinase A itself is a substrate for PDK1-dependent phosphorylation, and activates Akt in a PI3K-independent manner by a mechanism that does not result in Akt phosphorylation on serine 473 (54, 55). Other reports have shown PI3K-independent activation of Akt by oxidative stress (56) and MAPKAP kinase-2, a stress-activated kinase that is able to phosphorylate Akt on serine 473 *in vitro* (57). Recent studies, however, contradict earlier findings by underscoring the importance of growth factor signalling and PI3K activity in stress-mediated Akt activation (58). Finally, several different heat shock proteins including hsp25/27 and hsp90 associate with Akt, suggesting a role in the reported activation of Akt following exposure of cells to heat shock and other stresses, perhaps by altering the accessibility of Akt to inactivating protein phosphatases (59, 60).

6. Negative regulators of PI3K/Akt signalling

Phosphorylation is critical in Akt activation by upstream kinases and important for the maintenance of its activity. Thus, it is not surprising that several protein phosphatases inactivate Akt by dephosphorylation. Non-specific inhibition of endogenous phosphatase activity efficiently activates Akt (61) and phosphatase activity is involved in mediating effects of extracellular stresses on Akt signal transduction (62, 63). Dephosphorylation of Akt is also observed after increased ceramide levels, possibly due to inhibition of PI3K or other upstream signalling molecules (64, 65). Insights into the importance of phosphatases for the regulation of Akt activity have also been obtained from the study of phosphatases that not only dephosphorylate proteins, but also the phosphoinositide products of PI3K. Of these, the SH2-containing inositol phosphatases SHIP1/2 and the PTEN phosphatase (also known as MMAC1 and TEP1) are critical mediators in determining Akt activity. SHIP1 dephosphorylates inositides and phosphoinositides on the 5′-position and regulates B cell/myeloid cell function (66–68), whereas SHIP-2 controls the intracellular levels of Akt-activating phosphoinositides in non-haematopoietic cell types (69).

The PTEN gene was originally identified as a tumour suppressor gene and is frequently mutated or deleted in both inherited and spontaneous human cancer diseases (for review see refs 16 and 70). Although substrate specificity predictions based on structural comparisons initially predicted that PTEN might be a phosphotyrosine-specific phosphatase, Maehama and Dixon (71) showed that PTEN is a phosphoinositide phosphatase that efficiently dephosphorylates PI3K products at the 3′-position. A great number of papers have analysed the expression and function of PTEN in human cancer cell lines and in primary human tumours (for review see refs 16 and 70). Since PTEN dephosphorylates the phosphoinositide products of PI3K that otherwise trigger the activation of Akt and other signalling pathways downstream of activated PI3K, inactivating mutations or loss of PTEN expression lead to increased levels of PI3K products in cells. These increased levels of second messenger molecules result in a constitutive signal leading to enhanced Akt activity and to increased cell cycle progression, apoptosis resistance, and oncogenic transformation (for review see ref. 16). These findings explain the tumour suppressor activity of PTEN: PTEN counteracts the pro-oncogenic effects of elevated PI3K activity by decreasing the intracellular levels of Akt-activating phospholipids (see also Fig. 1). Additional support for the importance of PTEN in normal cell function is obtained from transgenic animal studies (72, 73) and from studies that examine the function of PTEN in controlling the threshold of Akt signalling in *C. elegans* (24).

7. Apoptosis suppression by Akt

An important function of activated PI3K in cells is the inhibition of programmed cell death (74, 75), and Akt is a good candidate for mediating these PI3K-dependent cell survival responses. The initial evidence that Akt acted as an anti-apoptotic signalling molecule was shown in cerebellar granule neurons after trophic factor withdrawal (76) and in fibroblasts after exogenous induction of c-Myc (77). Subsequent work in many laboratories has further established the principle role of Akt in the regulation of cell survival, consistent with its ubiquitous expression pattern. Akt has been implicated as an anti-apoptotic in many different cell types and death paradigms, including withdrawal of extracellular signalling factors, oxidative and osmotic stress, irradiation and treatment of cells with chemotherapeutic drugs, and ischaemic shock (for review see refs 3 and 22). Multiple studies supporting the role of Akt in apoptosis suppression have connected Akt to cell death regulation either by demonstrating its down-regulation following pro-apoptotic insults, or by using gene transfer experiments that transduce both activated, anti-apoptotic and inactive, pro-apoptotic mutants of Akt.

Taken together, these observations suggest that Akt may play a critical role both in the function of cancer cells and in the pathogenesis of degenerative diseases. By promoting cell survival of mutated, damaged, or transformed cells even under adverse conditions, Akt can promote cancer cell growth by protecting cells that would otherwise be eliminated by apoptosis. To experimentally prove the importance of Akt kinases in oncogenic transformation, Peter Vogt and colleagues recently demonstrated that a transformed cellular phenotype could be reverted in a cell model for PI3K-dependent oncogenesis when dominant-negative mutants of Akt were expressed concomitantly (78). Akt is also likely to play a significant role in degenerative diseases where excessive or inappropriate cell death occurs, possibly because proper trophic factor support is lacking. The relevance of Akt signalling in neurodegenerative disease is supported by recent studies that examine its activity and function in Alzheimer disease models (79, 80). A role for Akt has also been suggested in other models of human degenerative diseases including cardiac failure (81), diseases where there is increased and chronic loss of cells (for review see ref. 82). Finally, an increasing amount of work has implicated Akt in apoptotic processes that are caused by a loss of directed cell–cell interaction and not by soluble factors (for review see ref. 83). However, not all survival signals necessarily require the PI3K/Akt pathway. Additional studies are required to characterize survival pathways that are independent of PI3K/Akt (84–86).

8. Protein substrates of Akt

The first molecular insights into the function of Akt were derived from studies directed at its role in insulin-dependent metabolic responses. Thus, the very first molecular function associated with proto-oncogenic Akt kinase turned out to be within the glucose metabolism. When searching for kinases that could regulate glycogen synthase kinase-3 (GSK3), the groups of Brian Hemmings and Phil Cohen realized that Akt inhibited GSK3 activity downstream of insulin-activated PI3K by direct phosphorylation of an N-terminal regulatory serine residue (87). GSK3 is also involved in consequences of Akt signalling that are not primarily related to metabolic responses (87): GSK3 is involved in regulating anti-apoptotic pathways (88, 89), and in regulating cell cycle progression by phosphorylation of cyclin D1 (90). However, other kinases besides Akt have been implicated in GSK3 regulation (91).

The identification of GSK3 as an Akt substrate helped to define the residues adjacent to serine or threonine residues that are required for Akt phosphorylation to occur (92). By systematically permutating the amino acid sequence surrounding the Akt phosphorylation site in GSK3, Alessi *et al.* (92) derived an optimal peptide sequence for Akt phosphorylation. Indeed, most of the known Akt substrates contain a phosphorylation site similar to that found in GSK3 (see also Table 1). To date

Table 1 Akt substrates and their phosphorylation sites[a]

Substrate		Phosphorylation site		Function
(1a) Enzymatic substrates that are activated by Akt phosphorylation (Group Ia)				
PFK2	460	**VRMRRNSFT**	468	Activation in insulin-dependent metabolic responses
PFK2	477	**RRPRNYSVG**	485	
hTERT	221	**ARRRGGSAS**	229	DNA damage control and regulation of genomic ageing
hTERT	818	**VRIRGKSYV**	826	
eNOS	1171	**SRIRTQSFS**	1179	Nitric oxide production
PDE	267	**IRPRRRSSC**	275	Regulation of intracellular cAMP levels
IKKα	17	**MRERLGTGG**	25	Targeting IκB for degradation
(1b) Enzymatic substrates that are inhibited by Akt phosphorylation (Group Ib)				
GSK3α	15	**GRARTSSFA**	23	Inhibition of glycogen synthesis
GSK3β	3	**GRPRTTSFA**	11	
Caspase-9	190	**LRRRFSSLH**	198	Inhibition of cytochrome *c*-induced cell death in cancer cells
(2) Akt substrates that are regulated by binding to 14-3-3 proteins (Group II)				
mBAD	106	**TRSRHSSYP**	114	Promotion of Bcl-x_L-dependent cell survival
mBAD	130	**FRGRSRSAP**	138	
huBAD	69	**IRSRHSSYP**	77	
huBAD	93	**FRGRSRSAP**	101	
AFX	187	**PRRRAASMD**	195	Cell cycle progression
AFX	252	**FRPRSSSNA**	260	
FKHRL1	26	**SRPRSCTWP**	34	Inhibition of cell death by reducing transcription of Fas-ligand
FKHRL1	247	**PRRRAVSMD**	255	
(3) Other substrates				
CREB	129	**SRRPSYR**	135	Induction of CREB-dependent transcription
Histone 2B	31	**KRSRKESYS**	40	Unknown
Akt peptide	1	**RPRAATF**	7	Optimized Akt substrate peptide

[a] Akt activation results in the direct phosphorylation of specific protein substrates. Two different groups of substrates can be distinguished: (I) Akt either directly activates or inhibits the enzymatic activity of a protein substrate, or (II) substrate function is affected indirectly by binding to 14-3-3 molecules that determine the intracellular localization of phosphorylated substrates. Experimentally confirmed Akt substrates are listed and their Akt phosphorylation site(s) are included. The relative positions of the flanking amino acids within the phosphorylation motif of the (poly)-peptide are listed. Residues that agree with the principal motif (RXRXXS/T) defined by Alessi *et al.* (92) are shaded. Consequences of Akt phosphorylation of these proteins for the regulation of specific biological functions are mentioned where applicable.

several additional Akt substrates have been identified and even more have been predicted and await experimental verification (for review see ref. 93). Although recent research has confirmed the accurateness of the original phosphorylation motif (94), the power of predicting possible downstream substrates solely based on a phosphorylation motif remains questionable as constraints other than the primary peptide sequence can influence kinase–substrate interactions.

Several Akt substrates are involved in downstream metabolic consequences of insulin stimulation in cells, including enzymes such as 6-phosphofructo-2-kinase (95) as well as winged-helix, a forkhead-related transcription factor involved in the regulation of insulin-dependent transcription (96, 97). In *C. elegans*, the forkhead-related transcription factor DAF16 is a substrate for Akt in an insulin-like receptor pathway that is involved in regulation of the exit from the *dauer* stage (20). Akt activity in insulin-dependent responses is also involved in the regulation of translocation and/or expression of glucose transporters including GLUT4 or GLUT1 (98, 99). Akt has also been implicated in the regulation of phosphodiesterase by direct phosphorylation of phosphodiesterase 3B (100), regulation of telomerase activity (101), induction of nitric oxide synthase activity (102–104), and in several other biological functions that are related to metabolic control (for review see ref. 3).

9. Substrate regulation by Akt-dependent phosphorylation

Since protein kinases act on downstream protein substrates by adding phosphate groups to serine/threonine, tyrosine, or histidine residues, it is not very surprising that the identification of such downstream substrates for Akt has received a great deal of attention by many researchers. Over 40 Akt downstream substrates have been uncovered to date and still more are being added to this list (for review see ref. 93). We do not plan to review every substrate here, but to focus on a selected number of substrates that are primarily involved in regulation of programmed cell death.

Substrates of Akt can be classified into two general categories (here called Group I and Group II) depending upon the consequences of Akt-dependent phosphorylation for their biological function. For Group I substrates, Akt phosphorylation either increases (Group Ia) or decreases (Group Ib) the activity of enzymatic protein substrates. For Group II substrates, Akt phosphorylation increases the affinity of the substrate for interaction with 14-3-3 proteins. 14-3-3 proteins are abundantly expressed in the cytoplasm and specifically bind phosphoserine/threonine-containing polypeptides, and retain phosphorylated Akt substrates in the cytosol.

Examples of Akt phosphorylation events that inhibit substrate enzymatic activity include the inhibition of different GSK3 isoforms (87) and cell death protease caspase-9 in human cells (105). Examples of downstream enzymes that are activated by Akt include the activation of PFK2 (95) and nitric oxide synthase (102–104). Furthermore, increased FRAP/mTOR kinase activity after Akt stimulation leads to increased 4E-BP1 phosphorylation and changes in the translation of specific RNA species (106, 107). Increased telomerase activity has also been linked to Akt activation (101) as well as increased phosphodiesterase activity (108, 109). Finally, the regulation of NF-κB transcriptional activity is regulated by PI3K/Akt (110–113) and involves direct phosphorylation and regulation of IκB-kinase (IKK)-α activity by Akt (114).

In addition, specific kinases in mitogen-activated protein kinase (MAPK) and p38/JunK-type stress-regulated kinase cascades including Raf-1, B-Raf, and apoptosis signal-regulating kinase 1 (ASK1) are functionally inhibited following Akt-dependent phosphorylation (115–118). It is not clear, however, whether Akt phosphorylation directly inhibits the activity of these kinases, or whether Akt determines their cellular function indirectly by affecting their cellular localization and/or their interaction with accessory molecules. For the regulation of the serine/threonine kinase Raf-1, Akt-dependent inhibition plays an important role in the cellular coordination and integration of different branches of Ras-dependent signalling, especially in relation to cell differentiation (for review see ref. 119).

10. A BAD kinase makes good

One of the first targets of Akt identified that has direct implications for regulating cell survival is the pro-apoptotic Bcl-2-family member BAD. When it is not phosphorylated, BAD will inhibit Bcl-X_L and other anti-apoptotic Bcl-2 family members by direct binding (for review see ref. 120). Once phosphorylated, however, the phosphoserine residues of BAD form high affinity binding sites for 14-3-3 molecules, thus localizing phosphorylated BAD to the cytosol and effectively neutralizing its pro-apoptotic activity (121). In addition, phosphorylation of BAD changes its affinity for interaction with Bcl-X_L molecules, an idea that is supported by the observation that a BAD phosphorylation site mutant resembling a constitutive phosphorylation event is unable to inhibit Bcl-X_L function, even though it is not bound to 14-3-3 proteins. Two of the phosphorylated serine residues in BAD that are involved in mediating its interactions with 14-3-3 resemble an Akt phosphorylation motif (Group II substrate). Two studies initially demonstrated that Akt was able to phosphorylate BAD both in IL-3-dependent myeloid cells (122), and also in fibroblasts and primary neuronal cells (123). These studies suggested that Akt was directly involved in the regulation of survival by phosphorylation of BAD (for review see ref. 124). Additional studies have subsequently linked Akt activity to BAD phosphorylation (53, 65, 84, 125–127), but other studies have pointed out a dissociation of Akt activity and BAD phosphorylation in certain signalling models (128–130). In fact, a recent report suggests that kinases other than Akt actually mediate BAD phosphorylation in some IL-3-dependent cell models (131). Today, the list of kinases that have been shown to regulate BAD phosphorylation includes calcium/calmodulin-dependent protein kinase kinase (CaM-KK), MAPK/ERK kinase (MEK), protein kinase A, protein kinase C, p21-activated kinases (PAK1 and PAK4), cytokine-independent survival kinase (CISK), and Rsks (130–139). Not all of these kinases phosphorylate BAD on the 14-3-3 binding sites; indeed, an additional phosphorylation site is present in BAD and not involved in binding to 14-3-3 proteins (140). Rather, its phosphorylation facilitates binding to 14-3-3 proteins, possibly by forcing the BAD protein into an optimal protein conformation for 14-3-3 binding (141). Thus, even though Akt was the first BAD kinase to be identified, apoptosis regulation by BAD phosphorylation has emerged as prime example of a physiological mechanism that integrates kinase cascades with apoptosis regulation. In that, it is not too surprising that several different signalling pathways exist that lead to BAD phosphorylation.

One could argue that BAD phosphorylation has emerged as a general anti-apoptotic regulatory mechanism that co-ordinates the cellular activation state with cell survival. Both PI3K-dependent pathways, as well as PI3K-independent pathways (such as those that lead to increased levels of intracellular cAMP or increased intracellular PAK activity) impinge on BAD phosphorylation. In addition, phosphorylation of BAD as a regulatory mechanism appears to be evolutionary conserved as its BH3 domain and the 14-3-3 consensus binding sites are present in all known BAD homologues (T. Franke, unpublished). Finally, phosphatases have been identified including calcineurin that inhibit Akt-dependent cell survival by dephosphorylation of BAD (142).

11. Other substrates of Akt in apoptosis regulation

Although the mechanism of BAD phosphorylation by Akt provides an elegant model of apoptosis regulation, it is most likely not the only mechanism by which Akt can promote cell survival. One reason is that Akt expression is fairly ubiquitous, whereas BAD expression itself is fairly restricted (143). Furthermore, several cell types have been examined that do not express BAD protein, but in which Akt still efficiently suppresses apoptosis. Additional targets of Akt in cell survival have been postulated and several candidate protein substrates have already been identified and characterized (for review see ref. 93) (a selection is shown in Fig. 3).

One set of targets may include proteins involved in the mitochondrial pathway of apoptosis. Akt not only plays a role in keeping cytochrome *c* in the mitochondria, but its activity also inhibits the response of cells to cytochrome *c* after it is released from the mitochondria (105, 144). Although caspase-9 appears to be a target of Akt in human cells (145), and could explain this cytochrome *c* resistance (105), it is not the only, or most important, target, as Akt-dependent cytochrome *c* resistance is observed in other vertebrate cell models in which caspase-9 lacks an Akt phosphorylation site (146, 147).

An important subset of Akt substrates that is, like BAD, regulated by phosphorylation-dependent binding to 14-3-3 proteins, includes the forkhead-related family of mammalian transcription factors. The *C. elegans* forkhead-related DAF-16 protein transduces insulin receptor-like signals in a PI3K/Akt-homologous signalling cascade (20). In mammalian cells, forkhead-related transcription factors are retained within the cytoplasm when phosphorylated and bound to 14-3-3. Loss of Akt activity results in their dephosphorylation and translocation to the nucleus, where they then induce transcriptional events (93, 96, 97, 148–152). It has been suggested that transcriptional events depending on forkhead-related transcription factors are involved in pro-apoptotic responses. Indeed, several studies have already demonstrated how Akt inhibits the transcription of specific pro-apoptotic genes by retaining forkhead-related protein in the cytoplasm (93, 151). Furthermore, studies in primary human tumours and cell lines suggest that nuclear exclusion of forkhead-related transcription factors is frequent and often associated with oncogenic transformation (18, 153–155). It remains to be determined whether this exclusion is caused primarily by increased Akt activity or by other kinases that also phosphorylate these proteins (156).

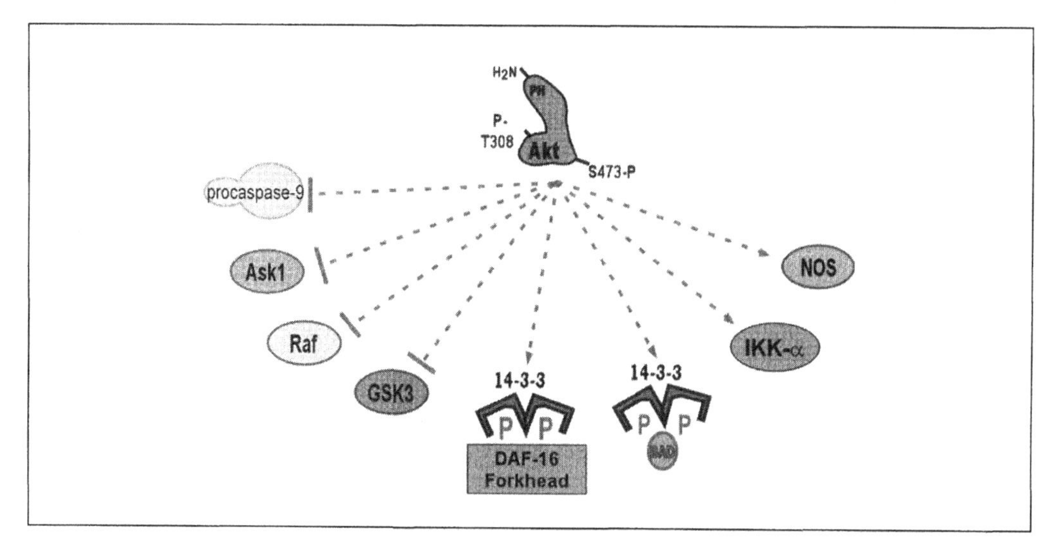

Fig. 3 Several Akt protein substrates have been defined (see also Table 1). Three classes of substrates can be distinguished. They include enzymes that are (Ia) activated or (Ib) inhibited following Akt phosphorylation, and (II) substrates that, when phosphorylated, bind to 14-3-3 proteins in the cytoplasm. The last mechanism is particularly important for the regulation of the pro-apoptotic protein BAD and for nuclear exclusion of Akt-dependent transcription factors that are related to DAF-16 forkhead-related transcription factors.

Another important class of Akt targets is those proteins involved in the stress-activated/mitogen-activated protein kinase (SAPK/MAPK) cascades. Growing experimental evidence points to a close functional interrelation between the Akt survival pathway and SAPK/MAPK cascades that are activated by various cellular stresses and linked to apoptosis. Increased Akt activity has been shown to suppress the JNK and p38 pathways in a number of cell systems (157–159). Recently, it has been shown that ASK1 is regulated by Akt and contains an Akt-specific phosphorylation site (118). Thus, ASK1 is likely to be one of the points of convergence between PI3K/Akt signalling and stress-activated kinases cascades, although probably not the only one. For example, Akt also phosphorylates and regulates the small G protein Rac1, an activator of the JNK pathway (160).

12. Conclusions

When considering the current understanding of all the signals leading to and from Akt, we face a growing complexity that is in part compounded by the intersection of different signalling cascades. Many substrates of Akt are shared with other kinases that have similar specificities, but alternative activation mechanisms. Moreover, signals originating from activated Akt do not simply lead to changes in the biological activity of specific downstream substrates, but affect whole signalling cascades, including changes in the activity of transcription factors. In mammalian cells, known points of convergence between different signalling cascades include BAD, GSK3, CREB, and forkhead-related transcription factors. In invertebrate genetic models of

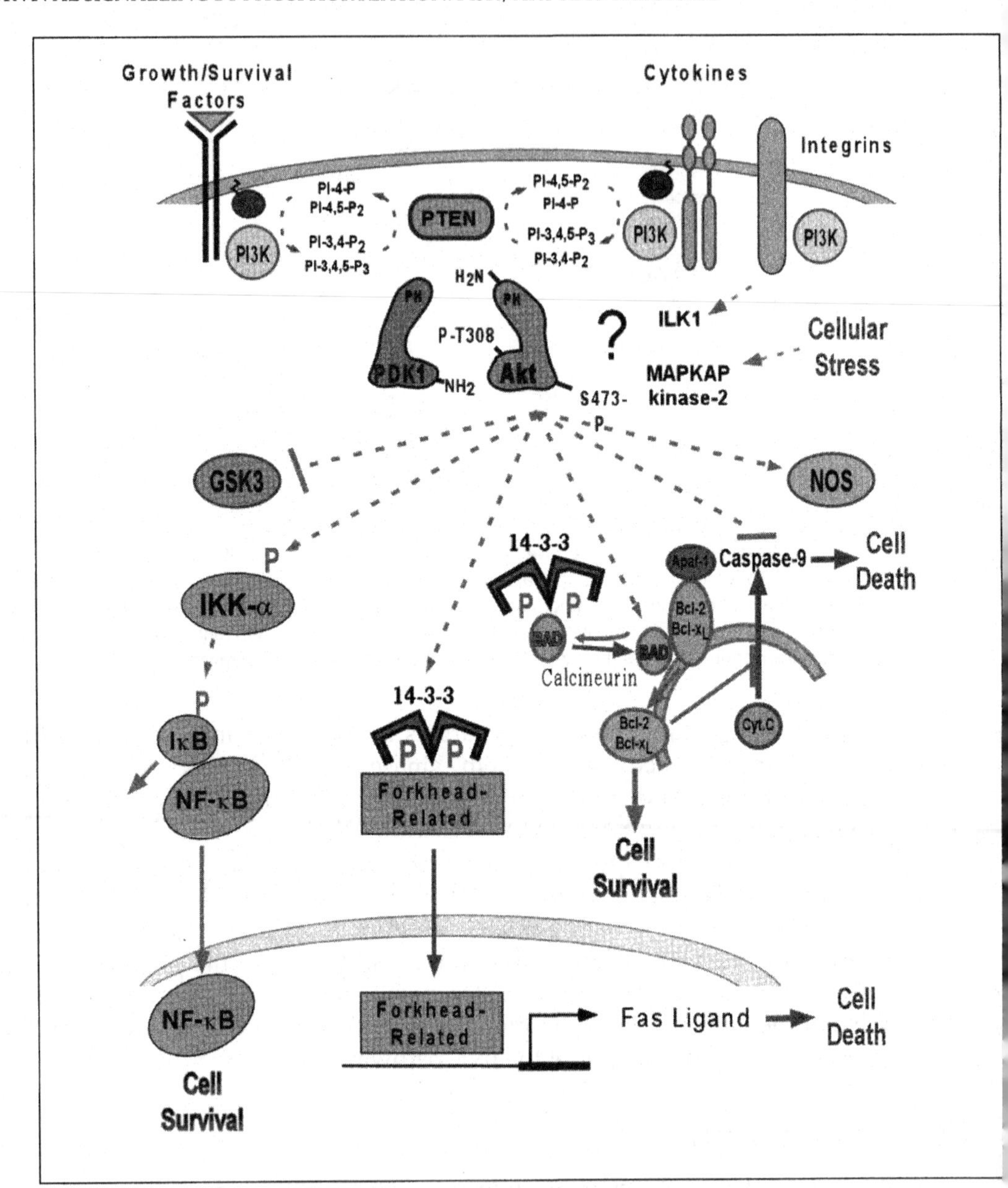

Fig. 4 This figure summarizes our current understanding of Akt activation and the downstream consequences of its activation on cell growth and survival. Extracellular signals that include growth factors, cytokines, integrin-mediated signals, and other survival factors induce the activity of intracellular phosphoinositide 3-kinase (PI3K), leading to the generation of phospholipid products that act as second messenger molecules (PI-3,4-P_2 and PI-3,4,5-P_3). These products activate PDK1 and Akt. Activated Akt phosphorylates enzymatic substrates including GSK3, caspase-9, NOS, and IKK-α, resulting in inhibition (GSK3, caspase-9) or activation (NOS, IKK-α) of enzymatic activity. Phosphorylation of the pro-apoptotic factor BAD on serine residues results in dissociation from Bcl-x_L and association with 14-3-3. This allows Bcl-x_L to assume its function of suppressing cell death. Phosphorylation of serine and threonine residues in forkhead-related transcription factors also induces association with 14-3-3 proteins, resulting in nuclear exclusion. See text for details.

signal transduction cascades, however, findings of overlapping and intersecting pathways are rare (for review see ref. 161). It thus remains a possibility that some of the current complexity of PI3K/Akt signal transduction in vertebrates may be an artifact of *in vitro* studies (see also Fig. 4).

The growing number of Akt substrates discovered complicates our understanding. It is therefore important to determine which of these substrates are genuine, which are the more important, and how Akt independently regulates each one. As it stands now, it appears that multiple substrates are phosphorylated simultaneously in any given system, suggesting that Akt regulates multiple downstream functions in parallel. How Akt does this, however, is not clear.

Pharmacological inhibitors of Akt are likely to become important tools in examining Akt function in cells and modulating Akt-dependent downstream cell responses. Recent studies that have identified endogenous inhibitors of PI3K/Akt signal transduction have significantly fuelled and accelerated the effort to identify and characterize synthetic small molecule compounds. Such reagents, along with more detailed analysis of Akt signal transduction *in vivo*, should help to clarify the role of Akt in transducing survival signals.

Acknowledgements

Dr Franke is supported in part through Career Development Award DAMD17-0-1-0214.

References

1. Staal, S. P., Hartley, J. W., and Rowe, W. P. (1977). Isolation of transforming murine leukemia viruses from mice with a high incidence of spontaneous lymphoma. *Proc. Natl. Acad. Sci. USA*, 74, 3065–7.
2. Alessi, D. R. and Cohen, P. (1998). Mechanism of activation and function of protein kinase B. *Curr. Opin. Genet. Dev.*, **8**, 55–62.
3. Downward, J. (1998). Mechanisms and consequences of activation of protein kinase B/Akt. *Curr. Opin. Cell Biol.*, **10**, 262–7.
4. Chan, T. O., Rittenhouse, S. E., and Tsichlis, P. N. (1999). AKT/PKB and other D3 phosphoinositide-regulated kinases: kinase activation by phosphoinositide-dependent phosphorylation. *Annu. Rev. Biochem.*, **68**, 965–1014.
5. Bellacosa, A., Testa, J. R., Staal, S. P., and Tsichlis, P. N. (1991). A retroviral oncogene, *akt*, encoding a serine-threonine kinase containing an SH2-like region. *Science*, **254**, 274–7.
6. Bellacosa, A., Franke, T. F., Gonzales-Portal, M. E., Datta, K., Taguchi, T., Gardner, J., *et al.* (1993). Structure, expression and chromosomal mapping of c-*akt*: relationship to v-*akt* and its implications. *Oncogene*, **8**, 745–54.
7. Coffer, P. J. and Woodgett, J. R. (1991). Molecular cloning and characterisation of a novel putative protein-serine kinase related to the cAMP-dependent and protein kinase C families. *Eur. J. Biochem.*, **201**, 475–81.
8. Jones, P. F., Jakubowicz, T., Pitossi, F. J., Maurer, F., and Hemmings, B. A. (1991). Molecular cloning and identification of a serine/threonine protein kinase of the second-messenger subfamily. *Proc. Natl. Acad. Sci. USA*, **88**, 4171–5.

9. Franke, T. F., Tartof, K. D., and Tsichlis, P. N. (1994). The SH2-like *Akt* homology (AH) domain of c-*akt* is present in multiple copies in the genome of vertebrate and invertebrate eucaryotes. Cloning and characterization of the *Drosophila melanogaster* c-*akt* homolog *Dakt1*. *Oncogene*, **9**, 141–8.
10. Jones, P. F., Jakubowicz, T., and Hemmings, B. A. (1991). Molecular cloning of a second form of *rac* protein kinase. *Cell Reg.*, **2**, 1001–9.
11. Konishi, H., Shinomura, T., Kuroda, S.-i., Ono, Y., and Kikkawa, U. (1994). Molecular cloning of rat rac protein kinase α and β and their association with protein kinase C ζ. *Biochem. Biophys. Res. Commun.*, **205**, 817–25.
12. Nakatani, K., Sakaue, H., Thompson, D. A., Weigel, R. J., and Roth, R. A. (1999). Identification of a human Akt3 (protein kinase B gamma) which contains the regulatory serine phosphorylation site. *Biochem. Biophys. Res. Commun.*, **257**, 906–10.
13. Franke, T. F. (2000). Assays for Akt. In *Methods in enzymology* (ed. J. C. Reed), Vol. 322, pp. 400–10. Academic Press, San Diego.
14. Cheng, J. Q., Ruggeri, B., Klein, W. M., Sonoda, G., Altomare, D. A., Watson, D. K., *et al.* (1996). Amplification of *AKT2* in human pancreatic cancer cells and inhibition of *AKT2* expression and tumorigenicity by antisense RNA. *Proc. Natl. Acad. Sci. USA*, **93**, 3636–41.
15. Nakatani, K., Thompson, D. A., Barthel, A., Sakaue, H., Liu, W., Weigel, R. J., *et al.* (1999). Up-regulation of Akt3 in estrogen receptor-deficient breast cancers and androgen-independent prostate cancer lines. *J. Biol. Chem.*, **274**, 21528–32.
16. Cantley, L. C. and Neel, B. G. (1999). New insights into tumor suppression: PTEN suppresses tumor formation by restraining the phosphoinositide 3-kinase/AKT pathway. *Proc. Natl. Acad. Sci. USA*, **96**, 4240–5.
17. Bellacosa, A., de Feo, D., Godwin, A. K., Bell, D. W., Cheng, J. Q., Altomare, D. A., *et al.* (1995). Molecular alterations of the AKT2 oncogene in ovarian and breast carcinomas. *Int. J. Cancer*, **64**, 280–5.
18. Graff, J. R., Konicek, B. W., McNulty, A. M., Wang, Z., Houck, K., Allen, S., *et al.* (2000). Increased AKT activity contributes to prostate cancer progression by dramatically accelerating prostate tumor growth and diminishing p27Kip1 expression. *J. Biol. Chem.*, **275**, 24500–5.
19. Cheng, J. Q., Godwin, A. K., Bellacosa, A., Taguchi, T., Franke, T. F., Hamilton, T. C., *et al.* (1992). *AKT2*, a putative oncogene encoding a member of a subfamily of protein-serine/threonine kinases, is amplified in human ovarian carcinomas. *Proc. Natl. Acad. Sci. USA*, **89**, 9267–71.
20. Paradis, S. and Ruvkun, G. (1998). *Caenorhabditis elegans* Akt/PKB transduces insulin receptor-like signals from AGE-1 PI3 kinase to the DAF-16 transcription factor. *Genes Dev.*, **12**, 2488–98.
21. Meili, R., Ellsworth, C., Lee, S., Reddy, T. B., Ma, H., and Firtel, R. A. (1999). Chemoattractant-mediated transient activation and membrane localization of Akt/PKB is required for efficient chemotaxis to cAMP in Dictyostelium. *EMBO J.*, **18**, 2092–105.
22. Franke, T. F., Kaplan, D. R., and Cantley, L. C. (1997). PI3K: Downstream Aktion blocks apoptosis. *Cell*, **88**, 435–7.
23. Staveley, B. E., Ruel, L., Jin, J., Stambolic, V., Mastronardi, F. G., Heitzler, P., *et al.* (1998). Genetic analysis of protein kinase B (AKT) in *Drosophila*. *Curr. Biol.*, **8**, 599–602.
24. Ogg, S. and Ruvkun, G. (1998). The *C. elegans* PTEN homolog, DAF-18, acts in the insulin receptor-like metabolic signaling pathway. *Mol. Cell*, **2**, 887–93.

25. Taub, J., Lau, J. F., Ma, C., Hahn, J. H., Hoque, R., Rothblatt, J., *et al.* (1999). A cytosolic catalase is needed to extend adult lifespan in *C. elegans* daf-C and clk-1 mutants. *Nature*, **399**, 162–6.
26. Finkel, T. and Holbrook, N. J. (2000). Oxidants, oxidative stress and the biology of ageing. *Nature*, **408**, 239–47.
27. Kapeller, R. and Cantley, L. C. (1994). Phosphatidylinositol 3-kinase. *Bioessays*, **16**, 565–76.
28. Carpenter, C. L. and Cantley, L. C. (1996). Phosphoinositide kinases. *Curr. Opin. Cell Biol.*, **8**, 153–8.
29. Franke, T. F., Yang, S.-I., Chan, T. O., Datta, K., Kazlauskas, A., Morrison, D. K., *et al.* (1995). The protein kinase encoded by the *Akt* proto-oncogene is a target of the PDGF-activated phosphatidylinositol 3-kinase. *Cell*, **81**, 727–36.
30. Burgering, B. M. T. and Coffer, P. J. (1995). Protein kinase B (c-Akt) in phosphatidylinositol-3-OH kinase signal transduction. *Nature*, **376**, 599–602.
31. Kohn, A. D., Kovacina, K. S., and Roth, R. A. (1995). Insulin stimulates the kinase activity of RAC-PK, a pleckstrin homology domain containing ser/thr kinase. *EMBO J.*, **14**, 4288–95.
32. Rameh, L. E. and Cantley, L. C. (1999). The role of phosphoinositide 3-kinase lipid products in cell function. *J. Biol. Chem.*, **274**, 8347–50.
33. Franke, T. F., Kaplan, D. R., Cantley, L. C., and Toker, A. (1997). Direct regulation of the *Akt* proto-oncogene by phosphatidylinositol-3,4-bisphosphate. *Science*, **275**, 665–8.
34. Frech, M., Andjelkovic, M., Ingley, E., Reddy, K. K., Falck, J. R., and Hemmings, B. A. (1997). High affinity binding of inositol phosphates and phosphoinositides to the pleckstrin homology domain of RAC/protein kinase B and their influence on kinase activity. *J. Biol. Chem.*, **272**, 8474–81.
35. Andjelkovic, M., Alessi, D. R., Meier, R., Fernandez, A., Lamb, N. J., Frech, M., *et al.* (1997). Role of translocation in the activation and function of protein kinase B. *J. Biol. Chem.*, **272**, 31515–24.
36. Stephens, L., Anderson, K., Stokoe, D., Erdjument-Bromage, H., Painter, G. F., Holmes, A. B., *et al.* (1998). Protein kinase B kinases that mediate phosphatidylinositol 3,4,5-trisphosphate- dependent activation of protein kinase B. *Science*, **279**, 710.
37. Alessi, D. R., James, S. R., Downes, C. P., Holmes, A. B., Gaffney, P. R. J., Reese, C. B., *et al.* (1997). Characterization of a 3-phosphoinositide-dependent protein kinase which phosphorylates and activates protein kinase Bα. *Curr. Biol.*, **7**, 261–9.
38. Anderson, K. E., Coadwell, J., Stephens, L. R., and Hawkins, P. T. (1998). Translocation of PDK-1 to the plasma membrane is important in allowing PDK-1 to activate protein kinase B. *Curr. Biol.*, **8**, 684–91.
39. Filippa, N., Sable, C. L., Hemmings, B. A., and Van Obberghen, E. (2000). Effect of phosphoinositide-dependent kinase 1 on protein kinase B translocation and its subsequent activation. *Mol. Cell. Biol.*, **20**, 5712–21.
40. Belham, C., Wu, S., and Avruch, J. (1999). Intracellular signalling: PDK1–a kinase at the hub of things. *Curr. Biol.*, **9**, R93–6.
41. Balendran, A., Casamayor, A., Deak, M., Paterson, A., Gaffney, P., Currie, R., *et al.* (1999). PDK1 acquires PDK2 activity in the presence of a synthetic peptide derived from the carboxyl terminus of PRK2. *Curr. Biol.*, **9**, 393–404.
42. Biondi, R. M., Cheung, P. C., Casamayor, A., Deak, M., Currie, R. A., and Alessi, D. R. (2000). Identification of a pocket in the PDK1 kinase domain that interacts with PIF and the C-terminal residues of PKA. *EMBO J.*, **19**, 979–88.

43. Williams, M. R., Arthur, J. S., Balendran, A., van der Kaay, J., Poli, V., Cohen, P., *et al.* (2000). The role of 3-phosphoinositide-dependent protein kinase 1 in activating AGC kinases defined in embryonic stem cells. *Curr. Biol.*, **10**, 439–48.
44. Toker, A. and Newton, A. C. (2000). Akt/protein kinase B is regulated by autophosphorylation at the hypothetical PDK-2 site. *J. Biol. Chem.*, **275**, 8271–4.
45. Dutil, E. M., Toker, A., and Newton, A. C. (1998). Regulation of conventional protein kinase C isozymes by phosphoinositide-dependent kinase 1 (PDK-1). *Curr. Biol.*, **8**, 1366–75.
46. Yoganathan, T. N., Costello, P., Chen, X., Jabali, M., Yan, J., Leung, D., *et al.* (2000). Integrin-linked kinase (ILK): a 'hot' therapeutic target. *Biochem. Pharmacol.*, **60**, 1115–19.
47. Persad, S., Attwell, S., Gray, V., Delcommenne, M., Troussard, A., Sanghera, J., *et al.* (2000). Inhibition of integrin-linked kinase (ILK) suppresses activation of protein kinase B/Akt and induces cell cycle arrest and apoptosis of PTEN-mutant prostate cancer cells. *Proc. Natl. Acad. Sci. USA*, **97**, 3207–12.
48. Attwell, S., Roskelley, C., and Dedhar, S. (2000). The integrin-linked kinase (ILK) suppresses anoikis. *Oncogene*, **19**, 3811–15.
49. D'Amico, M., Hulit, J., Amanatullah, D. F., Zafonte, B. T., Albanese, C., Bouzahzah, B., *et al.* (2000). The integrin-linked kinase regulates the cyclin D1 gene through glycogen synthase kinase 3beta and cAMP-responsive element-binding protein-dependent pathways. *J. Biol. Chem.*, **275**, 32649–57.
50. Lynch, D. K., Ellis, C. A., Edwards, P. A., and Hiles, I. D. (1999). Integrin-linked kinase regulates phosphorylation of serine 473 of protein kinase B by an indirect mechanism. *Oncogene*, **18**, 8024–32.
51. Pekarsky, Y., Koval, A., Hallas, C., Bichi, R., Tresini, M., Malstrom, S., *et al.* (2000). Tcl1 enhances Akt kinase activity and mediates its nuclear translocation. *Proc. Natl. Acad. Sci. USA*, **97**, 3028–33.
52. Laine, J., Kunstle, G., Obata, T., Sha, M., and Noguchi, M. (2000). The protooncogene TCL1 is an Akt kinase coactivator. *Mol. Cell*, **6**, 395–407.
53. Yano, S., Tokumitsu, H., and Soderling, T. R. (1998). Calcium promotes cell survival through CaM-K kinase activation of the protein-kinase-B pathway. *Nature*, **396**, 584–7.
54. Sable, C. L., Filippa, N., Hemmings, B., and Van Obberghen, E. (1997). cAMP stimulates protein kinase B in a Wortmannin-insensitive manner. *FEBS Lett.*, **409**, 253–7.
55. Filippa, N., Sable, C. L., Filloux, C., Hemmings, B., and Van Obberghen, E. (1999). Mechanism of protein kinase B activation by cyclic AMP-dependent protein kinase. *Mol. Cell. Biol.*, **19**, 4989–5000.
56. Konishi, H., Matsuzaki, H., Tanaka, M., Ono, Y., Tokunaga, C., Kuroda, S. i., *et al.* (1996). Activation of RAC-protein kinase by heat shock and hyperosmolarity stress through a pathway independent of phosphatidylinositol 3-kinase. *Proc. Natl. Acad. Sci. USA*, **93**, 7639–43.
57. Clifton, A. D., Young, P. R., and Cohen, P. (1996). A comparison of the substrate specificity of MAPKAP kinase-2 and MAPKAP kinase-3 and their activation by cytokines and cellular stress. *FEBS Lett.*, **392**, 209–14.
58. Shaw, M., Cohen, P., and Alessi, D. R. (1998). The activation of protein kinase B by H_2O_2 or heat shock is mediated by phosphoinositide 3-kinase and not by mitogen-activated protein kinase- activated protein kinase-2. *Biochem. J.*, **336**, 241–6.

59. Konishi, H., Matsuzaki, H., Tanaka, M., Takemura, Y., Kuroda, S., Ono, Y., *et al.* (1997). Activation of protein kinase B (Akt/RAC-protein kinase) by cellular stress and its association with heat shock protein Hsp27. *FEBS Lett.*, **410**, 493–8.
60. Sato, S., Fujita, N., and Tsuruo, T. (2000). Modulation of Akt kinase activity by binding to Hsp90. *Proc. Natl. Acad. Sci. USA*, **97**, 10832–7.
61. Andjelkovic, M., Jakubowicz, T., Cron, P., Ming, X.-F., Han, J.-W., and Hemmings, B. A. (1996). Activation and phosphorylation of a pleckstrin homology domain containing protein kinase (RAC-PK/PKB) promoted by serum and protein phosphatase inhibitors. *Proc. Natl. Acad. Sci. USA*, **93**, 5699–704.
62. Chen, D., Fucini, R. V., Olson, A. L., Hemmings, B. A., and Pessin, J. E. (1999). Osmotic shock inhibits insulin signaling by maintaining Akt/protein kinase B in an inactive dephosphorylated state. *Mol. Cell. Biol.*, **19**, 4684–94.
63. Meier, R., Thelen, M., and Hemmings, B. A. (1998). Inactivation and dephosphorylation of protein kinase Balpha (PKBalpha) promoted by hyperosmotic stress. *EMBO J.*, **17**, 7294–303.
64. Zhou, H., Summers, S. A., Birnbaum, M. J., and Pittman, R. N. (1998). Inhibition of Akt kinase by cell-permeable ceramide and its implications for ceramide-induced apoptosis. *J. Biol. Chem.*, **273**, 16568–75.
65. Zundel, W. and Giaccia, A. (1998). Inhibition of the anti-apoptotic PI(3)K/Akt/Bad pathway by stress. *Genes Dev.*, **12**, 1941–6.
66. Aman, M. J., Lamkin, T. D., Okada, H., Kurosaki, T., and Ravichandran, K. S. (1998). The inositol phosphatase SHIP inhibits Akt/PKB activation in B cells. *J. Biol. Chem.*, **273**, 33922–8.
67. Liu, Q., Sasaki, T., Kozieradzki, I., Wakeham, A., Itie, A., Dumont, D. J., *et al.* (1999). SHIP is a negative regulator of growth factor receptor-mediated PKB/Akt activation and myeloid cell survival. *Genes Dev.*, **13**, 786–91.
68. Astoul, E., Watton, S., and Cantrell, D. (1999). The dynamics of protein kinase B regulation during B cell antigen receptor engagement. *J. Cell Biol.*, **145**, 1511–20.
69. Taylor, V., Wong, M., Brandts, C., Reilly, L., Dean, N. M., Cowsert, L. M., *et al.* (2000). 5′ phospholipid phosphatase SHIP-2 causes protein kinase B inactivation and cell cycle arrest in glioblastoma cells. *Mol. Cell. Biol.*, **20**, 6860–71.
70. Myers, M. P. and Tonks, N. K. (1997). PTEN: sometimes taking it off can be better than putting it on. *Am. J. Hum. Genet.*, **61**, 1234–8.
71. Maehama, T. and Dixon, J. E. (1998). The tumor suppressor, PTEN/MMAC1, dephosphorylates the lipid second messenger, phosphatidylinositol 3,4,5-trisphosphate. *J. Biol. Chem.*, **273**, 13375–8.
72. Di Cristofano, A., Pesce, B., Cordon-Cardo, C., and Pandolfi, P. P. (1998). Pten is essential for embryonic development and tumour suppression. *Nature Genet.*, **19**, 348–55.
73. Stambolic, V., Suzuki, A., de la Pompa, J. L., Brothers, G. M., Mirtsos, C., Sasaki, T., *et al.* (1998). Negative regulation of PKB/Akt-dependent cell survival by the tumor suppressor PTEN. *Cell*, **95**, 29–39.
74. Yao, R. and Cooper, G. M. (1996). Growth factor-dependent survival of rodent fibroblasts requires phosphatidylinositol 3-kinase but is independent of pp70S6k activity. *Oncogene*, **13**, 343–51.
75. Yao, R. and Cooper, G. M. (1995). Requirement for phosphatidylinositol-3 kinase in the prevention of apoptosis by nerve growth factor. *Science*, **267**, 2003–6.

76. Dudek, H., Datta, S. R., Franke, T. F., Birnbaum, M. J., Yao, R., Cooper, G. M., *et al.* (1997). Regulation of neuronal survival by the serine-threonine protein kinase Akt. *Science*, **275**, 661–5.
77. Kauffmann-Zeh, A., Rodriguez-Viciana, P., Ulrich, E., Gilbert, C., Coffer, P., Downward, J., *et al.* (1997). Suppression of c-Myc-induced apoptosis by Ras signalling through PI(3)K and PKB. *Nature*, **385**, 544–8.
78. Aoki, M., Batista, O., Bellacosa, A., Tsichlis, P., and Vogt, P. K. (1998). The akt kinase: molecular determinants of oncogenicity. *Proc. Natl. Acad. Sci. USA*, **95**, 14950–5.
79. Hong, M. and Lee, V. M. (1997). Insulin and insulin-like growth factor-1 regulate tau phosphorylation in cultured human neurons. *J. Biol. Chem.*, **272**, 19547–53.
80. Weihl, C. C., Ghadge, G. D., Kennedy, S. G., Hay, N., Miller, R. J., and Roos, R. P. (1999). Mutant presenilin-1 induces apoptosis and downregulates Akt/PKB. *J. Neurosci.*, **19**, 5360–9.
81. Matsui, T., Li, L., del Monte, F., Fukui, Y., Franke, T. F., Hajjar, R. J., *et al.* (1999). Adenoviral gene transfer of activated phosphatidylinositol 3′-kinase and Akt inhibits apoptosis of hypoxic cardiomyocytes *in vitro*. *Circulation*, **100**, 2373–9.
82. Reed, J. C. and Paternostro, G. (1999). Postmitochondrial regulation of apoptosis during heart failure. *Proc. Natl. Acad. Sci. USA*, **96**, 7614–16.
83. Kundra, V., Escobedo, J. A., Kazlauskas, A., Kim, H. K., Rhee, S. G., Williams, L. T., *et al.* (1994). Regulation of chemotaxis by the platelet-derived growth factor receptor-β. *Nature*, **367**, 474–6.
84. Kulik, G. and Weber, M. J. (1998). Akt-dependent and -independent survival signaling pathways utilized by insulin-like growth factor I. *Mol. Cell. Biol.*, **18**, 6711–18.
85. Carson, J. P., Kulik, G., and Weber, M. J. (1999). Antiapoptotic signaling in LNCaP prostate cancer cells: a survival signaling pathway independent of phosphatidylinositol 3′-kinase and Akt/protein kinase B. *Cancer Res.*, **59**, 1449–53.
86. Philpott, K. L., McCarthy, M. J., Klippel, A., and Rubin, L. L. (1997). Activated phosphatidylinositol 3-kinase and Akt kinase promote survival of superior cervical neurons. *J. Cell Biol.*, **139**, 809–15.
87. Cross, D. A. E., Alessi, D. R., Cohen, P., Andjelkovich, M., and Hemmings, B. A. (1995). Inhibition of glycogen synthase kinase-3 by insulin mediated by protein kinase B. *Nature*, **378**, 785–9.
88. Ikeda, S., Kishida, S., Yamamoto, H., Murai, H., Koyama, S., and Kikuchi, A. (1998). Axin, a negative regulator of the Wnt signaling pathway, forms a complex with GSK-3beta and beta-catenin and promotes GSK-3beta- dependent phosphorylation of beta-catenin. *EMBO J.*, **17**, 1371–84.
89. Pap, M. and Cooper, G. M. (1998). Role of glycogen synthase kinase-3 in the phosphatidylinositol 3- kinase/Akt cell survival pathway. *J. Biol. Chem.*, **273**, 19929–32.
90. Diehl, J. A., Cheng, M., Roussel, M. F., and Sherr, C. J. (1998). Glycogen synthase kinase-3beta regulates cyclin D1 proteolysis and subcellular localization. *Genes Dev.*, **12**, 3499–511.
91. Shaw, M. and Cohen, P. (1999). Role of protein kinase B and the MAP kinase cascade in mediating the EGF-dependent inhibition of glycogen synthase kinase 3 in Swiss 3T3 cells. *FEBS Lett.*, **461**, 120–4.
92. Alessi, D. R., Caudwell, F. B., Andjelkovic, M., Hemmings, B. A., and Cohen, P. (1996). Molecular basis for the substrate specificity of protein kinase B; comparison with MAPKAP kinase-1 and p70 S6 kinase. *FEBS Lett.*, **399**, 333–8.

93. Datta, S. R., Brunet, A., and Greenberg, M. E. (1999). Cellular survival: a play in three Akts. *Genes Dev.*, **13**, 2905–27.
94. Obata, T., Yaffe, M. B., Leparc, G. G., Piro, E. T., Maegawa, H., Kashiwagi, A., *et al.* (2000). Peptide and protein library screening defines optimal substrate motifs for AKT/PKB. *J. Biol. Chem.*, **275**, 36108–15.
95. Deprez, J., Vertommen, D., Alessi, D. R., Hue, L., and Rider, M. H. (1997). Phosphorylation and activation of heart 6-phosphofructo-2-kinase by protein kinase B and other protein kinases of the insulin signaling cascades. *J. Biol. Chem.*, **272**, 17269–75.
96. Nakae, J., Park, B. C., and Accili, D. (1999). Insulin stimulates phosphorylation of the forkhead transcription factor FKHR on serine 253 through a wortmannin-sensitive pathway. *J. Biol. Chem.*, **274**, 15982–5.
97. Guo, S., Rena, G., Cichy, S., He, X., Cohen, P., and Unterman, T. (1999). Phosphorylation of serine 256 by protein kinase B disrupts transactivation by FKHR and mediates effects of insulin on insulin-like growth factor-binding protein-1 promoter activity through a conserved insulin response sequence. *J. Biol. Chem.*, **274**, 17184–92.
98. Barthel, A., Okino, S. T., Liao, J., Nakatani, K., Li, J., Whitlock, J. P., Jr., *et al.* (1999). Regulation of GLUT1 gene transcription by the serine/threonine kinase Akt1. *J. Biol. Chem.*, **274**, 20281–6.
99. Kohn, A. D., Summers, S. A., Birnbaum, M. J., and Roth, R. A. (1996). Expression of a constitutively active Akt Ser/Thr kinase in 3T3-L1 adipocytes stimulates glucose uptake and glucose transporter 4 translocation. *J. Biol. Chem.*, **271**, 31372–8.
100. Wijkander, J., Landstrom, T. R., Manganiello, V., Belfrage, P., and Degerman, E. (1998). Insulin-induced phosphorylation and activation of phosphodiesterase 3B in rat adipocytes: possible role for protein kinase B but not mitogen- activated protein kinase or p70 S6 kinase. *Endocrinology*, **139**, 219–27.
101. Kang, S. S., Kwon, T., Kwon, D. Y., and Do, S. I. (1999). Akt protein kinase enhances human telomerase activity through phosphorylation of telomerase reverse transcriptase subunit. *J. Biol. Chem.*, **274**, 13085–90.
102. Dimmeler, S., Fleming, I., Fisslthaler, B., Hermann, C., Busse, R., and Zeiher, A. M. (1999). Activation of nitric oxide synthase in endothelial cells by Akt-dependent phosphorylation. *Nature*, **399**, 601–5.
103. Fulton, D., Gratton, J. P., McCabe, T. J., Fontana, J., Fujio, Y., Walsh, K., *et al.* (1999). Regulation of endothelium-derived nitric oxide production by the protein kinase Akt. *Nature*, **399**, 597–601.
104. Michell, B. J., Griffiths, J. E., Mitchelhill, K. I., Rodriguez Crespo, I., Tiganis, T., Bozinovski, S., *et al.* (1999). The Akt kinase signals directly to endothelial nitric oxide synthase. *Curr. Biol.*, **12**, 845–8.
105. Cardone, M. H., Roy, N., Stennicke, H. R., Salvesen, G. S., Franke, T. F., Stanbridge, E., *et al.* (1998). Regulation of cell death protease caspase-9 by phosphorylation. *Science*, **282**, 1318–21.
106. Scott, P. H., Brunn, G. J., Kohn, A. D., Roth, R. A., and Lawrence, J. C., Jr. (1998). Evidence of insulin-stimulated phosphorylation and activation of the mammalian target of rapamycin mediated by a protein kinase B signaling pathway. *Proc. Natl. Acad. Sci. USA*, **95**, 7772–7.
107. Gingras, A. C., Kennedy, S. G., O'Leary, M. A., Sonenberg, N., and Hay, N. (1998). 4E-BP1, a repressor of mRNA translation, is phosphorylated and inactivated by the Akt(PKB) signaling pathway. *Genes Dev.*, **12**, 502–13.

108. Ahmad, F., Gao, G., Wang, L. M., Landstrom, T. R., Degerman, E., Pierce, J. H., *et al.* (1999). IL-3 and IL-4 activate cyclic nucleotide phosphodiesterases 3 (PDE3) and 4 (PDE4) by different mechanisms in FDCP2 myeloid cells. *J. Immunol.*, **162**, 4864–75.
109. Kitamura, T., Kitamura, Y., Kuroda, S., Hino, Y., Ando, M., Kotani, K., *et al.* (1999). Insulin-induced phosphorylation and activation of cyclic nucleotide phosphodiesterase 3B by the serine-threonine kinase Akt. *Mol. Cell. Biol.*, **19**, 6286–96.
110. Beraud, C., Henzel, W. J., and Baeuerle, P. A. (1999). Involvement of regulatory and catalytic subunits of phosphoinositide 3- kinase in NF-kappaB activation. *Proc. Natl. Acad. Sci. USA*, **96**, 429–34.
111. Sizemore, N., Leung, S., and Stark, G. R. (1999). Activation of phosphatidylinositol 3-kinase in response to interleukin- 1 leads to phosphorylation and activation of the NF-kappaB p65/RelA subunit. *Mol. Cell. Biol.*, **19**, 4798–805.
112. Romashkova, J. A. and Makarov, S. S. (1999). NF-kappaB is a target of AKT in anti-apoptotic PDGF signalling. *Nature*, **401**, 86–90.
113. Kane, L. P., Shapiro, V. S., Stokoe, D., and Weiss, A. (1999). Induction of NF-kappaB by the Akt/PKB kinase. *Curr. Biol.*, **9**, 601–4.
114. Ozes, O. N., Mayo, L. D., Gustin, J. A., Pfeffer, S. R., Pfeffer, L. M., and Donner, D. B. (1999). NF-kappaB activation by tumour necrosis factor requires the Akt serine-threonine kinase. *Nature*, **401**, 82–5.
115. Guan, K. L., Figueroa, C., Brtva, T. R., Zhu, T., Taylor, J., Barber, T. D., *et al.* (2000). Negative regulation of the serine/threonine kinase B-Raf by Akt. *J. Biol. Chem.*, **275**, 27354–9.
116. Zimmermann, S. and Moelling, K. (1999). Phosphorylation and regulation of Raf by Akt (protein kinase B). *Science*, **286**, 1741–4.
117. Rommel, C., Clarke, B. A., Zimmermann, S., Nunez, L., Rossman, R., Reid, K., *et al.* (1999). Differentiation stage-specific inhibition of the Raf-MEK-ERK pathway by Akt. *Science*, **286**, 1738–41.
118. Kim, A. H., Khursigara, G., Sun, X., Franke, T. F., and Chao, M. V. (2001). Akt phosphorylates and negatively regulates apoptosis signal-regulating kinase 1. *Mol. Cell. Biol.*, **21**, 893–901.
119. Scheid, M. P. and Woodgett, J. R. (2000). Protein kinases: six degrees of separation? *Curr. Biol.*, **10**, R191–4.
120. Gajewski, T. F. and Thompson, C. B. (1996). Apoptosis meets signal transduction: elimination of a BAD influence. *Cell*, **87**, 589–92.
121. Zha, J., Harada, H., Yang, E., Jockel, J., and Korsmeyer, S. J. (1996). Serine phosphorylation of death agonist BAD in response to survival factor results in binding to 14-3-3 not BCL-X_L. *Cell*, **87**, 619–28.
122. del Paso, L., Gonzales-Garcia, M., Page, C., Herrera, R., and Nunez, G. (1997). Interleukin-3-induced phosphorylation of BAD through the protein kinase Akt. *Science*, **278**, 687–9.
123. Datta, S. R., Dudek, H., Tao, X., Masters, S., Fu, H., Gotoh, Y., *et al.* (1997). Akt phosphorylation of BAD couples survival signals to the cell-intrinsic death machinery. *Cell*, **91**, 231–41.
124. Franke, T. F. and Cantley, L. C. (1997). Apoptosis. A Bad kinase makes good. *Nature*, **390**, 116–17.
125. Blume-Jensen, P., Janknecht, R., and Hunter, T. (1998). The kit receptor promotes cell survival via activation of PI 3-kinase and subsequent Akt-mediated phosphorylation of Bad on Ser136. *Curr. Biol.*, **8**, 779–82.

126. Pastorino, J. G., Tafani, M., and Farber, J. L. (1999). Tumor necrosis factor induces phosphorylation and translocation of BAD through a phosphatidylinositide-3-OH kinase-dependent pathway. *J. Biol. Chem.*, **274**, 19411–16.
127. Mok, C. L., Gil-Gomez, G., Williams, O., Coles, M., Taga, S., Tolaini, M., *et al.* (1999). Bad can act as a key regulator of T cell apoptosis and T cell development. *J. Exp. Med.*, **189**, 575–86.
128. Hinton, H. J. and Welham, M. J. (1999). Cytokine-induced protein kinase B activation and Bad phosphorylation do not correlate with cell survival of hemopoietic cells. *J. Immunol.*, **162**, 7002–9.
129. Craddock, B. L., Orchiston, E. A., Hinton, H. J., and Welham, M. J. (1999). Dissociation of apoptosis from proliferation, protein kinase B activation, and BAD phosphorylation in interleukin-3-mediated phosphoinositide 3-kinase signaling. *J. Biol. Chem.*, **274**, 10633–40.
130. Scheid, M. P. and Duronio, V. (1998). Dissociation of cytokine-induced phosphorylation of Bad and activation of PKB/akt: involvement of MEK upstream of Bad phosphorylation. *Proc. Natl. Acad. Sci. USA*, **95**, 7439–44.
131. Harada, H., Becknell, B., Wilm, M., Mann, M., Huang, L. J., Taylor, S. S., *et al.* (1999). Phosphorylation and inactivation of BAD by mitochondria-anchored protein kinase A. *Mol. Cell*, **3**, 413–22.
132. Bonni, A., Brunet, A., West, A. E., Datta, S. R., Takasu, M. A., and Greenberg, M. E. (1999). Cell survival promoted by the Ras-MAPK signaling pathway by transcription-dependent and -independent mechanisms. *Science*, **286**, 1358–62.
133. Scheid, M. P., Schubert, K. M., and Duronio, V. (1999). Regulation of bad phosphorylation and association with Bcl-x(L) by the MAPK/Erk kinase. *J. Biol. Chem.*, **274**, 31108–13.
134. Schurmann, A., Mooney, A. F., Sanders, L. C., Sells, M. A., Wang, H. G., Reed, J. C., *et al.* (2000). p21-activated kinase 1 phosphorylates the death agonist bad and protects cells from apoptosis. *Mol. Cell. Biol.*, **20**, 453–61.
135. Lizcano, J. M., Morrice, N., and Cohen, P. (2000). Regulation of BAD by cAMP-dependent protein kinase is mediated via phosphorylation of a novel site, Ser155. *Biochem. J.*, **349**, 547–57.
136. Tan, Y., Ruan, H., Demeter, M. R., and Comb, M. J. (1999). p90(RSK) blocks bad-mediated cell death via a protein kinase C-dependent pathway. *J. Biol. Chem.*, **274**, 34859–67.
137. Bertolotto, C., Maulon, L., Filippa, N., Baier, G., and Auberger, P. (2000). Protein kinase C theta and epsilon promote T-cell survival by a rsk- dependent phosphorylation and inactivation of BAD. *J. Biol. Chem.*, **275**, 37246 50.
138. Shimamura, A., Ballif, B. A., Richards, S. A., and Blenis, J. (2000). Rsk1 mediates a MEK-MAP kinase cell survival signal. *Curr. Biol.*, **10**, 127–35.
139. Liu, D., Yang, X., and Songyang, Z. (2000). Identification of CISK, a new member of the SGK kinase family that promotes IL-3-dependent survival. *Curr. Biol.*, **10**, 1233–6.
140. Tan, Y., Demeter, M. R., Ruan, H., and Comb, M. J. (2000). BAD Ser-155 phosphorylation regulates BAD/Bcl-XL interaction and cell survival. *J. Biol. Chem.*, **275**, 25865–9.
141. Datta, S. R., Katsov, A., Hu, L., Petros, A., Fesik, S. W., Yaffe, M. B., *et al.* (2000). 14-3-3 proteins and survival kinases cooperate to inactivate BAD by BH3 domain phosphorylation. *Mol. Cell*, **6**, 41–51.
142. Wang, H.-G., Pathan, N., Shibasaki, F., McKeon, F., Franke, T. F., and Reed, J. C. (1999). Dephosphorylation of BAD by calcineurin promotes cell death. *Science*, **284**, 339–43.

143. Kitada, S., Krajewska, M., Zhang, X., Scudiero, D., Zapata, J. M., Wang, H. G., *et al.* (1998). Expression and location of pro-apoptotic Bcl-2 family protein BAD in normal human tissues and tumor cell lines. *Am. J. Pathol.*, **152**, 51–61.
144. Kennedy, S. G., Kandel, E. S., Cross, T. K., and Hay, N. (1999). Akt/Protein kinase B inhibits cell death by preventing the release of cytochrome c from mitochondria. *Mol. Cell. Biol.*, **19**, 5800–10.
145. Kelley, T. W., Graham, M. M., Doseff, A. I., Pomerantz, R. W., Lau, S. M., Ostrowski, M. C., *et al.* (1999). Macrophage colony-stimulating factor promotes cell survival through Akt/protein kinase B. *J. Biol. Chem.*, **274**, 26393–8.
146. Fujita, E., Jinbo, A., Matuzaki, H., Konishi, H., Kikkawa, U., and Momoi, T. (1999). Akt phosphorylation site found in human caspase-9 is absent in mouse caspase-9. *Biochem. Biophys. Res. Commun.*, **264**, 550–5.
147. Zhou, H., Li, X. M., Meinkoth, J., and Pittman, R. N. (2000). Akt regulates cell survival and apoptosis at a postmitochondrial level. *J. Cell Biol.*, **151**, 483–94.
148. Kops, G. J., de Ruiter, N. D., De Vries-Smits, A. M., Powell, D. R., Bos, J. L., and Burgering, B. M. (1999). Direct control of the Forkhead transcription factor AFX by protein kinase B. *Nature*, **398**, 630–4.
149. Biggs, W. H., 3rd, Meisenhelder, J., Hunter, T., Cavenee, W. K., and Arden, K. C. (1999). Protein kinase B/Akt-mediated phosphorylation promotes nuclear exclusion of the winged helix transcription factor FKHR1. *Proc. Natl. Acad. Sci. USA*, **96**, 7421–6.
150. Rena, G., Guo, S., Cichy, S. C., Unterman, T. G., and Cohen, P. (1999). Phosphorylation of the transcription factor forkhead family member FKHR by protein kinase B. *J. Biol. Chem.*, **274**, 17179–83.
151. Tang, E. D., Nunez, G., Barr, F. G., and Guan, K. L. (1999). Negative regulation of the forkhead transcription factor FKHR by Akt. *J. Biol. Chem.*, **274**, 16741–6.
152. Eder, A. M., Dominguez, L., Franke, T. F., and Ashwell, J. D. (1998). Phosphoinositide 3-kinase regulation of T cell receptor-mediated interleukin-2 gene expression in normal T cells [In Process Citation]. *J. Biol. Chem.*, **273**, 28025–31.
153. Jackson, J. G., Kreisberg, J. I., Koterba, A. P., Yee, D., and Brattain, M. G. (2000). Phosphorylation and nuclear exclusion of the forkhead transcription factor FKHR after epidermal growth factor treatment in human breast cancer cells. *Oncogene*, **19**, 4574–81.
154. Nakamura, N., Ramaswamy, S., Vazquez, F., Signoretti, S., Loda, M., and Sellers, W. R. (2000). Forkhead transcription factors are critical effectors of cell death and cell cycle arrest downstream of PTEN. *Mol. Cell. Biol.*, **20**, 8969–82.
155. Hutchinson, J., Jin, J., Cardiff, R. D., Woodgett, J. R., and Muller, W. J. (2001). Activation of Akt (protein kinase B) in mammary epithelium provides a critical cell survival signal required for tumor progression. *Mol. Cell. Biol.*, **21**, 2203–12.
156. Brunet, A., Park, J., Tran, H., Hu, L. S., Hemmings, B. A., and Greenberg, M. E. (2001). Protein kinase SGK mediates survival signals by phosphorylating the forkhead transcription factor FKHRL1 (FOXO3a). *Mol. Cell. Biol.*, **21**, 952–65.
157. Cerezo, A., Martinez, A. C., Lanzarot, D., Fischer, S., Franke, T. F., and Rebollo, A. (1998). Role of Akt and c-Jun N-terminal kinase 2 in apoptosis induced by interleukin-4 deprivation. *Mol. Biol. Cell*, **9**, 3107–18.
158. Berra, E., Diaz-Meco, M. T., and Moscat, J. (1998). The activation of p38 and apoptosis by the inhibition of Erk is antagonized by the phosphoinositide 3-kinase/Akt pathway. *J. Biol. Chem.*, **273**, 10792–7.

159. Okubo, Y., Blakesley, V. A., Stannard, B., Gutkind, S., and Le Roith, D. (1998). Insulin-like growth factor-I inhibits the stress-activated protein kinase/c-Jun N-terminal kinase. *J. Biol. Chem.*, **273**, 25961–6.
160. Kwon, T., Kwon, D. Y., Chun, J., Kim, J. H., and Kang, S. S. (2000). Akt protein kinase inhibits Rac1-GTP binding through phosphorylation at serine 71 of Rac1. *J. Biol. Chem.*, **275**, 423–8.
161. Noselli, S. and Perrimon, N. (2000). Signal transduction. Are there close encounters between signaling pathways? *Science*, **290**, 68–9.

10 Viruses and apoptosis

DAVID L. VAUX

1. Introduction

Like the best human relationships, the association between viruses and cell suicide is intimate and of long duration. Apoptosis is used to remove surplus cells during development and the same molecular processes are used for cellular defence. Indeed, the mechanisms of apoptosis may have arisen initially to protect against viruses, conceivably even in single-celled organisms. In an ongoing battle between viruses and their hosts, viruses have developed a large number of strategies to inhibit early cell death. Viral anti-apoptosis proteins have been useful tools to investigate the mechanisms of cell death and have provided clues to the identification of cellular homologues that regulate apoptosis in host cells. In mammals the main protection against viruses comes from cytotoxic T cells (CTL). While these cells can often kill infected targets and use some mechanisms that are related to those used for cell suicide, viruses have also developed strategies to avoid killing of the host cell by CTL.

2. Origins of apoptosis

In animals, apoptosis is a physiological process of cell death used for morphogenesis during development, to maintain homeostasis of cell number, and to remove unwanted cells such as those with deleterious mutations or those infected by micro-organisms (1). As certain apoptosis effector genes can function in animals as diverse as nematodes, insects, and vertebrates, it is clear that the mechanisms of apoptosis are highly conserved. The conservation of function raises the question of how these mechanisms arose during evolution. It is possible that the apoptosis mechanisms used by vertebrates initially evolved in single-celled organisms as a defence against pathogens.

In order to reproduce, viruses need to subvert the host cell's synthetic machinery. Therefore, the simplest event that could limit virus production is a rapid, early death of the host cell. An altruistic suicide strategy would prevent replication and subsequent spread of viruses from one cell to another. In many tissues it is easy to replace cells by cell division, therefore such a strategy would not be too costly to a multicellular organism and indeed, has been observed in many cases. Interestingly, there are also instances of cellular suicide being used as a defence against viral infection even among single-celled organisms. For example, in many *E. coli* strains the Lit protein allows exclusion of T4 phage by causing proteolysis of translation elongation factor Tu, thereby blocking translation. While phage multiplication is prevented, the bacterium also dies (2). Death of the bacterium would prevent it propagating the genes it carried and therefore might seem to select against genes for its suicidal mechanism, but this is not necessarily so. If, by committing suicide, one bacterium could increase the likelihood of survival and reproduction of relatives carrying the same genes, cell death could be an adaptive behaviour. The thought of a single-celled organism committing suicide initially seems counter-intuitive, but as long as the death of one individual enhanced the ability of its relatives to survive and propagate, the genes that encode the suicide process would persist; the ultimate in 'selfish genes' (Fig. 1).

Indeed, the need to defend against pathogens and adverse circumstances such as scarce nutrients may be the reason cell suicide first arose. Single-celled organisms could not have evolved an obligatory cell suicide mechanism, because all individuals carrying it would, by definition, have killed themselves. Single-celled organisms would have no need for cell death during development ('programmed' cell death) but would still be at risk from infection by micro-organisms.

Therefore, if the mechanism of apoptosis used by most metazoans was inherited from their single-celled ancestors, it must have initially been used as a contingent strategy, presumably activated for defence. With the evolution of multicellular organisms, the same mechanisms were put to use for morphogenesis. Such a scenario

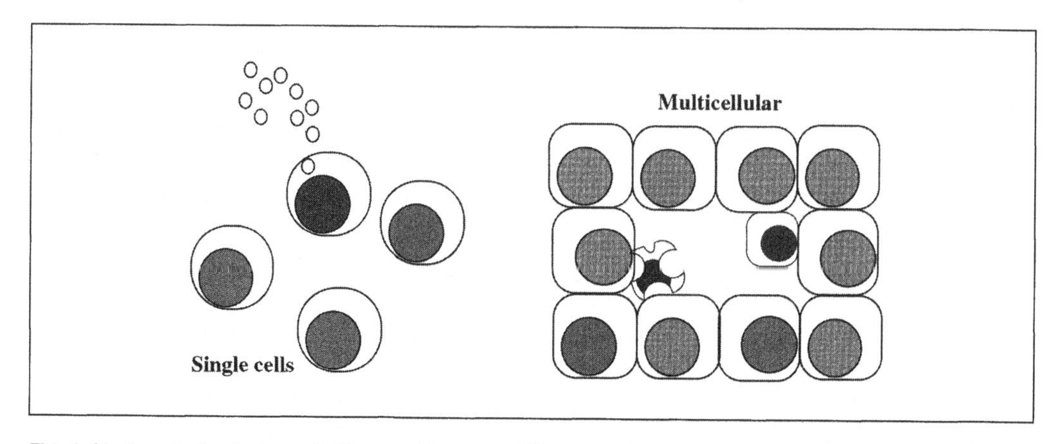

Fig. 1 Single-celled organisms (*left*) cannot have an obligatory cell death 'programme', but they could use cell death as a contingent strategy to prevent infection of related individuals. Multicellular organisms (*right*) use cell death both as a part of a developmental programme as well as to defend against infection. The infected cell may recognize the viral nucleic acid as being foreign, or may detect metabolic changes, such as unscheduled DNA replication. In metazoans, cells may be alerted to the presence of the virus by inflammatory cytokines such as tumour necrosis factor alpha (TNFα) or interferon (IFN). As the same apoptosis effector mechanisms are used for both defensive cell death and cell death during development, these process are likely to have a common evolutionary origin.

would explain why in some cases (e.g. development) apoptosis is not associated with inflammation, whereas in others (e.g. defence) it is.

3. Detection of viruses

How does a cell become aware that it has been infected in order that it can combat the virus? The answer to this question is not known in any detail, but there are a number of possibilities that have some experimental support.

Viral nucleic acid can be recognized as being foreign. In *E. coli*, for example, the restriction enzyme/DNA methylation system is used to distinguish self from foreign DNA, which is specifically degraded (3). Mammalian cells can respond to polyI-polyC RNA, which is thought to mimic viral double-stranded RNA, by producing IL-6 and interferon (4), provoking an immune response. p53 is up-regulated by single-stranded and broken host DNA, and may also be involved in the recognition of foreign DNA, although how it does so has not been established (5).

Another means by which cells may be alerted to the presence of a virus is by detecting changes in the expression of cell cycle regulators. Viruses that infect resting (G0) cells often endeavour to turn on the host cell's machinery for DNA replication, which is usually only expressed during S phase. A resting cell that detected expression of genes that are normally only turned on during proliferation would have a pretty good indication that something was amiss. For example, adenoviruses encode a transcription regulator, E1A, which turns on cellular genes whose products, such as DNA polymerase, are required for virus replication. Inappropriate expression of these genes, or perhaps the presence of unreplicated DNA during G0, may be what alerts the cell to the presence of the virus.

In addition to these cell autonomous means of detecting viruses, vertebrates are able to use their adaptive immune systems to recognize virally-infected cells. Primary roles of the major histocompatibility complex (MHC) proteins present on the surfaces of all cells, and of the T lymphocytes that recognize MHC molecules, are to defend against intracellular infections. The proteins made by a cell are constantly being digested and presented on the surface of cells bound to MHC class I molecules. If these are recognized by the highly variable T cell receptors on the surface of T cells, specifically a subset of T cells known as cytotoxic T lymphocytes (CTLs), they activate and attack the infected cell. CTLs then kill the infected cell extremely rapidly (within minutes of engagement) by apoptotic mechanisms (explained more fully in Section 8).

4. Viral pro-apoptotic genes?

The frequent close association of viral infection and apoptosis has led to suggestions that some viruses encode proteins for the purpose of inducing host cell death, rather than apoptosis being a defence response on the part a cell that detects a viral infection.

Whether viruses carry genes in order to activate apoptosis is controversial. It is has been suggested that viruses may induce host cell apoptosis after they have replicated, so that when the cell breaks up into apoptotic bodies and is engulfed, infectious particles will be carried into a new cell, thereby allowing the infection to spread. However, when cells are engulfed following apoptosis, they are enclosed within lysosomal vacuoles whose contents are rapidly degraded by enzymes in a low pH environment.

So far, all of the candidate pro-apoptotic viral proteins discovered also have other important roles for viral reproduction. Examples include adenovirus E1A (6), chicken anaemia virus apoptin (7), and HTLV-1 tax protein (8). It seems far more likely that these genes evolved to carry out some function essential for viral replication, which provokes apoptosis of the host cell, rather than having evolved specifically for the purpose of inducing cell death. As adenovirus encodes specific cell death inhibitors (E1b55kD and E1b19kD; see below) it is hard to imagine why it would also express a protein such as E1A for the purpose of inducing cell death. In fact, the picture that emerges from studying viral pro- and anti-apoptotic proteins suggests that the pro-apoptotic proteins are essential for viral functions such as replication and that the anti-apoptotic viral proteins function to prevent or to delay the host cell's innate defence mechanism, namely suicide.

5. Apoptosis as a defensive response to viral infection

In a large number of cases, infection with a virus particle results in apoptosis of the infected cell, both *in vitro* and *in vivo*. Thus infection of chicken blood cells by avian anaemia virus (9), infection of insect cells by baculoviruses (10), and infection of neurons by Sindbis virus (11) all result in apoptosis and a reduction in virus replication. Cells exposed to toxins or environmental stresses often respond in a similar way, namely by undergoing apoptosis and in many cases this apoptotic response to stress engages the same molecular mechanisms as the apoptotic response to virally-induced metabolic changes. A cell responding to radiation or staurosporine, for example, may be behaving by activating the cell death strategy that evolved originally to combat viruses.

That stress should commonly provoke an apoptotic response is quite understandable, given that a damaged cell may be ineffective or even harmful, and that new cells can often easily be replaced by mitosis. However, viral infection does not always result in apoptosis of the host cell, since in many cases apoptosis appears to be thwarted. Just as viral infections have put a selective pressure on organisms to come up with ways to detect and defend against them (for example by cell suicide), in an ongoing arms race viruses have themselves developed ways to subvert host cell defences, including cell death.

6. Viral inhibition of apoptosis

Viruses have been found to encode proteins that specifically target the host cell's death programme and the study of such proteins and their targets has provided information on critical aspects of the cell death machinery, as well as the importance of apoptosis in the defence against viruses. Virally encoded inhibitors of apoptosis include caspase inhibitors, Bcl-2-like proteins, p53 inhibitors, inhibitors of MHC expression, and inhibitors of intracellular signalling pathways activated by viral infection (Fig. 2).

6.1 CrmA

The products of the cytokine response modifier (*crmA*) gene from cowpox and the closely related *Spi-2* gene from Vaccinia directly inhibit caspases (12, 13). Structurally, these proteins resemble serine protease inhibitors (serpins), but rather than inhibiting serine proteases they target caspases. CrmA is able to bind to and inhibit caspase-1 (interleukin 1β converting enzyme, ICE) (12) and caspase-8 with inhibitory constants in the picomolar range (14, 15). In this way CrmA is able to block production of IL-1β and cell death induced via the CD95 and TNF pathways (16), which use caspase-8 to induce apoptosis. In the chick membrane model of cowpox infection, viruses lacking CrmA do not produce haemorrhagic pocks that normally mediate greater viral infection (12). It is not clear whether inhibition of cytokine production or inhibition of cell death is more important for viral pathogenesis.

CrmA binds to activated caspases via its pseudosubstrate site. While some CrmA molecules are cleaved by the caspase and released, intact CrmA molecules remain coupled to activated caspase, thereby inhibiting it (17). In this respect CrmA may act

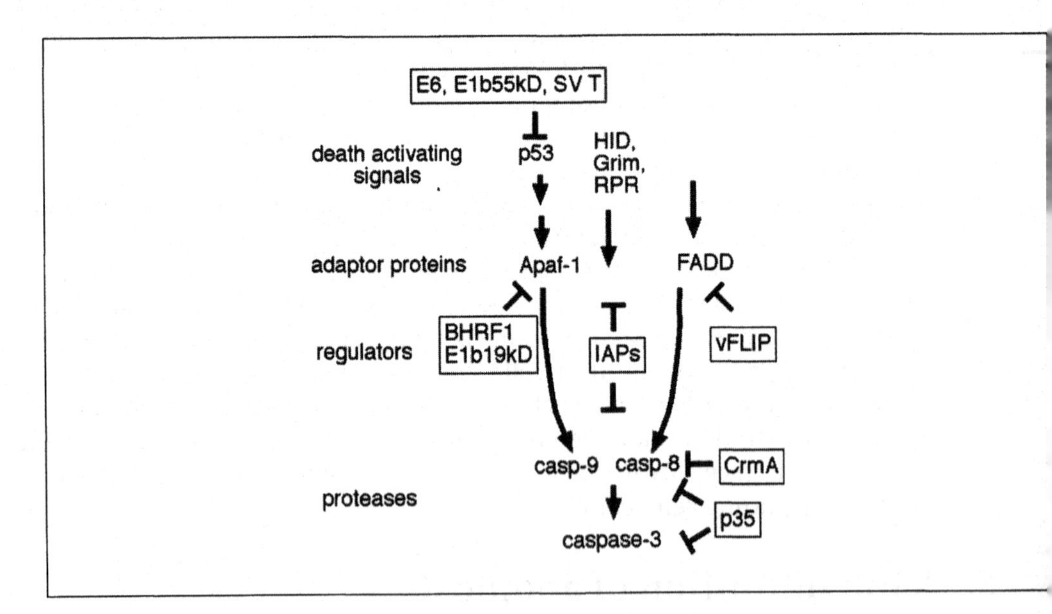

Fig. 2 Virally induced changes within the cell, such as unscheduled DNA replication, alert the cell to the presence of an infecting virus particle. Alternatively, apoptosis may be triggered by signals from cytokine receptors. Viruses encode inhibitors of both apoptosis activating pathways, such as homologues of Bcl-2 (BHRF1 and E1b19kD) and vFLIPs, or they may carry inhibitors of caspases, the effector proteases of apoptosis.

differently from some other serpins, which upon cleavage by their cognate serine protease undergo a stressed to relaxed conformational change and form an SDS-stable inhibitory complex with the protease.

6.2 p35 and IAP

Baculoviruses (such as *Autographa californica* nuclear polyhedrosis virus [AcNPV], *Bombyx mori* NPV, and *Spodoptera littoralis* NPV) encode a protein termed p35. AcNPV p35 was initially identified because it was mutated in an 'annihilator' stain of baculovirus—a strain that induced massive apoptosis in host cells, resulting in poor viral replication (10).

The p35 protein from AcNPV is able to bind to, and thereby inhibit, a variety of activated caspases with low nanomolar and picomolar inhibition constants (K_i) (14). As well as blocking insect caspases, p35 can potently inhibit mammalian caspases-3, -6, and -8 and can block the activity of the CED-3 caspase from *C. elegans* (18). In addition, expression of p35 in both neurons and oligodendrocytes in transgenic mice significantly inhibits the death of these cells in response to a number of different stimuli, demonstrating that p35 is a potent inhibitor of apoptosis in genetically diverse organisms (19, 20). p35 binds to activated caspases and is cleaved by them but remains bound in a stable complex (21). So far, no homologues of p35 have been identified, either in viruses other than baculoviruses or in animal cells.

It is interesting to note that not all insect caspases are blocked by p35. The *Drosophila* caspase DRONC is an upstream caspase activated by the death regulating proteins Reaper and Hid (22) (see Chapter 3). Active DRONC cleaves downstream caspases such as drICE, DCP-1, and DCP-2/DREDD, all of which are inhibited by p35. Instead, the activation of DRONC is prevented by another family of anti-apoptosis genes first characterized in *Baculovirus*, or inhibitors of apoptosis proteins (IAPs).

IAP proteins were identified as baculovirus genes that could prevent death of cells infected with baculoviruses with mutated p35 (23). IAPs bear no structural relationships to p35 or the serpins, but instead bear one or more motifs termed baculoviral IAP repeats (BIRs), a zinc binding fold. Genetic and biochemical evidence suggests the baculoviral IAPs and some of their insect and mammalian cellular homologues inhibit apoptosis by binding to pro- and active caspases (24–28). Caspase release from IAP molecules appears to require another set of unrelated IAP binding molecules. The insect apoptosis activating proteins Reaper (Rpr), Grim, and Head Involution Defective (HID) can promote caspase activation by binding to drIAP1 and removing it from the procaspase DRONC (22, 25). A similar system may also function in mammalian cells with the recent identification of Smac/DIABLO (29, 30). This protein is released from mitochondria and binds to IAPs such as XIAP and c-IAP2, inhibiting their binding to both upstream and downstream caspases.

The BIR motifs are required for the inhibition of active downstream caspases and inactive upstream ones in *Drosophila* and mammalian cells, but their biochemical function is still unclear. Recent data suggests that the BIRs may mediate homo-oligomerization of the IAPs and that this is required for functional inhibition of apoptosis (31). Several IAPs also have a ring finger motif that enables interaction with the ubiquitin-mediated pathway of protein degradation. IAPs may be subject to ubiquitin-mediated regulation and may also target downstream caspases for a similar fate, again blocking a pro-apoptotic signal (32).

Curiously, some baculoviruses bear IAPs for which no anti-apoptotic activity has been found. For example, in addition to encoding the caspase inhibitor p35, AcNPV also encodes an IAP that does not inhibit apoptosis in insect or mammalian systems (33). Similar IAPs can be found in BmNPV and OpNPV. Intriguingly, the latter bears several IAP genes, one encoding a cell death inhibitor, one whose function is yet to be determined, and possibly two more (34). The significance of these multiple IAP genes has yet to be determined, but several IAP homologues also have functions that are not directly involved with the activation of the apoptotic machinery. Both the human IAP homologue survivin and the *C. elegans* homologue BIR2 appear to regulate events at the mitotic and meiotic spindle (35, 36). Thus, IAPs may also function to regulate division as well as death.

6.3 FLIPs

Viral FLICE-inhibitory proteins (v-FLIPs) are encoded by several γ-herpesviruses, such as human herpesvirus-8 and human molluscipoxvirus, and inhibit apoptosis signalled through TNF receptor family members. Viral FLIPs and their cellular counterparts bear two death effector domains (DEDs) by which they bind to FADD, the adaptor molecule that activates caspase-8 (see Chapter 8). In this way FLIPs can prevent the recruitment and activation of caspase-8 following ligation of receptors such as CD95 (Fas/APO-1). Like cells expressing CrmA, cells expressing v-FLIPs are resistant to apoptosis induced by TNF receptor family members (37, 38). However, unlike CrmA, v-FLIP does not interact with active caspases. Instead, v-FLIP has many potential protein partners involved in mediating TNFα and FasL effects including members of the TRAF family (39). Moreover, v-FLIP is able to mediate NF-κB activation, which may activate anti-apoptotic pathways. But, given the multitude of downstream pathways mediated by NF-κB, it is likely that v-FLIP functions to subvert signalling pathways other than those associated with apoptosis (39, 40). Indeed, transgenic expression of v-FLIP in thymocytes decreases thymocyte numbers compared to controls. This decrease in cell number is independent of the Fas pathway and T cell receptor-mediated events, implying that an uncharacterized FLIP-sensitive pathway is required for normal thymocyte development (41).

Interestingly, equine herpes virus-2 encodes another FLIP-like molecule, E10 (42). Instead of containing two upstream DED domains, E10 contains a CARD domain (caspase activation and recruitment domain; see Chapter 4), a domain involved in protein–protein interactions. E10 also interacts with selective TRAF family members and c-JUN N-terminal kinase (JNK) and has a cellular homologue, Bcl-10. The role of E10 in equine herpes infection is unknown, as is the function of its cellular homologue.

6.4 Anti-p53

p53 protein is a transcription factor activated by DNA damage that can induce growth arrest or apoptosis. Mice and humans lacking one or both copies of p53 develop tumours in multiple cell lineages. Levels of p53 protein increase due to both transcriptional and post-translational mechanisms following double-stranded DNA breaks to chromatin, although it is not known how this damage is detected.

Since p53 has effects in addition to promoting cell death, such as causing cell cycle arrest and influencing expression of some cell cycle regulated genes, these viral p53 antagonists do not have identical effects to the viral Bcl-2 homologues (see below). Thus E1b55kD protein from adenovirus, which binds to and antagonizes p53, will not only inhibit p53-dependent apoptosis, but will inhibit p53-induced cycle arrest, whereas E1b19kD, also from adenovirus, resembles Bcl-2 and can prevent p53-induced apoptosis, but does not prevent p53-induced arrest (43, 44). Other viral p53 antagonists include SV40 virus large T antigen, which can bind and sequester p53, and human papilloma virus E6 protein, which can target p53 for ubiquitination and degradation by the proteasome (45, 46).

The fact that many mammalian viruses carry direct antagonists of p53 raises the possibility that it may have evolved to detect and defend against viruses, rather than to protect against oncogenesis.

6.5 Bcl-2 homologues

Many viruses encode functional Bcl-2 homologues, such as E1b19kD (adenovirus), BHRF1/BLAF1 (Epstein–Barr virus), and HSV (Kaposi associated Herpes virus 8) (43, 45–50). Presumably these proteins act like Bcl-2, namely to inhibit apoptosis induced via caspase-activating adaptor proteins such as Apaf-1 (51). The details of how cellular Bcl-2 family proteins function are described in Chapter 5.

Perhaps the most comprehensively studied viral Bcl-2 homologue is the adenovirus gene E1b19kD. While many of its functions appear to be comparable to Bcl-2, there are subtle differences. The recognition of adenovirus-infected host cells triggers an immune response that results in the release of many inflammatory cytokines including TNFα. Cells infected with either adenovirus or ectopic E1b19kD are resistant to TNFα-mediated apoptosis, a result not always seen in cells expressing either Bcl-2 or Bcl-x_L (52–54). The reason for this difference is unclear, but one could speculate that the ability to inhibit TNFα-mediated apoptosis would give adenovirus a specific replicative advantage by evading one arm of the immune response.

Another potential difference between E1b19kD is in the association with Bad, a BH3-only Bcl-2 family member (55). Bad binds to and mitigates the anti-apoptotic activity of Bcl-2. However, no such interaction between E1b19kD and Bad is seen, suggesting that E1b19kDs' function is enhanced by its inability to be functionally suppressed by Bad. This hypothesis would potentially imply that all BH3-only proteins would be unable to interact with E1b19kD, but this is not the case. Bik, another BH3-only protein, interacts with E1b19kD, but does not apparently mitigate its anti-apoptotic function (56, 57). This in turn suggests either a functional difference between Bad and Bik, or that Bad does not function in relation to viral-mediated apoptosis.

The BH3 domain is evident in both pro- and anti-apoptotic family members, a domain that E1b19kD lacks. In fact, E1b19kD possesses only the BH1 domain potentially involved in pore formation (see Chapter 5). E1b19kD binds to Bax, but not Bcl-2. However, swapping the BH3 domain of each protein reverses the binding capabilities of E1b19kD. This implies that E1b19kD may act to suppress the function of pro-apoptotic family members and not as an anti-apoptotic Bcl-2-like protein, as it was first described (58–60).

7. Inhibition of signal transduction pathways involved in defence

As well as blocking the host cell's apoptosis effector mechanisms, viruses have evolved clever ways of interfering with signal transduction pathways involved in the host defences. As some of these immune system responses can also lead to apoptosis, they are briefly mentioned here. Some viruses encode TNF receptor-like proteins that may act as 'decoys' to block all the effects of TNF, including activation of NF-κB and AP-1 transcription factors as well as caspase-8-dependent apoptosis. Thus the cowpox virus products CrmB and CrmC (61, 62) and the T2 proteins from Shope fibroma and myxoma viruses (63, 64) act as TNF receptor decoys to block all TNF activities (Fig. 3). In contrast, CrmA (see above), which acts downstream to directly inhibit caspase-8, leaves NF-κB and AP-1 signalling intact.

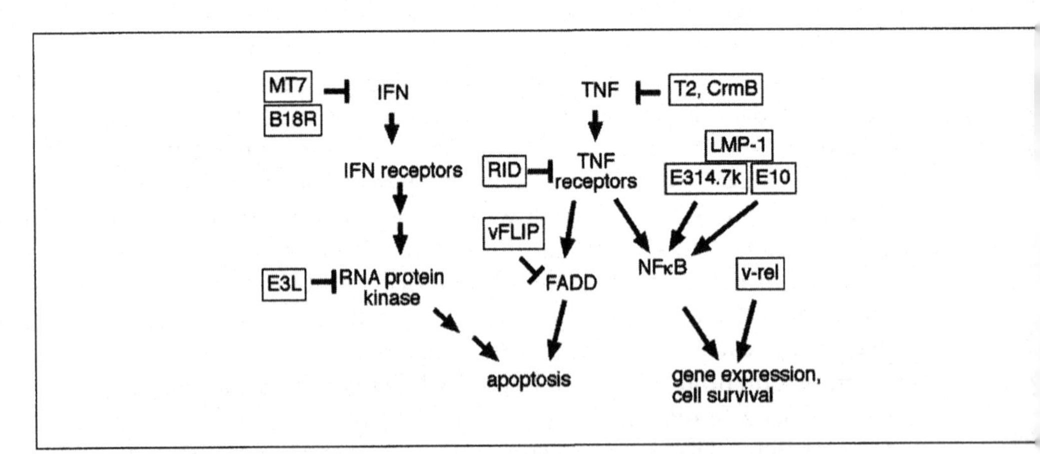

Fig. 3 Viruses encode proteins that inhibit defensive pathways such as those activated by cytokines, including interferons and TNFα. Apoptosis of the infected cell is often the outcome of activation of these pathways, but there are other effects. As the transcription factor NF-κB can stimulate production of proteins useful to the virus including cell death inhibitors, viruses may block the apoptosis pathway stemming from TNF receptors, but leave NF-κB activation unimpeded, or even stimulate it directly.

The adenoviral receptor internalization and degradation protein complex (RID), can inhibit apoptosis caused by CD95 by causing internalization of receptors on the cell's surface, which are then destroyed by lysosomes (65). RID (composed of E3 10.4 and 14.5 kD proteins) may prevent the destruction of adenovirus-infected cells by cytotoxic T cells by preventing expression of CD95 on the surface of the infected cell. However, RID does not prevent the death of infected cells killed via perforin-mediated pathways (see below).

E3-14.7k protein from adenovirus is also able to inhibit TNF-induced apoptosis, but acts intracellularly. While its precise mechanism of action is not certain, it probably does so by promoting signalling by NF-κB (66, 67), which is a transcriptional regulator of a number of genes, some of which promote cell survival (68, 69).

There are a several other examples of viral proteins that activate NF-κB, such as LMP-1 from Epstein–Barr virus, which associates with signalling/adaptor molecules such as TRAFs and TRADD to activate NF-κB (70), and E10 from equine herpesvirus-2 (42, 71). The avian reticuloendotheliosis virus strain T encodes *v-rel*, an activated form of NF-κB (72).

It is likely that these viruses activate NF-κB in order to promote host cell survival, and/or promote expression of host genes that are needed for viral replication. Expression of NF-κB can also be oncogenic; indeed, LMP-1 expression and EBV infections are associated with Burkitt's lymphoma, *v-rel* is an oncogene, and E10 is a homologue of Bcl10, an oncogene translocated in MALT lymphomas (71).

TNF is not the only cytokine targeted by viruses. As interferons are used in antiviral defence, viral proteins also target them. Viruses may carry interferon α/β or γ receptor decoys (e.g. myxoma virus M-T7, Vaccinia virus B8R), or inhibitors of IFN signalling (reviewed in ref. 73). Viruses also carry inhibitors of IFN-induced antiviral proteins, such as Vaccinia virus E3L protein, a dsRNA-binding protein that inhibits activation of IFN-induced RNA-dependent protein kinase (74).

8. Cytotoxic T cells

Viruses seem to have evolved effective mechanisms to prevent apoptosis of the infected cell, yet host organisms still manage to rid themselves of many viral infections. This is because cells of the adaptive immune system—specifically, cytotoxic T lymphocytes (CTL)—can induce apoptosis of virally-infected cells in a non-cell autonomous fashion while evading viral anti-apoptotic proteins. CTL manage to accomplish this by means of two parallel apoptosis-inducing mechanisms. One is by activation of CD95 on the target cell surface (see Chapter 8); the other is by perforin-mediated granule exocytosis (75), in which caspase-activating and other proteases (called granzymes) are directly introduced into the target cell. By directly introducing these powerful proteases, the CTL can activate the death programme downstream of most viral anti-apoptotic proteins.

In defence against viral infections, the perforin/granzyme mechanism used by CTL (and related NK cells) to directly kill infected targets appears to be more important, since mice lacking the gene for perforin are more immunocompromised than *lpr* mice, which do not express CD95 (75, 76). The cytolytic granules contain, in addition to perforin, a number of serine proteases. Granzyme B, the granule protease most responsible for rapid apoptosis of the target cells, is an Aspase capable of directly cleaving and activating a number of caspases, and may cleave some of the same substrates (77–79). In fact, granzyme B is the only mammalian serine protease that cleaves after aspartic acid residues and thus mimics the caspases in substrate specificity. Granzyme B is capable of cleaving many of the same substrates as the caspases and can cleave and thereby activate some of the caspases themselves. Proteolysis by granzyme B therefore explains why targets of CTL display classical apoptotic morphology. However, as CTL lacking both granzyme B and granzyme A can still kill target cells better than CTL lacking perforin, the granules must contain further cytotoxic components (80).

8.1 Viral defence against CTL and other immune system defences

By now it should be no surprise that viruses have come up with ways of avoiding the killing of their host cells by cells of the immune system. One strategy is to reduce expression of MHC, so that the infected cells aren't seen by the CTL. Adenovirus E319kD, for example, inhibits MHC class I maturation, and human cytomegalovirus proteins US3 and US2 prevent progression of MHC through the ER or induce it to be transported back from the ER to the cytosol (reviewed in ref. 81).

Another strategy is to avoid antibody-dependent cytotoxicity. Herpes simplex virus type 1 encodes glycoproteins that can bind to the Fc domain of immunoglobulins (82), thereby preventing them from being seen by Fc receptors on cytolytic cells. Other viruses have evolved proteins to inhibit complement-mediated lysis, such as homologues of complement regulatory proteins (83).

9. Conclusions

The relationship between viruses and apoptosis has been a long one, possible predating the evolution of multicellular metazoans. In fact, there is a similar battle between viruses and their host cells in plants and even prokaryotes (see ref. 1). But the battle is not over and every new strategy will select for a new counter-strategy. Studying the viral inhibitors of cell death has been especially fruitful, leading to the identification of the IAP family of proteins and the caspase inhibitors crmA and p35. Moreover, the study of the viral Bcl-2 homologues has shed some light on the function and interaction of this group of related proteins. Perhaps, had cell death not been studied in *C. elegans* and had not Bcl-2 been so frequently translocated in follicular lymphomas, then maybe all of the cell death mechanisms we know today could have been discovered by studying viruses. While this may be an exaggeration, I think few would argue that viruses are likely to reveal even more about cell death in the future.

References

1. Vaux, D. L. and Korsmeyer, S. J. (1999). Cell death in development. *Cell*, **96**, 245.
2. Georgiou, T., Yu, Y., Ekunwe, S., Buttner, M. J., Zuurmond, A. M., Kraal, B., *et al.* (1998). Specific peptide-activated proteolytic cleavage of *Escherichia coli* elongation factor tu. *Proc. Natl. Acad. Sci. USA*, **95**, 2891.
3. Naito, T., Kusano, K., and Kobayashi, I. (1995). Selfish behavior of restriction-modification systems. *Science*, **267**, 897.
4. Matsuyama, T., Kimura, T., Kitagawa, M., Pfeffer, K., Kawakami, T., Watanabe, N., *et al.* (1993). Targeted disruption of IRF-1 or IRF-2 results in abnormal type-I IFN gene induction and aberrant lymphocyte development. *Cell*, **75**, 83.
5. Lane, D. P. (1992). p53, guardian of the genome. *Nature*, **358**, 15.
6. Teodoro, J. G., Shore, G. C., and Branton, P. E. (1995). Adenovirus E1a proteins induce apoptosis by both p53-dependent and p53-independent mechanisms. *Oncogene*, **11**, 467.
7. Noteborn, M. H. M., Todd, D., Verschueren, C. A. J., De Gauw, F. M., Curran, W. L., Veldcamp, S., *et al.* (1993). A single chicken anemia virus protein induces apoptosis. *J. Virol.*, **68**, 346.
8. Yamada, T., Yamaoka, S., Goto, T., Nakai, M., Tsujimoto, Y., and Hatanaka, M. (1994). The human T-cell leukemia virus type I Tax protein induces apoptosis which is blocked by the Bcl-2 protein. *J. Virol.*, **68**, 3374.
9. Jeurissen, S. H., Wagenaar, F., Pol, J. M., van d. E. A., and Noteborn, M. H. (1992). Chicken anemia virus causes apoptosis of thymocytes after *in vivo* infection and of cell lines after *in vitro* infection. *J. Virol.*, **66**, 7383.
10. Clem, R. J., Fechheimer, M., and Miller, L. K. (1991). Prevention of apoptosis by a baculovirus gene during infection of insect cells. *Science*, **254**, 1388.
11. Griffin, D. E., Levine, B., Ubol, S., and Hardwick, J. M. (1994). The effects of alphavirus infection on neurones. *Ann. Neurol.*, **35**, 23.
12. Ray, C. A., Black, R. A., Kronheim, S. R., Greenstreet, T. A., Sleath, P. R., Salvesen, G. S., *et al.* (1992). Viral inhibition of inflammation: cowpox virus encodes an inhibitor of the interleukin-1 beta converting enzyme. *Cell*, **69**, 597.
13. Kettle, S., Alcami, A., Khanna, A., Ehret, R., Jassoy, C., and Smith, G. L. (1997). Vaccinia virus serpin B13r (SPI-2) inhibits interleukin-1-beta-converting enzyme and protects virus-infected cells from TNF- and Fas-mediated apoptosis, but does not prevent IL-1-beta-induced fever. *J. Gen. Virol.*, **78**, 667.
14. Zhou, Q., Snipas, S., Orth, K., Muzio, M., Dixit, V. M., and Salvesen, G. S. (1997). Target protease specificity of the viral serpin crmA—analysis of five caspases. *J. Biol. Chem.*, **272**, 7797.
15. Garcia-Calvo, M., Peterson, E. P., Leiting, B., Ruel, R., Nicholson, D. W., and Thornberry, N. A. (1998). Inhibition of human caspases by peptide-based and macromolecular inhibitors. *J. Biol. Chem.*, **273**, 32608.
16. Tewari, M., Beidler, D. R., and Dixit, V. M. (1995). CrmA-inhibitable cleavage of the 70-kDa protein component of the U1 small nuclear ribonucleoprotein during as- and tumor necrosis factor-induced apoptosis. *J. Biol. Chem.*, **270**, 18738.
17. Komiyama, T., Ray, C. A., Pickup, D. J., Howard, A. D., Thornberry, N. A., Peterson, E. P., *et al.* (1994). Inhibition of interleukin-1-beta converting enzyme by the Cowpox virus serpin CrmA—an example of cross-class inhibition. *J. Biol. Chem.*, **269**, 19331.

18. Sugimoto, A., Friesen, P. D., and Rothman, J. H. (1994). Baculovirus p35 prevents developmentally programmed cell death and rescues a *ced 9* mutant in the nematode *Caenorhabditis elegans*. *EMBO J.*, **13**, 2023.
19. Viswanth, V., Wu, Z., Fonck, C., Wei, Q., Boonplueang, R., and Anderson, J. K. (2000). Transgenic mice neuronally expressing baculoviral p35 are resistant to diverse types of induced apoptosis including seizure-associated neurodegeneration. *Proc. Natl. Acad. Sci. USA*, **97**, 2270.
20. Hisahara, S., Araki, T., Sugiyama, F., Yagami, Ki., Suzuki, M., Abe, K., *et al.* (2000). Targeted expression of baculovirus p35 caspase inhibitor in oligodendrocytes protects mice against autoimmune-mediated demyelination. *EMBO J.*, **19**, 341.
21. Bump, N. J., Hackett, M., Hugunin, M., Seshagiri, S., Brady, K., Chen, P., *et al.* (1995). Inhibition of ICE family proteases by baculovirus antiapoptotic protein p35. *Science*, **269**, 1885.
22. Meier, P., Silke, J., Leevers, S., and Evan, G. I. (2000). The *Drosophila* caspase DRONC is regulated by DIAP1. *EMBO J.*, **19**, 598.
23. Crook, N. E., Clem, R. J., and Miller, L. K. (1993). An apoptosis inhibiting baculovirus gene with a zinc finger like motif. *J. Virol.*, **67**, 2168.
24. Deveraux, Q. L., Roy, N., Stennicke, H. R., Vanarsdale, T., Zhou, Q., Srinivasula, S. M., *et al.* (1998). IAPs block apoptotic events induced by caspase-8 and cytochrome c by direct inhibition of distinct caspases. *EMBO J.*, **17**, 2215.
25. Kaiser, W. J., Vucic, D., and Miller, L. K. (1998). The *Drosophila* inhibitor of apoptosis D-IAP1 suppresses cell death induced by the caspase drICE. *FEBS Lett.*, **440**, 243.
26. Vucic, D., Kaiser, W. J., and Miller, L. K. (1998). A mutational analysis of the baculovirus inhibitor of apoptosis Op-IAP. *J. Biol. Chem.*, **273**, 33915.
27. Deveraux, Q. L., Takahashi, R., Salvesen, G. S., and Reed, J. C. (1997). X-linked IAP is a direct inhibitor of cell-death proteases. *Nature*, **388**, 300.
28. Roy, N., Deveraux, Q. L., Takahashi, R., Salvesen, G. S., and Reed, J. C. (1997). The c-IAP-1 and c-IAP-2 proteins are direct inhibitors of specific caspases. *EMBO J.*, **16**, 6914.
29. Du, C., Fang, M., Li, Y., Li, L., and Wang, X. (2000). Smac, a mitochondrial protein that promotes cytochrome c-dependent caspase activation by eliminating IAP inhibition. *Cell*, **102**, 33.
30. Verhagen, A. M., Ekert, P. G., Pakusch, M., Silke, J., Connolly, L. M., Reid, G. E., *et al.* (2000). Identification of DIABLO, a mammalian protein that promotes apoptosis by binding to and antagonizing IAP proteins. *Cell*, **102**, 43.
31. Hozak, R. R., Manji, G. A., and Friesen, P. D. (2000). The BIR motifs mediate dominant interference and oligomerisation of inhibitor of apoptosis Op-IAP. *Mol. Cell. Biol.*, **20**, 1877.
32. Huang, H-K., Joazeiro, C. A., Bonfoco, E., Kamada, S., Leverson, J. D., and Hunter, T. (2000). The inhibitor of apoptosis, cIAP2, functions as a ubiquitin-protein ligase and promotes *in vitro* monoubiquitination of caspases 3 and 7. *J. Biol. Chem.*, **275**, 26661.
33. Clem, R. J. and Miller, L. K. (1994). Control of programmed cell death by the baculovirus genes p35 and IAP. *Mol. Cell. Biol.*, **14**, 5212.
34. Uren, A. G., Coulson, E. J., and Vaux, D. L. (1998). Conservation of baculovirus inhibitor of apoptosis repeat proteins (BIRps) in viruses, nematodes, vertebrates and yeasts. *Trends Biochem. Sci.*, **23**, 159.
35. Fraser, A., James, C., Evan, G. I., and Hengartner, M. O. (1999). *Caenorhabditis elegans* inhibitor of apoptosis protein (IAP) homologue BIR1 plays a conserved role in cytokinesis. *Curr. Biol.*, **9**, 292.

36. Reed, J. C. and Reed, S. I. (1999). Survivin' cell-separation anxiety. *Nature Cell Biol.*, **1**, E199.
37. Thome, M., Schneider, P., Hofmann, K., Fickenscher, H., Meinl, E., Neipel, F., *et al.* (1997). Viral FLICE-inhibitory proteins (FLIPs) prevent apoptosis induced by death receptors. *Nature*, **386**, 517.
38. Irmler, M., Thome, M., Hahne, M., Schneider, P., Hofmann, B., Steiner, V., *et al.* (1997). Inhibition of death receptor signals by cellular FLIP. *Nature*, **388**, 190.
39. Chaudhary, P. M., Jasmin, A., Eby, M. T., and Hood, L. (1999). Modulation of the NFkB pathway by virally encoded death effector domains containing proteins. *Oncogene*, **18**, 5738.
40. Kataoka, T., Budd, R. C., Holler, N., Thome, M., Martinou, F., Irmler, M., *et al.* (2000). The caspase-8 inhibitor FLIP promotes activation of NF-kappaB and erk signalling pathways. *Curr. Biol.*, **10**, 640.
41. Ohayama, T., Tsukumo, S., Yajima, N., Sakamaki, K., and Yonehara, S. (2000). Reduction of thymocyte numbers in transgenic mice expressing viral FLICE-inhibitory protein in a Fas-independent manner. *Microbiol. Immunol.*, **44**, 289.
42. Thome, M., Maritnon, F., Hofmann, K., Rubio, V., Steiner, V., Schneider, P., *et al.* (1999). Equine herpesvirus-2 E10 gene product, but not its cellular homologue activates NF-kappaB transcription factor and c-Jun N-terminal kinase. *J. Biol. Chem.*, **274**, 9962.
43. Rao, L., Debbas, M., Sabbatini, P., Hockenbery, D., and Korsmeyer White, E. (1992). The adenovirus E1A proteins induce apoptosis, which is inhibited by the E1B 19-kDa and Bcl-2 proteins. *Proc. Natl. Acad. Sci. USA*, **89**, 7742.
44. Debbas, M. and White, E. (1993). Wild-type p53 mediates apoptosis by E1A, which is inhibited by E1B. *Genes Dev.*, **7**, 546.
45. Jay, G., Khoury, G., DeLeo, A. B., Dippold, W. C., and Old, L. J. (1981). p53 transformation-related protein: detection of an associated phosphotransferase activity. *Proc. Natl. Acad. Sci. USA*, **78**, 2932.
46. Scheffner, M., Huibregtse, J. M., Vierstra, R. D., and Howley, P. M. (1993). The HPV-16 E6 and E6-AP complex functions as a ubiquitin-protein ligase in the ubiquitination of p53. *Cell*, **75**, 495.
47. Henderson, S., Huen, D., Rowe, M., Dawson, C., Johnson, G., and Rickinson, A. (1993). Epstein Barr virus coded BHRF1 protein, a viral homolog of Bcl 2, protects human B cells from programmed cell death. *Proc. Natl. Acad. Sci. USA*, **90**, 8479.
48. Chiou, S. K., Tseng, C. C., Rao, L., and White, E. (1994). Functional complementation of the adenovirus E1B 19-kilodalton protein with Bcl-2 in the inhibition of apoptosis in infected cells. *J. Virol.*, **68**, 6553.
49. Russo, J. J., Bohenzky, R. A., Chien, M. C., Chen, J., Yan, M., Maddalena, D., *et al.* (1996). Nucleotide sequence of the Kaposi sarcoma-associated herpesvirus (HHV8). *Proc. Natl. Acad. Sci. USA*, **93**, 14862.
50. Virgin, H. W., Presti, R. M., Li, X. Y., Liu, C., and Speck, S. H. (1999). Three distinct regions of the murine gammaherpesvirus 68 genome are transcriptionally active in latently infected mice. *J. Virol.*, **73**, 2321.
51. Zou, H., Henzel, W. J., Liu, X. S., Lutschg, A., and Wang, X. D. (1997). Apaf-1, a human protein homologous to c-elegans ced-4, participates in cytochrome c-dependent activation of caspase-3. *Cell*, **90**, 405.
52. Gooding, L. R., Aquino, L., Duerksen-Hughes, P. J., Day, D., Horton, T. M., Yei, S. P., *et al.* (1991). The E1B 19,000-molecular-weight protein of group C adenoviruses prevents tumor necrosis factor cytolysis of human cells but not of mouse cells. *J. Virol.*, **65**, 3083.

53. Hashimoto, S., Ishii, A., and Yonehara, S. (1991). The E1b oncogene of adenovirus confers cellular resistance to cytotoxicity of tumor necrosis factor and monoclonal anti-Fas antibody. *Int. Immunol.*, **3**, 343.
54. White, E., Sabbatini, P., Debbas, M., Wold, W. S., Kusher, D. I., and Gooding, L. R. (1992). The 19-kilodalton adenovirus E1B transforming protein inhibits programmed cell death and prevents cytolysis by tumor necrosis factor alpha. *Mol. Cell. Biol.*, **12**, 2570.
55. Chen, G., Branton, P. E., Yang, E., Korsmeyer, S. J., and Shore, G. C. (1996). Adenovirus E1B 19-kDa death suppressor protein interacts with Bax but not with Bad. *J. Biol. Chem.*, **271**, 24221.
56. Boyd, J. M., Gallo, G. J., Elangovan, B., Houghton, A. B., Malstrom, S., Avery, B. J., *et al.* (1995). Bik, a novel death-inducing protein shares a distinct sequence motif with Bcl-2 family proteins and interacts with viral and cellular survival-promoting proteins. *Oncogene*, **11**, 1921.
57. Han, J., Sabbatini, P., and White, E. (1996). Induction of apoptosis by human Nbk/Bik, a BH3-containing protein that interacts with E1B 19K. *Mol. Cell. Biol.*, **16**, 5857.
58. Han, J., Sabbatini, P., Perez, D., Rao, L., Modha, D., and White, E. (1996). The E1B 19K protein blocks apoptosis by interacting with and inhibiting the p53-inducible and death-promoting Bax protein. *Genes Dev.*, **10**, 461.
59. Marshall, W. L., Yim, C., Gustafson, E., Graf, T., Sage, D. R., Hanify, K., *et al.* (1999). Epstein–Barr virus encodes a novel homolog of the bcl-2 oncogene that inhibits apoptosis and associates with Bax and Bak. *J. Virol.*, **73**, 5181.
60. Han, J., Modha, D., and White, E. (1998). Interaction of E1B 19K with Bax is required to block Bax-induced loss of mitochondrial membrane potential and apoptosis. *Oncogene*, **17**, 2993.
61. Hu, F. Q., Smith, C. A., and Pickup, D. J. (1994). Cowpox virus contains two copies of an early gene encoding a soluble secreted form of the type II TNF receptor. *Virology*, **204**, 343.
62. Smith, C. A., Hu, F. Q., Smith, T. D., Richards, C. L., Smolak, P., Goodwin, R. G., *et al.* (1996). Cowpox virus genome encodes a second soluble homologue of cellular TNF receptors, distinct from CrmB, that binds TNF but not LT alpha. *Virology*, **223**, 132.
63. Upton, C., Macen, J. L., Schreiber, M., and McFadden, G. (1991). Myxoma virus expresses a secreted protein with homology to the tumor necrosis factor receptor gene family that contributes to viral virulence. *Virology*, **184**, 370.
64. Smith, C. A., Davis, T., Wignall, J. M., Din, W. S., Farrah, T., Upton, C., *et al.* (1991). T2 open reading frame from the Shope fibroma virus encodes a soluble form of the TNF receptor. *Biochem. Biophys. Res. Commun.*, **176**, 335.
65. Tollefson, A. E., Hermiston, T. W., Lichtenstein, D. L., Colle, C. F., Tripp, R. A., Dimitrov, T., *et al.* (1998). Forced degradation of Fas inhibits apoptosis in adenovirus-infected cells. *Nature*, **392**, 726.
66. Li, Y. G., Kang, J., Friedman, J., Tarassishin, L., Ye, J. J., Kovalenko, A., *et al.* (1999). Identification of a cell protein (FIP-3) as a modulator of NF-kappa B activity and as a target of an adenovirus inhibitor of tumor necrosis factor alpha-induced apoptosis. *Proc. Natl. Acad. Sci. USA*, **96**, 1042.
67. Dimitrov, T., Krajcsi, P., Hermiston, T. W., Tollefson, A. E., Hannink, M., and Wold, W. S M. (1997). Adenovirus E3-10.4k/14.5k protein complex inhibits tumor necrosis factor-induced translocation of cytosolic phospholipase A(2) to membranes. *J. Virol.*, **71**, 2830.
68. Beg, A. A. and Baltimore, D. (1996). An essential role for NF-kappa-B in preventing TNF-alpha-induced cell death. *Science*, **274**, 782.

69. Wang, C. Y., Mayo, M. W., and Baldwin, A. S. (1996). TNF- and cancer therapy-induced apoptosis potentiation by inhibition of NF-kappa-B. *Science*, **274**, 784.
70. Devergne, O., McFarland, E. C., Mosialos, G., Izumi, K. M., Ware, C. F., and Kieff, E. (1998). Role of the TRAF binding site and NF-kappa-B activation in Epstein–Barr virus latent membrane protein 1-induced cell gene expression. *J. Virol.*, **72**, 7900.
71. Willis, T. G., Jadayel, D. M., Du, M. Q., Peng, H. Z., Perry, A. R., Abdul-Rauf, M., *et al.* (1999). Bcl10 is involved in t(1;14) (p22;q32) of MALT B cell lymphoma and mutated in multiple tumor types..*Cell*, **96**, 35.
72. Chen, I. S., Wilhelmsen, K. C., and Temin, H. M. (1983). Structure and expression of c-rel, the cellular homolog to the oncogene of reticuloendotheliosis virus strain T. *J. Virol.*, **45**, 104.
73. Krajcsi, P. and Wold, W. S. M. (1998). Inhibition of tumor necrosis factor and interferon triggered responses by DNA viruses. *Semin. Cell Dev. Biol.*, **9**, 351.
74. Rivas, C., Gil, J., Melkova, Z., Esteban, M., and Diazguerra, M. (1998). Vaccinia virus E3l protein is an inhibitor of the interferon (IFN)-induced 2–5a synthetase enzyme. *Virology*, **243**, 406.
75. Kagi, D., Vignaux, F., Ledermann, B., Burki, K., Depraetere, V., Nagata, S., *et al.* (1994). Fas and perforin pathways as major mechanisms of T cell-mediated cytotoxicity. *Science*, **265**, 528.
76. Kagi, D., Ledermann, B., Burki, K., Seiler, P., Odermattolsen, K. J., Podack, E. R., *et al.* (1994). Cytotoxicity mediated by T cells and natural killer cells is greatly impaired in perforin deficient mice. *Nature*, **369**, 31.
77. Heusel, J. W., Wesselschmidt, R. L., Shresta, S., Russell, J. H., and Ley, T. J. (1994). Cytotoxic lymphocytes require granzyme b for the rapid induction of DNA fragmentation and apoptosis in allogeneic target cells. *Cell*, **76**, 977.
78. Andrade, F., Roy, S., Nicholson, D., Thornberry, N., Rosen, A., and Casciolarosen, L. (1998). Granzyme B directly and efficiently cleaves several downstream caspase substrates—implications for CTL-induced apoptosis. *Immunity*, **8**, 451.
79. Harris, J. L., Peterson, E. P., Hudig, D., Thornberry, N. A., and Craik, C. S. (1998). Definition and redesign of the extended substrate specificity of granzyme B. *J. Biol. Chem.*, **273**, 27364.
80. Simon, M. M., Hausmann, M., Tran, T., Ebnet, K., Tschopp, J., Thahla, R., *et al.* (1997). *In vitro-* and *ex vivo*-derived cytolytic leukocytes from granzyme A x B double knockout mice are defective in granule-mediated apoptosis but not lysis of target cells. *J. Exp. Med.*, **186**, 1781.
81. Farrell, H. E. and Davis Poynter, N. J. (1998). From sabotage to camouflage—viral evasion of cytotoxic T lymphocyte and natural killer cell-mediated immunity. *Semin. Cell Dev. Biol.*, **9**, 369.
82. Dubin, G., Socolof, E., Frank, I., and Friedman, H. M. (1991). Herpes simplex virus type 1 Fc receptor protects infected cells from antibody-dependent cellular cytotoxicity. *J. Virol.*, **65**, 7046.
83. Fishelson, Z. (1994). Complement-related proteins in pathogenic organisms. *Springer Semin. Immunopathol.*, **15**, 345.

11 | Cell death in the nervous system

MICHAEL D. JACOBSON and LOUISE BERGERON

1. Cell death in the nervous system

Programmed cell death (PCD) is important in animal development, tissue homeostasis, and disease pathogenesis. Although the role that cell death plays in development has long been recognized by embryologists, only recently has the idea that animal cells have a built-in suicide programme become generally accepted (reviewed in refs 1–3). For neurobiologists, however, the study of cell death has a long history, possibly because it is much easier to observe cell death among nerve cells, which cannot replace themselves, than among dividing cells, which can hide their losses by cell division.

The modern study of cell death in the nervous system starts with the seminal discovery by Viktor Hamburger and Rita Levi-Montalcini that the growth and survival of developing neurons depends on the availability of their innervating target (4). This and subsequent findings led to the neurotrophin theory—the idea that neurons compete for limiting amounts of extracellular signalling molecules secreted by the targets that they innervate in order to survive (5, 6). This theory has since been generalized to encompass all animal cells, which are thought to have a built-in, default death programme that is suppressed by survival signals from neighbouring cells (7). Although the neurotrophin theory explains the requirement of neurons for survival signals, it does not explain exactly how neurons die in their absence. In the last several years, our understanding of the molecular basis of this intracellular death programme has been greatly expanded. This chapter reviews the intracellular pathways that regulate and execute the cell death programme in neurons, and the contribution of this cell death machinery to the pathogenesis of neurological diseases.

2. Neurotrophins: extracellular survival and death signals to neurons

The first target-derived factor to be discovered was nerve growth factor (NGF), based largely on research by Cohen, Levi-Montalcini, and Hamburger in the 1950s. Other neurotrophins were discovered in the 1980s and 90s, including brain-derived neurotrophic factor (BDNF), ciliary neurotrophic factor (CNTF), neurotrophins-3 (NT3) and -4/5 (NT4/5), glial-derived neurotrophic factor (GDNF), and cardiotrophin-1 (CT-1). All of these ligands have survival-promoting properties on specific populations of neurons that are conveyed by their cognate cell surface receptors. Experiments with these factors generally confirm the concept that neuronal survival depends on competition for limiting amounts of one or a combination of target-derived factors. Target tissues, however, are not the only important influence on neuronal survival, as afferent connections (i.e. those received from other neurons), neighbouring glial cells, and neuronal activity can all contribute. More recent observations suggest that neurons can also be directly triggered to die by ligands that activate death signalling receptors (DRs; see Chapter 8).

3. Signal transduction

Neuron survival depends on the integration of multiple signals, including the presence of neurotrophins, afferent excitation, death receptor agonists, and environmental stress. These signals are transmitted and integrated within each cell, ultimately resulting in its survival or death. The intracellular signalling pathways responsible for integrating these environmental signals are the subjects of intensive research. In this section we examine the intracellular signalling pathways that mediate the actions of these extracellular signals (see also refs 8 and 9).

3.1 Survival signalling

Survival factors are thought to act mainly by suppressing intrinsic death effector molecules. Neurotrophins signal survival by binding with high affinity to neuron-specific tyrosine kinase receptors, TrkA, TrkB, and TrkC. NGF binds to TrkA (but not TrkB or TrkC) and induces dimerization and autophosphorylation of the receptor on tyrosine residues. The domain containing the phosphorylated residue serves as a docking site for a variety of factors, including proteins that regulate the activities of the Ras/phosphatidyl inositol-3 (PI3)-kinase and Ras/MAPK(ERK) pathways. These two signalling pathways are responsible for the majority of survival signalling in neurons (Fig. 1). Although Ras is not the only means by which these pathways are activated in cells, Ras plays a major role in translating Trk-mediated survival signals in neurons.

3.1.1 PI3K/Akt

Activation of the PI3 kinase pathway is critical for the survival of many cell types. For example, PI3 kinase activity is necessary for the survival activity of insulin-like growth factor-1 (IGF-1) on fibroblasts (10, 11) and cerebellar granule neurons (12), for the survival activity of IL-3 on haematopoietic cells (13), and for the survival of NGF-dependent PC12 cells (14). The importance of PI3K as the major survival-promoting pathway for neurons is supported by numerous studies which show that PI3 kinase

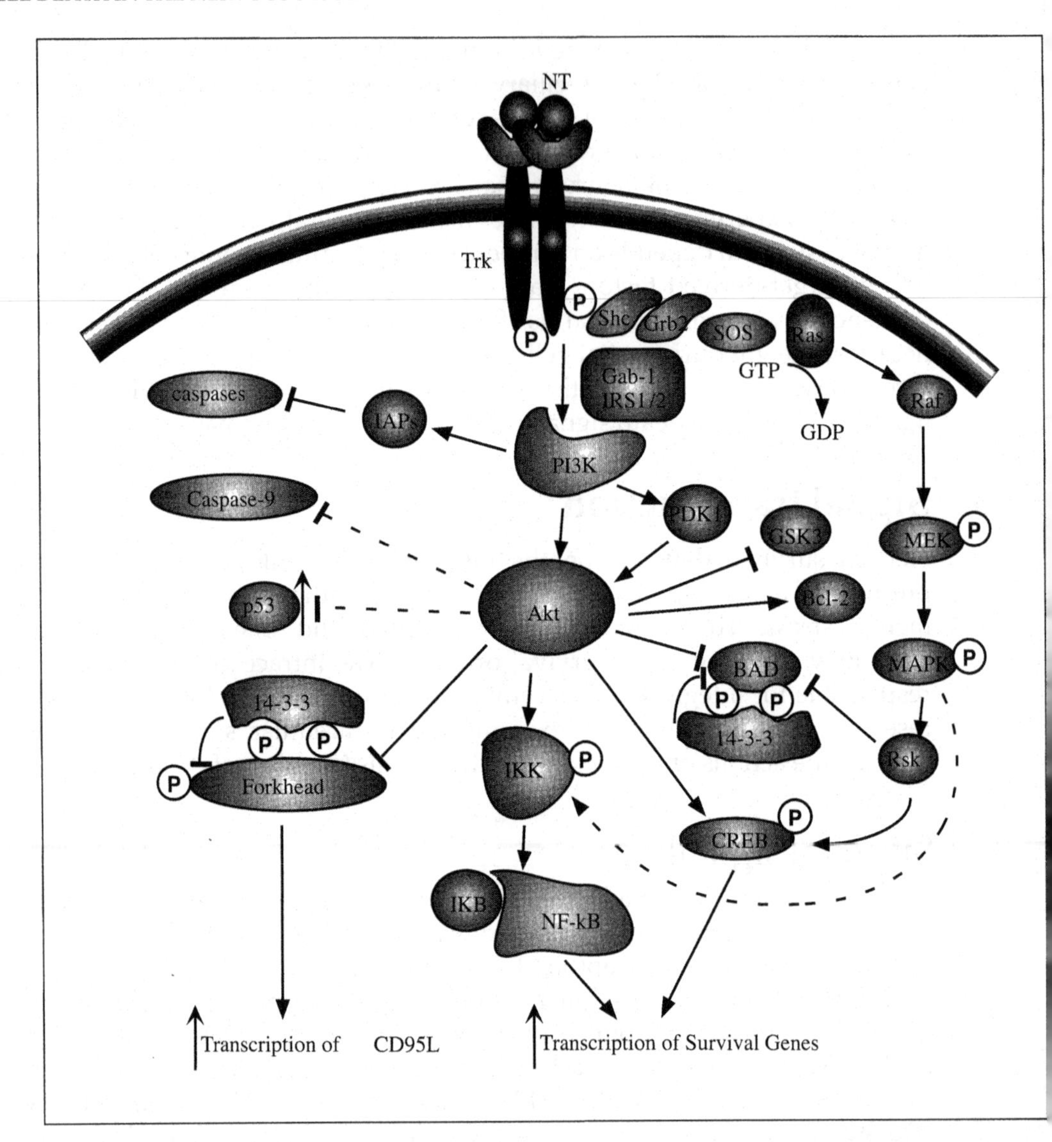

Fig. 1 Neurotrophin-stimulated survival pathways in neurons. Neurotrophins and other survival factors, such as IGF-1, stimulate survival via activation of receptor tyrosine kinases. Activated receptor tyrosine kinases (those for neurotrophins are known as Trks) autophosphorylate themselves and recruit factors that mediate the stimulation of multiple downstream signalling pathways. The most important of these for mediating neuronal survival are the PI3K/Akt and Ras/Raf/MEK pathways, which phosphorylate various downstream targets involved in the regulation or execution of programmed cell death. These targets include both transcriptional and post-transcriptional regulatory proteins, as described in the text. Many targets are not shown, including those involved in cell growth and metabolism, which are likely to influence cell survival indirectly. Other targets of PI3K besides Akt (not shown) may be involved in the regulation of cell survival, but have not been well characterized. Activation pathways are shown with arrowheads, inhibition with bars. Dashed lines indicate pathways that are less clear or are indirect.

is necessary for most of the survival signalling in various types of neurons (reviewed in ref. 8).

Neurotrophin binding to Trk stimulates receptor transphosphorylation, resulting in the recruitment of a number of signalling proteins to docking sites on the receptor. Trk activates Ras via Shc, Grb-2, and SOS; Ras directly interacts with PI3K (15) and is a critical effector of Trk-mediated activation of PI3K (16, 17). Trk also activates PI3K through the Gab-1 / IRS-1 / IRS-2 family of adaptor proteins (18–20). PI3K activity activates PDK1 and PDK2, which in turn activate Akt (21). The receptor for IGF-1, important for the survival of many cell types, also appears to signal survival by these pathways.

The survival function of PI3 kinase has been studied in many cells types, including neurons, by inhibiting its kinase activity. Interestingly, inhibition of PI3 kinase does not cause superior cervical ganglion (SCG) neurons to die in the presence of NGF, indicating that the PI3K pathway is not required for survival when NGF is present in these cells (22). However, in PC12 cells, PI3 kinase is important for survival signalling in the presence of NGF. Moreover, in cerebellar granule neurons maintained either by depolarization, IGF-1, or BDNF, inhibition of PI3K activity causes cell death even in the presence of the survival stimulus (23, 24). Therefore, it appears that the role of PI3K in signalling survival varies depending on neuronal cell type and the nature of the survival stimulus. Although PI3K is not required for NGF to signal survival in primary neurons, constitutive activation of the pathway by overexpression of PI3K or its downstream effector, the serine / threonine kinase Akt, blocks the death of sympathetic neurons after NGF is removed (22). Therefore, inhibition of the PI3K pathway may be required for the induction of apoptosis.

PI3K stimulates a number of kinases, including the serine / threonine kinase Akt (also known as protein kinase B, or PKB). Akt has been indentified as critical in a number of cell types for the survival response mediated by PI3K. Greenberg and colleagues first reported a requirement for Akt in neuronal survival by showing that a dominant-negative mutant Akt was cytotoxic when overexpressed in cerebellar granule neurons, whereas wild-type Akt markedly reduces the apoptotic response in these cells (12). Subsequent studies have similarly shown a requirement of Akt for survival of sympathetic neurons induced by NGF (25–27), as well as by depolarization (26, 28). Indeed, the PI3K / Akt pathway appears to be a convergence point for survival signals by both neurotrophins and depolarization (26). Although Akt is a critical target of PI3K, it is not the only one. Inhibition of PI3K, for example, is often more effective than inhibition of Akt for suppression of survival responses.

Other potentially important targets of the PI3 kinase pathway in neurons are the IAPs, (inhibitor of apoptosis proteins; see Chapter 3), a family of caspase inhibitors. Overexpression of ITA, a chicken IAP homologue, supported survival of chick sensory and sympathetic neurons in the absence of NGF, whereas introduction of ITA antisense RNA reduced their survival (29). Activation of PI3 kinase was required for the NGF-induced increase in ITA expression, suggesting that ITA might mediate part of the survival activity of NGF signalling, through TrkA and PI3 kinase.

In neurons, Akt has only been shown to regulate survival, and all of the targets identified so far have been proteins that regulate survival: Bad, procaspase-9,

Forkhead-1, and CREB. Upon activation, Akt phosphorylates these targets at the consensus sequence RXRXXS/T (reviewed in refs 21, 30, and 31). Bad and caspase-9 are cytoplasmic components of the intrinsic cell death machinery, and phosphorylation by Akt suppresses their pro-apoptotic activity (32, 33). Akt phosphorylates Bad at Ser 136, inducing it to bind the protein 14-3-3 and so preventing Bad's pro-apoptotic association with Bcl-2 and Bcl-x_L (32, 34). Moreover, transfection with constitutively-active Akt, or treatment with IGF-1, in CGNs inhibits the induction of apoptosis induced by transfection of wild-type Bad but not by Bad mutated at Ser 136 (32). Interestingly however, neurons from Bad-deficient mice do not show apparent alterations in survival (35), whereas studies of Bax knockout mice indicate a striking requirement for this protein in mediating PCD in both CGNs and sympathetic neurons (36). It may be that inhibition of Bad contributes to Akt-mediated survival, but that Bax mediates a pandemic death signal that is not regulated by Akt-mediated mechanisms.

Akt-induced phosphorylation of procaspase-9 inhibits its cleavage and activation, making procaspase-9 a similarly attractive target for Akt-mediated survival signalling (33). However, these studies are based on evidence from exogenous overexpression experiments and there are no reports of phosphorylation of endogenous procaspase-9 in neurons. Critically, the Akt phosphorylation site in human procaspase-9, Ser 196, is absent in mouse procaspase-9 (37).

Akt can also modulate the function of three families of transcription factors—Forkhead, cAMP response-element binding protein (CREB), and NF-κB—all of which are implicated in regulating cell survival (38–41). Genetic studies in *C. elegans* suggest that the Forkhead family member DAF-16 (Dauer-formation-16) is a downstream target of Akt (42), and activated Akt has been shown to translocate to the nucleus, suggesting that it may phosphorylate nuclear proteins. In CGNs, phosphorylated FKHRL1, a forkhead family member, is relocated to the cytoplasm where it binds to 14-3-3 protein and is inactive (41). When survival signals are in short supply, the PI3 kinase/Akt pathway is inactivated, FKHRL1 is unphosphorylated and accumulates in the nucleus where it has been shown to mediate the transcription of the FasL gene and possibly other pro-apoptotic genes (41). In these studies, overexpression of FKHRL1 mutated at its Akt phosphorylation sites (T32 and S315), which renders it constitutively active, promoted PCD in a manner that required Fas:FasL receptor/ligand interaction.

3.1.2 MEK/MAPK

The MEK/MAPK pathway is another survival pathway important in neurons. MAP kinases have multiple roles in neurons, including differentiation and neurite outgrowth as well as survival (reviewed in ref. 43). The Ras/MAPK pathway does not appear to be required for normal neurotrophic factor-dependent survival in several types of neurons (44–47) and differentiated PC12 cells (48)—a role ascribed mainly to the Ras/PI3K pathway. However, this pathway does appear to have an important role in protecting against injury and toxicity. The Ras/PI3 kinase/Akt pathway is required, for example, for NGF-dependent survival of sympathetic neurons, whereas the Ras/ERK pathway is selective for NGF-dependent protection against PCD induced by cytosine arabinoside, a DNA damaging agent (49, 50). Potassium depolarization also activates both the MAP kinase and PI3 kinase pathways in cerebellar granule neurons (23); both are required for neurotrophin-mediated protection against certain oxidative stresses (24).

An important target of the MAPK pathway is CREB, a transcription factor that is also activated by Akt (38). CREB is phosphorylated as a result of NGF-mediated activation of ERK/RSK and p38-MAP kinase pathways (51). CREB-mediated transcription has been found to be both necessary and sufficient for NGF-dependent survival of sympathetic neurons (40). Activation of CREB has also been observed to contribute to survival of cerebellar granule neurons, along with MAPK-stimulated phosphorylation of Bad (52). One of the targets of CREB-activated transcription is Bcl-2 (40, 53). Thus, activation of the MAP kinase pathway stimulates the transcription of pro-survival genes, via phosphorylation and activation of CREB, as well as the inhibition of pro-apopotitc proteins such as Bad.

3.1.3 NF-κB

The transcription factor NF-κB has been implicated in survival signalling in neurons and is activated indirectly by Akt (39). In cultured NGF-dependent sympathetic and sensory neurons, NGF results in the activation of NF-κB. Inhibition of NF-κB induces PCD in the presence of NGF, whereas overexpression of NF-κB subunits enhanced neuronal survival after NGF withdrawal (54, 55). Furthermore, in mice deficient in the p65 subunit of NF-κB there is increased neuronal apoptosis during embryogenesis in the trigeminal ganglia. Moreover, these cultured trigeminal neurons show a reduced survival response to NGF (55).

NF-κB plays an important role in inflammatory signalling—a process that, in the brain, involves microglia and macrophages, endothelial cells, and neurons in the pathogenesis of brain injury and neurodegenerative disease (see Section 6). The direct effects of NF-κB in these processes, however, are complex and not well understood. NF-κB has been implicated, for example, in mediating survival against excitotoxic injury (56, 57), but also appears to promote cell death in a model of focal ischaemia (58).

3.2 Death signals

3.2.1 Death receptors

Receptors with the capacity to induce apoptosis have been characterized extensively in dividing cells (see Chapter 8), but little in neuronal cells. Death domain containing receptors are expressed in neurons, suggesting that they might function to trigger neuronal cell death in certain circumstances. Two examples are the p75 neurotrophin receptor ($p75^{NTR}$) and Fas (CD95). Increased expression of the Fas receptor has been observed in the brains of Alzheimer's patients in both neurons and glial cells from the cortical and white matter regions damaged by degeneration (59, 60). Furthermore, withdrawal of NGF from PC12 cells or removing extracellular potassium from cerebellar granule neurons leads to the up-regulation of Fas ligand (61). In the latter case, blocking the action of Fas ligand with a decoy reagent (Fas-Fc), or by using neurons from FasL-deficient (*gld*) mice, caused a reduction in apoptosis, indicating a functional role for the Fas pathway in neuronal apoptosis. As discussed above (Section 3.1.1), the induction of Fas ligand in granular neurons deprived of survival factors appears to be mediated at least in part by the activation of the Forkhead transcription factor (41).

3.2.2 The p75 neurotrophin receptor

The first receptor shown to transduce a ligand-dependent death signal in neurons was the p75 neurotrophin receptor ($p75^{NTR}$), a low affinity NGF receptor and a member of the Fas/TNF receptor family. The major role previously assigned to $p75^{NTR}$, which binds all the neurotrophins (NGF, BDNF, NT3, and NT4/5) with low affinity, was to enhance survival signalling by the tyrosine kinase neurotrophin receptors TrkA, TrkB, and TrkC (reviewed in refs 8 and 62). However, $p75^{NTR}$ can also mediate PCD in response to NGF signalling in several neuronal cell types including oligodendrocytes (63), Schwann cells (64), sympathetic neurons (65), motor neurons (66, 67), and sensory neurons (68, 69) (see Fig. 2). The first strong evidence for such a role *in vivo* came from work by Frade *et al.* (70), which showed that, in early chick development, NGF caused death of retinal neuronal precursors that express $p75^{NTR}$ in the absence of TrkA. Knockout studies in mice indicate that endogenous $p75^{NTR}$ was necessary for naturally occurring sympathetic neuron death *in vivo* (65). Four general conclusions can be drawn from these studies (8):

(a) $p75^{NTR}$-mediated PCD is ligand dependent.

(b) The apoptotic signal mediated by $p75^{NTR}$ is independent of Trk activation.

(c) This signal is only active when Trk-signalling is reduced or absent, and is inhibited by Trk activation.

(d) In some cells, $p75^{NTR}$ is required for PCD.

$p75^{NTR}$ is important for PCD during development. Knockout studies in mice indicate that $p75^{NTR}$ and NGF are important for the death of Islet-1 immunoreactive cells along axon tracts in the developing optic nerve and spinal cord (71). $p75^{NTR}$-deficient mice also show delayed developmental death of sympathetic neurons (65). The specificity of survival responses to different neurotrophins may also be modulated by the presence of $p75^{NTR}$. Sympathetic neurons, from $p75^{NTR}$-deficient mice, which normally depend on NGF but not NT3 for their survival, can use NT3 as a survival factor both *in vivo* (72) and *in vitro* (73). This apparent modulation of neurotrophin responsiveness by $p75^{NTR}$ is thought to occur because of antagonistic signalling by $p75^{NTR}$ (which activates PCD pathways) and TrkA (which activates survival pathways). Presumably, binding of NT3 to $p75^{NTR}$ in normal neurons results in a dominant death signal.

$p75^{NTR}$ may also be involved in responses to neuronal injury. Transgenic expression of the intracellular domain of $p75^{NTR}$ increased the sensitivity of facial motor-

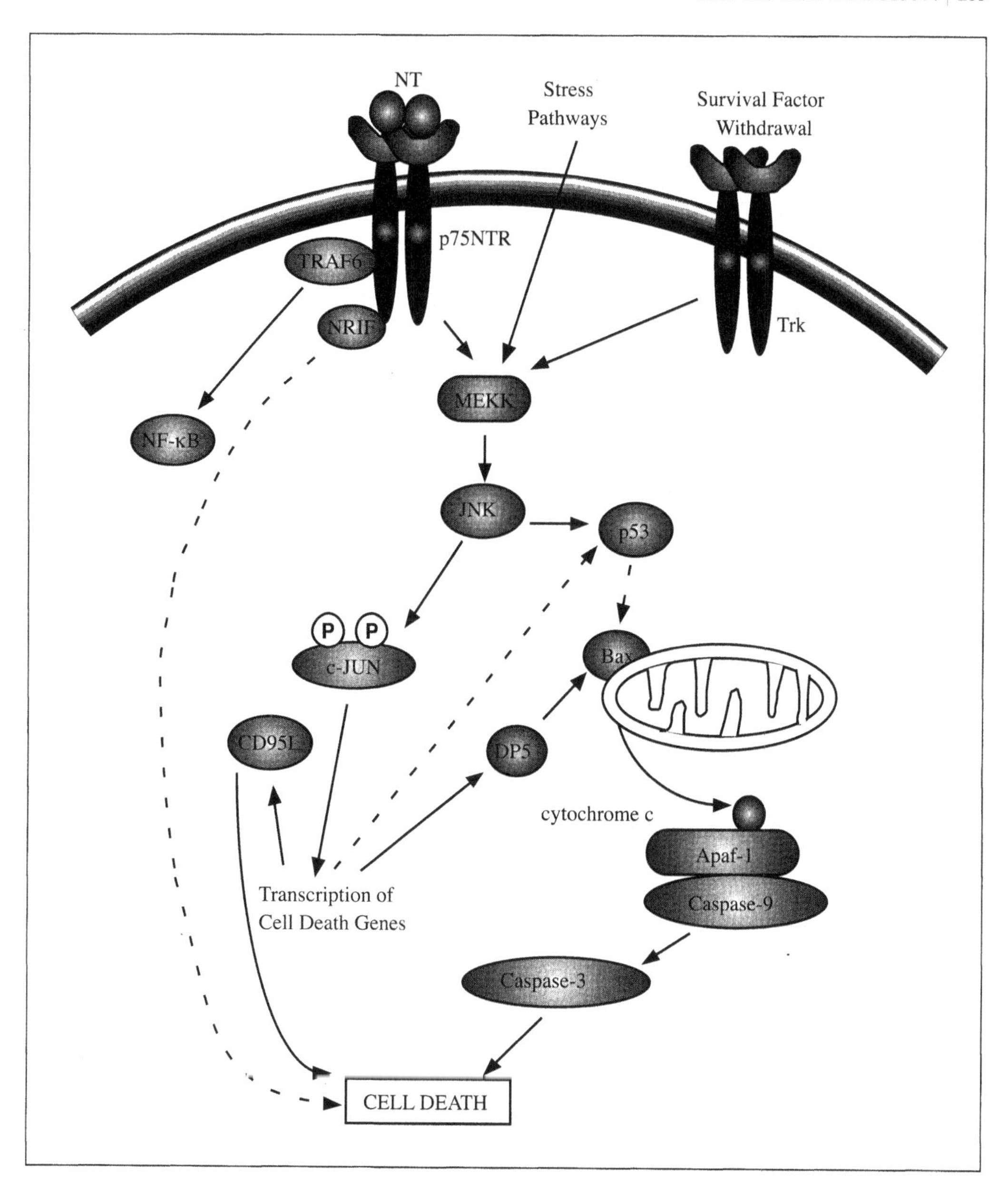

Fig. 2 Activation of programmed cell death in sympathetic neurons. Sympathetic neurons have been a model system for studying the activation of programmed cell death in neurons. PCD can be induced in several ways: (1) by NGF-mediated activation of its low affinity receptor, $p75^{NTR}$, in the absence of Trk-mediated survival activation; (2) by withdrawal of survival factors (which will remove the survival pathways shown in Fig. 1); and (3) by environmental stresses such as the action of toxins, reactive oxygen species, and irradiation. These pathways converge on the c-jun N-terminal kinases (JNKs), which are critical mediators of PCD in neurons. Bax is another critical mediator of PCD in sympathetic neurons, as loss of Bax function (e.g. in knockout mice) results in neurons that are resistant to induction of PCD. Relatively little is known about other cell death regulatory pathways activated by $p75^{NTR}$ via NRIF or TRAF6. The mechanisms by which p53 is transcriptionally regulated, or how p53 activates Bax, are also not settled.

neurons to death after axotomy (74). $p75^{NTR}$ levels have been reported to increase in motorneurons and Schwann cells of newborn mice after facial motorneuron injury (67, 75), and in apoptotic CNS neurons of adult rats after seizure (76), suggesting that $p75^{NTR}$ may contribute to apoptotic signalling after neuronal injury.

The mechanisms by which $p75^{NTR}$ activates apoptotic signalling pathways are not completely understood, but current evidence points to an important role of the JNK-p53-Bax pathway, with p53 as a critical component (77) (see Fig. 2). TrkA silences this JNK-p53 pathway in sympathetic neurons by activation of the Ras pathway through both PI3K/Akt and MEK signalling cascades (16). Other pathways are also likely to be involved, as $p75^{NTR}$ has been found to interact with a number of proteins *in vitro*. These proteins include NRIF (neurotrophin receptor interacting factor), NADE (neurotrophin receptor associated death executor), TRAF6 (tumour necrosis factor receptor associated factor-6), SC-1, and NRAGE (neurotrophin receptor interacting MAGE [melanoma antigen gene] homologue). Mice with a deletion in the NRIF gene show alterations in cell death *in vivo* that are indistinguishable from those of p75(–/–) and NGF(–/–) mice (78). NADE induced NGF-dependent PCD when overexpressed in cultured cells, but only when $p75^{NTR}$ was co-expressed (79), whereas TRAF6 appears to be important for $p75^{NTR}$-mediated activation of NF-κB (80). SC-1 and NRAGE appear to have roles in inducing growth arrest (81, 82).

$P75^{NTR}$ signalling activates multiple pathways, including apoptotic pathways, NF-κB activation, and inhibition of neuronal growth (8). Trk signalling silences $p75^{NTR}$ apoptotic pathways such as JNK-p53, but not activation of NF-κB signalling (83). Indeed, the $p75^{NTR}$-mediated activation of NF-κB appears to be survival-promoting, and to reinforce the survival activity of the Trk pathway (54, 55). In Schwann cells, $p75^{NTR}$-mediated activation of NF-κB can be blocked by dominant-negative TRAF6, a TNF receptor interacting protein that also binds to $p75^{NTR}$ (80). TRAF6 forms a signalling complex with TRANCE in non-neuronal cells to activate Akt (84), suggesting that $p75^{NTR}$ and Trk may co-operate to promote survival via the Akt pathway.

3.2.3 JNK

In the two best-studied paradigms of neuronal cell death *in vitro*—sympathetic and cerebellar granule neurons deprived of survival signals—the induction of apoptosis requires *de novo* protein synthesis. In these cases there must be inducible transcription factors that co-ordinate the expression of other genes that regulate the death machinery. The proto-oncogene c-Jun and its upstream kinase JNK (Jun N-terminal kinase), which is activated in response to various environmental stresses, appear to contribute directly to the demise of the cell (85). JNK is activated in many cell types by the triggering of death receptors as well as by environmental stresses such as UV irradiation and oxidative stress. However, while JNK activation is often associated with, but not necessary for, apoptosis in dividing cells, it is often required for neuronal apoptosis to take place. Elevated c-Jun expression is associated with neuronal degeneration caused by stroke and by neurodegenerative disorders such as Alzheimer's disease and amyotrophic lateral sclerosis (86). Microinjection of anti-c-

Jun antibodies and dominant-negative c-Jun (87, 88), or overexpression of the endogenous JNK inhibitor, JIP (JNK interacting protein) (89), can block neuronal death caused by trophic factor removal. JIP, which is highly expressed in the brain and may play a role in modulating death signalling, binds to JNK and blocks c-Jun phosphorylation. When expressed in pheochromocytoma (PC12) cells, JIP is a very efficient inhibitor of apoptosis induced by NGF deprivation.

Conclusive evidence for the involvement of JNK in receptor-mediated neuronal cell death has come from a study of targeted disruption of the Jnk-3 gene. Yang *et al.* (90) reported that deletion of the Jnk-3 gene caused inhibition of neuronal death in the hippocampus induced by treatment with kainate, an agonist of the excitotoxic glutamate receptor. This finding indicates that JNK signalling is an important component of the neurotoxic response to glutamate. Compound mutants lacking both Jnk1 and Jnk2 (but not either one alone) were embryonic lethal and showed reduced levels of apoptosis in the brain, indicating that JNK signalling is required for PCD during early brain development (91).

One way that JNK may induce apoptosis in neurons is through a p53-Bax-cytochrome *c* pathway. p53 appears to be essential for the death of sympathetic neurons following NGF withdrawal *in vitro*, and for their physiological death *in vivo* (77). One of the transcriptional targets of p53 is Bax, which is also required for neuronal death in several paradigms (discussed below). Examination of fibroblasts from compound JNK-deficient animals indicates that stress-induced activation of the cytochrome *c*-mediated cell death pathway requires JNK activation (92), consistent with a JNK-p53-Bax-cytochrome *c* cell death pathway. The JNK protein kinases are activated by two dual-specificity protein kinses (MAP KKs), MKK4 and MKK7, which in turn are activated by a large group of MAP kinase kinase kinase (KKK). One such MAPKKK, Ask1, appears to be involved in the death of SCG and PC12 neurons and in JNK activation following NGF withdrawal (93).

3.3 Summary: convergence of survival and death signals

All cells receive both survival and death signals, from neighbouring cells and from the environment, and it is the balance of such signals that determines whether a cell undergoes PCD. The most important intracellular integrators of survival signals are the kinases PI3K and Akt. Akt phosphorylates multiple target proteins, including known components of the cell death machinery, and can antagonize cell death signalling. Cell death signalling in neurons, by the p75 neurotrophin receptor as well as in response to stress pathways, converge onto the JNK pathway. Downstream of JNK, activation of Bax is required for cell death in both sympathetic and cerebellar granule neurons, the two most-studied neuronal cell death paradigms. Competition between cell death and survival signals is well illustrated by studies of the p75 and Trk neurotrohpin receptors (8, 83, 94). To succeed, cell death signalling must override concurrent survival signals (and vice versa). A striking example is the caspase-mediated cleavage of the p65 subunit of NF-κB, resulting in the production of a dominant-negative NF-κB inhibitor and consequent supression of NF-κB survival signalling (95).

4. Bcl-2 family and mitochondria

The functions of Bcl-2 family proteins appear to be the same in neurons as they are in many other cells. For example, in transgenic mice overexpressing Bcl-2 (96, 97) or Bcl-x_L (98) targeted to the CNS, both of these proteins inhibit cell death and are neuroprotective in ischaemia and axotomy models. Similarly, Bax overexpression in sympathetic neurons induces cell death, which can be blocked by co-expression of Bcl-x_L or a caspase inhibitor (99). Bax, Bcl-2, and Bcl-x_L proteins are all expressed in the nervous systems of developing and neonatal rats (99, 100), suggesting that these proteins regulate neuronal cell death during development. Postnatally, however, Bax levels decrease (99), whereas Bcl-x_L expression is maintained at high levels in the CNS (101, 102), suggesting that Bcl-x_L may be important for the survival of mature neurons.

Studies of knockout mice have been particularly informative of the role of endogenous Bcl-2 family proteins in neuronal survival. Although *bcl-2* knockout mice do not show any obvious gross abnormalities in nervous system development (103), more subtle effects can be seen on closer inspection. For instance, *bcl-2*-deficient cerebellar granule neurons are reduced in number and are more sensitive to cell death inducing treatments (104). *bcl-x_L* knockout mice, however, show more obvious nervous system defects; examination of these mice indicate that Bcl-x_L is required for neuronal survival during brain development (105). But as these knockout mice do not survive beyond about embryonic day 13.5, an assessment cannot be made of Bcl-x_L's importance in the mature nervous system.

Among Bcl-2 family members, Bax seems to be particularly important in neuronal cell death. Bax knockout mice have shown striking effects on neuronal survival, demonstrating a requirement of Bax for many neuronal cell deaths. During embryonic development, when most neuronal cell death normally occurs, Bax-deficient mice exhibited virtually no PCD between E11.5 and P1 in most peripheral ganglia, motor pools in spinal cord, and trigeminal brainstem nuclear complex. There was also a reduction in cell death throughout the developing cerebellum, some layers of retina, and in hippocampus (106, 107). SCG and facial motor neurons from *bax*-deficient mice survived NGF withdrawal and axotomy, respectively (107). Strikingly, cultured SCG neurons from these mice survived completely without NGF for at least one month. Although these neurons atrophied, they responded to trophic factor re-addition by regaining their normal size and neurite outgrowth. Similarly, cerebellar granule neurons from Bax-deficient mice were resistant to death in low extracellular potassium, although they were not protected from glutamate excitotoxicity induced by NMDA treatment (36). Bax-deficient cortical neurons, however, exhibited significant protection from both excitotoxic and DNA damage-induced killing (108). In this study, glutamate was shown to induce elevated Bax protein in wild-type neurons, which was dependent on the presence of p53. Although adenovirus transfection of p53 alone is not sufficient to induce Bax expression (108), Bax deficiency has been shown to protect against cell death induced by adenovirus transfected p53 expression, at least in cerebellar granule neurons (109).

In pathways of cell death signalling in neurons, Bax acts between early signalling events and the activation of downstream caspases (Fig. 2). Studies of the death pathway induced by NGF withdrawal in Bax-deficient SCG sympathetic neurons have provided especially detailed information, and suggest the following model:

(a) Following NGF withdrawal, early changes occur. These include decreased protein synthesis and glucose uptake, increased levels of c-jun mRNA, and phosphorylation of c-Jun, all of which depend on protein synthesis but are independent of Bax (110).

(b) Subsequent to these events, Bax relocalizes from the cytosol to mitochondria. This relocalization may involve a BH3-only protein such as Bid (111) or DP5/Hrk (see Chapter 5).

(c) Bax localization to mitochondria is followed by release of cytochrome *c* from mitochondria into the cytosol by processes that are protein synthesis-dependent but caspase-independent (112, 113).

Cytochrome *c* release appears to be necessary, but not sufficient, for the death of sympathetic neurons after NGF withdrawal. Although microinjection of anti-cytochrome *c* antibodies can block cell death (114), microinjection of cytochrome *c* was insufficient to induce cell death in NGF-maintained neurons (112, 114). Cytochrome *c* was able to induce caspase-dependent cell death, however, in either Bax-deficient or cycloheximide-treated neurons that had been deprived of NGF for 15–20 hours, suggesting that NGF withdrawal induced another pathway that is independent of both protein synthesis and Bax (112). This finding suggests that NGF withdrawal initiates this second pathway that, together with cytochrome *c* release, triggers cell death.

It is not clear, however, whether cytochrome *c* is required for normal neuron development *in vivo*. Surprisingly, the Apaf-1/cytochrome *c* pathway does not appear to be required in later embryonic development, as the brains of surviving adult Apaf-1 knockout mice develop normally (of the 5% that survive; the remainder die *in utero* from excess numbers of neuronal progenitor cells) (115). Neuron development in cytochrome *c*-deficient mice has not been evaluated, as these mice also die *in utero* and none have been reported to survive postnatally (116).

The point at which sympathetic neurons are committed to die appears to be at the point of caspase activation. In the presence of a caspase inhibitor, cells remained alive (albeit atrophied) in the absence of NGF even after cytochrome *c* was released; indeed, these neurons recovered their size and cytochrome *c* content by a process requiring *de novo* protein synthesis, once NGF was added back (117). Once mitochondrial potential was lost, following cytochrome *c* release, caspase inhibition failed to allow rescue by NGF (118).

The role of Bax in cerebellar granule neurons appears to be similar to that in sympathetic neurons. Programmed cell death in CGNs induced by 5 mM potassium did not occur in Bax-deficient mice and the requirement of Bax can be placed downstream of early changes in macromolecular synthesis and kinase signalling, but upstream of caspase activation (36).

We have discussed the special roles that members of the Bcl-2 and Bax subfamily proteins might play in neurons; the remaining subfamily consists of the BH3-only proteins (see Chapter 5). What little is known about the functions of these proteins in neurons tends to confirm what is already known about BH3-only proteins in other cells (32, 111, 119). It is as yet unclear, however, whether proteins such as Bid or Bad are required for neuronal PCD. The analysis of different BH3 knockout and mutant transgenic mice should help to answer this question.

5. Caspases

The neuronal phenotypes of several caspase knockout mice have been examined, and confirm the critical importance of caspases in neuronal development and response to injury (see Chapter 4). The most severe phenotypes are in the caspase-3- and caspase-9-deficient animals, which are embryonic or perinatal lethal and display a lack of apoptosis in neuroepithelial progenitor cells during development (120–122). Animals deficient in caspase-1, -2, -11, and -12 all develop normally, but show resistance to certain toxic stimuli. Caspase-1 (123) and -11 (124) knockout mice are both resistant to ischaemic brain injury, with a decreased incidence and severity in experimental autoimmune encephalomyelitis reported for the caspase-1-deficient animals. Animals deficient in caspase-2 (125) and -12 (126) were resistant to amyloid-β toxicity.

6. Neurodegenerative diseases and stroke

There are two important questions regarding the role of PCD in neurodegenerative diseases and stroke: first, do cells die by PCD (versus necrosis) and secondly, is PCD involved in or required for disease pathogenesis? These questions have important practical implications for treating these diseases because they determine whether anti-apoptotic therapies are likely to be effective. In principle, such an approach has been proven to work, at least in the fruitfly *Drosophila*, where prevention of apoptosis by expression of the baculovirus caspase inhibitor p35 blocks retinal degeneration (127). The extent to which caspase inhibition can suppress neuron loss in chronic neurodegenerative diseases in humans, where the process takes place over years or decades, remains to be seen. Here we shall discuss the evidence for a role of programmed cell death in the pathogenesis of ischaemia, glutamine (CAG) repeat diseases, Alzheimer's disease, amyotrophic lateral sclerosis (ALS), and spinal muscular atrophy (SMA) (see also ref. 9).

We first discuss the role of PCD in ischaemia, the main source of damage in stroke, as the evidence for the importance of PCD here is best. We next discuss the role of PCD in four chronic neurodegenerative diseases—glutamine repeat diseases (e.g. Huntingtin's disease), Alzheimer's disease, amyotrophic lateral sclerosis (ALS), and spinal muscular atrophy (SMA). For the first three of these, their pathogenesis, like that of Parkinson's disease and the prion diseases, involves altered protein clearance (128). Although defining the role of PCD in any of these chronic diseases remains contentious, we shall discuss what is known about the role of the molecular machinery of PCD in each of them. Not only does this reveal important insights into the nature of the cell death programme, but it is highly relevant to the design of therapeutic strategies.

6.1 Ischaemic stroke

The neuronal loss in ischaemic stroke can be divided into two main phases, based on timing and location. The first phase occurs in a core infarct region that is immediately affected by the loss of circulation. This involves rapid necrotic death of neurons that

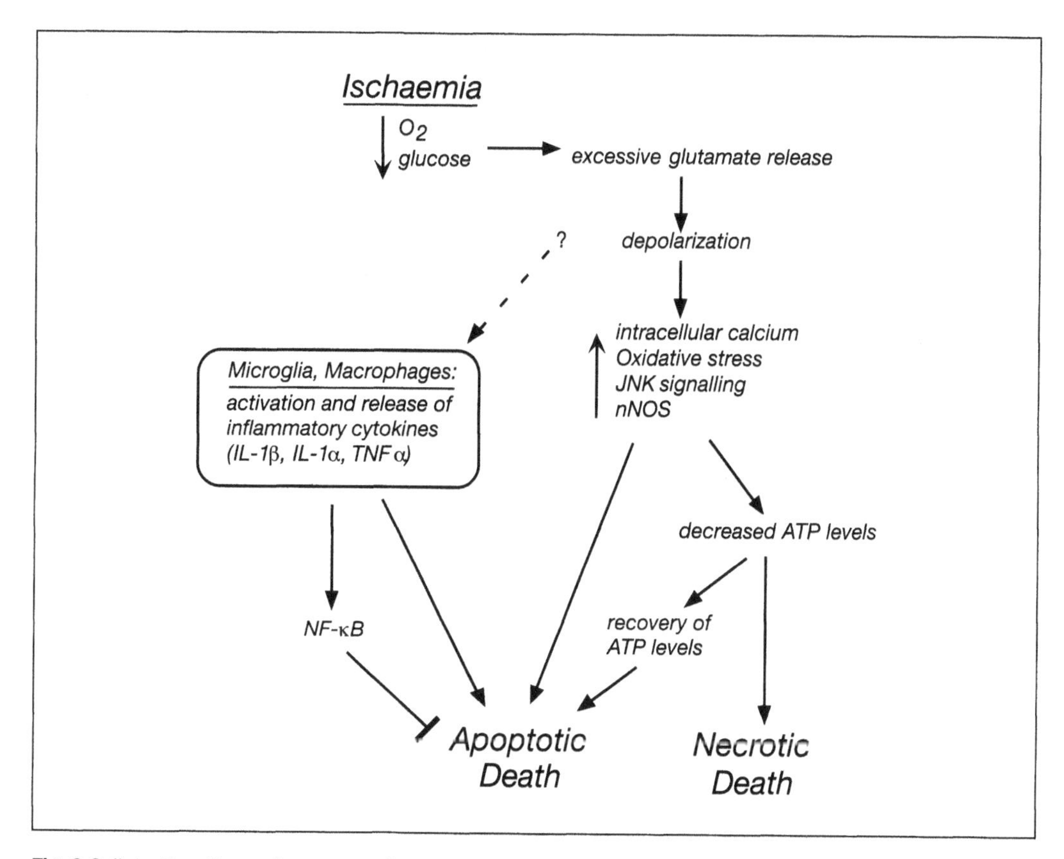

Fig. 3 Cell death pathways in neurons after ischaemic stroke. Transient or focal ischaemia to the brain results in a temporary and/or localized loss of blood circulation bringing oxygen and glucose. Although the neurons immediately affected within the ischaemic zone generally die rapidly by necrosis, those in the surrounding region (the 'penumbra') die later, mainly by apoptosis. The proximal cause of these delayed neuronal deaths is thought to be triggered by excessive levels of glutamate released by neurons and not adequately removed by neighbouring astrocyte glial cells. Downstream events within the neurons (increased intracellular calcium, oxidative stress, and JNK activation), as well as release of cytokines from activated microglia and macrophages, contribute to the activation of apoptotic pathways within the neuronal cells. Excitotoxic levels of glutamate also result in decreased ATP levels within nerve cells. If ATP levels drop substantially, the cell dies by necrosis; otherwise, if its ATP levels recover and the cell dies, it does so by apoptosis.

have been severely deprived of oxygen and nutrients. The second phase occurs in a surrounding region called the penumbra and involves the delayed apoptotic death of neurons that were not immediately killed by the ischaemic event. This delay lasts usually about 12–24 hours or more, during which time several biochemical and cellular processes ensue (Fig. 3). These processes include (in order of timing): an immediate excitotoxicity caused by excess release of excitatory amino acids such as glutamate; depolarization of neurons surrounding the infarct; inflammation; and lastly programmed cell death. In this section we shall focus on the evidence that such delayed neuronal deaths are apoptotic and on the preceding processes involved in triggering them.

6.1.1 PCD in ischaemia

Early studies that support a role for PCD in both global and focal ischaemia in rat brains show evidence of DNA fragmentation characteristic of apoptosis (129, 130). These observations have been confirmed and extended in several different animal models of brain ischaemia using TUNEL, nuclear morphology, and measures of DNA fragmentation (131–135). Activation of caspase-3, involved in the execution phase of PCD, has also been detected in neurons following ischaemic injury (136–141). The strongest evidence for the involvement of apoptosis in brain ischaemia, however, comes from findings of reduced neuron death in transgenic mice that overexpress Bcl-2 or Bcl-x_L in CNS neurons and in mice and rats treated with caspase inhibitors. Infarct size in a permanent ischaemia model was reduced by about 50% in mice that overexpressed neuron-targeted Bcl-2 compared to control mice (97). Neuronal Bcl-x_L overexpression has been shown to have the same effect, reducing tissue loss by 39% to 61.5%, depending on brain region, compared to controls (98). Inhibition of caspases, using both broad-spectrum as well as more subclass-specific (YVAD- and DEVD-specific) peptide inhibitors, reduces brain injury and neuronal loss, as well as subsequent behavioural deficits, in mice (142, 143) and rats (137–139, 144).

Interestingly, these caspase inhibitor studies suggest the involvement of caspases in two processes in ischaemia: the processing of cytokines such as pro-IL-1beta by caspase-1 and the execution of the cell death programme by activated caspase-3. IL-1beta signalling is involved in the inflammatory response that follows the initial ischaemic event and blocking this pathway can inhibit ischaemic neuronal damage (145–152). Consistent with IL-1beta as an inflammatory mediator, caspase-1 protein and mRNA levels have been observed to increase in microglia after ischaemic damage and this correlates with the appearance of apoptotic neurons (153). IL-1beta signalling may also induce programmed cell death in target cells directly (154). The observation that sympathetic neurons cultured from caspase-1-deficient mice are resistant to PCD induced by trophic factor withdrawal supports this contention (150). Most importantly, mice deficient in either caspase-1 (123) or caspase-11 (124) are resistant to apoptosis induced by ischaemic brain injury.

Caspase-3 is a critical component of the execution phase of programmed cell death and is activated following ischaemia, so blocking its activity would be expected to prevent cell death without affecting cytokine processing. Nevertheless, isolating its specific involvement is not straightforward. The early embryonic lethality of mice deficient in caspases-3 and -9 makes their importance in brain ischaemia difficult to assess. Furthermore, although peptide inhibitors based on the sequence DEVD are relatively specific for execution caspases such as caspase-3, they also inhibit caspase-8, an upstream signalling caspase activated by death signalling receptors such as CD95 (Fas) (see Chapter 8). Indeed, caspase-8 is expressed by cortical neurons and its activation has been detected in neurons undergoing delayed cell death after focal stroke (155). Virally-mediated overexpression of the endogenous anti-apoptotic protein XIAP has been shown to block caspase-3 activation and neuronal damage in cerebral ischaemia (140), but the specificity of this protein when overexpressed is not certain. Despite the difficulty in completely separating out inflammatory cytokine signalling from the execution of the cell death programme, the evidence is fairly compelling that the delayed neuronal death following ischaemia is apoptotic and mainly caspase-dependent.

6.1.2 Excitotoxicity

One of the earliest effects of ischaemia is the release of excess L-glutamate and L-aspartate, amino acids used as excitatory neurotransmitters in the brain but which are toxic to neurons at high concentrations (reviewed in ref. 156). Inhibition of NMDA-type glutamate receptors can reduce infarct volume in animal models if these antagonists are given before or immediately after ischaemia. Although this treatment does not completely prevent the caspase-dependent death of neurons that occurs later, the combination of caspase and NMDA inhibitors is more effective than either alone in some models (139, 143).

The neurotoxicity that follows exposure to high extracellular concentrations of glutamate can be divided into two types of cell death, one that occurs early and appears necrotic, and another that occurs later and is apoptotic (reviewed in ref. 156). The proportion of apoptotic versus necrotic death depends on concentration: low concentrations of L-glutamate cause only apoptotic cell death, whereas high micromolar to millimolar concentrations initially induce cell swelling, loss of mitochondrial membrane potential, and reduced ATP levels. The neurons that survive this initial necrotic phase and recover, however, then proceed to die later by programmed cell death (157).

Some of the programmed cell death induced by L-glutamate can be attributed to any one of several sequelae. Activation of ionotropic glutamate receptors induces depolarization and calcium influx in neurons. Much of the toxicity of glutamate is attributed to this excessive increase in intracellular calcium, which activates neuronal nitric oxide synthase (nNOS), mitochondrial respiratory chain dysfunction and ATP depletion (discussed above), and activation of specific enzymes such as phospholipase A_2, tissue plasminogen activator (tPA), and calpain. These biochemical changes produce oxidative stress and activate downstream MAP kinase pathways, leading to programmed cell death. JNK in particular has been implicated in excitotoxin-induced PCD in neurons. Disruption of the gene encoding Jnk3 in mice caused the mice to be resistant to the excitotoxic glutamate receptor agonist kainic acid: these mice showed a reduction in seizure activity and prevention of apoptosis in hippocampal neurons (90). Prolonged N-terminal phosphorylation of c-Jun has been detected in neurons after transient ischaemia along with expression of CD95(Fas)-ligand and TUNEL-labelled nuclei (158), suggesting that JNK activation participates in an ischaemia-induced PCD pathway that may also involve CD95 signalling.

In addition to the activating glutamate receptors and consequent calcium influx, glutamate also inhibits cysteine uptake by the glutamate/cysteine antiporter. Intracellular cysteine is limiting for the synthesis of glutathione, an important cellular antioxidant that helps to inhibit the toxicity of reactive oxygen species. Depletion of glutathione has been shown to lead to apoptotic death in neurons in culture (159, 160). These two mechanisms may both contribute to the delayed apoptotic death of neurons after ischaemia.

6.1.3 NF-κB and inflammatory cytokines

In addition to IL-1beta, TNFα and IL-1α are produced by microglia and invading macrophages shortly after ischaemia as part of the inflammatory response. TNFα induces the apoptotic death of a variety of neurons (161), particularly in cerebral ischaemia (162). TNFα induces cell death by activation of its p55 receptor (Chapter 8). TNFα also induces NF-κB activation, which can suppress apoptosis induced by TNFα (163). In cerebellar granule neurons, TNFα has also been shown to inhibit signalling by the survival peptide IGF-1. TNFα promotes IGF-1 receptor resistance through suppression of IGF-1-induced tyrosine phosphorylation of insulin receptor substrate 2 (IRS-2) and subsequent activation of PI3 kinase (164).

All of the stimuli that are activated in ischaemic stroke, including glutamate excitotoxicity, TNFα, hypoxia, reactive oxygen species, and IL-1α, also activate NF-kappa B (165). The role of NF-kappa B in ischaemia is somewhat complex as it has both protective and pro-apoptotic actions in neurons (review in ref. 166). Whether NF-kappa B is ultimately protective or pro-apoptotic is thought to depend on the timing and intensity of the stimulus.

There also appears to be feed-forward and feedback interactions between NF-kappa B and other signalling pathways. In endothelial cells deprived of growth factors, for example, the surviving cells exhibit increased NF-kappa B activity, whereas the apoptotic cells show caspase-mediated cleavage of the NF-kappa B RelA subunit (95). This cleavage inactivates the transcriptional activity of the p65 molecule and produces a dominant-negative inhibitor of NF-kappa B that promotes apoptosis.

6.2 Polyglutamine repeat diseases

A group of adult-onset neurodegenerative diseases is caused by proteins with expanded polyglutamine tracts (167). These diseases include Huntington's disease (HD) and several forms of spinal cerebellar ataxia (SCA) and are characterized by the loss of select subpopulations of neurons. The causative genes in these disorders, which are inherited in an autosomal dominant fashion, have been identified and include huntingtin in HD, ataxin-1 in spinocerebellar ataxia type 1 (SCA1), ataxin-2 in SCA2, ataxin-3 in SCA3 (Machado Joseph disease), ataxin-7 in SCA7, atrophin-1 in dentalorubralpallidoluydian atrophy (DRPLA), and the androgen receptor in spino-bulbar muscular atrophy (SMBA). In all of these illnesses except one, the affected proteins have no homology to one another outside of their polyglutamine repeats and all exhibit a gain-of-function mutation. A possible exception is SCA6, which has smaller trinucleotide expansion and the mutation may act via a different route (168, but see ref. 169). Although the target genes are widely expressed in the brain as well as in other tissues, each are associated with the dysfunction and loss of selected subpopulations of neurons. In all of these disorders except SCA6, a threshold of 35–40 glutamines provokes the clinical manifestation of the disease and longer repeats are associated with more severe pathology and an earlier age of clinical onset.

Studies of transgenic animals that express mutant proteins with expanded glutamine repeats provide some insight into possible mechanisms of pathogenesis. Transgenic mice that express mutant ataxin-1 (170) and Huntingtin (171) proteins with expanded glutamine repeats develop neurological phenotypes and neuronal pathologies similar to those observed in SCA-1 and HD. A prominent histological feature observed in both the human diseases and the mouse models is the appearance of neuronal intranuclear inclusions formed by aggregates of the mutant protein. Expression of mutant proteins containing polyglutamine tracts in *Drosophila* and *C. elegans* also results in neuronal degradation and intranuclear inclusions (172–174), suggesting that the pathogenesis of these diseases is a result of an inherent neurotoxicity of expanded polyglutamine containing proteins.

Is the glutamine repeat the critical mediator of neuronal toxicity, or is the context of the specific protein in which the repeat resides somehow important? To answer this question, Ordway *et al.* (175) produced mice that express a long glutamine repeat inserted into the mouse hypoxanthine-guanine phosphoribosyltransferase (HPRT) gene, a housekeeping gene not associated with neurological disease. These animals progressively developed several neurological abnormalities, exhibited intranulcear inclusions in affected neurons, and died prematurely, suggesting that long poly-glutamine stretches themselves may be responsible for the neurotoxicity.

The consistent appearance of intranuclear inclusions in polyglutamine repeat diseases and their animal models suggests that the neurotoxicity of polyglutamine stretches may relate to their aggregation. Proteins with polyglutamine stretches can aggregate *in vitro* to form amyloid-like fibrils (176), and aggregates of mutant protein are observed in neurons of patients with HD, SCA-1 and -3, and DRPLA (177). Furthermore, in HD, in which a larger number of cases have been studied, the length of the glutamine expansion is directly proportional to the density of the inclusions (178). The intranuclear aggregates contain ubiquitinated derivatives of the mutant protein, suggesting that ineffective clearance by the ubiquitin pathway might contribute to the formation of intranuclear inclusions (128).

The significance of intranuclear inclusions, however, is not yet clear. In transgenic mice that express exon 1 of huntingtin containing an expanded glutamine repeat, nuclear inclusions containing huntingtin and ubiquitin appeared prior to the neurological phenotype and changes in neuronal morphology, consistent with a causative role for nuclear inclusions (179). When mutant huntingtin was transfected into cultured striatal neurons, however, the formation of inclusions did not correlate with huntingtin-induced neuronal death (180). Surprisingly, when the formation of nuclear aggregates was inhibited, by preventing ubiquitination of huntingtin, the rate of cell death was accelerated. This raises the possibility that aggregate formation is a by-product of a process to clear and de-toxify the mutant protein. In another study, the formation of nuclear aggregates was not required for mutant ataxin-1-mediated pathogenesis in transgenic mice (181). Interestingly, in both studies nuclear localization of the mutant proteins was required for pathogenesis.

Apoptotic cell death has been implicated in HD with the observation of increased levels of DNA strand breaks in brains of HD patients, and by the involvement of caspases. The expression of an expanded polyglutamine could induce small aggregates and neuronal PCD in primary cultures of rat neurons via the recruitment of FADD and caspase-8 (182). Moreover, the death of these neurons could be inhibited by a dominant-negative form of FADD, as well as by Bcl-2, Bcl-x_L, and the caspase inhibitor CrmA, implicating a caspase cascade that involves caspase-8. Caspase-8 was also found localized to huntingtin-containing aggregates in the brains of patients that had died of HD. Direct evidence for a role of caspases in a CAG repeat disease comes from studies of a transgenic mouse model of HD. Crossing these mice with a dominant-negative caspase-1 mutant or intra-cerebroventricular injection of a caspase inhibitor delayed disease onset and progression of pathology, and prevented the appearance of huntingtin cleavage product (183).

The contribution of apoptosis to the observed neurological deficits remains unresolved. In early stages of HD, the characteristic motor deficits are observed without evidence of the striatal atrophy which becomes prominent later in the disease (184). These findings are mimicked in a transgenic mouse model of HD in which neuronal PCD is also not seen (171). Furthermore, in a conditional HD transgenic mouse, both the nuclear inclusions and neuronal deficits can be reversed by turning off expression of the transgene (185). It is therefore possible that expression of expanded polyglutamine repeats causes neuronal dysfunction before neuronal cell death occurs, and that this dysfunction—and not cell death—is responsible for symptoms early in the disease.

6.3 Alzheimer's disease

Alzheimer's disease (AD) is a neurodegenerative disease associated with neuronal loss and specific pathological features including the formation of neurofibrillary tangles (NFTs) and the accumulation of β-amyloid in extracellular deposits called plaques (reviewed in ref. 186). Although apoptotic cell death has been reported as a pathological feature of AD post-mortem brains (187–189) and of neurons treated with Aβ *in vitro* (reviewed in ref. 190), the extent to which apoptosis contributes to the pathogenesis of AD is difficult to assess. Since AD is a chronic disease, only a few cells are likely to be observed dying at any time. Furthermore, TUNEL staining, an important source of histological evidence for apoptosis (187, 188), must be interpreted with caution as it is subject to artefactual staining and can label non-apoptotic cells that have been damaged by oxidative stress or during post-mortem preparation of tissue (191). Direct measurements of cells with apoptotic morphology, coupled with immunohistochemical staining of activated caspase-3, suggest levels of apoptosis compatible with the progression of neuronal degeneration seen in AD (192), although apoptosis may not be the only mechanism of cell death in AD. Although the proximal cause of neurodegeneration is not settled, genetic studies of inherited forms of AD have implicated several proteins: Amyloid-β protein, the presenilins (PS-1 and PS-2), Tau, and ApoE.

6.3.1 APP and Aβ

Amyloid precursor protein (APP) is a membrane-spanning protein that is normally processed by proteolytic cleavage to produce secreted peptide fragments, Aβ and p3. The identified familial Alzheimer's disease (FAD) mutations in APP lead to increased production of amyloid-β peptide (Aβ), particularly of its amyloidogenic form, Aβ(1–42), which accumulates in senile plaques of both sporadic and familial forms of AD. Although mutations in APP are rare, the more frequent mutations observed in the PS1 and PS2 genes all result in increased levels of secreted Aβ(1–42), supporting a central role for Aβ in the pathogenesis of Alzheimer's disease. Further support for a role of Aβ in AD pathogenesis comes from APP transgenic mice, which overexpresses mutant forms of human APP and progressively develop many of the neuropathological hallmarks of AD (193–195). The development of AD-like neuropathology in one such APP mouse model was shown to be prevented by immunization of young mice against $A\beta_{42}$ (196). But does Aβ induce apoptosis and is this essential for the neuropathological features of AD?

Aβ peptide, when presented in an aggregated form, is neurotoxic in culture and *in vivo* and induces programmed cell death (197, 198). Although the mechanism of Aβ toxicity is not settled, it is known to induce oxidative stress (199), increased intracellular calcium (200), and apoptosis (201) (see Fig. 4). Aβ may activate signalling pathways by interacting with neuronal receptors such as RAGE (receptor for advanced glycosylation end-products), $p75^{NTR}$, and APP. RAGE may mediate free radical production (202), while neuronal cell death may be induced by $p75^{NTR}$ (203) and APP (204). The most direct evidence that Aβ activates cell death signalling pathways is the requirement of specific caspases. Cortical neurons derived from mice deficient in caspase-12 (126) or treated with antisense oligonucleotides to caspase-2 or a caspase inhibitor (125) are resistant to Aβ-induced apoptosis. Intriguingly, caspase-3 can cleave APP in its cytoplasmic domain to produce a truncated form of APP that, when transfected into cells, produces more Aβ than unprocessed APP, suggesting that the activation of caspases promotes the cleavage of APP and facilitates the generation of Aβ (205). However, there is a substantial amount of evidence against this hypothesis, not least that caspases do not always localize with cleaved APP and that the aspartate cleavage seen in APP does not conform to a classic caspase cleavage reaction. Indeed, caspase activation and Aβ production could be occurring simultaneously and independently.

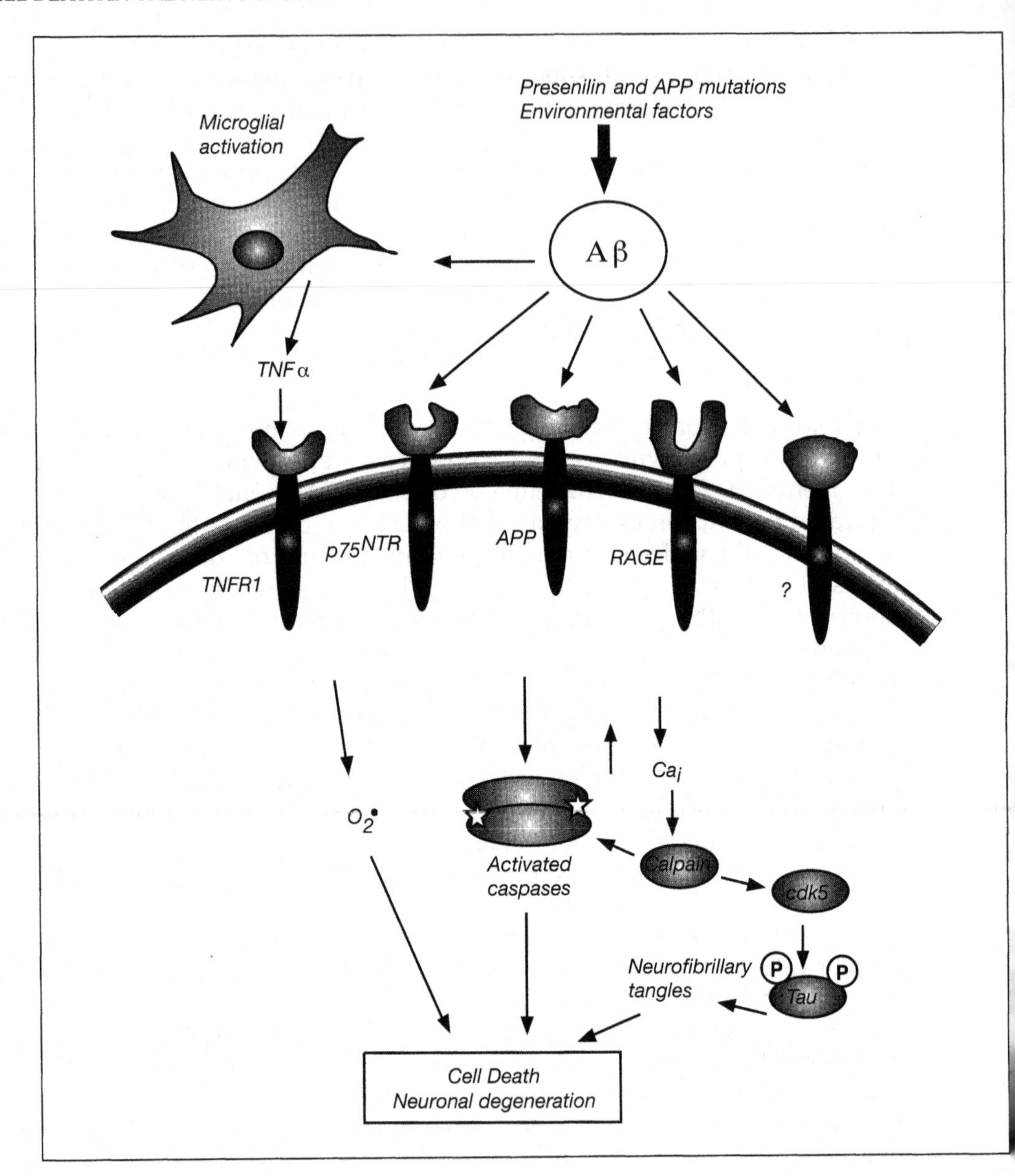

Fig. 4 Cellular pathways of amyloid-β-mediated neurotoxicity in Alzheimer's disease. Aggregated forms of amyloid-β protein interact with several cell surface receptors and can activate microglia, which in turn can generate reactive oxygen species, activate caspases, and induce increases in intracellular calcium. Activated calpain can activate certain caspases and cleave cdk5 into a form that hyperphosphorylates Tau, leading to neurofibrillary tangles. These pathways have all been implicated in neurodegeneration and neuronal cell death, although their relative contributions to the pathogenesis of Alzheimer's disease has not been settled. Figure adapted from ref. 9.

6.3.2 The presenilins

Mutations in presenilin genes are a major cause of early-onset FAD (206–208). Recent studies suggest that presenilins may be gamma-secretases, proteases involved in the processing of APP into Aβ (209–211). Mutations in presenilin genes are associated with increased production of Aβ(1–42) in the brains of AD patients (212–215), suggesting that these mutations alter normal APP processing. Presenilin mutations increase neuronal vulnerability to apoptosis (reviewed in ref. 216), although their contribution to neuronal death in AD remains to be established. The presenilins appear to be caspase substrates and can be regulated by phosphorylation (217). In addition, presenilins interact with other proteins that may be involved in signalling pathways linked to PCD (218–220), although the relevance of these interactions to apoptosis in AD is not known.

6.3.3 Tau phosphorylation and neurofibrillary tangles

Neurofibrillary tangles (NFTs) are an obvious feature in the brains of patients with AD and are made up of aggregates of a hyperphosphorylated form of the protein Tau. Although NFTs are not required, *per se*, for the neuronal cell loss observed in AD (191), the hyperphosphorylation of Tau may be explained by the overactivation of cyclin-dependent kinase 5 (cdk5) in affected neurons. Cdk5 is regulated by a cyclin-like protein, p35, which has been found to be truncated in neurons from AD patients. This truncation constitutively activates Cdk5 and results in hyperphosphorylated Tau and possibly the resulting NFTs. Importantly, expression of the truncated Cdk5 protein causes apoptosis in cultured primary neurons, suggesting that the rogue activation of this protein is important in the pathogenesis of AD (221). The finding that calpain is responsible for cleaving p35 into its truncated, toxic form suggests that Aβ may exert some of its toxicity in neurons by increasing intracellular calcium, activating calpain, and thereby activating Cdk5 (222).

6.3.4 Microglial activation

Microglia are a prominent feature of inflammatory processes in the brain, and their activation can contribute to neuronal death. In addition to their involvement in ischaemia (see Section 6.1), activation of microglia is a feature of several neurological disorders including HIV dementia, MS, and AD (223), and has been implicated in the neurotoxicity of a form of prion protein (224). Microglia are associated with amyloid plaques and can be induced by Aβ (225). Aβ-induced microglial activation induces the secretion of TNFα and other cytokines that can induce neuronal PCD (226), suggesting that microglia can contribute indirectly to the neurotoxicity of Aβ.

6.4 Amyotrophic lateral sclerosis (ALS)

Amyotrophic lateral sclerosis (ALS) is a disease characterized by a progressive loss of spinal motorneurons leading to muscle weakness, paralysis, and finally death. The onset of symptoms generally starts in middle age and invariably proceeds to death in about five years. About 5–10% of ALS cases are inherited (familial ALS, or FALS), in an autosomal dominant fashion, of which about 25% are due to mutations in the gene encoding Cu/Zn-superoxide dismutase (SOD1). Other than this clue (227), the aetiology of ALS remains unclear. Nevertheless, the question of whether motorneurons ultimately die by apoptosis in ALS is an important one because it may help in the design of anti-apoptotic therapies to delay or prevent the progression of the disease.

The best evidence for the involvement of apoptosis in ALS may come from studies of transgenic mice that express FALS-mutant human SOD-1 protein and show progressive motorneuron degeneration similar to the human disease (reviewed in ref. 228). By contrast, mice that are deficient in SOD-1, or that overexpress the wild-type protein, do not develop motorneuron disease, indicating that the disease is caused by a gain-of-function in the mutant form of SOD-1. Although the mechanism of mSOD-1 toxicity is not established, the protein can form intracellular aggregates and induce oxidative stress. When overexpressed in neuronal cell lines, mSOD-1 is pro-apoptotic (229, 230)

So far, experiments that cross these SOD-1 mutant mice with mice that express anti-apoptotic proteins in neurons are suggestive, albeit not conclusive, of a role for apoptosis in ALS. Bcl-2 is protective in mSOD-1 (G93A) transgenic mice, resulting in delayed onset of ALS symptoms and mortality and attenuated motoneuron death in the spinal cord, although without altering the progression of the disease (231). In another study (232), a mouse line that expresses a dominant-negative form of ICE (caspase-1) in neurons (150) was crossed with the G93R SOD mutant mouse (233). These mice showed increased length of disease and slower mortality, but no change in onset, compared to the founder SOD mice, suggesting a role for caspase-1-related caspases in the progression—but not the onset—of ALS. In another study, in which G93A SOD-1 mice (234) were intra-cerebroventricularly administered zVAD-fmk, a broad-spectrum caspase inhibitor, both disease onset and mortality were delayed (235), suggesting that caspase activity is important in the initiation of disease pathogenesis. Indeed, activated caspase-1 and caspase-3 have been detected in the spinal cords of ALS patients and in mSOD-1 transgenic mice (229, 235). Given the prominent role of caspase-1 and closely related caspases in cytokine processing, it seems likely that inflammatory cytokines may contribute to the progression of ALS, possibly through the involvement of microglia. But whether apoptosis is required remains uncertain.

6.5 Spinal muscular atrophy (SMA)

Spinal muscular atrophy (SMA) is a motor neuron disease characterized by degeneration of the anterior horn cells of the spinal cord. It is a common fatal autosomal recessive disorder and linkage studies have identified two candidate genes, SMN and NAIP, both on chromosome 5q13. SMN is believed to be the causative gene in SMA (236), whereas NAIP is thought to play a modulatory role, exacerbating the severity of the disease (237).

The SMN gene is deleted or interrupted on both chromosomes in nearly all SMA patients. SMN shows no significant sequence similarity to any other known protein, but is thought to be important for the biogenesis of snRNPs, suggesting that defects in SMN might cause SMA by interfering with spliceosome synthesis and RNA splicing (238, 239). How such a defect might lead to programmed cell death in motor neurons—if indeed the motor neurons die by PCD—is not yet known. One suggestion is that SMN promotes the anti-apoptotic activity of Bcl-2. SMN interacts with Bcl-2; when co-transfected the two proteins are synergistically anti-apoptotic, whereas SMN alone is only weakly anti-apoptotic and SMA-mutant SMN proteins are not effective at synergizing with Bcl-2 at all (240).

NAIP is an anti-apoptotic protein, related to the IAP proteins first discovered in baculovirus. Partial deletions in NAIP have been found in a significant portion of patients with SMA (241), but the role of NAIP in SMA has been difficult to separate from that of SMN, because of the physical proximity of these two genes. Ectopic overexpression of NAIP in the CNS of transgenic mice attenuated neuronal damage in a transient ischaemia model, suggesting that NAIP is neuroprotective (242). NAIP-deleted mice develop normally, and do not show any symptoms of SMA, demonstrating that deletion of this gene is not necessary for disease (243). However, these mice were more vulnerable to kainic acid-induced injury, supporting the hypothesis that NAIP is neuroprotective, and explaining how deletions in this gene would exacerbate the severity of SMA caused by SMN deletions.

7. Conclusions

Analysis of the molecular machinery involved in neuronal cell death reinforces the concept of a core cell death mechanism. Molecules critical for neuronal cell death include key components of the cell death machinery—Apaf-1, the Bcl-2 family, and caspases. Bcl-2 family proteins and caspases play a major role in the developing brain, as demonstrated by the neuronal phenotypes observed in knockout mice. Neurotrophins play a particularly important role in mediating cell–cell signalling in the developing nervous system, regulating both survival and death of neurons and thereby regulating their overall numbers and patterns of connectivity. Neurotrophin receptors regulate critical kinase pathways that regulate cell survival, most importantly the PI3K/Akt and MAPK pathways. In human neurodegenerative diseases as well as stroke, caspases play an important role, both as essential components of the cell death pathway and as mediators of inflammatory cytokine signalling. Although apoptosis appears to play a prominent part in many neurodegenerative diseases, it is not known exactly how toxic stimuli trigger programmed cell death in neurons, or how important PCD pathways are in the pathogenesis of many of these diseases. In particular, it remains to be seen whether rescuing neurons by inhibiting these cell death pathways in humans will retain sufficient neuronal function to be an effective strategy against neurodegenerative disease.

References

1. Glucksmann, A. (1951). Cell deaths in normal vertebrate ontogeny. *Biol. Rev.*, **26**, 59–86.
2. Clarke, P. G. H. and Clarke, S. (1996). Nineteenth century research on naturally occurring cell death and related phenomena. *Anat. Embryol.*, **193**, 81–99.
3. Jacobson, M. D., Weil, M., and Raff, M. C. (1997). Programmed cell death in animal development. *Cell*, **88**, 347–54.
4. Hamburger, V. and Levi-Montalcini, R. (1949). Proliferation, differentiation and degeneration in the spinal ganglia of the chick embryo under normal and experimental conditions. *J. Exp. Zool.*, **111**, 457–501.
5. Purves, D. (1988). *Body and brain: a trophic theory of neural connections.* Harvard Press, Cambridge, Mass.
6. Barde, Y. A. (1989). Trophic factors and neuronal survival. *Neuron*, **2**, 1525–34.
7. Raff, M. C. (1992). Social controls on cell survival and cell death. *Nature*, **356**, 397–400.
8. Kaplan, D. R. and Miller, F. D. (2000). Neurotrophin signal transduction in the nervous system. *Curr. Opin. Neurobiol.*, **10**, 381–91.
9. Yuan, J. and Yankner, B. A. (2000). Apoptosis in the nervous system. *Nature*, **407**, 802–9.
10. Rohn, J. L., Hueber, A. O., McCarthy, N. J., Lyon, D., Navarro, P., Burgering, B. M., *et al.* (1998). The opposing roles of the Akt and c-Myc signalling pathways in survival from CD95-mediated apoptosis. *Oncogene*, **17**, 2811–18.
11. Kulik, G., Klippel, A., and Weber, M. J. (1997). Antiapoptotic signalling by the insulin-like growth factor I receptor, phosphatidylinositol 3-kinase, and Akt. *Mol. Cell. Biol.*, **17**, 1595–606.
12. Dudek, H., Datta, S. R., Franke, T. F., Birnbaum, M. J., Yao, R., Cooper, G. M., *et al.* (1997). Regulation of neuronal survival by the serine-threonine protein kinase Akt. *Science*, **275**, 661–5.
13. Songyang, Z., Baltimore, D., Cantley, L. C., Kaplan, D. R., and Franke, T. F. (1997). Interleukin 3-dependent survival by the Akt protein kinase. *Proc. Natl. Acad. Sci. USA*, **94**, 11345–50.
14. Yao, R. and Cooper, G. M. (1995). Requirement for phosphatidylinositol-3 kinase in the prevention of apoptosis by nerve growth factor. *Science*, **267**, 2003–6.
15. Rodriguez-Viciana, P., Warne, P. H., Dhand, R., Vanhaesebroeck, B., Gout, I., Fry, M. J., *et al.* (1994). Phosphatidylinositol-3-OH kinase as a direct target of Ras. *Nature*, **370**, 527–32.
16. Mazzoni, I. E., Said, F. A., Aloyz, R., Miller, F. D., and Kaplan, D. (1999). Ras regulates sympathetic neuron survival by suppressing the p53- mediated cell death pathway. *J. Neurosci.*, **19**, 9716–27.
17. Klesse, L. J. and Parada, L. F. (1998). p21 ras and phosphatidylinositol-3 kinase are required for survival of wild-type and NF1 mutant sensory neurons. *J. Neurosci.*, **18**, 10420–8.
18. Holgado-Madruga, M., Moscatello, D. K., Emlet, D. R., Dieterich, R., and Wong, A. J. (1997). Grb2-associated binder-1 mediates phosphatidylinositol 3-kinase activation and the promotion of cell survival by nerve growth factor. *Proc. Natl. Acad. Sci. USA*, **94**, 12419–24.
19. Korhonen, J. M., Said, F. A., Wong, A. J., and Kaplan, D. R. (1999). Gab1 mediates neurite outgrowth, DNA synthesis, and survival in PC12 cells. *J. Biol. Chem.*, **274**, 37307–14.
20. Yamada, M., Ohnishi, H., Sano, S., Nakatani, A., Ikeuchi, T., and Hatanaka, H. (1997). Insulin receptor substrate (IRS)-1 and IRS-2 are tyrosine- phosphorylated and associated

with phosphatidylinositol 3-kinase in response to brain-derived neurotrophic factor in cultured cerebral cortical neurons. *J. Biol. Chem.*, **272**, 30334–9.

21. Vanhaesebroeck, B. and Alessi, D. R. (2000). The PI3K-PDK1 connection: more than just a road to PKB. *Biochem. J.*, **346 Pt 3**, 561–76.
22. Philpott, K. L., McCarthy, M. J., Klippel, A., and Rubin, L. L. (1997). Activated phosphatidylinositol 3-kinase and Akt kinase promote survival of superior cervical neurons. *J. Cell Biol.*, **139**, 809–15.
23. Miller, T. M., Tansey, M. G., Johnson, E. M., Jr., and Creedon, D. J. (1997). Inhibition of phosphatidylinositol 3-kinase activity blocks depolarization- and insulin-like growth factor I-mediated survival of cerebellar granule cells. *J. Biol. Chem.*, **272**, 9847–53.
24. Skaper, S. D., Floreani, M., Negro, A., Facci, L., and Giusti, P. (1998). Neurotrophins rescue cerebellar granule neurons from oxidative stress- mediated apoptotic death: selective involvement of phosphatidylinositol 3-kinase and the mitogen-activated protein kinase pathway. *J. Neurochem.*, **70**, 1859–68.
25. Crowder, R. J. and Freeman, R. S. (1998). Phosphatidylinositol 3-kinase and Akt protein kinase are necessary and sufficient for the survival of nerve growth factor-dependent sympathetic neurons. *J. Neurosci.*, **18**, 2933–43.
26. Vaillant, A. R., Mazzoni, I., Tudan, C., Boudreau, M., Kaplan, D. R., and Miller, F. D. (1999). Depolarization and neurotrophins converge on the phosphatidylinositol 3-kinase-Akt pathway to synergistically regulate neuronal survival. *J. Cell Biol.*, **146**, 955–66.
27. Virdee, K., Xue, L., Hemmings, B. A., Goemans, C., Heumann, R., and Tolkovsky, A. M. (1999). Nerve growth factor-induced PKB/Akt activity is sustained by phosphoinositide 3-kinase dependent and independent signals in sympathetic neurons. *Brain Res.*, **837**, 127–42.
28. Crowder, R. J. and Freeman, R. S. (1999). The survival of sympathetic neurons promoted by potassium depolarization, but not by cyclic AMP, requires phosphatidylinositol 3-kinase and Akt. *J. Neurochem.*, **73**, 466–75.
29. Wiese, S., Digby, M. R., Gunnersen, J. M., Gotz, R., Pei, G., Holtmann, B., *et al.* (1999). The anti-apoptotic protein ITA is essential for NGF-mediated survival of embryonic chick neurons. *Nature Neurosci.*, **2**, 978–83.
30. Datta, S. R., Brunet, A., and Greenberg, M. E. (1999). Cellular survival: a play in three Akts. *Genes Dev.*, **13**, 2905–27.
31. Chan, T. O., Rittenhouse, S. E., and Tsichlis, P. N. (1999). AKT/PKB and other D3 phosphoinositide-regulated kinases: kinase activation by phosphoinositide-dependent phosphorylation. *Annu. Rev. Biochem.*, **68**, 965–1014.
32. Datta, S. R., Dudek, H., Tao, X., Masters, S., Fu, H., Gotoh, Y., *et al.* (1997). Akt phosphorylation of BAD couples survival signals to the cell- intrinsic death machinery [In Process Citation]. *Cell*, **91**, 231–41.
33. Cardone, M. H., Roy, N., Stennicke, H. R., Salvesen, G. S., Franke, T. F., Stanbridge, E., *et al.* (1998). Regulation of cell death protease caspase-9 by phosphorylation. *Science*, **282**, 1318–21.
34. del Peso, L., Gonzalez-Garcia, M., Page, C., Herrera, R., and Nunez, G. (1997). Interleukin-3-induced phosphorylation of BAD through the protein kinase Akt. *Science*, **278**, 687–9.
35. Shindler, K. S., Yunker, A. M., Cahn, R., Zha, J., Korsmeyer, S. J., and Roth, K. A. (1998). Trophic support promotes survival of bcl-x-deficient telencephalic cells *in vitro*. *Cell Death Differ.*, **5**, 901–10.

36. Miller, T. M., Moulder, K. L., Knudson, C. M., Creedon, D. J., Deshmukh, M., Korsmeyer, S. J., *et al.* (1997). Bax deletion further orders the cell death pathway in cerebellar granule cells and suggests a caspase-independent pathway to cell death. *J. Cell Biol.*, **139**, 205–17.
37. Fujita, E., Jinbo, A., Matuzaki, H., Konishi, H., Kikkawa, U., and Momoi, T. (1999). Akt phosphorylation site found in human caspase-9 is absent in mouse caspase-9. *Biochem. Biophys. Res. Commun.*, **264**, 550–5.
38. Du, K. and Montminy, M. (1998). CREB is a regulatory target for the protein kinase Akt/PKB. *J. Biol. Chem.*, **273**, 32377–9.
39. Kane, L. P., Shapiro, V. S., Stokoe, D., and Weiss, A. (1999). Induction of NF-kappaB by the Akt/PKB kinase. *Curr. Biol.*, **9**, 601–4.
40. Riccio, A., Ahn, S., Davenport, C. M., Blendy, J. A., and Ginty, D. D. (1999). Mediation by a CREB family transcription factor of NGF-dependent survival of sympathetic neurons. *Science*, **286**, 2358–61.
41. Brunet, A., Bonni, A., Zigmond, M. J., Lin, M. Z., Juo, P., Hu, L. S., *et al.* (1999). Akt promotes cell survival by phosphorylating and inhibiting a Forkhead transcription factor. *Cell*, **96**, 857–68.
42. Paradis, S. and Ruvkun, G. (1998). *Caenorhabditis elegans* Akt/PKB transduces insulin receptor-like signals from AGE-1 PI3 kinase to the DAF-16 transcription factor. *Genes Dev.*, **12**, 2488–98.
43. Grewal, S. S., York, R. D., and Stork P. J. (1999). Extracellular-signal-regulated kinase signalling in neurons. *Curr. Opin. Neurobiol.*, **9**, 544–53.
44. Creedon, D. J., Johnson, E. M., and Lawrence, J. C. (1996). Mitogen-activated protein kinase-independent pathways mediate the effects of nerve growth factor and cAMP on neuronal survival. *J. Biol. Chem.*, **271**, 20713–18.
45. Virdee, K. and Tolkovsky, A. M. (1995). Activation of p44 and p42 MAP kinases is not essential for the survival of rat sympathetic neurons. *Eur. J. Neurosci.*, **7**, 2159–69.
46. Virdee, K. and Tolkovsky, A. M. (1996). Inhibition of p42 and p44 mitogen-activated protein kinase activity by PD98059 does not suppress nerve growth factor-induced survival of sympathetic neurones. *J. Neurochem.*, **67**, 1801–5.
47. Gunn-Moore, F. J., Williams, A. G., Toms, N. J., and Tavare, J. M. (1997). Activation of mitogen-activated protein kinase and p70S6 kinase is not correlated with cerebellar granule cell survival. *Biochem. J.*, **324**, 365–9.
48. Klesse, L. J., Meyers, K. A., Marshall, C. J., and Parada, L. F. (1999). Nerve growth factor induces survival and differentiation through two distinct signaling cascades in PC12 cells. *Oncogene*, **18**, 2055–68.
49. Anderson, C. N. G. and Tolkovsky, A. M. (1999). A role for MAPK/ERK in sympathetic neuron survival: protection against a p53-dependent, JNK-independent induction of apoptosis by cytosine arabinoside. *J. Neurosci.*, **19**, 664–73.
50. Xue, L., Murray, J. H., and Tolkovsky, A. M. (2000). The Ras/phosphatidylinositol 3-kinase and Ras/ERK pathways function as independent survival modules each of which inhibits a distinct apoptotic signaling pathway in sympathetic neurons. *J. Biol. Chem.*, **275**, 8817–24.
51. Xing, J., Kornhauser, J. M., Xia, Z., Thiele, E. A., and Greenberg, M. E. (1998). Nerve growth factor activates extracellular signal-regulated kinase and p38 mitogen-activated protein kinase pathways to stimulate CREB serine 133 phosphorylation. *Mol. Cell. Biol.*, **18**, 1946–55.
52. Bonni, A., Brunet, A., West, A. E., Datta, S. R., Takasu, M. A., and Greenberg, M. E. (1999). Cell survival promoted by the ras-MAPK signaling pathway by transcription-dependent and -independent mechanisms. *Science*, **286**, 1358–62.

53. Liu, Y. Z., Boxer, L. M., and Latchman, D. S. (1999). Activation of the Bcl-2 promoter by nerve growth factor is mediated by the p42/p44 MAPK cascade. *Nucleic Acids Res.*, **27**, 2086–90.
54. Maggirwar, S. B., Sarmiere, P. D., Dewhurst, S., and Freeman, R. S. (1998). Nerve growth factor-dependent activation of NF-kappaB contributes to survival of sympathetic neurons. *J. Neurosci.*, **18**, 10356–65.
55. Hamanoue, M., Middleton, G., Wyatt, S., Jaffray, E., Hay, R. T., and Davies, A. M. (1999). p75-mediated NF-kappaB activation enhances the survival response of developing sensory neurons to nerve growth factor. *Mol. Cell. Neurosci.*, **14**, 28–40.
56. Yu, Z., Zhou, D., Bruce-Keller, A. J., Kindy, M. S., and Mattson, M. P. (1999). Lack of the p50 subunit of nuclear factor-kappaB increases the vulnerability of hippocampal neurons to excitotoxic injury. *J. Neurosci.*, **19**, 8856–65.
57. Qin, Z. H., Wang, Y., Nakai, M., and Chase, T. N. (1998). Nuclear factor-kappa B contributes to excitotoxin-induced apoptosis in rat striatum. *Mol. Pharmacol.*, **53**, 33–42.
58. Schneider, A., Martin-Villalba, A., Weih, F., Vogel, J., Wirth, T., and Schwaninger, M. (1999). NF-kappaB is activated and promotes cell death in focal cerebral ischemia. *Nature Med.*, **5**, 554–9.
59. Seidl, R., Fang-Kircher, S., Bidmon, B., Cairns, N., and Lubec, G. (1999). Apoptosis-associated proteins p53 and APO-1/Fas (CD95) in brains of adult patients with Down syndrome. *Neurosci. Lett.*, **260**, 9–12.
60. de la Monte, S. M., Sohn, Y. K., and Wands, J. R. (1997). Correlates of p53- and Fas (CD95)-mediated apoptosis in Alzheimer's disease. *J. Neurol. Sci.*, **152**, 73–83.
61. Le-Niculescu, H., Bonfoco, E., Kasuya, Y., Claret, F. X., Green, D. R., and Karin, M. (1999). Withdrawal of survival factors results in activation of the JNK pathway in neuronal cells leading to Fas ligand induction and cell death. *Mol. Cell. Biol.*, **19**, 751–63.
62. Carter, B. D. and Lewin, G. R. (1997). Neurotrophins live or let die: does p75NTR decide? *Neuron*, **18**, 187–90.
63. Casaccia Bonnefil, P., Carter, B. D., Dobrowsky, R. T., and Chao, M. V. (1996). Death of oligodendrocytes mediated by the interaction of nerve growth factor with its receptor p75. *Nature*, **383**, 716–19.
64. Soilu-Hanninen, M., Ekert, P., Bucci, T., Syroid, D., Bartlett, P. F., and Kilpatrick, T. J. (1999). Nerve growth factor signaling through p75 induces apoptosis in Schwann cells via a Bcl-2-independent pathway. *J. Neurosci.*, **19**, 4828–38.
65. Bamji, S. X., Majdan, M., Pozniak, C. D., Belliveau, D. J., Aloyz, R., Kohn, J., *et al.* (1998). The p75 neurotrophin receptor mediates neuronal apoptosis and is essential for naturally occurring sympathetic neuron death. *J. Cell Biol.*, **140**, 911–23.
66. Sedel, F., Bechade, C., and Triller A. (1999). Nerve growth factor (NGF) induces motoneuron apoptosis in rat embryonic spinal cord *in vitro*. *Eur J. Neurosci.*, **11**, 3904–12.
67. Wiese, S., Metzger, F., Holtmann, B., and Sendtner, M. (1999). The role of p75NTR in modulating neurotrophin survival effects in developing motoneurons. *Eur J. Neurosci.*, **11**, 1668–76.
68. Barrett, G. L. and Bartlett, P. F. (1994). The p75 nerve growth factor receptor mediates survival or death depending on the stage of sensory neuron development. *Proc. Natl. Acad. Sci. USA*, **91**, 6501–5.
69. Davey, F. and Davies, A. M. (1998). TrkB signalling inhibits p75-mediated apoptosis induced by nerve growth factor in embryonic proprioceptive neurons. *Curr. Biol.*, **8**, 915–18.

70. Frade, J. M., Rodríguez-Tébar, A., and Barde, Y.-A. (1996). Induction of cell death by endogenous nerve growth factor through its p75 receptor. *Nature*, **383**, 166–8.
71. Frade, J. M. and Barde, Y. A. (1999). Genetic evidence for cell death mediated by nerve growth factor and the neurotrophin receptor p75 in the developing mouse retina and spinal cord. *Development*, **126**, 683–90.
72. Brennan, C., Rivas-Plata, K., and Landis, S. C. (1999). The p75 neurotrophin receptor influences NT-3 responsiveness of sympathetic neurons *in vivo*. *Nature Neurosci.*, **2**, 699–705.
73. Lee, K. F., Bachman, K., Landis, S., and Jaenisch, R. (1994). Dependence on p75 for innervation of some sympathetic targets. *Science*, **263**, 1447–9.
74. Majdan, M., Lachance, C., Gloster, A., Aloyz, R., Zeindler, C., Bamji, S., *et al.* (1997). Transgenic mice expressing the intracellular domain of the p75 neurotrophin receptor undergo neuronal apoptosis. *J. Neurosci.*, **17**, 6988–98.
75. Ferri, C. C., Moore, F. A., and Bisby, M. A. (1998). Effects of facial nerve injury on mouse motoneurons lacking the p75 low-affinity neurotrophin receptor. *J. Neurobiol.*, **34**, 1–9.
76. Roux, P. P., Colicos, M. A., Barker, P. A., and Kennedy, T. E. (1999). p75 neurotrophin receptor expression is induced in apoptotic neurons after seizure. *J. Neurosci.*, **19**, 6887–96.
77. Aloyz, R. S., Bamji, S. X., Pozniak, C. D., Toma, J. G., Atwal, J., Kaplan, D. R., *et al.* (1998). p53 is essential for developmental neuron death as regulated by the TrkA and p75 neurotrophin receptors. *J. Cell Biol.*, **143**, 1691–703.
78. Casademunt, E., Carter, B. D., Benzel, I., Frade, J. M., Dechant, G., and Barde, Y. A. (1999). The zinc finger protein NRIF interacts with the neurotrophin receptor p75(NTR) and participates in programmed cell death. *EMBO J.*, **18**, 6050–61.
79. Mukai, J., Hachiya, T., Shoji-Hoshino, S., Kimura, M. T., Nadano, D., Suvanto, P., *et al.* (2000). NADE, a p75NTR-associated cell death executor, is involved in signal transduction mediated by the common neurotrophin receptor p75NTR. *J. Biol. Chem.*, **275**, 17566–70.
80. Khursigara, G., Orlinick, J. R., and Chao, M. V. (1999). Association of the p75 neurotrophin receptor with TRAF6. *J. Biol. Chem.*, **274**, 2597–600.
81. Chittka, A. and Chao, M. V. (1999). Identification of a zinc finger protein whose subcellular distribution is regulated by serum and nerve growth factor. *Proc. Natl. Acad. Sci. USA*, **96**, 10705–10.
82. Salehi, A. H., Roux, P. P., Kubu, C. J., Zeindler, C., Bhakar, A., Tannis, L. L., *et al.* (2000). NRAGE, a novel MAGE protein, interacts with the p75 neurotrophin receptor and facilitates nerve growth factor-dependent apoptosis. *Neuron*, **27**, 279–88.
83. Yoon, S. O., Casaccia-Bonnefil, P., Carter, B., and Chao, M. V. (1998). Competitive signaling between TrkA and p75 nerve growth factor receptors determines cell survival. *J. Neurosci.*, **18**, 3273–81.
84. Wong, B. R., Besser, D., Kim, N., Arron, J. R., Vologodskaia, M., Hanafusa, H., *et al.* (1999). TRANCE, a TNF family member, activates Akt/PKB through a signaling complex involving TRAF6 and c-Src. *Mol. Cell*, **4**, 1041–9.
85. Davis, R. J. (2000). Signal transduction by the JNK group of MAP kinases. *Cell*, **103**, 239–52.
86. Herdegen, T., Skene, P., and Bahr, M. (1997). The c-Jun transcription factor–bipotential mediator of neuronal death, survival and regeneration. *Trends Neurosci.*, **20**, 227–31.

87. Estus, S., Zaks, W. J., Freeman, R. S., Gruda, M., Bravo, R., and Johnson, E. M., Jr. (1994). Altered gene expression in neurons during programmed cell death: identification of c-jun as necessary for neuronal apoptosis. *J. Cell Biol.*, **127**, 1717–27.
88. Ham, J., Babij, C., Whitfield, J., Pfarr, C. M., Lallemand, D., Yaniv, M., *et al.* (1995). A c-Jun dominant negative mutant protects sympathetic neurons against programmed cell death. *Neuron*, **15**, 927–39.
89. Dickens, M., Rogers, J. S., Cavanagh, J., Raitano, A., Xia, Z., Halpern, J. R., *et al.* (1997). A cytoplasmic inhibitor of the JNK signal transduction pathway. *Science*, **277**, 693–6.
90. Yang, D. D., Kuan, C. Y., Whitmarsh, A. J., Rincon, M., Zheng, T. S., Davis, R. J., *et al.* (1997). Absence of excitotoxicity-induced apoptosis in the hippocampus of mice lacking the Jnk3 gene. *Nature*, **389**, 865–70.
91. Kuan, C. Y., Yang, D. D., Samanta Roy, D. R., Davis, R. J., Rakic, P., and Flavell, R. A. (1999). The Jnk1 and Jnk2 protein kinases are required for regional specific apoptosis during early brain development. *Neuron*, **22**, 667–76.
92. Tournier, C., Hess, P., Yang, D. D., Xu, J., Turner, T. K., Nimnual, A., *et al.* (2000). Requirement of JNK for stress-induced activation of the cytochrome c-mediated death pathway. *Science*, **288**, 870–4.
93. Kanamoto, T., Mota, M., Takeda, K., Rubin, L. L., Miyazono, K., Ichijo, H., *et al.* (2000). Role of apoptosis signal-regulating kinase in regulation of the c-Jun N- terminal kinase pathway and apoptosis in sympathetic neurons. *Mol. Cell. Biol.*, **20**, 196–204.
94. Casaccia-Bonnefil, P., Kong, H., and Chao, M. V. (1998). Neurotrophins: the biological paradox of survival factors eliciting apoptosis. *Cell Death Differ.*, **5**, 357–64.
95. Levkau, B., Scatena, M., Giachelli, C. M., Ross, R., and Raines, E. W. (1999). Apoptosis overrides survival signals through a caspase-mediated dominant-negative NF-kappa B loop. *Nature Cell Biol.*, **1**, 227–33.
96. Dubois Dauphin, M., Frankowski, H., Tsujimoto, Y., Huarte, J., and Martinou, J. C. (1994). Neonatal motoneurons overexpressing the bcl-2 protooncogene in transgenic mice are protected from axotomy-induced cell death. *Proc. Natl. Acad. Sci. USA*, **91**, 3309–13.
97. Martinou, J. C., Dubois Dauphin, M., Staple, J. K., Rodriguez, I., Frankowski, H., Missotten, M., *et al.* (1994). Overexpression of BCL-2 in transgenic mice protects neurons from naturally occurring cell death and experimental ischemia. *Neuron*, **13**, 1017–30.
98. Parsadanian, A. S., Cheng, Y., Keller-Peck, C. R., Holtzman, D. M., and Snider, W. D. (1998). Bcl-xL is an antiapoptotic regulator for postnatal CNS neurons. *J. Neurosci.*, **18**, 1009–19.
99. Vekrellis, K., McCarthy, M. J., Watson, A., Whitfield, J., Rubin, L. L., and Ham J. (1997). Bax promotes neuronal cell death and is downregulated during the development of the nervous system. *Development*, **124**, 1239–49.
100. Merry, D. E., Veis, D. J., Hickey, W. F., and Korsmeyer, S. J. (1994). bcl-2 protein expression is widespread in the developing nervous system and retained in the adult PNS. *Development*, **120**, 301–11.
101. Gonzalez Garcia, M., Perez Ballestero, R., Ding, L., Duan, L., Boise, L. H., Thompson, C. B., *et al.* (1994). bcl-XL is the major bcl-x mRNA form expressed during murine development and its product localizes to mitochondria. *Development*, **120**, 3033–42.
102. Gonzalez Garcia, M., Garcia, I., Ding, L., O'Shea, S., Boise, L. H., Thompson, C. B., *et al.* (1995). bcl-x is expressed in embryonic and postnatal neural tissues and functions to prevent neuronal cell death. *Proc. Natl. Acad. Sci. USA*, **92**, 4304–8.
103. Nakayama, K., Nakayama, K., Negishi, I., Kuida, K., Sawa, H., and Loh, D. Y. (1994). Targeted disruption of Bcl-2 alpha beta in mice: occurrence of gray hair, polycystic kidney disease, and lymphocytopenia. *Proc. Natl. Acad. Sci. USA*, **91**, 3700–4.

104. Tanabe, H., Eguchi, Y., Kamada, S., Martinou, J. C., and Tsujimoto, Y. (1997). Susceptibility of cerebellar granule neurons derived from Bcl-2-deficient and transgenic mice to cell death. *Eur. J. Neurosci.*, **9**, 848–56.
105. Motoyama, N., Wang, F., Roth, K. A., Sawa, H., Nakayama, K., Nakayama, K., *et al.* (1995). Massive cell death of immature hematopoietic cells and neurons in Bcl-x-deficient mice. *Science*, **267**, 1506–10.
106. White, F. A., Keller-Peck, C. R., Knudson, C. M., Korsmeyer, S. J., and Snider, W. D. (1998). Widespread elimination of naturally occurring neuronal death in Bax-deficient mice. *J. Neurosci.*, **18**, 1428–39.
107. Deckwerth, T. L., Elliott, J. L., Knudson, C. M., Johnson, E. M., Jr., Snider, W. D., and Korsmeyer, S. J. (1996). BAX is required for neuronal death after trophic factor deprivation and during development. *Neuron*, **17**, 401–11.
108. Xiang, H., Kinoshita, Y., Knudson, C. M., Korsmeyer, S. J., Schwartzkroin, P. A., and Morrison, R. S. (1998). Bax involvement in p53-mediated neuronal cell death. *J. Neurosci.*, **18**, 1363–73.
109. Cregan, S. P., MacLaurin, J. G., Craig, C. G., Robertson, G. S., Nicholson, D. W., Park, D. S., *et al.* (1999). Bax-dependent caspase-3 activation is a key determinant in p53-induced apoptosis in neurons. *J. Neurosci.*, **19**, 7860–9.
110. Deckwerth, T. L., Easton, R. M., Knudson, C. M., Korsmeyer, S. J., and Johnson, E. M., Jr. (1998). Placement of the BCL2 family member BAX in the death pathway of sympathetic neurons activated by trophic factor deprivation. *Exp. Neurol.*, **152**, 150–62.
111. Desagher, S., Osen-Sand, A., Nichols, A., Eskes, R., Montessuit, S., Lauper, S., *et al.* (1999). Bid-induced conformational change of Bax is responsible for mitochondrial cytochrome c release during apoptosis. *J. Cell Biol.*, **144**, 891–901.
112. Deshmukh, M. and Johnson, E. M., Jr. (1998). Evidence of a novel event during neuronal death: development of competence-to-die in response to cytoplasmic cytochrome c. *Neuron*, **21**, 695–705.
113. Putcha, G. V., Deshmukh, M., and Johnson, E. M., Jr. (1999). BAX translocation is a critical event in neuronal apoptosis: regulation by neuroprotectants, BCL-2, and caspases. *J. Neurosci.*, **19**, 7476–85.
114. Neame, S. J., Rubin, L. L., and Philpott, K. L. (1998). Blocking cytochrome c activity within intact neurons inhibits apoptosis. *J. Cell Biol.*, **142**, 1583–93.
115. Honarpour, N., Du, C., Richardson, J. A., Hammer, R. E., Wang, X., and Herz, J. (2000). Adult Apaf-1-deficient mice exhibit male infertility. *Dev. Biol.*, **218**, 248–58.
116. Li, K., Li, Y., Shelton, J. M., Richardson, J. A., Spencer, E., Chen, Z. J., *et al.* (2000). Cytochrome c deficiency causes embryonic lethality and attenuates stress-induced apoptosis. *Cell*, **101**, 389–99.
117. Martinou, I., Desagher, S., Eskes, R., Antonsson, B., Andre, E., Fakan, S., *et al.* (1999). The release of cytochrome c from mitochondria during apoptosis of NGF-deprived sympathetic neurons is a reversible event. *J. Cell Biol.*, **144**, 883–9.
118. Deshmukh, M., Kuida, K., and Johnson, E. M., Jr. (2000). Caspase inhibition extends the commitment to neuronal death beyond cytochrome c release to the point of mitochondrial depolarization. *J. Cell Biol.*, **150**, 131–44.
119. Wang, H. G., Pathan, N., Ethell, I. M., Krajewski, S., Yamaguchi, Y., Shibasaki, F., *et al.* (1999). Ca^{2+}-induced apoptosis through calcineurin dephosphorylation of BAD. *Science*, **284**, 339–43.
120. Kuida, K., Haydar, T. F., Kuan, C. Y., Gu, Y., Taya, C., Karasuyama, H., *et al.* (1998). Reduced apoptosis and cytochrome c-mediated caspase activation in mice lacking caspase 9. *Cell*, **94**, 325–37.

121. Kuida, K., Zheng, T. S., Na, S., Kuan, C.-Y., Yang, D., Karasuyama, H., *et al.* (1996). Decreased apoptosis in the brain and premature lethality in CPP32-deficient mice. *Nature*, **384**, 368–72.
122. Hakem, R., Hakem, A., Duncan, G. S., Henderson, J. T., Woo, M., Soengas, M. S., *et al.* (1998). Differential requirement for caspase 9 in apoptotic pathways *in vivo*. *Cell*, **94**, 339–52.
123. Liu, X. H., Kwon, D., Schielke, G. P., Yang, G. Y., Silverstein, F. S., and Barks, J. D. (1999). Mice deficient in interleukin-1 converting enzyme are resistant to neonatal hypoxic-ischemic brain damage. *J. Cereb. Blood Flow Metab.*, **19**, 1099–108.
124. Kang, S. J., Wang, S., Hara, H., Peterson, E. P., Namura, S., Amin-Hanjani, S., *et al.* (2000). Dual role of caspase-11 in mediating activation of caspase-1 and caspase-3 under pathological conditions. *J. Cell Biol.*, **149**, 613–22.
125. Troy, C. M., Rabacchi, S. A., Friedman, W. J., Frappier, T. F., Brown, K., and Shelanski, M. L. (2000). Caspase-2 mediates neuronal cell death induced by beta-amyloid. *J. Neurosci.*, **20**, 1386–92.
126. Nakagawa, T., Zhu, H., Morishima, N., Li, E., Xu, J., Yankner, B. A., *et al.* (2000). Caspase-12 mediates endoplasmic-reticulum-specific apoptosis and cytotoxicity by amyloid-beta. *Nature*, **403**, 98–103.
127. Davidson, F. F. and Steller, H. (1998). Blocking apoptosis prevents blindness in *Drosophila* retinal degeneration mutants. *Nature*, **391**, 587–91.
128. Orr, H. T. and Zoghbi, H. Y. (2000). Reversing neurodegeneration: a promise unfolds. *Cell*, **101**, 1–4.
129. Linnik, M. D., Zobrist, R. H., and Hatfield, M. D. (1993). Evidence supporting a role for programmed cell death in focal cerebral ischemia in rats. *Stroke*, **24**, 2002–8; Discussion 2008–9.
130. MacManus, J. P., Buchan, A. M., Hill, I. E., Rasquinha, I., and Preston, E. (1993). Global ischemia can cause DNA fragmentation indicative of apoptosis in rat brain. *Neurosci. Lett.*, **164**, 89–92.
131. Ferrer, I., Tortosa, A., Macaya, A., Sierra, A., Moreno, D., Munell, F., *et al.* (1994). Evidence of nuclear DNA fragmentation following hypoxia-ischemia in the infant rat brain, and transient forebrain ischemia in the adult gerbil. *Brain Pathol.*, **4**, 115–22.
132. MacManus, J. P., Hill, I. E., Huang, Z. G., Rasquinha, I., Xue, D., and Buchan, A. M. (1994). DNA damage consistent with apoptosis in transient focal ischaemic neocortex. *Neuroreport*, **5**, 493–6.
133. Mehmet, H., Yue, X., Squier, M. V., Lorek, A., Cady, E., Penrice, J., *et al.* (1994). Increased apoptosis in the cingulate sulcus of newborn piglets following transient hypoxia-ischaemia is related to the degree of high energy phosphate depletion during the insult. *Neurosci. Lett.*, **181**, 121–5.
134. Hill, I. E., MacManus, J. P., Rasquinha, I., and Tuor, U. I. (1995). DNA fragmentation indicative of apoptosis following unilateral cerebral hypoxia-ischemia in the neonatal rat. *Brain Res.*, **676**, 398–403.
135. Nitatori, T., Sato, N., Waguri, S., Karasawa, Y., Araki, H., Shibanai, K., *et al.* (1995). Delayed neuronal death in the CA1 pyramidal cell layer of the gerbil hippocampus following transient ischemia is apoptosis. *J. Neurosci.*, **15**, 1001–11.
136. Gillardon, F., Bottiger, B., Schmitz, B., Zimmermann, M., and Hossmann, K. A. (1997). Activation of CPP-32 protease in hippocampal neurons following ischemia and epilepsy. *Brain Res. Mol. Brain Res.*, **50**, 16–22.

137. Chen, J., Nagayama, T., Jin, K., Stetler, R. A., Zhu, R. L., Graham, S. H., *et al.* (1998). Induction of caspase-3-like protease may mediate delayed neuronal death in the hippocampus after transient cerebral ischemia. *J. Neurosci.*, **18**, 4914–28.
138. Cheng, Y., Deshmukh, M., D'Costa, A., Demaro, J. A., Gidday, J. M., Shah, A., *et al.* (1998). Caspase inhibitor affords neuroprotection with delayed administration in a rat model of neonatal hypoxic-ischemic brain injury. *J. Clin. Invest.*, **101**, 1992–9.
139. Schulz, J. B., Weller, M., Matthews, R. T., Heneka, M. T., Groscurth, P., Martinou, J. C., *et al.* (1998). Extended therapeutic window for caspase inhibition and synergy with MK-801 in the treatment of cerebral histotoxic hypoxia. *Cell Death Differ.*, **5**, 847–57.
140. Xu, D., Bureau, Y., McIntyre, D. C., Nicholson, D. W., Liston, P., Zhu, Y., *et al.* (1999). Attenuation of ischemia-induced cellular and behavioral deficits by X chromosome-linked inhibitor of apoptosis protein overexpression in the rat hippocampus. *J. Neurosci.*, **19**, 5026–33.
141. Namura, S., Zhu, J., Fink, K., Endres, M., Srinivasan, A., Tomaselli, K. J., *et al.* (1998). Activation and cleavage of caspase-3 in apoptosis induced by experimental cerebral ischemia. *J. Neurosci.*, **18**, 3659–68.
142. Hara, H., Friedlander, R. M., Gagliardini, V., Ayata, C., Fink, K., Huang, Z., *et al.* (1997). Inhibition of interleukin 1beta converting enzyme family proteases reduces ischemic and excitotoxic neuronal damage. *Proc. Natl. Acad. Sci. USA*, **94**, 2007–12.
143. Endres, M., Namura, S., Shimizu-Sasamata, M., Waeber, C., Zhang, L., Gomez-Isla, T., *et al.* (1998). Attenuation of delayed neuronal death after mild focal ischemia in mice by inhibition of the caspase family. *J. Cereb. Blood Flow Metab.*, **18**, 238–47.
144. Loddick, S. A., MacKenzie, A., and Rothwell, N. J. (1996). An ICE inhibitor, z-VAD-DCB attenuates ischaemic brain damage in the rat. *Neuroreport*, **7**, 1465–8.
145. Relton, J. K. and Rothwell, N. J. (1992). Interleukin-1 receptor antagonist inhibits ischaemic and excitotoxic neuronal damage in the rat. *Brain Res. Bull.*, **29**, 243–6.
146. Garcia, J. H., Liu, K. F., and Relton, J. K. (1995). Interleukin-1 receptor antagonist decreases the number of necrotic neurons in rats with middle cerebral artery occlusion. *Am. J. Pathol.*, **147**, 1477–86.
147. Loddick, S. A. and Rothwell, N. J. (1996). Neuroprotective effects of human recombinant interleukin-1 receptor antagonist in focal cerebral ischaemia in the rat. *J. Cereb. Blood Flow Metab.*, **16**, 932–40.
148. Yang, G. Y., Zhao, Y. J., Davidson, B. L., and Betz, A. L. (1997). Overexpression of interleukin-1 receptor antagonist in the mouse brain reduces ischemic brain injury. *Brain Res.*, **751**, 181–8.
149. Stroemer, R. P. and Rothwell, N. J. (1997). Cortical protection by localized striatal injection of IL-1ra following cerebral ischemia in the rat. *J. Cereb. Blood Flow Metab.*, **17**, 597–604.
150. Friedlander, R. M., Gagliardini, V., Hara, H., Fink, K. B., Li, W., MacDonald, G., *et al.* (1997). Expression of a dominant negative mutant of interleukin-1 beta converting enzyme in transgenic mice prevents neuronal cell death induced by trophic factor withdrawal and ischemic brain injury. *J. Exp. Med.*, **185**, 933–40.
151. Hara, H., Fink, K., Endres, M., Friedlander, R. M., Gagliardini, V., Yuan, J., *et al.* (1997). Attenuation of transient focal cerebral ischemic injury in transgenic mice expressing a mutant ICE inhibitory protein. *J. Cereb. Blood Flow Metab.*, **17**, 370–5.
152. Schielke, G. P., Yang, G. Y., Shivers, B. D., and Betz, A. L. (1998). Reduced ischemic brain injury in interleukin-1 beta converting enzyme-deficient mice. *J. Cereb. Blood Flow Metab.*, **18**, 180–5.

153. Bhat, R. V., DiRocco, R., Marcy, V. R., Flood, D. G., Zhu, Y., Dobrzanski, P., *et al.* (1996). Increased expression of IL-1beta converting enzyme in hippocampus after ischemia: selective localization in microglia. *J. Neurosci.*, **16**, 4146–54.
154. Friedlander, R. M., Gagliardini, V., Rotello, R. J., and Yuan, J. (1996). Functional role of interleukin 1 beta (IL-1 beta) in IL-1 beta- converting enzyme-mediated apoptosis. *J. Exp. Med.*, **184**, 717–24.
155. Velier, J. J., Ellison, J. A., Kikly, K. K., Spera, P. A., Barone, F. C., and Feuerstein, G. Z. (1999). Caspase-8 and caspase-3 are expressed by different populations of cortical neurons undergoing delayed cell death after focal stroke in the rat. *J. Neurosci.*, **19**, 5932–41.
156. Michaelis, E. K. (1998). Molecular biology of glutamate receptors in the central nervous system and their role in excitotoxicity, oxidative stress and aging. *Prog. Neurobiol.*, **54**, 369–415.
157. Ankarcrona, M., Dypbukt, J. M., Bonfoco, E., Zhivotovsky, B., Orrenius, S., Lipton, S. A., *et al.* (1995). Glutamate-induced neuronal death: a succession of necrosis or apoptosis depending on mitochondrial function. *Neuron*, **15**, 961–73.
158. Herdegen, T., Claret, F. X., Kallunki, T., Martin-Villalba, A., Winter, C., Hunter, T., *et al.* (1998). Lasting N-terminal phosphorylation of c-Jun and activation of c-Jun N- terminal kinases after neuronal injury. *J. Neurosci.*, **18**, 5124–35.
159. Ratan, R. R., Murphy, T. H., and Baraban, J. M. (1994). Oxidative stress induces apoptosis in embryonic cortical neurons. *J. Neurochem.*, **62**, 376–9.
160. Ratan, R. R., Lee, P. J., and Baraban, J. M. (1996). Serum deprivation inhibits glutathione depletion-induced death in embryonic cortical neurons: evidence against oxidative stress as a final common mediator of neuronal apoptosis. *Neurochem. Int.*, **29**, 153–7.
161. Munoz-Fernandez, M. A. and Fresno, M. (1998). The role of tumour necrosis factor, interleukin 6, interferon-gamma and inducible nitric oxide synthase in the development and pathology of the nervous system. *Prog. Neurobiol.*, **56**, 307–40.
162. Barone, F. C., Arvin, B., White, R. F., Miller, A., Webb, C. L., Willette, R. N., *et al.* (1997). Tumor necrosis factor-alpha. A mediator of focal ischemic brain injury. *Stroke*, **28**, 1233–44.
163. Van Antwerp, D. J., Martin, S. J., Kafri, T., Green, D. R., and Verma, I. M. (1996). Suppression of TNF-alpha-induced apoptosis by NF-kappaB. *Science*, **274**, 787–9.
164. Venters, H. D., Tang, Q., Liu, Q., VanHoy, R. W., Dantzer, R., and Kelley, K. W. (1999). A new mechanism of neurodegeneration: a proinflammatory cytokine inhibits receptor signaling by a survival peptide. *Proc. Natl. Acad. Sci. USA*, **96**, 9879–84.
165. Baeuerle, P. A. and Henkel, T. (1994). Function and activation of NF-kappa B in the immune system. *Annu. Rev. Immunol.*, **12**, 141–79.
166. Lipton, S. A. (1997). Janus faces of NF-kappa B: neurodestruction versus neuroprotection [News]. *Nature Med.*, **3**, 20–2.
167. Bates, G. (1996). Expanded glutamines and neurodegeneration–a gain of insight. *Bioessays*, **18**, 175–8.
168. Matsuyama, Z., Wakamori, M., Mori, Y., Kawakami, H., Nakamura, S., and Imoto, K. (1999). Direct alteration of the P/Q-type Ca^{2+} channel property by polyglutamine expansion in spinocerebellar ataxia 6. *J. Neurosci.* (Online), **19**, RC14.
169. Ishikawa, K., Fujigasaki, H., Saegusa, H., Ohwada, K., Fujita, T., Iwamoto, H., *et al.* (1999). Abundant expression and cytoplasmic aggregations of [alpha]1A voltage-dependent calcium channel protein associated with neurodegeneration in spino-cerebellar ataxia type 6. *Hum. Mol. Genet.*, **8**, 1185–93.

170. Burright, E. N., Clark, H. B., Servadio, A., Matilla, T., Feddersen, R. M., Yunis, W. S., *et al.* (1995). SCA1 transgenic mice: a model for neurodegeneration caused by an expanded CAG trinucleotide repeat. *Cell*, **82**, 937–48.
171. Mangiarini, L., Sathasivam, K., Seller, M., Cozens, B., Harper, A., Hetherington, C., *et al.* (1996). Exon 1 of the HD gene with an expanded CAG repeat is sufficient to cause a progressive neurological phenotype in transgenic mice. *Cell*, **87**, 493–506.
172. Jackson, G. R., Salecker, I., Dong, X., Yao, X., Arnheim, N., Faber, P. W., *et al.* (1998). Polyglutamine-expanded human huntingtin transgenes induce degeneration of *Drosophila* photoreceptor neurons. *Neuron*, **21**, 633–42.
173. Warrick, J. M., Paulson, H. L., Gray-Board, G. L., Bui, Q. T., Fischbeck, K. H., Pittman, R. N., *et al.* (1998). Expanded polyglutamine protein forms nuclear inclusions and causes neural degeneration in *Drosophila*. *Cell*, **93**, 939–49.
174. Faber, P. W., Alter, J. R., MacDonald, M. E., and Hart, A. C. (1999). Polyglutamine-mediated dysfunction and apoptotic death of a *Caenorhabditis elegans* sensory neuron. *Proc. Natl. Acad. Sci. USA*, **96**, 179–84.
175. Ordway, J. M., Tallaksen-Greene, S., Gutekunst, C. A., Bernstein, E. M., Cearley, J. A., Wiener, H. W., *et al.* (1997). Ectopically expressed CAG repeats cause intranuclear inclusions and a progressive late onset neurological phenotype in the mouse. *Cell*, **91**, 753–63.
176. Scherzinger, E., Lurz, R., Turmaine, M., Mangiarini, L., Hollenbach, B., Hasenbank, R., *et al.* (1997). Huntingtin-encoded polyglutamine expansions form amyloid-like protein aggregates *in vitro* and *in vivo*. *Cell*, **90**, 549–58.
177. Lunkes, A. and Mandel, J. L. (1997). Polyglutamines, nuclear inclusions and neurodegeneration [News]. *Nature Med.*, **3**, 1201–2.
178. DiFiglia, M., Sapp, E., Chase, K. O., Davies, S. W., Bates, G. P., Vonsattel, J. P., *et al.* (1997). Aggregation of huntingtin in neuronal intranuclear inclusions and dystrophic neurites in brain. *Science*, **277**, 1990–3.
179. Davies, S. W., Turmaine, M., Cozens, B. A., DiFiglia, M., Sharp, A. H., Ross, C. A., *et al.* (1997). Formation of neuronal intranuclear inclusions underlies the neurological dysfunction in mice transgenic for the HD mutation. *Cell*, **90**, 537–48.
180. Saudou, F., Finkbeiner, S., Devys, D., and Greenberg, M. E. (1998). Huntingtin acts in the nucleus to induce apoptosis but death does not correlate with the formation of intranuclear inclusions. *Cell*, **95**, 55–66.
181. Klement, I. A., Skinner, P. J., Kaytor, M. D., Yi, H., Hersch, S. M., Clark, H. B., *et al.* (1998). Ataxin-1 nuclear localization and aggregation: role in polyglutamine- induced disease in SCA1 transgenic mice. *Cell*, **95**, 41–53.
182. Sanchez, I., Xu, C. J., Juo, P., Kakizaka, A., Blenis, J., and Yuan, J. (1999). Caspase-8 is required for cell death induced by expanded polyglutamine repeats. *Neuron*, **22**, 623–33.
183. Ona, V. O., Li, M., Vonsattel, J. P., Andrews, L. J., Khan, S. Q., Chung, W. M., *et al.* (1999). Inhibition of caspase-1 slows disease progression in a mouse model of Huntington's disease. *Nature*, **399**, 263–7.
184. Vonsattel, J. P., Myers, R. H., Stevens, T. J., Ferrante, R. J., Bird, E. D., and Richardson, E. P., Jr. (1985). Neuropathological classification of Huntington's disease. *J. Neuropathol. Exp. Neurol.*, **44**, 559–77.
185. Yamamoto, A., Lucas, J. J., and Hen, R. (2000). Reversal of neuropathology and motor dysfunction in a conditional model of Huntington's disease. *Cell*, **101**, 57–66.
186. Tanzi, R. E. (1998). The molecular genetics of alzheimer's disease. In *Scientific American molecular neurobiology* (ed. J. B. Marton), pp. 55–75. Scientific American, New York.

187. Su, J. H., Anderson, A. J., Cummings, B. J., and Cotman, C. W. (1994). Immunohistochemical evidence for apoptosis in Alzheimer's disease. *Neuroreport*, **5**, 2529–33.
188. Smale, G., Nichols, N. R., Brady, D. R., Finch, C. E., and Horton, W. E., Jr. (1995). Evidence for apoptotic cell death in Alzheimer's disease. *Exp. Neurol.*, **133**, 225–30.
189. Troncoso, J. C., Sukhov, R. R., Kawas, C. H., and Koliatsos, V. E. (1996). *In situ* labeling of dying cortical neurons in normal aging and in Alzheimer's disease: correlations with senile plaques and disease progression. *J. Neuropathol. Exp. Neurol.*, **55**, 1134–42.
190. Cotman, C. W. and Anderson, A. J. (1995). A potential role for apoptosis in neurodegeneration and Alzheimer's disease. *Mol. Neurobiol.*, **10**, 19–45.
191. Perry, G., Nunomura, A., Lucassen, P., Lassmann, H., and Smith, M. A. (1998). Apoptosis and Alzheimer's disease. *Science*, **282**, 1268–9.
192. Stadelmann, C., Deckwerth, T. L., Srinivasan, A., Bancher, C., Bruck, W., Jellinger, K., *et al.* (1999). Activation of caspase-3 in single neurons and autophagic granules of granulovacuolar degeneration in Alzheimer's disease. Evidence for apoptotic cell death. *Am. J. Pathol.*, **155**, 1459–66.
193. Games, D., Adams, D., Alessandrini, R., Barbour, R., Berthelette, P., Blackwell, C., *et al.* (1995). Alzheimer-type neuropathology in transgenic mice overexpressing V717F beta-amyloid precursor protein. *Nature*, **373**, 523–7.
194. Johnson-Wood, K., Lee, M., Motter, R., Hu, K., Gordon, G., Barbour, R., *et al.* (1997). Amyloid precursor protein processing and A beta42 deposition in a transgenic mouse model of Alzheimer disease. *Proc. Natl. Acad. Sci. USA*, **94**, 1550–5.
195. Sturchler-Pierrat, C., Abramowski, D., Duke, M., Wiederhold, K. H., Mistl, C., Rothacher, S., *et al.* (1997). Two amyloid precursor protein transgenic mouse models with Alzheimer disease-like pathology. *Proc. Natl. Acad. Sci. USA*, **94**, 13287–92.
196. Schenk, D., Barbour, R., Dunn, W., Gordon, G., Grajeda, H., Guido, T., *et al.* (1999). Immunization with amyloid-beta attenuates Alzheimer-disease-like pathology in the PDAPP mouse. *Nature*, **400**, 173–7.
197. Yankner, B. A. (1996). Mechanisms of neuronal degeneration in Alzheimer's disease. *Neuron*, **16**, 921–32.
198. Geula, C., Wu, C. K., Saroff, D., Lorenzo, A., Yuan, M., and Yankner, B. A. (1998). Aging renders the brain vulnerable to amyloid beta-protein neurotoxicity. *Nature Med.*, **4**, 827–31.
199. Behl, C., Davis, J. B., Lesley, R., and Schubert, D. (1994). Hydrogen peroxide mediates amyloid beta protein toxicity. *Cell*, **77**, 817–27.
200. Mattson, M. P., Tomaselli, K. J., and Rydel, R. E. (1993). Calcium-destabilizing and neurodegenerative effects of aggregated beta-amyloid peptide are attenuated by basic FGF. *Brain Res.*, **621**, 35–49.
201. Loo, D. T., Copani, A., Pike, C. J., Whittemore, E. R., Walencewicz, A. J., and Cotman, C. W. (1993). Apoptosis is induced by beta-amyloid in cultured central nervous system neurons. *Proc. Natl. Acad. Sci. USA*, **90**, 7951–5.
202. Yan, S. D., Chen, X., Fu, J., Chen, M., Zhu, H., Roher, A., *et al.* (1996). RAGE and amyloid-beta peptide neurotoxicity in Alzheimer's disease. *Nature*, **382**, 685–91.
203. Yaar, M., Zhai, S., Pilch, P. F., Doyle, S. M., Eisenhauer, P. B., Fine, R. E., *et al.* (1997). Binding of beta-amyloid to the p75 neurotrophin receptor induces apoptosis. A possible mechanism for Alzheimer's disease. *J. Clin. Invest.*, **100**, 2333–40.
204. Lorenzo, A., Yuan, M., Zhang, Z., Paganetti, P. A., Sturchler-Pierrat, C., Staufenbiel, M., *et al.* (2000). Amyloid beta interacts with the amyloid precursor protein: a potential toxic mechanism in Alzheimer's disease. *Nature Neurosci.*, **3**, 460–4.

205. Gervais, F. G., Thornberry, N. A., Ruffolo, S. C., Nicholson, D. W., and Roy, S. (1998). Caspases cleave focal adhesion kinase during apoptosis to generate a FRNK-like polypeptide. *J. Biol. Chem.*, **273**, 17102–8.
206. Sherrington, R., Rogaev, E. I., Liang, Y., Rogaeva, E. A., Levesque, G., Ikeda, M., *et al.* (1995). Cloning of a gene bearing missense mutations in early-onset familial Alzheimer's disease. *Nature*, **375**, 754–60.
207. Levy-Lahad, E., Wasco, W., Poorkaj, P., Romano, D. M., Oshima, J., Pettingell, W. H., *et al.* (1995). Candidate gene for the chromosome 1 familial Alzheimer's disease locus. *Science*, **269**, 973–7.
208. Price, D. L., Tanzi, R. E., Borchelt, D. R., and Sisodia, S. S. (1998). Alzheimer's disease: genetic studies and transgenic models. *Annu. Rev. Genet.*, **32**, 461–93.
209. Wolfe, M. S., Xia, W., Ostaszewski, B. L., Diehl, T. S., Kimberly, W. T., and Selkoe, D. J. (1999). Two transmembrane aspartates in presenilin-1 required for presenilin endoproteolysis and gamma-secretase activity. *Nature*, **398**, 513–17.
210. Kimberly, W. T., Xia, W., Rahmati, T., Wolfe, M. S., and Selkoe, D. J. (2000). The transmembrane aspartates in presenilin 1 and 2 are obligatory for gamma-secretase activity and amyloid beta-protein generation. *J. Biol. Chem.*, **275**, 3173–8.
211. Li, Y. M., Xu, M., Lai, M. T., Huang, Q., Castro, J. L., DiMuzio-Mower, J., *et al.* (2000). Photoactivated gamma-secretase inhibitors directed to the active site covalently label presenilin 1. *Nature*, **405**, 689–94.
212. Lemere, C. A., Lopera, F., Kosik, K. S., Lendon, C. L., Ossa, J., Saido, T. C., *et al.* (1996). The E280A presenilin 1 Alzheimer mutation produces increased A beta 42 deposition and severe cerebellar pathology. *Nature Med.*, **2**, 1146–50.
213. Duff, K., Eckman, C., Zehr, C., Yu, X., Prada, C. M., Perez-tur, J., *et al.* (1996). Increased amyloid-beta42(43) in brains of mice expressing mutant presenilin 1. *Nature*, **383**, 710–13.
214. Borchelt, D. R., Thinakaran, G., Eckman, C. B., Lee, M. K., Davenport, F., Ratovitsky, T., *et al.* (1996). Familial Alzheimer's disease-linked presenilin 1 variants elevate Abeta1–42/1–40 ratio *in vitro* and *in vivo*. *Neuron*, **17**, 1005–13.
215. Scheuner, D., Eckman, C., Jensen, M., Song, X., Citron, M., Suzuki, N., *et al.* (1996). Secreted amyloid beta-protein similar to that in the senile plaques of Alzheimer's disease is increased *in vivo* by the presenilin 1 and 2 and APP mutations linked to familial Alzheimer's disease. *Nature Med.*, **2**, 864–70.
216. Mattson, M. P., Guo, Q., Furukawa, K., and Pedersen, W. A. (1998). Presenilins, the endoplasmic reticulum, and neuronal apoptosis in Alzheimer's disease. *J. Neurochem.*, **70**, 1–14.
217. Walter, J., Schindzielorz, A., Grunberg, J., and Haass, C. (1999). Phosphorylation of presenilin-2 regulates its cleavage by caspases and retards progression of apoptosis. *Proc. Natl. Acad. Sci. USA*, **96**, 1391–6.
218. Stabler, S. M., Ostrowski, L. L., Janicki, S. M., and Monteiro, M. J. (1999). A myristoylated calcium-binding protein that preferentially interacts with the Alzheimer's disease presenilin 2 protein. *J. Cell Biol.*, **145**, 1277–92.
219. Buxbaum, J. D., Choi, E. K., Luo, Y., Lilliehook, C., Crowley, A. C., Merriam, D. E., *et al.* (1998). Calsenilin: a calcium-binding protein that interacts with the presenilins and regulates the levels of a presenilin fragment. *Nature Med.*, **4**, 1177–81.
220. Zhang, Z., Hartmann, H., Do, V. M., Abramowski, D., Sturchler-Pierrat, C., Staufenbiel, M., *et al.* (1998). Destabilization of beta-catenin by mutations in presenilin-1 potentiates neuronal apoptosis. *Nature*, **395**, 698–702.

221. Patrick, G. N., Zukerberg, L., Nikolic, M., de la Monte, S., Dikkes, P., and Tsai, L. H. (1999). Conversion of p35 to p25 deregulates Cdk5 activity and promotes neurodegeneration. *Nature*, **402**, 615–22.
222. Lee, M. S., Kwon, Y. T., Li, M., Peng, J., Friedlander, R. M., and Tsai, L. H. (2000). Neurotoxicity induces cleavage of p35 to p25 by calpain. *Nature*, **405**, 360–4.
223. Gonzalez-Scarano, F. and Baltuch, G. (1999). Microglia as mediators of inflammatory and degenerative diseases. *Annu. Rev. Neurosci.*, **22**, 219–40.
224. Brown, D. R., Schmidt, B., and Kretzschmar, H. A. (1996). Role of microglia and host prion protein in neurotoxicity of a prion protein fragment. *Nature*, **380**, 345–7.
225. Giulian, D., Haverkamp, L. J., Yu, J. H., Karshin, W., Tom, D., Li, J., *et al.* (1996). Specific domains of beta-amyloid from Alzheimer plaque elicit neuron killing in human microglia. *J. Neurosci.*, **16**, 6021–37.
226. Tan, J., Town, T., Paris, D., Mori, T., Suo, Z., Crawford, F., *et al.* (1999). Microglial activation resulting from CD40-CD40L interaction after beta- amyloid stimulation. *Science*, **286**, 2352–5.
227. Rosen, D. R., Siddique, T., Patterson, D., Figlewicz, D. A., Sapp, P., Hentati, A., *et al.* (1993). Mutations in Cu/Zn superoxide dismutase gene are associated with familial amyotrophic lateral sclerosis [Published erratum appears in *Nature* (1993) Jul 22, 364 (6435), 362]. *Nature*, **362**, 59–62.
228. Cleveland, D. W. (1999). From Charcot to SOD1: mechanisms of selective motor neuron death in ALS. *Neuron*, **24**, 515–20.
229. Pasinelli, P., Borchelt, D. R., Houseweart, M. K., Cleveland, D. W., and Brown, R. H., Jr. (1998). Caspase-1 is activated in neural cells and tissue with amyotrophic lateral sclerosis-associated mutations in copper-zinc superoxide dismutase [Published erratum appears in *Proc. Natl. Acad. Sci. USA* (1999) Mar 16, 96 (6), 3330]. *Proc. Natl. Acad. Sci. USA*, **95**, 15763–8.
230. Rabizadeh, S., Gralla, E. B., Borchelt, D. R., Gwinn, R., Valentine, J. S., Sisodia, S., *et al.* (1995). Mutations associated with amyotrophic lateral sclerosis convert superoxide dismutase from an antiapoptotic gene to a proapoptotic gene: studies in yeast and neural cells. *Proc. Natl. Acad. Sci. USA*, **92**, 3024–8.
231. Kostic, V., Jackson-Lewis, V., de Bilbao, F., Dubois-Dauphin, M., and Przedborski, S. (1997). Bcl-2: prolonging life in a transgenic mouse model of familial amyotrophic lateral sclerosis. *Science*, **277**, 559–62.
232. Friedlander, R. M., Brown, R. H., Gagliardini, V., Wang, J., and Yuan, J. (1997). Inhibition of ICE slows ALS in mice. *Nature*, **388**, 31.
233. Gurney, M. E., Pu, H., Chiu, A. Y., Dal Canto, M. C., Polchow, C. Y., Alexander, D. D., *et al.* (1994). Motor neuron degeneration in mice that express a human Cu,Zn superoxide dismutase mutation. *Science*, **264**, 1772–5.
234. Dal Canto, M. C. and Gurney, M. E. (1994). Development of central nervous system pathology in a murine transgenic model of human amyotrophic lateral sclerosis. *Am. J. Pathol.*, **145**, 1271–9.
235. Li, M., Ona, V. O., Guegan, C., Chen, M., Jackson-Lewis, V., Andrews, L. J., *et al.* (2000). Functional role of caspase-1 and caspase-3 in an ALS transgenic mouse model. *Science*, **288**, 335–9.
236. Lefebvre, S., Burlet, P., Liu, Q., Bertrandy, S., Clermont, O., Munnich, A., *et al.* (1997). Correlation between severity and SMN protein level in spinal muscular atrophy. *Nature Genet.*, **16**, 265–9.

237. Gambardella, A., Mazzei, R., Toscano, A., Annesi, G., Pasqua, A., Annesi, F., *et al.* (1998). Spinal muscular atrophy due to an isolated deletion of exon 8 of the telomeric survival motor neuron gene. *Ann. Neurol.*, **44**, 836–9.
238. Fischer, U., Liu, Q., and Dreyfuss, G. (1997). The SMN-SIP1 complex has an essential role in spliceosomal snRNP biogenesis. *Cell*, **90**, 1023–9.
239. Liu, Q., Fischer, U., Wang, F., and Dreyfuss, G. (1997). The spinal muscular atrophy disease gene product, SMN, and its associated protein SIP1 are in a complex with spliceosomal snRNP proteins. *Cell*, **90**, 1013–21.
240. Iwahashi, H., Eguchi, Y., Yasuhara, N., Hanafusa, T., Matsuzawa, Y., and Tsujimoto, Y. (1997). Synergistic anti-apoptotic activity between Bcl-2 and SMN implicated in spinal muscular atrophy. *Nature*, **390**, 413–17.
241. Roy, N., Mahadevan, M. S., McLean, M., Shutler, G., Yaraghi, Z., Farahani, R., *et al.* (1995). The gene for neuronal apoptosis inhibitory protein is partially deleted in individuals with spinal muscular atrophy. *Cell*, **80**, 167–78.
242. Xu, D. G., Crocker, S. J., Doucet, J. P., St-Jean, M., Tamai, K., Hakim, A. M., *et al.* (1997). Elevation of neuronal expression of NAIP reduces ischemic damage in the rat hippocampus. *Nature Med.*, **3**, 997–1004.
243. Holcik, M., Thompson, C. S., Yaraghi, Z., Lefebvre, C. A., MacKenzie, A. E., and Korneluk, R. G. (2000). The hippocampal neurons of neuronal apoptosis inhibitory protein 1 (NAIP1)-deleted mice display increased vulnerability to kainic acid- induced injury. *Proc. Natl. Acad. Sci. USA*, **97**, 2286–90.

Index

Page numbers in italic, e.g. *216*, refer to figures. Page numbers in bold, e.g. **245**, signify entries in tables.

Printed in the United States
By Bookmasters